Springer Series in Optical Sciences

Volume 176

For further volumes:
http://www.springer.com/series/624

Springer Series in Optical Sciences

The Springer Series in Optical Sciences, under the leadership of Editor-in-Chief William T. Rhodes, Georgia Institute of Technology, USA, provides an expanding selection of research monographs in all major areas of optics: lasers and quantum optics, ultrafast phenomena, optical spectroscopy techniques, optoelectronics, quantum information, information optics, applied laser technology, industrial applications, and other topics of contemporary interest.

With this broad coverage of topics, the series is of use to all research scientists and engineers who need up-to-date reference books.

The editors encourage prospective authors to correspond with them in advance of submitting a manuscript. Submission of manuscripts should be made to the Editor-in-Chief or one of the Editors. See also www.springer.com/series/624

Shigemasa Suga · Akira Sekiyama

Photoelectron Spectroscopy

Bulk and Surface Electronic Structures

Shigemasa Suga
Institute of Scientific and Industrial Research
Osaka University
Osaka
Japan

Akira Sekiyama
Graduate School of Engineering Science
Osaka University
Osaka
Japan

ISSN 0342-4111
ISSN 1556-1534 (electronic)
ISBN 978-3-642-37529-3
ISBN 978-3-642-37530-9 (eBook)
DOI 10.1007/978-3-642-37530-9
Springer Heidelberg New York Dordrecht London

Library of Congress Control Number: 2013937610

Printed on acid-free paper

Springer is part of Springer Science+Business Media (www.springer.com)

Preface

The phenomenon of photoelectron emission was found more than a century ago and yielded the concept of photons. The relation between the photon energies and the kinetic energies of photoelectrons was soon clarified. The usefulness of photoelectron spectroscopy (PES) for materials sciences was widely recognized in the last century. Electronic structures of gases and solids were mostly studied at low photon energies and chemical analyses were mainly performed in the X-ray region. Besides the development of high performance electron analyzers, light sources have in parallel been intensively improved till date. The use of synchrotron radiation has promoted the photoelectron spectroscopy since ~1975 due to the tunability of photon energies. The development of undulator light sources enabled the high energy resolution photoelectron spectroscopy even in the regions of soft and hard X-rays. Now, synchrotron radiation can provide photons with energy ($h\nu$) resolutions less than a few meV below 10 eV, 40 meV around 800 eV, and less than 300 μeV at 14 keV by use of good monochromators. Meanwhile, the resolution $\leq$1 meV can be achieved in laboratories by use of very low energy (~7 eV) quasi-CW, high-repetition lasers as well as high resolution rare gas (Xe, Kr, Ar) lamps.

By utilizing the angle resolving power of the electron energy analyzer, the angle resolved photoelectron spectroscopy (ARPES) became very popular to probe the band dispersions of solids as well as the fermi surface topology in metals. In strongly correlated electron systems, where the surface electronic structures as well as surface band dispersions are often noticeably different from those of the bulk electronic structures, the high energy photoelectron spectroscopy with high bulk sensitivity and enough energy and momentum resolutions is now extensively performed. Since the inelastic mean free path of photoelectrons becomes again increased in some materials below the kinetic energies of a few eV, extremely high resolution photoelectron spectroscopy is under development below $h\nu$ ~ 10 eV. For a complete understanding of electronic structures of strongly correlated electron systems, studies in wide photon energies in the range of 10 keV down to a few eV will be highly desired. By utilizing the high brilliance of the synchrotron radiation, studies of micro- and nano-materials are progressing rapidly. The scanning photoelectron microscopy and the photoelectron emission microscopy are such examples. In addition, spin polarized measurements are widely used to study the spin states.

Complementary techniques to probe materials electronic structures are also under intensive development. For example, absorption and reflectivity spectroscopy as well as the resonance inelastic X-ray scattering are very powerful techniques with high bulk sensitivity. Inverse photoemission spectroscopy is a useful tool to probe the unoccupied electronic states. On the other hand, photoelectron diffraction can provide the structural information down to sub nm range. The scanning tunneling spectroscopy by use of scanning tunneling microscope can also probe the occupied and unoccupied electronic states with extremely high spatial resolution down to the atomic resolution. By means of the Fourier transform, the information in the momentum space can also be obtained. Thus we intended to include up-to-date achievements in photoelectron spectroscopy and frontiers of some typical complementary techniques in this book.

Osaka

Shigemasa Suga
Akira Sekiyama

Acknowledgments

The photoelectron spectroscopy with use of such a bright light source as synchrotron radiation has already a history of more than four decades. One of the authors S. S. has first experienced synchrotron radiation photoelectron spectroscopy in 1975 at DESY with F. J. Himpsel. The initial interest of S. S. in this field was stimulated by M. Cardona of Max-Planck-Institute für Festkörperforschung in Stuttgart in the last stage of his stay in Germany between 1973 and 1976. S. S. was then engaged in the instrumentation, spectroscopy, and materials' sciences in the synchrotron radiation laboratory of the Institute for Solid State Physics (ISSP), The University of Tokyo, with a dedicated synchrotron light source, SOR-RING (0.38 GeV), where he was responsible for the national use of this facility over 13 years together with H. Kanzaki and T. Ishii. There he cooperated with M. Taniguchi and S. Shin in photoelectron spectroscopy. He also collaborated with A. Fujimori on XPS and S.-J. Oh on resonance photoelectron spectroscopy. He was also engaged in VUV-IPES and X-BIS with H. Namatame and S. Ogawa. During this period he also joined spin polarized electron experiments at BESSY-I and KFA, Jülich in collaboration with J. Kirschner. He also collaborated with A. Kakizaki, K. Soda, M. Fujisawa, and T. Kinoshita to construct two beam lines of the ISSP at Photon Factory (2.5 GeV) in Tsukuba.

He then moved to Osaka University in 1989 to join the SPring-8 (8 GeV) project as an active outside user. He devoted the full potential of his old and new laboratories to the construction of the first soft X-ray beam line, BL25SU, of SPring-8 with a twin helical undulator (designed by H. Kitamura) together with H. Daimon, S. Imada, and Y. Saitoh. Excellent performance of the soft X-ray monochromator equipped with varied line spacing plane gratings was confirmed at BL25SU in 2000 by the author's group. Till the full commissioning of this new beam line with three different experimental stations (soft X-ray photoelectron spectroscopy, soft X-ray absorption magnetic circular dichroism, and photoelectron diffraction with two-dimensional electron analyzer) in 1998, he continued such experiments at Photon Factory. He met with the other author A. S. in 1995, who was a Ph.D. student under supervision of A. Fujimori in The University of Tokyo. A. S. partly collaborated with S. S. on the experiments at Photon Factory, and then moved to Osaka University after taking his Ph.D. degree to join S. S. lab. Both authors did extensive PES studies on rare earth compounds. During this period, they became acquainted with P. Baltzer with whom they discussed quite a lot on electron analyzers and the future of bulk sensitive photoelectron spectroscopy. It must be mentioned here that four more soft X-ray beam lines were later constructed up to now in SPring-8, demonstrating that an 8 GeV storage ring is very useful for soft X-ray spectroscopy.

Through various research collaborations making use of this high performance soft X-ray BL25SU, the authors profited immensely from discussions with J. W. Allen, D. Vollhardt, V. I. Anisimov, R. Claessen, M. Sing, G. Güntherodt, J. Osterwalder, J.-S. Kang, and W. Kuch. In our recent activity in the hard X-ray region, the authors are much obliged to T. Ishikawa. S. S. is also much obliged to C. M. Schneider and L. Plucinski for collaboration in the field of extremely low energy photoelectron spectroscopy. The authors would like to thank all of these people for fruitful collaborations and discussions. This book has benefitted quite a lot from continuous cooperation and discussions with many foreign and domestic friends as well as colleagues in the field of synchrotron radiation spectroscopy.

The content of this book is based on the lectures, which S. S. gave from 1980 in the Graduate School of Applied Physics, The University of Tokyo and later from 1989 in the Graduate School of Engineering Science, Osaka University. It is also based on the lectures given by A. S. from 2010 in the latter graduate school. The very recent studies performed together with A. Yamasaki, H. Fujiwara, and many other students in the graduate school are included in this book and the authors acknowledge them all.

Over a long period of his life, S. S. acknowledges his wife Sachiha for her patience and appreciation of his hard scientific job while taking care of their four children. A. S. acknowledges his wife Haruko for supporting his researches till late at night while taking care of two children. S. S. is much obliged to the Japan Society for Promotion of Science for twice supporting the Japan–Germany research collaborations and the Ministry of Education, Culture, Sports, Science and Technology, MEXT, Japan, for financial supports through COE, 21COE, and Grant-in-Aids for Creative Scientific Research (15GS0213). A. S. thanks the Japan Society for promotion of science for financial support through Global COE (G10) and scientific research on innovative areas "Heavy Electrons" (20102003). S. S. also thanks the Helmholtz Association and the Alexander von Humboldt Foundation for supporting the Japan–Germany research collaborations.

The authors are much obliged to A. Yamasaki of Konan University, Japan, M. Sing of University of Würzburg, Germany, K. Sakamoto of Chiba University, and T Okuda of Hiroshima University Japan, for their careful reading of the first manuscript and many useful comments for further improvement. The stimulating discussion with A. Winkelmann of the Max-Planck Institute for Microstructure Physics, Germany, is highly acknowledged. This book would not have been completed without the patience and continuous encouragement of C. Ascheron of Springer-Verlag.

Contents

Acronyms

2PP	Two-photon photoemission
AB	Alloy band
ABS	Antibonding state
AES	Auger electron spectroscopy
AFI	Antiferromagnetic insulator (AFMI is also used in the same meaning)
AFM	Atomic force microscope (microscopy) or antiferromagnetic
AFMI	Antiferromagnetic insulator
AFMM	Antiferromagnetic metal
AFPM	Antiferroparamagnetic
AP-PES	Ambient pressure photoelectron spectroscopy
ARPES	Angle resolved photoelectron spectroscopy
BS	Bonding state
BZ	Brillouin zone
CB	Conduction band
CCD	Charge coupled device
CD	Circular dichroism
CDA	Cylindrical deflector analyzer
CDMFT	Cluster extension of dynamical mean field theory
CDW	Charge density wave
CEF	Crystal electric field
CFS	Constant final state (spectrum)
CIS	Constant initial state (spectrum)
CMA	Cylindrical mirror analyzer
COI	Charge ordered insulator
CPA	Coherent potential approximation
DAC	Diamond anvil cell
DFT	Density functional theory
dHvA	De Haas-van Alphen
DMFT	Dynamical mean field theory
DOS	Density of states
EAL	Effective attenuation length
E_B	Binding energy

EDC	Energy distribution curve
ELEPES	Extremely low energy photoelectron spectroscopy in comparison with HAXPES
ELEARPES	Extremely low energy angle-resolved photoelectron spectroscopy
ESCA	Electron spectroscopy for chemical analysis
EUV	Extreme ultraviolet
FD	Fermi-Dirac distribution function
FEL	Free electron laser
FLAPW	Full-potential linearized augmented plane wave
FM	Ferromagnetic
FMM	Ferromagnetic metal
FS	Fermi surface
FT-STS	Fourier transform scanning tunneling spectroscopy
FWHM	Full width at half maximum
FZP	Fresnel zone plate
HAXPES	Hard X-ray photoelectron spectroscopy
HAXARPES	Hard X-ray angle resolved photoelectron spectroscopy
HF	Heavy Fermion or Hartree–Fock
HOPG	Highly oriented pyrolytic graphite
HTSC	High temperature superconductors or superconductivity
I	Insulator
IDEA	Imaging double energy analyzer
INS	Inelastic neutron scattering
IPS	Image potential state
IPES	Inverse photoemission (spectroscopy)
IR	Infrared
ITS	Inelastic tunneling spectroscopy
JDOS	Joint density of states
K–K	Kramers–Kronig
LD	Linear dichroism
LDA	Local density approximation
LEED	Low energy electron diffraction
LEEM	Low energy electron microscope or microscopy
LHB	Lower Hubbard band
LMTO	Linear muffin-tin orbital
M	Metal
MCD	Magnetic circular dichroism
MCP	Micro channel plate
MDC	Momentum distribution curve
MFM	Magnetic force microscope
MIT	Metal to insulator transition
MLD	Magnetic linear dichroism
NB	Non-bonding
NCA	Non crossing approximation

NEA	Negative electron affinity
OSA	Order sorting aperture
PDOS	Partial density of states
PED	Photoelectron diffraction
PEEM	Photoelectron emission microscope or microscopy. Photoemission electron microscope has the same meaning
PES	Photoelectron spectroscopy or photoelectron spectrum
PFY	Partial fluorescence yield
PI	Paramagnetic insulator
PICS	Photoionization cross section
PLD	Pulsed laser deposition
PM	Paramagnetic or paramagnetic metal
PMA	Plane mirror analyzer
PMI	Paramagnetic insulator
PMM	Paramagnetic metal
PY	Partial yield
QMC	Quantum Monte Carlo
QP	Quasi-particle
QWR	Quantum well resonance
QWS	Quantum well state
RE	Rare earth
RF	Radio frequency wave
RHEED	Reflection high energy electron diffraction
RIXS	Resonance inelastic X-ray scattering
RPES	Resonance photoelectron spectroscopy
RVB	Resonance valence bond
SBZ	Surface Brillouin zone
SC	Superconductive
SCES	Strongly correlated electron system
S_{eff}	Effective Sherman function
SEM	Scanning electron microscope or microscopy
SEMPA	Scanning electron microscope with polarization analysis
SIAM	Single impurity Anderson model
SP-ARPES	Spin polarized (or resolved) angle resolved photoelectron spectroscopy
SPEA	Scattering pattern extraction algorithm
SP-ELEPES	Spin polarized (or resolved) extremely low energy photoelectron spectroscopy
SP-ELEARPES	Spin polarized extremely low energy angle resolved photoelectron spectroscopy
SPEM	Scanning photoelectron microscope or scanning photoelectron microscopy
SP-IPES	Spin polarized inverse photoemission spectroscopy
SPLEED	Spin polarized LEED
SP-PES	Spin polarized (or resolved) photoelectron spectroscopy

SR	Synchrotron radiation
STM	Scanning tunneling microscope or microscopy
STS	Scanning tunneling spectroscopy
SW	Standing wave
SX	Soft X-ray
SXPES	Soft X-ray photoelectron spectroscopy
SX-RIXS	Soft X-ray resonant inelastic scattering
SXARPES	Soft X-ray angle resolved photoelectron spectroscopy
SXSW	Soft X-ray standing wave
TEY	Total electron yield
TFY	Total fluorescence yield
TLL	Tomonaga-Luttinger liquid
TM	Transition metal
TOF	Time of flight
TR-2PP	Time resolved two-photon photoemission
UHB	Upper Hubbard band
UHV	Ultra high vacuum
UPS	Ultraviolet photoelectron spectroscopy
VB	Valence band
VLEED	Very-low-energy-electron-diffraction
VLSPG(M)	Varied line spacing plane grating (monochromator)
VUV	Vacuum ultraviolet
XAS	X-ray core absorption spectroscopy
XMCD	X-ray absorption magnetic circular dichroism
XPS	X-ray photoelectron spectroscopy or spectra
XSW	X-ray standing wave
ZRS	Zhang-Rice singlet

Symbols

a, b, c	Lattice constant of a crystal
1D, 2D, 3D	One-dimensional, two-dimensional, and three-dimensional
B	Magnetic field
$\underline{c}$	Core hole
d	Period length of grooves of a grating or lattice plane distance of crystals
d_{hkl}	Interplane distance between nearest neighbor (*hkl*) planes
E	Electron energy
E_K	Kinetic energy of photoelectron in vacuum
E_F	Fermi level
E_k	Kinetic energy of photoelectron in solid
E_V	Energy of the vacuum level
e	A constant 2.71828 or electron or electron charge
F_{hkl}	Structure factor of crystal
(hkl)	Miller index
J	Exchange coupling or total angular momentum
k_F	Fermi wave number
L	Langmuir or ligand
$\underline{L}$	Ligand hole
M	Magnification or mass of an ion (atom)
m	integer or electron mass
n	Refractive index or an integer
R	Radius of curvature of a mirror
s	Surface layer thickness
ss	Sub-surface layer thickness
T	Temperature
T_C	Curie temperature
T_{CDW}	Critical temperature of charge density wave
T_c	Critical temperature of superconductivity or metal-insulator transition
T_K	Kondo temperature
T_N	Neel temperature
T_V	Verwey transition temperature

T_v	Valence transition temperature
t	Transfer energy or T/T*, where T is the sample temperature and T* is the pseudogap opening temperature
U	Electron correlation energy (on site Coulomb repulsive energy)
V_0	Inner potential
V_b	Bulk valence
V_s	Surface valence
V_{ss}	Sub-surface valence
Γ	(0,0,0) point of the BZ
$\Gamma_{e\text{-}e}$	Damping by electron–electron interaction
$\Gamma_{e\text{-}ph}$	Damping by electron–phonon interaction
$\Gamma_{e\text{-}imp}$	Damping by electron–impurity scattering
α	Incidence angle or asymmetry parameter or value of power-law
β	Exit angle
λ	Photon wavelength
λ_{mp}	Inelastic mean free path of photoelectrons
λ_u	Period length of the periodic magnet composing an undulator or a wiggler
ϕ	Work function or azimuthal angle
θ_B	Bragg angle
θ	Polar angle
θ_{blaze}	Blazing angle of a blazed grating
ρ	Density (g/cm^3) or radius of electron orbital motion
τ	Lifetime
$h\nu$	Photon energy
$\|\|$	Parallel
$\perp$	Perpendicular
γ	Lorentz factor, namely E/m_0c^2, energy divided by the electron rest mass energy. The electron specific heat coefficient is also expressed as γ

Chapter 1
Introduction

A detailed understanding of electronic structures of materials is very important from the view points of fundamental science as well as application. Among various means to probe electronic structures, the photon-in and electron-out measurements seem to be very useful since the energies, wave vectors numbers, polarizations and time structures are well defined for photons and electrons. Historically speaking, the photoelectric effects were first observed more than a century ago [1] when the spark between two electrodes was facilitated by illumination of the negative electrode by ultraviolet light. Electrons were discovered in 1897 [2] and the above mentioned photoelectric effect was found to be associated with the emission of electrons from the metal under ultraviolet light irradiation [3]. From the dependence of the emitted electron current on the light intensity and the electron velocity on the light frequency, Einstein proposed in 1905 the concept of photons [4]. The concept of the work function was also recognized around this time. Many metals were later carefully studied and the validity of the Einstein's photoelectric equation was confirmed.

The question of whether the photoelectric effect is a pure surface effect or due to a bulk effect was raised and unsettled for a long time because of poor vacuum condition. Considerable efforts were made to determine the work functions of many materials. It was then found that the work function of tungsten was significantly lowered by Cs coverage. From the application point of view, the lowering of the work function by means of Cs deposition was applied in Ag–O-Cs cathodes to detect infrared radiation. A similar technique is now often used for negative electron affinity cathodes to have sensitivity for infrared radiation down to $h\nu \sim 0.9$ eV or to prepare the spin polarized electron source from GaAs by circularly polarized laser light excitation. The photoelectric effect is now used for photomultipliers in a wide $h\nu$ region.

By the late 1950s the understanding of electronic structures of simple solids was noticeably advanced. For example, the photoelectron spectrum of Si was compared with the experimentally obtained optical spectra and band structures obtained by band structure calculations. Then a lot of experimental works were made on tetrahedral semiconductors. Correspondingly, development of theoretical

S. Suga and A. Sekiyama, *Photoelectron Spectroscopy*,
Springer Series in Optical Sciences 176, DOI: 10.1007/978-3-642-37530-9_1,

works followed. The concept of k-conserving direct transitions and non-conserving transitions was intensively discussed in the early 1960s [5] in photoelectron spectroscopy (PES).

The very early PES was performed below $h\nu \sim 6$ eV limited by the absorption by air. The PES in higher $h\nu$ was performed in the vacuum ultraviolet (VUV) with use of hydrogen lamps separated from the experimental chamber by LiF windows with a cut off at $h\nu \sim 11.8$ eV in the mid 1960s [6, 7]. The use of synchrotron radiation (SR) for PES became a reality in late 1960s at Deutsches Elektronen-Synchrotron (DESY) in Hamburg, Germany. The use of the SR from electron or positron storage rings instead of electron synchrotrons became a major stream after the mid 1970s due to the beam stability (with respect to the intensity and the time dependence of the beam position) in storage rings. Even in the case of VUV lamp sources, the extension of the $h\nu$ toward higher energies became feasible by He discharge lamps, which provide a strong discrete line at $h\nu = 21.2$ eV (HeI) and a weaker line at $h\nu = 40.8$ eV (HeII) under low pressures. Due to the narrow line width of the strong HeI line source, even the molecular vibration levels were studied in the case of gaseous samples. Rare gas lamps using Xe, Kr, Ar and Ne were also used for VUV PES.

On the other hand, photoelectron spectroscopy with X-ray sources later called XPS was also widely used. The origin of XPS can be traced back to the early 1910s. However, the energy resolution around that time was very poor (> few eV) due to the broad X-ray lines and inadequate resolution of the analyzers. When X-ray absorption and emission spectroscopy became popular, XPS became less used. Four decades later, Siegbahn and collaborators started to realize a high resolution photoelectron spectrometer for XPS in 1951 [8]. It was soon found that sharp peaks were observed at the high kinetic energy end of several broad bands or inelastic backgrounds. The sharp peaks observed in XPS were corresponding to core levels and often much sharper than absorption and emission spectra. Consequently such a new applications as the electron spectroscopy for chemical analysis (ESCA) was opened [9]. The Kα lines of Al and Mg are nowadays most often used for conventional XPS. However, the width of these characteristic lines is $\sim$1 eV. By use of bent quartz crystals, a resolution down to 0.2 eV was achieved for the Al Kα line at the sacrifice of intensity. Nowadays XPS in the valence band region is thought to provide information on the density of states modified by the photoionization cross sections (PICS) of the constituent electronic states. Fine structures and satellites of core levels accessible by high resolution XPS are also useful to discuss the final state interactions and screening processes besides obtaining chemical information as discussed in Chaps. 5, 6, 7 and 8.

Later conventional PES systems in laboratories became often equipped with both He lamp and X-ray tube to study valence band as well as core levels with as good resolution as possible. The ultra high vacuum (UHV) conditions were satisfied to some extent and various surface cleaning techniques became available. The large $h\nu$ gap between the He lamp and the X-ray tube became gradually covered by SR source. SR has various useful properties compared with the radiation from the X-ray tubes and VUV radiation from the He, Ne and other rare gas lamps as discussed later in Sect. 3.1. First of all, the high brilliance of SR due to its low emittance enables one to focus high photon flux in a small spot ($\ll$1 mm) on the sample, facilitating the measurement of small-size samples. Tunability of hν by use of the photon monochromators is another advantage, which enables the efficient excitation of some particular core levels. The polarization, either linear (horizontal or vertical) or circular (left or right helicity), enables one to probe the specific symmetry of the orbitals due to the selection rules and also enables one the measurements of dichroism in PES. Moreover, the pulsed feature of SR facilitates the time-of-flight PES measurements.

From the mid 1970s the angle resolved photoelectron spectroscopy (ARPES) became feasible to study band dispersions utilizing the energy and momentum conservation, where the momentum parallel to the specular surface was thought to be conserved during the electron escape from the surface. The momentum perpendicular to the surface was thought not to be conserved in this case. For quasi-two-dimensional (2D) or one-dimensional (1D) systems with negligible dispersion to the perpendicular direction, the ARPES at one hν could provide information on 2D or 1D band dispersions when the sample was properly rotated and/or the detection angle was scanned. For three-dimensional (3D) materials, the band dispersions depend on the momenta along three directions. In this case, the dispersion perpendicular to the surface was evaluated by normal emission measurements with changing the hν if the final state dispersion was already known or approximated by a parabolic band. ARPES measurements could be performed by (1) rotating a small analyzer with enough angular resolution around the sample or (2) rotating the sample in front of the analyzer entrance slit or (3) using a 2D detector installed on a hemispherical analyzer. One example of (3) was to use a microchannel plate just behind the exit of the hemispherical analyzer, where the energy corresponds to one axis on the detector and the emission angle corresponds to the other axis parallel to the exit slit of the hemispherical analyzer. In addition to these approaches, (4) really two dimensional detectors are now used which can detect the full angular distribution simultaneously. Development of this technique is explained in Sects. 3.4, 3.6, Chaps. 11 and 12.

The use of SR made it possible to do the resonance photoemission (RPES) with tuning hν in a particular core absorption edge region, thereby enhancing the intensity of the state with a particular orbital character as a result of quantum interference between the direct photoemission and the direct recombination following the core absorption. Such states as f and d outer shell states buried in the valence band can be effectively enhanced and probed by this technique. The resonance enhancement seems to be strongly related to the degree of the localization of the electronic states.

Besides the energies and momenta, the spin is another important physical quantity to be probed by PES. In accordance with the development of spin detectors, spin-polarized and angle-integrated and -resolved PES measurements became feasible. In the case of ferromagnetic materials, the essence of the long range and short range spin exchange interactions is gradually clarified. The spin polarization of the emitted photoelectron is also observed for nonmagnetic materials when they are excited by the circularly polarized light as a result of spin–orbit interaction. In addition, spin polarization is observed in the cases of Rashba effects and topological insulators in non-magnetic materials as explained in Sect. 11.3.

While abundant information was obtained on occupied electronic states by PES, less is known for the unoccupied states. The inverse photoemission spectroscopy (IPES) became available for the study of unoccupied states in the early 1980s, where monochromatic electrons impinged onto the sample and relaxed to the lower unoccupied states above the Fermi level with the emission of photons. Angle resolved IPES could provide information on the band dispersion of the unoccupied states. When spin polarized electrons were used, the spin polarized and angle resolved IPES is feasible. These subjects are handled in Chap. 10 and Sect. 11.5.

In the PES studies the concept of the inelastic mean free path (λ_{mp}) of photoelectrons is very important. The electron–electron interaction induces the finite and rather short inelastic mean free path, which is often in the range of 3–5 Å between $\sim$15 and $\sim$200 eV of kinetic energies (E_K). Therefore PES in this energy region is generally surface sensitive. Although this quantity depends upon the individual materials, it is widely accepted that the λ_{mp} increases above $E_K \sim$ 200 eV. On the other hand its increase below $\sim$10 eV is known to be much more material specific. Therefore caution is required to discuss the bulk electronic structures based on the PES data for $E_K < 10$ eV.

In the case of strongly correlated electron systems, the surface electronic structures are noticeably different from the bulk electronic structures. In order to overcome this difficulty, high resolution PES above a few hundred eV is strongly favored. Such experiments are now progressing on several beam lines of SR facilities in the world by using bright undulator light sources, high transmittance and high resolution photon monochromators and high performance electron analyzers.

As for low energy high resolution PES, the resolution better than 1 meV has been already achieved by use of a quasi-CW high repetition pulsed lasers. Microwave-excited electron cyclotron resonance Xe, Kr and Ar lamps have comparable hν resolution and are nowadays employed for high resolution low energy PES and ARPES for $E_K < 12$ eV. Many examples of the soft X-ray, hard X-ray and low energy bulk sensitive high resolution PES are explained in Chaps. 7, 8 and 9. As for electron analyzers, performance of hemispherical analyzers has been amazingly improved in the last two decades, realizing the resolution better than 1 meV in the low energies and $\leq$50 meV in the region up to a few keV as explained in Sect. 3.4.

There are a lot of ARPES results for band mapping in the E_K range below 200 eV (Sect. 3.6 and Chap. 6), providing useful information on the momentum dependence of the electron energies. As is well known the momentum resolution is in proportion to the square root of E_K. Therefore ARPES is feasible even in the soft X-ray region of a few hundred eV as discussed in Sect. 7.2. However, still the λ_{mp} is at most of the order of 10 to 20 Å (1–2 nm) and some contribution from the surface is not fully negligible. Although higher bulk sensitivity is achieved at a few keV, ARPES becomes more difficult due to the worse momentum resolution and low photoelectron count rate. In addition, recoil effects on photoelectron emission have recently been recognized in various solid materials composed of non-heavy elements even though such effects are not observed in some solids. Therefore ARPES in the hard X-ray region (Sect. 8.7) is not so popular yet. In the case of extremely low energy ARPES for $E_K < 10$ eV, on the other hand, the matrix element effects are very strong and the measurement in a wide region of the Brillouin zone (BZ) is not so simple. One must be very careful to check whether the extremely-low-energy-ARPES and -PES are really bulk sensitive in the individual materials (Chap. 9).

In order to realize an extreme bulk sensitivity for the study of the momentum dependence of electronic structure, resonance inelastic X-ray scattering (RIXS) experiment can be employed as very simply explained in Sect. 13.3. In this case the probing depth can be easily >10 μm and one can cover a few BZ by slightly and properly rotating the sample. One of the advantages of RIXS is that even insulators could be studied without the problem of charging up. Therefore RIXS is a powerful complementary tool to PES to study, for example, metal–insulator transition (MIT) systems.

In accordance with the development of nanotechnology, PES of micro- and nano-materials is of strong interest. By focusing the high brilliance synchrotron radiation and scanning the sample, such materials can be studied by scanning photoelectron microscopy (SPEM) with the lateral resolution better than 100 nm. If only the secondary electrons are detected, photoelectron emission microscopy (PEEM) is applicable to nano-materials with the lateral resolution down to 10 nm. Such techniques are explained in Sects. 3.6.4 and 3.6.5. In the case of magnetic materials, magnetic circular and/or linear dichroism is utilized to realize the contrast of the magnetic domains. In the case of quantum well states (QWS) in thin films on substrate materials, the confinement is perpendicular to the surface and the dispersion parallel to the surface can be observed by angle resolved measurements as discussed in Sect. 6.7 in the case of ARPES and in Chaps. 10 and Chap. 11 in the case of IPES. PES utilizing standing waves is also a powerful means to study such systems and also interface systems as explained in Sect. 7.3. Further the time resolution is now utilized for PES experiments as explained in Sect. 3.4.4. The aspects of high resolutions in energy, momentum, space and time are treated as much as possible in this book to overview the state of the art technologies.

Although the $h\nu$ resolution was much improved in the last two decades in the wide $h\nu$ region from a few eV up to 10 keV, the efforts for realizing higher total resolution and higher bulk sensitivity are still going on. To clarify the difference between the surface and bulk properties is one of the urgent subjects in materials sciences. Combination of not only PES and ARPES in wide $h\nu$ regions but also various complementary techniques are thought to provide fruitful information on electronic and atomic structures of materials attracting wide interest from both fundamental science and application.

References

1. H. Hertz, Ann. Physik **31**, 983 (1887)
2. J.J. Thompson, Phil. Mag. **44**, 293 (1897)
3. P. Lenard, Ann. Physik, **2**, 359 (1900). ibid. **8**, 149 (1902)
4. A. Einstein, Ann. Physik **17**, 132 (1905)
5. W.E. Spicer, Phys. Rev. Lett. **11**, 243 (1963)
6. J.A. Samson, *Techniques of Vacuum Ultraviolet Spectroscopy* (Wiley, New York)
7. J.A. Samson, D.L. Ederer, *Vacuum Ultraviolet Spectroscopy* (Academic Press, London, 2000)
8. M. Cardona, L. Ley (eds) Detailed stories are given in Photoemission in Solids I, (Springer, Berlin, 1978)
9. H. Fellner-Feldegg, U. Gelius, B. Wannberg, A.G. Nilsson, E. Basilier, K. Siegbahn, J. Electron. Spectrosc. Rel. Phenom. **5**, 643 (1974)

Chapter 2
Theoretical Background

In this chapter, the theoretical background necessary for the discussion of electronic structures in solids based on the photoemission spectra is described. Although the photoemission process itself is not a main issue in this book, the processes taking place in solids should briefly be explained to help the understanding of the discussions given later. The photoemission process in solids can theoretically be described by the inverse Low-Energy Electron Diffraction (LEED) formalism [1, 2]. In this formalism, all many-body interactions (including photoelectron inelastic scattering process) are taken into account to quantitatively calculate the photoemission spectral weights. However, it is not practical to use this theory for the analysis of photoemission data except for simple metals since there is no straightforward way to apply this formalism in other systems. In most cases, the photoemission spectra of solids have successfully been analyzed on the basis of the "three-step model" for several decades although it is phenomenological. Thus, the three-step model is introduced first, and then several theoretical models for describing the electronic structure of solids in the initial and final photoemission states are discussed. Namely, the valence band photoexcitation process is first discussed for non-interacting systems then for strongly correlated electron systems. In the latter half of this chapter are discussed the core-level photoemission process for strongly correlated electron systems, matrix element effects and various theoretical models to describe the spectra of strongly correlated electron systems. Single impurity Anderson model, cluster model, Hubbard model, dynamical mean field theory are briefly described together with new directions and some remarks. For detailed discussions on the photoemission process (especially for the core-level photoemission spectral functions with asymmetric line shape due to the creation of electron-hole pairs, the satellite structure due to plasmon excitations etc.), see previously published books of photoemission [2, 3].

S. Suga and A. Sekiyama, *Photoelectron Spectroscopy*,
Springer Series in Optical Sciences 176, DOI: 10.1007/978-3-642-37530-9_2,

2.1 Three-Step Model

The three-step model is schematically shown in Fig. 2.1. In this model, it is assumed that the photoemission process can be divided into three steps:

1. Photoexcitation of an electron inside the solid (creation of a photoelectron),
2. Travel of the photoelectron to the sample surface,
3. Emission of the photoelectron into the vacuum.

Then the photocurrent or photoemission intensity as a function of the photoelectron kinetic energy in vacuum (E_K) and the excitation photon energy $h\nu$ is proportional to the product of the probabilities corresponding to each step. In fact, the wave vector (or momentum) **K** of the photoelectron in the vacuum is different from its wave vector **k** in the solid. The relation between **K** and **k** is given later for discussing the band dispersions. So far as its mutual difference is not essential, **K** is employed for simplicity. When the probabilities of the three steps are represented by $P(E_K, h\nu)$, $T(E_K, h\nu)$ and $D(E_K)$, respectively, the photoemission intensity is proportional to

$$P(E_K, h\nu)T(E_K, h\nu)D(E_K). \tag{2.1}$$

The information on the electronic states of solids in the initial state is included in the first term $P(E_K, h\nu)$. If the other functions do not depend on E_K within a scanned energy range (this assumption is practically reasonable at least for $E_K \gtrsim 100$ eV as discussed later), the photoemission spectra directly reflect $P(E_K, h\nu)$. The inelastic photoelectron scattering effects affect the term $T(E_K, h\nu)$ to some extent. In many cases, the so called satellite structures due to intrinsic plasmon, exciton and/or interband excitations (for instance, rare earth $5p \rightarrow 5d$ transitions with the energy of 20–30 eV [4]) taking place simultaneously with the photoexcitation are seen in the core-level photoemission spectra. But they are not considered for $P(E_K, h\nu)$ in the formulation of the photoemission for strongly

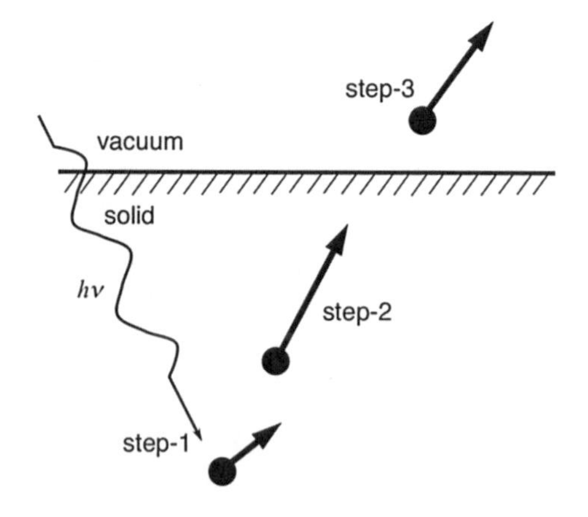

Fig. 2.1 Schematic representation of a photoemission process in the three-step model

correlated electron systems for simplicity. For the practical analysis of the photoemission spectra, it is convenient to additionally take these excitations together with the extrinsic excitations as inelastic scattering events (taking place during the traveling of the photoelectron to the surface, step-2) into account since the discrimination of the extrinsic excitations from the intrinsic ones is actually difficult. Hereafter these excitations are not discussed in this chapter although they could implicitly be considered in the photoemission process.

From the Fermi's golden rule, the photoexcitation probability in the N-electron system is proportional to

$$\sum_{f,j} \left|\langle f|\mathbf{A}\cdot\mathbf{p}_j|i\rangle\right|^2 \delta(E_f(N) - E_i(N) - h\nu), \tag{2.2}$$

where $|i\rangle(|f\rangle)$, $E_i(N)\,(E_f(N))$, $\mathbf{A}$ and $\mathbf{p}_j$ stand for the photoemission initial (final) state including the photon field, the total energy of the N-electron system in the initial (final) state, the quantized vector potential of the excitation light and the momentum operator for the j-th electron to be excited, respectively. Since the photoexcited electron energy is generally much larger than each one-electron energy in the remaining (N – 1)-electron system, it is reasonable to assume that the photoexcited electron does not interact with the N – 1 electrons. This assumption is equivalent to the situation that the time scale of the photoexcitation is much shorter than that of the electron interactions and thus the photoelectron is instantaneously created by the electron-photon interaction. This is called as "sudden approximation", which is applicable at least to the high-energy photoelectron limit. To date, a serious problem of using the sudden approximation has not been reported for the analyses of valence-band photoemission spectra. However, the applicability of the sudden approximation could be controversial for extremely low photon energies $h\nu < 10$ eV, which should be examined in the future.

Under the sudden approximation, the total N-electron energies $E_f(N)$ in the photoexcited (final) states are expressed as

$$E_f(N) = E_K + \phi + E_f(N-1), \tag{2.3}$$

where ϕ stands for the work function of the solid. The photoexcited states $|f\rangle$ with the photon field are represented by the direct product of the states in the subspaces of the photon field, the photoelectron, and the remaining (N – 1)-electrons as

$$|f\rangle = |n_{h\nu} - 1\rangle|E_k\rangle|E_f(N-1)\rangle, \tag{2.4}$$

where $n_{h\nu}$ denotes the number of incident photons with the energy of $h\nu$. By the same way, the initial state $|i\rangle$ is expressed as

$$|i\rangle = |n_{h\nu}\rangle|0_{PE}\rangle|E_i(N)\rangle, \tag{2.5}$$

where $|0_{PE}\rangle$ denotes the vacuum state in the photoelectron subspace. Since the photon annihilation operator is involved in the (quantized) term $\mathbf{A}\cdot\mathbf{p}_j$, the function $P(E_K, h\nu)$ is represented as

$$\begin{aligned} P(E_K, h\nu) &\propto n_{h\nu} \sum_{f,j} |\langle E_f(N-1)|\langle E_K|M_{Kj}a_K^{\dagger}a_j|0_{PE}\rangle|E_i(N)\rangle|^2 \\ &\quad \times \delta\big(E_f(N-1) - E_i(N) + E_K + \phi - h\nu\big) \\ &= n_{h\nu} \sum_{f,j} |M_{Kj}\langle E_f(N-1)|a_j|E_i(N)\rangle|^2 \delta(E_f(N-1) - E_i(N) + E_K + \phi - h\nu), \end{aligned} \tag{2.6}$$

where M_{Kj}, $a_K{}^{\dagger}$ and a_j stand for the matrix element including the one-electron photoexcitation process, the photoelectron creation operator, and the annihilation operator of the j-th electron, respectively.

The term of $T(E_K, h\nu)$ represents the probability of the photoelectron motion to the surface without serious inelastic scattering in step-2, being expressed by using the absorption coefficient $\alpha(h\nu)$ for the incident photon and the photoelectron inelastic mean free path $\lambda_{mp}(E_K)$. Since the bulk and surface contributions to the experimental spectra depend directly on $\lambda_{mp}(E_K)$, this term will be discussed in detail in Chap. 4. $1/\alpha(h\nu)$ is of the order of 100–1000 Å or more for $h\nu$ in the range of 6–10,000 eV, which is much longer than $\lambda_{mp}(E_K) \lesssim 100$ Å for most elemental solids as shown in Fig. 2.2 [5, 6], and $T(E_K, h\nu)$ is given as

$$T(E_K, h\nu) = \frac{\alpha(h\nu)\lambda_{mp}(E_K)}{1 + \alpha(h\nu)\lambda_{mp}(E_K)} \simeq \alpha(h\nu)\lambda_{mp}(E_K). \tag{2.7}$$

The λ_{mp} takes a minimum of ~3–5 Å at E_K of ~15–200 eV in many cases. This minimum length corresponds roughly to lattice constants of various solids. Therefore, the valence-band photoemission spectra with E_K of ~15–200 eV by using a He discharge lamp or synchrotron light source mainly reflect the surface electronic states of solids. A certain way to obtain the spectra predominantly reflecting the bulk electronic structure is to use higher $h\nu$ than ~500 eV. As discussed in Chap. 7, the bulk contribution in the spectra is actually reported to be >60 % at E_K of 600–1,000 eV for various strongly correlated materials. So far the analytical formula known under the acronym TPP-2 M was often used to calculate $\lambda_{mp}(E_K)$ at E_K ~50–2,000 eV [6]. According to this formula, $\lambda_{mp}(E_K)$ is generally at most 6 Å at $E_K < 100$ eV (hence the bulk contribution to the spectra at normal emission $I_B/I = \exp(-s/\lambda_{mp})$ is expected to be ~40 % for a sample with a "surface thickness" s of 5 Å) and > 15 Å at E_K >700 eV (in this case, I_B/I ~70 % for s of 5 Å), although it depends really on individual materials and is not so universal. The calculated values are fairly consistent with those obtained from optical data with a deviation of ~10 %. The λ_{mp} was expected to be longer at $E_K <$ ~10 eV according to Fig. 2.2, as nowadays believed by several scientists. However, from

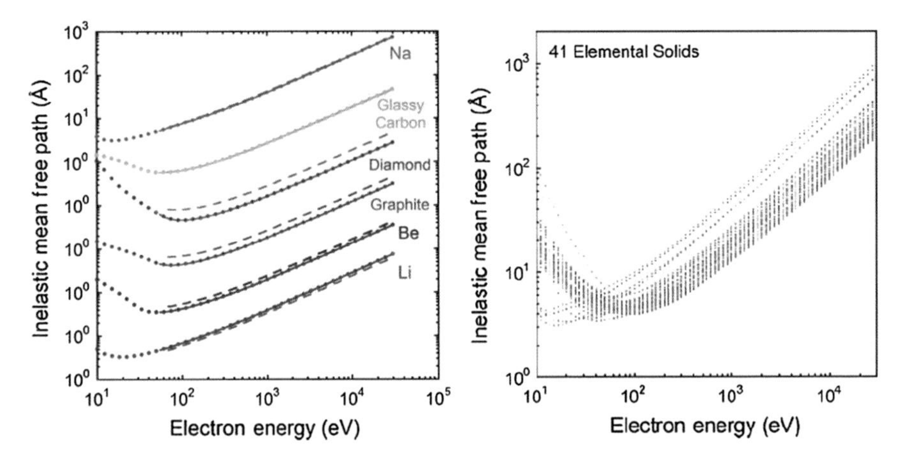

Fig. 2.2 Kinetic energy dependence of the photoelectron inelastic mean free path λ_{mp} (cited from [6]). So far the TPP-2M calculation was often used to predict λ_{mp}. Single-pole approximation was applied for electron energies $E_k \sim E_K$ higher than 300 eV, whereas λ_{mp} was calculated from optical data by using the full Penn algorithm for E_K up to 300 eV in Ref. [6]

the facts that prominent surface contributions have been seen in the spectra near the Fermi level (E_F) even at $h\nu < 10$ eV for silver [7], copper [8] as well as topological insulator [9], the "expected" long λ_{mp} comparable to that at $h\nu > 1{,}000$ eV is quite controversial. More detailed discussions on the bulk and surface sensitivity are given in Chaps. 4 and 9.

The term $D(E_K)$ can be calculated when the photoelectrons can be treated as nearly free electrons with a potential of depth $E_v - E_0 \equiv V_0$ (this is called the inner potential), where E_v denotes the vacuum level and E_0 stands for the bottom energy in a nearly free electron band. This approximation is appropriate since the photoelectron energy is much higher than that for bound electrons in solids. In the nearly free electron model, the kinetic energy of a photoelectron inside the solid is $E_K + V_0$ whereas it becomes E_K suddenly at the boundary. Since the "force" is applied to the photoelectron only perpendicular to the surface, the photoelectron momentum parallel to the sample surface is conserved on the emission into vacuum, which is one of the fundamental principles for angle-resolved photoelectron spectroscopy discussed below. To satisfy the condition that the perpendicular momentum component of the photoelectron emitted into the vacuum to be positive, $D(E_K)$ is calculated as

$$D(E_\mathbf{K}) = \frac{1}{2}\left(1 - \sqrt{\frac{V_0}{E_\mathbf{K} + V_0}}\right). \tag{2.8}$$

This function depends gently on E_K, and can be regarded as a constant when the recorded kinetic energy range is narrow enough compared with E_K. Therefore, it is hereafter assumed that the E_K dependence of the terms $T(E_K, h\nu)$ and $D(E_K)$ is negligible within the discussed kinetic energy range of one spectrum.

2.2 Valence-Band Photoexcitation Process for Non-interacting Systems

The one-electron binding energy for the j-th electron to be photoexcited is defined as

$$E_B(j) \equiv E_f(N-1) - E_i(N). \tag{2.9}$$

From the energy conservation law in Eq. (2.6) the relation

$$E_B(j) = h\nu - E_K - \phi > 0 \tag{2.10}$$

is obtained. In the initial states, the total N-electron energy $E_i(N)$ is expressed by the one-electron energy of the j-th electron ε_j and the remaining (N – 1)-electron energy $E_i(N-1)$ as

$$E_i(N) = \varepsilon_j + E_i(N-1). \tag{2.11}$$

If the photoexcitation process does not produce any change in the (N – 1)-electron system, which means that there is no orbital relaxation in the process, the (N – 1) electron energy is conserved as

$$E_f(N-1) = E_i(N-1). \tag{2.12}$$

From Eqs. (2.9), (2.11) and (2.12),

$$E_B(j) = -\varepsilon_j \tag{2.13}$$

is obtained (Koopmans' theorem), indicating that the one-electron energy $\varepsilon_j < 0$ can be measured by photoelectron spectroscopy. Usually, the concept of the binding energy is used also for the inner core electrons. It should be noted, however, that the experimentally measured core-level binding energy in solids is not directly related to its inner core orbital energy since such orbital relaxations as the electron-hole excitations (for metals) and screening by charge transfer (for strongly correlated electron systems) always take place due to the presence of the created core hole in the photoexcitation final states.

In the non-interacting systems in which the effects of electron–electron interactions are negligible and hence the Koopmans' theorem is satisfied, the photoemission spectral function $\rho(E_K, h\nu) \propto P(E_K, h\nu)$ can be rewritten as a function of one-electron energies. From Eqs. (2.6), (2.11) and (2.12),

$$\rho(E_K, h\nu) \propto \sum_{fj} \left|M_{Kj}\langle E_f(N-1)|a_j|E_j(N)\rangle\right|^2 \delta(E_K + \phi - h\nu - \varepsilon_j) \tag{2.14}$$

is obtained. If the term

$$\sum_{f} |M_{Kj}\langle E_f(N-1)|a_j|E_i(N)\rangle|^2 \tag{2.15}$$

has negligible energy and orbital dependence, the spectral function is expressed as

$$\rho(E_K, h\nu) = \rho(\omega) = \sum_j \delta(\omega - \varepsilon_j) \equiv N(\omega), \tag{2.16}$$

where

$$\omega \equiv E_K + \phi - h\nu \tag{2.17}$$

stands for the one-electron energy relative to E_F ($\omega < 0$ for the occupied energy side). N(ω) is equivalent to the density of states, and hence the valence-band (angle-integrated) photoemission spectra probe the density of states (DOS). It is experimentally well known that the photoionization cross-section (PICS), which is taken into account in (2.15), depends on the atomic orbital occupied by the j-th electron. Therefore, the actual spectral function is often approximated by using the coefficient C_l proportional to the cross-section and partial density of states $N_l(\omega)$ for the orbital l as

$$\rho(\omega) = \sum_l C_l N_l(\omega). \tag{2.18}$$

Figure 2.3 shows the photon energy dependence of the cross-sections [10–12] reflecting C_l for several orbitals. One can easily notice that the cross-sections drastically decrease on going from several tens eV to a higher excitation energy side, which suggests the fundamental difficulty in measuring the PES spectra at high-excitation energy (>a few hundreds eV) with comparable energy resolution and statistics to those at low-excitation energy (<100 eV).

For angle-resolved photoelectron spectroscopy (ARPES), momentum conservation must be taken into account in addition to energy conservation. As shown in Fig. 2.4 with **k** in the crystal, the photoelectron momentum $\mathbf{k}_f$ is the sum of the

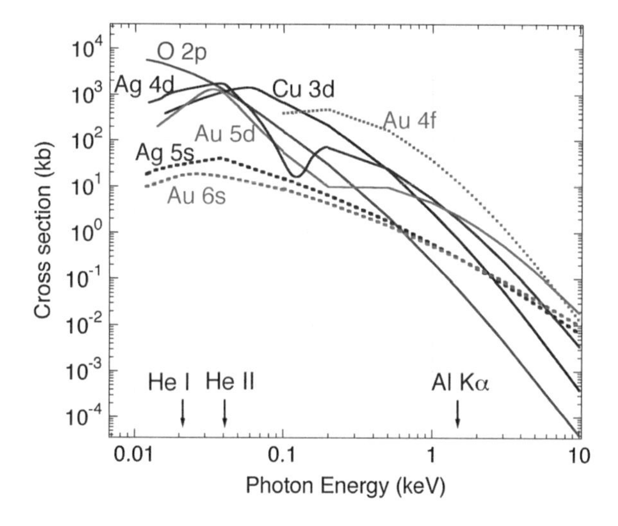

Fig. 2.3 Excitation photon energy dependence of the photoionization cross-sections (PICS) for several orbitals. The calculated data are from Ref. [10–12]

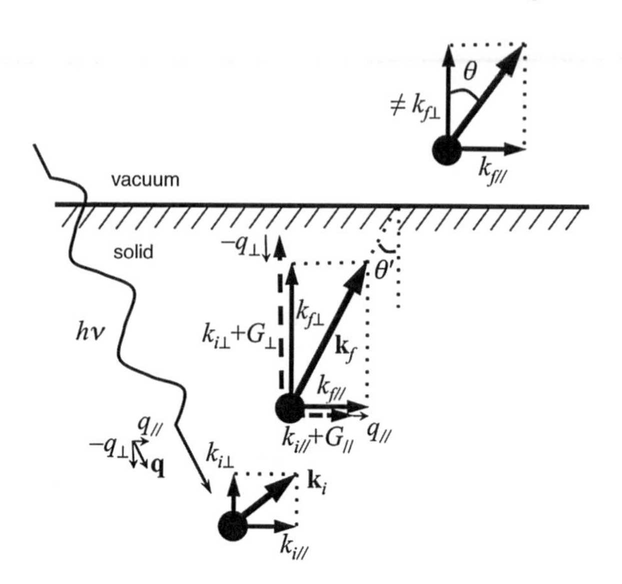

Fig. 2.4 Schematic representation of the momentum conservation at each step in the photoemission process in solids. In the figure, photon momentum normal to the sample surface is defined as $-q_\perp$ ($q_\perp > 0$). The $K_\perp$ in the vacuum is not equal to the $k_{f\perp}$ in the crystal

momenta of the electron in the initial state $\mathbf{k}_i$, the crystal $\mathbf{G}$, and the incident photon $\mathbf{q}$ as

$$k_{f//} = k_{i//} + G_{//} + q_{//}, \ : \ \text{parallel direction to the surface} \tag{2.19a}$$

$$k_{f\perp} = k_{i\perp} + G_\perp - q_\perp. \ : \ \text{normal direction to the surface} \tag{2.19b}$$

(Here the photon momentum normal to the sample surface is defined as $-q_\perp$.) In the case of low-energy ARPES at hν < 100 eV, $\mathbf{q}$ is often neglected since it is much smaller than $\mathbf{k}_i$ and $\mathbf{k}_f$. The photoelectron momentum component is also conserved along the surface parallel direction at step-3. Since the photoelectrons emitted into the vacuum are treated as free electrons, $K_{f//} = \mathbf{k}_{f//}$ can be obtained from E_K and the polar angle θ as

$$k_{f//} = \frac{\sqrt{2mE_K}}{\hbar}\sin\theta, \tag{2.20}$$

where m denotes the electron mass. Instead of (2.20), a practical formula

$$k_{f//}\left[\text{Å}\right]^{-1} = 0.5123\sqrt{E_K[\text{eV}]}\sin\theta \tag{2.21}$$

is often used. The momentum resolution is thus obtained as

$$\Delta k_{f||} = \frac{\sqrt{2mE_K}}{2\hbar}\cdot\frac{\Delta E_K}{E_K}\sin\theta + \frac{\sqrt{2mE_K}}{\hbar}\cos\theta\cdot\Delta\theta, \tag{2.22}$$

where ΔE_K denotes the energy resolution and $\Delta\theta$ stands for the acceptance angle of photoelectrons. In Eq. (2.22), the first term on the right side is negligibly smaller than the second term in general. More than a few decades ago, typical $\Delta\theta$ was

$2° = 0.035$ radian or larger. When the so-called XPS was performed by using an Al-Kα line (hν = 1,486.6 eV) with the above acceptance angle, $\Delta k_{f//} \sim 0.7\,\text{Å}^{-1}$ covers nearly the half of the Brillouin zone ($\pi/a \sim 0.8\,\text{Å}^{-1}$ for $a \sim 4$ Å), leading naturally to momentum-integrated valence-band photoemission. From Eqs. (2.19a) and (2.20),

$$k_{i//} = \frac{\sqrt{2mE_K}}{\hbar}\sin\theta - q_{//} - G_{//}. \tag{2.23}$$

Although the momentum is not conserved along the surface normal direction, $k_{i\perp}$ can be obtained when the photoelectrons can be treated as nearly free electrons in the solid by using the inner potential V_0 and Eq. (2.19b) as

$$k_{i\perp} = \frac{\sqrt{2m(E_K\cos^2\theta + V_0)}}{\hbar} + q_\perp - G_\perp. \tag{2.24}$$

It should be noted that Eqs. (2.23) and (2.24) are satisfied also for the strongly correlated electron systems. The momentum resolution along the normal direction $\Delta k_{f\perp}$ is not determined from (2.24) but depends on the $\lambda_{mp}(E_K)$ [13] as

$$\Delta k_{f\perp} \sim 1/\lambda_{mp}(E_K). \tag{2.25}$$

When an orbital (or band) l is used to express the ARPES spectra I(E_K, hν, θ) instead of j, it is represented as

$$\begin{aligned} I(E_K, h\nu, \theta) = \sum_l \delta(\omega - \varepsilon_l(\mathbf{k}_i))\delta\left(k_{i//} + q_{//} + G_{//} - \frac{\sqrt{2mE_K}}{\hbar}\sin\theta\right) \\ \times\,\delta\left(k_{i\perp} - q_\perp + G_\perp - \frac{\sqrt{2m(E_K\cos^2\theta + V_0)}}{\hbar}\right). \end{aligned} \tag{2.26}$$

Usually, it is more convenient to use the retarded Green's function while retaining the relations in Eqs. (2.21) and (2.24) for representing the ARPES spectra. Using the Dirac identity

$$\frac{1}{x + i\eta} = P\frac{1}{x} - i\pi\delta(x), \tag{2.27}$$

where η is a positive infinitesimal number, one obtains

$$\delta(\omega - \varepsilon_l(\mathbf{k}_i)) = -\frac{1}{\pi}\text{Im}\frac{1}{\omega - \varepsilon_l(\mathbf{k}_i) + i\eta}. \tag{2.28}$$

Since the retarded Green's function for the non-interacting systems is given by

$$G_l^0(\mathbf{k}_i, \omega) = \frac{1}{\omega - \varepsilon_l(\mathbf{k}_i) + i\eta}, \tag{2.29}$$

the ARPES spectral function for the band l, A_l ($\mathbf{k}_i$, ω), is thus represented by

$$A_l(\mathbf{k}_i, \omega) = -\frac{1}{\pi}\mathrm{Im}G_l^0(\mathbf{k}_i, \omega). \quad (2.30)$$

2.3 Valence-Band Photoexcitation Process for Strongly Correlated Electron Systems

For an electronic system called strongly correlated electron system in which the Coulomb repulsive interactions between the electrons are not negligible, the Koopmans' theorem (Eq. 2.14) is no longer satisfied as discussed below. Here the one-site $|4f^n\rangle$ electron system (the same as n-4f electron system) is considered, where the Coulomb repulsive (electron correlation) energy is defined as U_{ff} and the one-electron 4f orbital energy ε_f is supposed to have no momentum dependence. This discussion may be applied not only to 4f electron systems but also to other correlated electron systems. The total energy of a $|4f^n\rangle$ electron system $E_i(n)$ is described as

$$E_i(n) = n\varepsilon_f + \binom{n}{2}U_{ff} = n\varepsilon_f + \frac{n(n-1)}{2}U_{ff}. \quad (2.31)$$

If a photon with the energy hν impinges on the sample with this electronic state, the photoelectron is emitted while leaving the (n – 1)-4f electron system behind. Since

$$E_f(n-1) = (n-1)\varepsilon_f + \binom{n-1}{2}U_{ff} = (n-1)\varepsilon_f + \frac{(n-1)(n-2)}{2}U_{ff}, \quad (2.32)$$

the energy difference between the (n – 1)-4f final and n-4f initial states is given by (see Eq. 2.9)

$$E_B(4f) = E_f(n-1) - E_i(n) = -\varepsilon_f - (n-1)U_{ff} \neq -\varepsilon_f, \quad (2.33)$$

which does no longer satisfy the Koopmans' theorem.

Before going to the formulation of the ARPES spectra for the strongly correlated systems, the inverse photoemission process is considered to discuss the effects of U_{ff} on the photoemission spectra. In the inverse photoemission process, monochromatic electrons impinge on the sample and the electrons relax to lower unoccupied states with the emission of photons. Then the n-4f initial state is excited to the (n + 1)-4f final state. Since

$$E_f(n+1) = (n+1)\varepsilon_f + \binom{n+1}{2}U_{ff} = (n+1)\varepsilon_f + \frac{n(n+1)}{2}U_{ff}, \quad (2.34)$$

the energy difference is obtained from Eqs. (2.31) and (2.34) as

$$E_{\text{inv}}(4f) = E_f(n+1) - E_i(n) = \varepsilon_f + nU_{ff} = -E_B(4f) + U_{ff}. \tag{2.35}$$

From (2.33) and (2.35), one obtains

$$E_f(n+1) + E_f(n-1) - 2E_i(n) = U_{ff}, \tag{2.36}$$

which can be modified to $E_i(n) + E_i(n) + U_{ff} = E_f(n-1) + E_f(n-1)$. Accordingly, the energy difference between the photoemission and inverse photoemission peaks in the spectra for n-4f electron systems is equal to U_{ff}. If the energy term $E_B(4f)$ in Eq. (2.33) is very close to 0 eV, the valence fluctuation phenomenon is expected, where the ground state is expressed as a hybrid of both n-4f and (n – 1)-4f states and the number of the f electron is no more an integer. In this case, the (n – 2)-4f final state is observed in addition to the (n – 1)-4f final state in the photoemission spectrum and the n-4f final state in addition to the (n + 1)-4f state in the inverse photoemission spectrum. In both cases, the mutual energy splitting is close to U_{ff}.

We now turn to the question how the electron correlation effects as discussed above are taken into account in the ARPES spectral function. It is natural to expand the Green's retarded function for non-interacting systems to that for the strongly correlated electron systems. That is, the ARPES spectral function (2.30) should be modified as

$$A_l(\mathbf{k}_i, \omega) = -\frac{1}{\pi}\text{Im}G_l(\mathbf{k}_i, \omega) = -\frac{1}{\pi}\text{Im}\frac{1}{\omega - \varepsilon_l(\mathbf{k}_i) - \Sigma(\mathbf{k}_i, \omega)}, \tag{2.37}$$

where $\Sigma(\mathbf{k}_i, \omega)$ is a complex function called self-energy whose real and imaginary parts give the deviation of the one-electron energy between a bare band electron and a *dressed* quasi-particle, and the inverse life-time for the quasi-particle (QP), respectively.

The ARPES spectral function in Eq. (2.37) can also be derived from Eq. (2.6) if the matrix element M_{Kj} is assumed to have negligible energy dependence. When the valence-band electrons are labeled by band index l and momentum $\mathbf{k}_i$ instead of using j, one obtains

$$P(\mathbf{k}_i, \omega) \propto \sum_{f,l} |\langle E_f(N-1)|a_{l\text{ki}}|E_i(N)\rangle|^2 \delta(\omega + E_f(N-1) - E_i(N)). \tag{2.38}$$

Using the Dirac identity (2.27) and

$$\sum_f \big|E_f(N-1)\big\rangle\big\langle E_f(N-1)\big| = 1, \tag{2.39}$$

(2.38) is transformed into the form

$$P(\text{k}_i, \omega) \propto \sum_l -\frac{1}{\pi}\text{Im}\big\langle E_i(N)|a^{\dagger}_{l\text{k}i}\frac{1}{\omega + \mathcal{H} - E_i(N) + i\eta}a_{l\text{k}i}\big|E_{i(N)}\big\rangle, \tag{2.40}$$

where $\mathcal{H}$ stands for the Hamiltonian for the electron system. It is known that the right side of (2.40) is equivalent to the retarded Green's function shown in Eq. (2.37).

From Eq. (2.37), it is recognized that the quasi-particle peak near E_F is seen at $\omega = \varepsilon_l^*$, where ε_l^* satisfies the equation listed below:

$$\varepsilon_l^* = \varepsilon_l(\mathbf{k}_i) + \mathrm{Re}\Sigma(\mathbf{k}_i, \varepsilon_l^*), \tag{2.41}$$

with the reduced spectral weight $z(\varepsilon_l^*)$ as

$$z(\varepsilon_l^*) = \left[1 - \frac{\partial \mathrm{Re}\Sigma(\mathrm{k}_i,\ \omega)}{\partial \omega}\right]^{-1}\Bigg|_{\omega=\varepsilon_l^*} < 1. \tag{2.42}$$

From Eqs. (2.41) and (2.42), one can imagine that some portion $(z(\varepsilon_l^*))$ of the correlated electrons near E_F are observed as quasi-particles at energies ε_l^*, renormalized with respect to the bare band energies $\varepsilon_l(\mathbf{k}_\mathrm{i})$ (<0 for the occupied side) by $\mathrm{Re}\ \Sigma(\mathrm{k}_i, \varepsilon_l^*)$ in the photoemission process. This behaviour is schematically shown in Fig. 2.5, where the ARPES spectral functions for non-interacting and interacting cases are schematically shown by the dashed and solid lines. For the Fermi liquid, the self energy $\Sigma(\mathrm{k}_i, \omega)$ can be expanded as $-g_1\omega - \mathrm{i}g_2\omega^2$ $(g_1, g_2 > 0)$ in the vicinity of E_F if the momentum dependence of the self energy can be neglected. Here it is known that (2.37) may give a broad peak away from E_F called incoherent (since there is no counterpart in the non-interacting systems) when the electron correlation effects become large, whose spectral weight is $1 - z(\varepsilon_l^*)$. This broad peak corresponds to the (n – 1)-4f final state mentioned above or a so-called lower Hubbard band discussed later, reflecting the Coulomb repulsion energy U_{ff}.

In the self-energy, all other electron interactions besides the electro-electron interactions are taken into account, affecting the spectral function within respective energy scales. The electron–electron interactions, electron–phonon interactions, and electron-impurity interactions causing the electron scattering are typical in the photo-excitation process, whose energy scales are of the order of < several eV, <100 meV and <a few tens meV, respectively [14]. Among them, the impurity scattering just gives an energy- and momentum-independent finite value of $\mathrm{Im}\Sigma(\mathrm{k}, \omega)$, a broadening of the quasi-particle peak. For magnetic materials, the additional effects of electron-magnon interactions to the self-energy have been reported based on ARPES data [15].

2.4 Core-Level Photoemission Process for Strongly Correlated Electron Systems

In the case of the strongly correlated electron systems, the core-level photoemission is often very useful to get information on valence-band electronic states since they are affected by the inner core hole created in the photoemission final

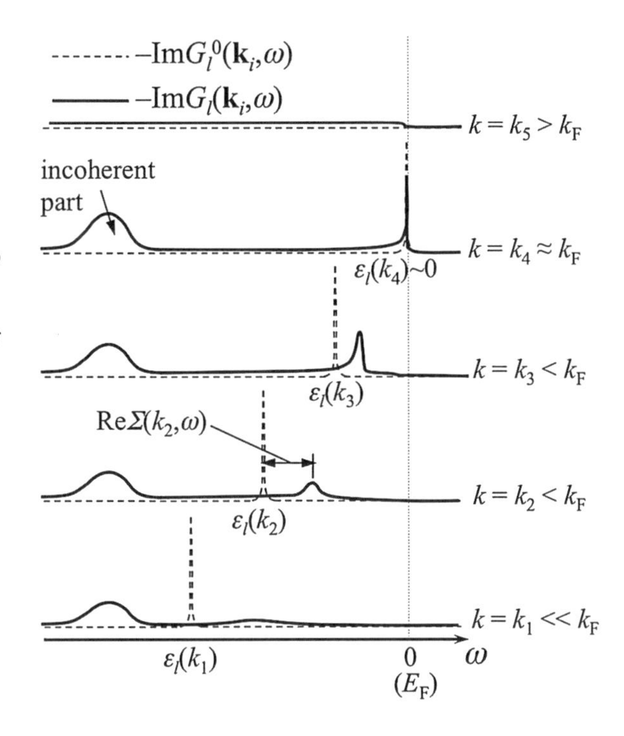

Fig. 2.5 Schematic representation of the comparison of ARPES spectral functions between the non-interacting and interacting (strongly correlated) electron systems. On going from k_1 to k_3, the electron energy approaches to E_F and further crosses E_F at $\sim k_4$, whereas the electron energy is above E_F at k_5 after the band crossing

states. In other words, the core-hole screening by the valence electrons can give a multiple-peak structure such as atomic multiplet, charge-transfer satellites and well-screened final-state structure in the core-level photoemission spectra. When the core hole is created by the incident photon, each valence-band electron feels a Coulomb attractive energy U_{fc}, which is empirically known to be larger than the Coulomb repulsive energy between the valence-band electrons U_{ff} ($U_{fc} \sim 1.5 - 2U_{ff}$). This is one of the orbital relaxation effects in the photoexcitation process.

It is convenient to divide the N-electron system into N_C-inner core and N_V-outer (valence) electron subsystems with $N = N_C + N_V$ to discuss the core-level photoemission process in solids following the formulation by Kotani [3] and Gunnarsson and Schönhammer [16]. Instead of (2.4), the final states are represented by the direct product of the states in the subspaces of the photon field, the photoelectron, the core hole, and the outer N_V-electrons as

$$|f\rangle = |n_{h\nu} - 1\rangle |E_K\rangle |c\rangle |E_f'(N_v)\rangle, \tag{2.43}$$

where $|E_f'(N_V)\rangle$ and $|c\rangle$ denote the N_V-outer electron subsystem with the given total energy in the presence of the core hole, and the core-level states, respectively. By the same way, the initial state $|i\rangle$ is expressed as

$$|i\rangle = |n_{h\nu}\rangle|0_{PE}\rangle|0_c\rangle|E_i(N_V)\rangle \tag{2.44}$$

Here, $E_i(N_V)$ is re-defined as the total energy of the N_V-outer electron subsystem in the initial state (without core hole) whereas $|0_{PE}\rangle$ and $|0_c\rangle$ denote the vacuum states in the photoelectron and core hole subspaces, respectively. Then $P(E_K, h\nu)$ is represented by

$$\begin{aligned} P(E_k, h\nu) &\propto n_{h\nu} \sum_{f,c} |\langle E_f'(N_V)|\langle c|\langle E_K|M_{Kc}a_K^\dagger a_c|0_{PE}\rangle|0_C\rangle|E_i(N_V)\rangle|^2 \\ &\quad \times \delta(E_f'(N_V) - E_i(N_V) + E_K + \phi - h\nu - \varepsilon_c) \\ &= n_{h\nu} \sum_{f,c} |M_{Kc}\langle E_f'(N_V)|E_i(N_V)\rangle|^2 \delta(E_f'(N_V) - E_i(N_V) + E_K + \phi - h\nu - \varepsilon_c), \end{aligned} \tag{2.45}$$

where M_{Kc}, a_c and ε_c stand for the matrix element including the one-electron photoexcitation process, the core-hole creation operator, and the one-electron core level, respectively. When M_{Kc} is assumed to have negligible energy dependence, which is much more realistic than in the case of the valence-band photoemission process, one measures the core-level photoemission spectral function

$$\rho_{\mathrm{core}}(\omega) \propto \sum_{f,c} |\langle E_f'(N_V)|E_i(N_V)\rangle|^2 \delta(\omega - \varepsilon_c + E_f'(N_V) - E_i(N_V)). \tag{2.46}$$

2.5 Matrix Element Effects

In a wide sense, the so-called matrix element effects include

1. Orbital and $h\nu$ dependence of the photoionization cross-sections (PICS),
2. Polarization dependence of the observed spectra with respect to the photoelectron angular distribution, reflecting the selection rule between the initial and final states based on the idea that the photoexcitation is essentially electric dipole transition,
3. $h\nu$ dependence of the observed spectral line shape due to the effects of the energy dependence of the photoexcited final states deviated from the nearly free-electron state mentioned above, and
4. Photoelectron diffraction effects reflecting the spatial configuration of neighboring sites (see Chap. 12).

Nowadays the term "matrix element effects" in a narrow sense used in papers on low-energy-excitation ARPES means the effects-2 and -3. For the angle-integrated photoemission of polycrystalline solids, the effect 4 is cancelled out. One should always pay attention to the effect 3 in the low-energy excitation photoemission with $h\nu < 100$ eV as reported a few decades ago [17]. The effect of the final-state electronic structures (energy dependence) becomes gradually negligible on going to higher-energy photoexcitation.

By using the effect-2 for the linearly polarized photoexcitations, one can perform "orbital-dependent photoemission", by which the orbital contributions to the valence-band spectra can be revealed as discussed in Sects. 6.1 and 8.8. It should be noted that the effect-2 is not fully cancelled out for ARPES even by using unpolarized excitation light since the anisotropy in the electric field of the incident photons is still remaining (please recall that the electric field is zero along the photon-propagation direction). It is empirically known that even when the expected ARPES spectral weight is negligible or very weak in a certain Brillouin zone (BZ), it might be clearly observed in the next BZ, as often experienced in the case of high-energy soft X-ray ARPES (SXARPES). Namely, please "Survey the next BZ if the spectral weight is weak in the BZ under the present scanning!" in the ARPES measurements.

2.6 Theoretical Models to Describe the Spectra of Strongly Correlated Electron Systems

In order to better understand the experimental photoemission spectra of strongly correlated electron systems and then to derive new physics from the data, appropriate theoretical models should be employed, in which the electron correlation effects are taken into account. The so-called periodic Anderson model (PAM) and the d-p model are schematically illustrated in Fig. 2.6, in which both translational symmetry for strongly correlated as well as itinerant sites and electron correlation are properly taken into account. These models could be good approximations for rare-earth compounds and transition-metal oxides. The Hamiltonian of the periodic Anderson model is described as

$$\mathcal{H} = \mathcal{H}_f + \mathcal{H}_c + \mathcal{H}_{cf}, \tag{2.47}$$

$$\mathcal{H}_f = \sum_{i,\mu,\sigma} \varepsilon_{fi}^{\mu} f_{i\mu\sigma}^{\dagger} f_{i\mu\sigma} + \frac{1}{2} \sum_{i} \sum_{(\mu,\sigma)\neq(\mu',\sigma')} U_{ff}^{\mu\mu'\sigma\sigma'} n_{fi\mu\sigma} n_{fi\mu'\sigma'}, \tag{2.47a}$$

$$\mathcal{H}_c = \sum_{\mathrm{k},l,\sigma} \varepsilon_{\mathrm{k}l} c_{\mathrm{k}l\sigma}^{\dagger} c_{\mathrm{k}l\sigma}, \tag{2.47b}$$

$$\mathcal{H}_{cf} = \sum_{\mathrm{k},l,i,\mu,\sigma} V_{cf}^{\mathrm{k}l\mu} \left(f_{i\mu\sigma}^{\dagger} c_{\mathrm{k}l\sigma} + c_{\mathrm{k}l\sigma}^{\dagger} f_{i\mu\sigma} \right). \tag{2.47c}$$

Here ε_{fi}^{μ} and $U_{ff}^{\mu\mu'\sigma\sigma'}$ in (2.47a) stand for the one-electron energy level for the strongly correlated orbital μ with the spin σ on the site i and the on-site Coulomb repulsive energy between the strongly correlated electrons, whereas $f_{i\mu\sigma}^{\dagger}(f_{i\mu\sigma})$ and

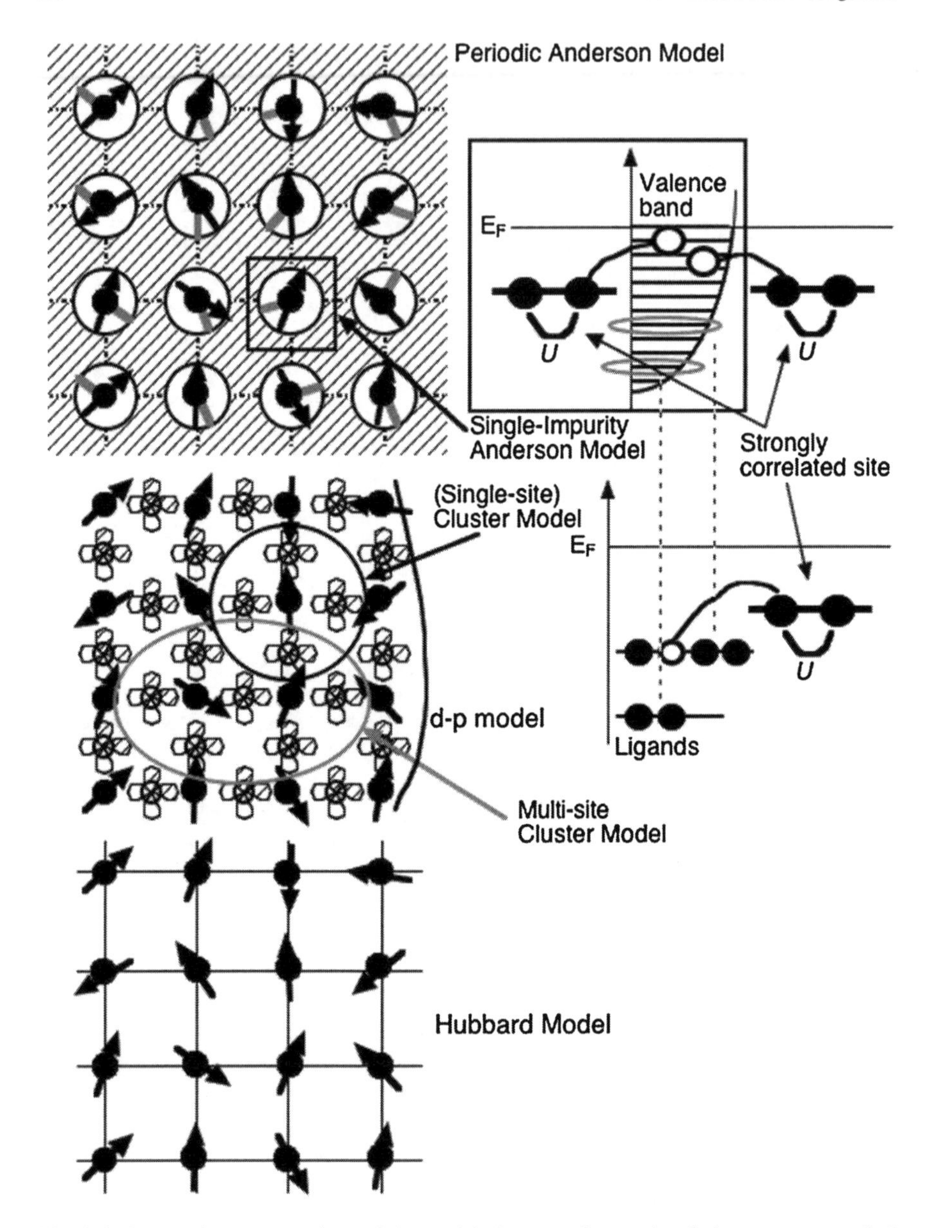

Fig. 2.6 Schematic representations of the models for strongly correlated electron systems. *Left* column: *Top*: Spatial description for the periodic Anderson model, where the hybridization is symbolically shown by thick *gray* bars. The description for the single-impurity Anderson model, in which only one strongly correlated site hybridized with the itinerant valence bands is considered, is shown in a *rectangle*. *Middle*: Spatial representation for the d-p model, where the single-site cluster model is surrounded by a *circle* and multi-site cluster model is surrounded by an *oval*. *Bottom*: Spatial description for the Hubbard model. *Right* column: *Top*: Energy diagrams for the periodic Anderson model. Single-impurity Anderson model is surrounded by a *rectangle*. *Middle*: Case for a single-site cluster model. The *dashed lines* show the correspondence of the ligand levels in the cluster model to the representative discrete levels extracted from the itinerant continuum bands in the single-impurity Anderson model, where both are hybridized with the strongly correlated states. In the *right* panels showing the energy diagrams, the filled (empty) *circles* denote the occupied electrons (*holes*) and the lines between the filled and empty *circles* represent a charge transfer

$n_{fi\mu\sigma} = f^{\dagger}_{i\mu\sigma} f_{i\mu\sigma}$ denote the creation (annihilation) and number operators for the strongly correlated electrons (either f or d state) $\varepsilon_{\mathbf{k}l}$ and $c^{\dagger}_{\mathbf{k}l\sigma}(c_{\mathbf{k}l\sigma})$ in (2.47b) denote the energy of one electron for the band $\imath$ at the momentum **k**, and its creation (annihilation) operator. $V^{\mathbf{k}l\mu}_{cf}$ represents the hybridization strength between the strongly correlated electron and the band electron.

For the d-p model, Eqs. (2.47b) and (2.47c) should be replaced by

$$\mathcal{H}_p = \sum_{j,l',\sigma} \varepsilon_{jl'} c^{\dagger}_{jl'\sigma} c_{jl'\sigma} \tag{2.47b$'$}$$

and

$$\mathcal{H}_{pf} = \sum_{(i,j)\mu,l',\sigma} V^{l'\mu}_{pf} \left(f^{\dagger}_{i\mu\sigma} c_{jl'\sigma} + c^{\dagger}_{jl'\sigma} f_{i\mu\sigma} \right), \tag{2.47c$'$}$$

respectively, where the momentum **k** in the band $\imath$ is replaced by the orbital $\imath'$ at the ligand (oxygen in the case of oxides) site j. In Eq. (2.47c′), the sum of the hybridization strength between the strongly correlated and ligand electrons is made on the neighboring (i, j) sites.

Of course the theoretical calculations based on the above models are still impossible even now for realistic materials without some simplification. Therefore, these models are employed to predict low-energy excitation properties for the electrons near E_F of strongly correlated materials by reducing the number of split 3d or 4f levels. However, such a theoretical calculation is not suitable for interpreting observed experimental photoemission results, for which the electron excitations are much higher at least by several eV. For instance, the 4f levels of Ce^{3+} ($4f^1$) are split by $\sim$300 meV into $4f_{7/2}$ and $4f_{5/2}$ levels due to the spin–orbit interaction, where the latter (also the former, though not discussed here) is further split into three levels for the tetragonal or lower symmetry by the crystal field with energy differences ranging from a few to several hundreds K (see Fig. 7.8). When one discusses such low-energy excitation phenomena as seen in thermodynamic measurements at low temperatures, the consideration of only the $4f_{5/2}$ levels can lead to excellent predictions consistent with the observed results. In the case of photoemission, however, the actual 4f spectra cannot be well reproduced unless all realistic 4f levels are taken into account in the theoretical models, since the charge transfer to the $4f_{7/2}$ states by the screening of the photoexcited 4f sites might give a prominent peak structure in the spectra [Indeed, the real quasi-particle spectral weight for the $4f^1_{5/2L}$ photoexcitation final state ($4f_{5/2L}$ denotes the lowest $4f_{5/2}$ level) could be very weak and not distinguishable from the other $4f^1_{5/2}$ final states for heavy fermion systems (see Sect. 7.1 for details)]. Therefore, somehow simplified models described below and/or phenomenological models based on the retarded Green's function are usually used for the actual analysis of the spectra. Here, a few models starting from the localized limit and from the itinerant limit (a dynamical mean field theory (DMFT)) are briefly introduced together with explanations of their basic concepts, advantages and limitations.

2.6.1 Single Impurity Anderson Model

The single impurity Anderson model (SIAM) expressed as

$$\begin{aligned}\mathcal{H}_{SIAM} = &\sum_{\mu,\sigma} \varepsilon_f^{\mu} f_{\mu\sigma}^{\dagger} f_{\mu\sigma} + \frac{1}{2} \sum_{(\mu,\sigma)\neq(\mu\prime,\sigma\prime)} U_{ff}^{\mu\mu'\sigma\sigma'} n_{f\mu\sigma} n_{f\mu'\sigma'} + \sum_{\mathbf{k},l,\sigma} \varepsilon_{\mathbf{k}l} c_{\mathbf{k}l\sigma}^{\dagger} c_{\mathbf{k}l\sigma} \\ &+ \sum_{\mathbf{k},l,\mu,\sigma} V_{cf}^{\mathbf{k}l\sigma} \left(f_{\mu\sigma}^{\dagger} c_{\mathbf{k}l\sigma} + c_{\mathbf{k}l\sigma}^{\dagger} f_{\mu\sigma} \right),\end{aligned} \tag{2.48}$$

in which only one site is taken into account, is often used for the analysis of the rare-earth core-level and valence-band spectra [3, 18, 19]. For the strongly correlated states in highly localized systems and the rare-earth core-level photoexcitation final states in strongly hybridized systems (including the core-levels even in the rather itinerant $CeRu_2$ [20]), the photoemission spectra can be well described by this model. The advantages in using the SIAM are as follows:

1. Since all the 4f orbitals can be taken into account, the valence-band spectral features near E_F due to the high-energy excitations can be correctly handled.
2. Rare-earth core-level photoemission and absorption spectra can also be well reproduced by considering the Coulomb attractive interactions between the strongly correlated electron and the core hole created by the photoxcitation as discussed below.
3. Due to the above advantages, the spectra of rare-earth core-level photoemission and absorption, valence-band photoemission, and inverse photoemission, resonance inverse photoemission and bremsstrahlung isochromat spectroscopy (BIS) can simultaneously be well reproduced by the same unique parameter set.

The item-3 is very important to obtain highly reliable $U_{ff}^{\mu\mu'\sigma\sigma'}$ and/or $V_{cf}^{\mathbf{k}l\mu}$ within the model [18]. For the core-level excitation process, the term of the Coulomb attractive interactions between the strongly correlated electron and the core hole

$$-\sum_{\mu,\sigma} U_{fc} n_{f\mu\sigma}(1 - n_c) \tag{2.49}$$

should be added in Eq. (2.48) for the core-level photoexcitation final states, where U_{fc} (>0) and n_c-represent the Coulomb interactions between the strongly correlated electron and the core hole and the number operator for the core electron level, respectively. (Here, the core-hole energy $-\varepsilon_c$ is not considered in the Hamiltonian since it is already taken into account in the δ-function in Eq. (2.46).) The energy dependence of the hybridization strength reflecting the density of states for the itinerant bands hybridized with the strongly correlated orbitals should be given before the calculation. Usually, this term is determined from the experimental valence-band spectra in which the spectral weight from the strongly correlated electrons is negligible, or from the partial density of states predicted from band-structure calculations.

From the 1980s, the Ce 3d core-level photoemission and absorption spectra as well as the BIS spectra have been analyzed by the SIAM based on the $1/N_F$-expansion (N_F denotes the degeneracy of the Ce 4f level) method (so-called GS model) with taking $4f^0$, $4f^1$ and $4f^2$ configurations into account in the initial states. The realistic 4f photoemission spectra can be calculated by the GS model, but the temperature is assumed to be zero [18]. On the other hand, the SIAM based on the non-crossing approximation (NCA) has been developed [21–23], which can calculate temperature-dependent 4f spectra as well as the Kondo temperature [24]. However, the NCA calculations usually assume $U_{ff} \rightarrow \infty$, namely, they omit the $4f^2$ components. Even now, it is very difficult to fully take the $4f^2$ states into account in the NCA calculations.

Within the SIAM, the calculation with taking into account all of the (1) itinerant bands contained, (2) 4f levels split by the spin-orbital interaction and crystal field, and (3) atomic multiplet structures is not practical even now. Generally, the atomic final-state multiplet structures are prominent in many rare-earth 3d core-level and valence-band 4f spectra except for the cases of the Ce^{4+} ($4f^0$) and Yb^{2+}($4f^{14}$) ions. When the atomic multiplet effects are taken into account to reproduce the spectra, the itinerant bands are often replaced by a few discrete levels (usually located near E_F) in the calculations [25, 26], which is equivalent to the cluster model discussed below.

From the 1990s, the applicability of the SIAM to the valence-band 4f spectra has been a matter of controversy. This problem was based on the fact that the 4f translational symmetry was not considered at all in the SIAM [27, 28]. Of course, the momentum dependence of the 4f spectra obtained by ARPES cannot be discussed by this model. On the other hand, the main issue of controversy was on the applicability to the angle-integrated spectra of the Ce and Yb systems. Although it had been revealed that most Ce and Yb 4f spectra could not be reproduced by the simple band-structure calculations except for, for example, $CeRu_2$, rigorous tests of the applicability required temperature-dependent data with high resolution (<100 meV) and negligible surface contributions. So far it was widely believed that the SIAM was applicable to the rather weakly hybridized Ce compounds (with low Kondo temperatures $< \sim 100$ K). For the Yb compounds, it was found that the temperature dependence of the Yb^{2+} and Yb^{3+} 4f peak positions was the key to check its applicability as discussed in Chap. 7.

2.6.2 Cluster Model

In a wide sense, the cluster model should be defined as a model in which the strongly correlated orbitals located at a finite number of sites hybridized with discrete levels are treated. At first, the single-site (meaning the strongly correlated orbitals located at one certain site) cluster model is introduced. As mentioned above, the SIAM with the itinerant bands reduced to only one or a few discrete levels is also a kind of a cluster model. On the other hand, a more realistic TMO_6 cluster model as schematically shown in Fig. 2.6 is often used to discuss the strongly correlated electronic structure of the transition metal oxides, where TM stands for the transition metal element.

Hereafter the configuration-interaction cluster model for a TMO_6 site (or a single TM site surrounded by several realistic ligands) [29, 30] is mainly described. In the same manner as for the SIAM, the cluster model can correctly calculate both TM core-level and valence-band photoemission spectra simultaneously by a unique parameter set. The other advantages of the realistic cluster model with respect to the SIAM are

1. As described below, the ground and photoexcited states are expressed as linear combinations of Slater determinants, and therefore the electronic states can be rigorously defined within the TMO_6 cluster,
2. Anisotropy in the hybridization is clearly taken into account since the wave functions of the ligand orbitals or the molecular orbitals can be well defined by the linear combination of the atomic orbitals.

For a strongly correlated n electron system, the electron correlation effects originally indicate the effects not taken into account in a single Slater determinant for the n electrons. One way to take the correlation effects into account is to consider the configuration interactions, namely, the interactions between the different Slater determinants, by representing the ground state $|\Psi_g\rangle$ by their linear combination as

$$|\Psi_g\rangle = a_n|d^n\rangle + a_{n+1}|d^{n+1}\underline{L}\rangle + a_{n+2}|d^{n+2}\underline{L}^2\rangle + \cdots + a_{10}|d^{10}\underline{L}^{10-n}\rangle, \qquad (2.50)$$

where $a_n^2 + a_{n+1}^2 + a_{n+2}^2 \cdots + a_{10}^2 = 1$ and $|d^{n+j}\underline{L}^j\rangle$ denotes the states in which j electrons are transferred from the ligand orbitals to the strongly correlated orbitals originally occupied by n electrons.

The cluster model has originally been applied to insulating transition metal oxides. Especially, the analysis of the photoemission and optical spectra of NiO by the cluster model tells us that the doped hole is mainly located at the O 2p sites hybridized with the Ni 3d orbitals, indicating that NiO is a charge-transfer type insulator and not a Mott–Hubbard type insulator [29]. Nowadays the insulating transition metal oxides are categorized into the Mott-Hubbard type, charge-transferred type and so on by the so-called Zaanen-Sawatzky-Allen diagram [31]. It has been found that the doped hole is also mainly located at the O 2p sites in the high-temperature superconducting cuprates according to the cluster model calculations [32].

Since the TMO_6 cluster model starts from the rigorously defined Slater determinants, it is difficult to describe the electronic structure of doped transition metal oxides such as $La_{1-x}Sr_xTMO_3$ with the nominal d-electron number deviating from an integer value by the cluster model. Further, spectral features, which cannot be reproduced by the single-site cluster model, have been recognized in the TM 2p core-level spectra. For the latter, multi-site cluster (TM_2O_{11}, TM_3O_{16},...) models have been developed [33–35]. From these theoretical studies, it has been found that the so-called non-local screening takes place in the oxides. In this case the doped holes in the photoexcitation process are no longer located at the excited TM or nearest neighbor O site, but transferred to the next TM site or further distant site. This screening process suggests the rather itinerant nature of the correlated electrons near E_F. One can imagine that the physics of the strongly correlated electron systems starting from the localized limit would smoothly continue to the physics of the itinerant picture if the itinerant nature of these electrons is properly introduced in the cluster model. At present, however, the effects of screening by the quasi-particles in the metallic transition metal oxides do not seem to be successfully taken into account even by these multi-site cluster models.

2.6.3 Hubbard Model

The Hubbard model is often used in the theoretical study of strongly correlated electron systems, being typically expressed as

$$\mathcal{H}_{MH} = \sum_{i \neq j, \sigma} t_{ij} d_{i\sigma}^{\dagger} d_{j\sigma} + \frac{1}{2} \sum_{i,\sigma} U_{dd} n_{i\sigma} n_{i-\sigma}, \tag{2.51}$$

in which the ligand states are removed from the d-p model, or the band states removed from the periodic Anderson model. Here, t_{ij} represents the hopping between neighboring sites via the ligand ion or itinerant bands, which are implicitly taken into account. $d_{i\sigma}{}^{\dagger}$ $(d_{i\sigma})$ denotes the creation (annihilation) operator of the strongly correlated electron with spin σ in the "d orbital" of the site i. It should be noted that the effects of the ligand orbitals or partially hybridized bands are implicitly taken into account in the strongly correlated but renormalized d-orbitals before discussing the electronic state based on the Hubbard model. For instance, when the electronic structure of high-temperature superconducting cuprates is handled, one should assume the so-called Zhang-Rice singlet [36] for the d orbital instead of the bare atomic orbital (Otherwise, the doped hole located in the oxygen sites would be missed). Consequently, the Coulomb repulsion energy U_{dd} in Eq. (2.51) is no longer the same as that in the SIAM or in the cluster model discussed above. U_{dd} should be recognized as the effective Coulomb repulsion energy between the renormalized d electrons.

Although the Hubbard model is in a sense a simplified version of the d-p model or periodic Anderson model (PAM), there are still difficulties in the simulations of the photoemission spectra for a realistically large size of a Hubbard lattice in two- and three-dimensions. The Hubbard model itself is hardly employed to reproduce the photoemission data by the experimentalists except for one-dimensional systems, though theoreticians are making efforts to obtain reliable spectral functions based on this model. On the other hand, the concept of the upper and lower Hubbard bands is derived owing to the Hubbard model, which should be always kept in mind when studying the strongly correlated electron systems by photoemission.

2.6.4 Dynamical Mean Field Theory

Dynamical mean field theory (DMFT) could be shortly described as a "Hubbard model with the infinity-dimension", in which the self-energy in the Green's function (2.37) has no longer momentum dependence [37]. This model is a good approximation for the strongly correlated electron sites surrounded by many ligands. A theoretical advantage of the DMFT with respect to the analysis of the valence-band photoemission spectra is that the calculations for realistic compounds are feasible by the combination with band-structure calculations using the local density approximation, known as LDA + DMFT [38], which may be formulated as a mapping of the lattice problem (strongly correlated sites with translational symmetry) onto an effective SIAM model coupled self-consistently to an effective conduction bandwidth [39]. LDA + DMFT can be recognized as the theoretical approach from the itinerant limit. It is known that the Coulomb repulsion energy due to the electron–electron interactions has already been taken into account in the LDA calculations within a static mean field approach. Therefore, a procedure of subtracting the Coulomb energy calculated within the LDA calculations is required in order to avoid a double-counting of the Coulomb interactions.

As discussed in Chap. 7, the LDA + DMFT using quantum Monte Carlo simulations (QMC) has successfully reproduced the experimental results of strongly correlated vanadium oxides in the metallic phases [39, 40]. So far, the valence-band photoemission spectra near E_F have been calculated for several light transition metal oxides by using this method [41]. On the other hand, the simulations of the spectral functions are limited to higher temperatures, i.e. more than a few hundred K, by LDA + DMFT (QMC). Other candidate compounds to apply the DMFT would be the Ce systems, where the study of the realistic Ce compounds is in progress [42].

2.6.5 New Directions and Some Remarks

As mentioned above, the combination of first-principle band structure calculations and the theoretical treatment of strongly correlated electrons has partly been achieved. One way to analyze the valence-band photoemission spectra is to compare them with the simulated spectra from the LDA + DMFT study, for which the collaboration between experimentalists and theoreticians is required. Meanwhile, high-resolution core-level photoelectron spectroscopy became feasible due to the dramatic developments in diverse apparatus. Their spectral line-shapes are sometimes very complicated resulting from the atomic multiplets, charge-transfer satellites, and the screening by the renormalized quasi-particles. So for a phenomenological approach such as the cluster model plus extrinsically introduced conduction-band states [43] has been applied to metallic transition metal oxides. This approach is effective to understand the origin of various peaks in the core-level spectra, but a more rigorous theoretical model should be established in order to really understand the physics of strongly correlated electron systems in the future.

Finally, some remarks should be noted here. The Coulomb repulsion energy U between the strongly correlated electrons depends on the specific model. Theoretical models from the localized limit like the SIAM or the cluster model define U as the bare Coulomb energy before effective screenings by the hybridization and/or the configuration interactions. On the other hand, the Hubbard model including DMFT defines U as the additional Coulomb energy introduced after the hybridized orbitals are formed. In this sense, we should not directly compare and discuss U among different theoretical models. Another remark is on the configuration dependence of the hybridization strength in the core-level photoexcitations [44, 45]. One can easily imagine that the outer orbitals would be shrunken in the photoexcitation final state due to the Coulomb attractive force from the core hole, and that they would be spatially expanded when the number of correlated electrons were increased due to the charge transfer. The former tends to a decrease in the hybridization strength and the latter tends to its enhancement. Such effects are often considered in theoretical simulations of the spectra, but description of the configuration dependence are sometimes lacking in published papers. When the core-level photoemission and absorption spectra are analyzed by using the simulations, one should pay careful attention to their configuration dependence.

References

1. P.J. Feibelman, D.E. Eastman, Phys. Rev. B **10**, 4932 (1974)
2. S. Hüfner, *Photoelectron Spectroscopy, Principles and Application* (Springer, Berlin, 2003)
3. A. Kotani, *Inner shell photoelectron process in solids in Handbook on Synchrotron Radiation,* vol. 2, ed. by G.V. Marr (North-Holland, 1987)

4. M. Matsunami, A. Chainani, M. Taguchi, R. Eguchi, Y. Ishida, Y. Takata, H. Okamura, T. Nanba, M. Yabashi, K. Tamasaku, Y. Nishino, T. Ishikawa, Y. Senba, H. Ohashi, N. Tsujii, A. Ochiai, S. Shin, Phys. Rev. B **78**, 195118 (2008)
5. C.R. Brundle, J. Vac. Sci. Technol. **11**, 212 (1974)
6. S. Tanuma, C.J. Powell, D.R. Penn, Surf. Interface Anal. **43**, 689 (2011)
7. T. Miller, W.E. McMahon, T.-C. Chiang, Phys. Rev. Lett. **77**, 1167 (1996)
8. S. Suga, A. Sekiyama, G. Funabashi, J. Yamaguchi, M. Kimura, M. Tsujibayashi, T. Uyama, H. Sugiyama, Y. Tomida, G. Kuwahara, S. Kitayama, K. Fukushima, K. Kimura, T. Yokoi, K. Murakami, H. Fujiwara, Y. Saitoh, L. Plucinski, C.M. Shneider, Rev. Sci. Instrum. **81**, 105111 (2010)
9. L. Plucinski, G. Mussler, J. Krumrain, A. Herdt, S. Suga, D. Grützmacher, C.M. Schneider, Appl. Phys. Lett. **98**, 222503 (2011)
10. J.J. Yeh, I. Lindau, At. Data Nucl. Data Tables **32**, 1 (1985)
11. M.B. Trzhaskovskaya, V.I. Nefedov, V.G. Yarzhemsky, At. Data Nucl. Data Table **77**, 97 (2001). ibid **82**, 257 (2002)
12. M.B. Trzhaskovskaya, V.K. Nukulin, V.I. Nefedov, V.G. Yarzhemsky, At. Data Nucl. Data Table **92**, 245 (2006)
13. H. Wadati, A. Chikamatsu, M. Takizawa, R. Hashimoto, H. Kumigashira, T. Yoshida, T. Mizokawa, A. Fujimori, M. Oshima, M. Lippmaa, M. Kawasaki, H. Koinuma, Phys. Rev. B **74**, 115114 (2006)
14. T. Valla, A.V. Fedorov, P.D. Johnson, S.L. Hulbert, Phys. Rev. Lett. **83**, 2085 (1999)
15. J. Schäfer, D. Schrupp, E. Rotenberg, K. Rossnagel, H. Koh, P. Blaha, R. Claessen, Phys. Rev. Lett. **92**, 097205 (2004)
16. O. Gunnarsson, K. Schönhammer, Phys. Rev. B **31**, 4815 (1985)
17. J. Freeouf, M. Erbudak, D.E. Eastman, Solid State Commun. **13**, 771 (1973)
18. J.W. Allen, S.J. Oh, O. Gunnarsson, K. Schönhammer, M.B. Maple, M.S. Torikachvili, I. Lindau, Adv. Phys. **35**, 275 (1986)
19. T. Jo, A. Kotani, J. Phys. Soc. Jpn. **55**, 2457 (1986)
20. A. Sekiyama, T. Iwasaki, K. Matsuda, Y. Saitoh, Y. Onuki, S. Suga, Nature **403**, 396 (2000)
21. Y. Kuramoto, Z. Phys. B: Condens. Matter **53**, 37 (1983)
22. N.E. Bickers, D.L. Cox, J.W. Wilkins, Phys. Rev. B **36**, 2036 (1987)
23. N.E. Bickers, Rev. Mod. Phys. **59**, 845 (1987)
24. A. Sekiyama, S. Imada, A. Yamasaki, S. Suga, in *Very high-resolution photoelectron spectroscopy* ed. by S. Hüfner, Lecture Notes in Physics vol. 715, (Springer, Berlin, 2007), p. 351
25. A. Tanaka, T. Jo, G. Sawatzky, J. Phys. Soc. Jpn. **61**, 2636 (1992)
26. A. Yamasaki, S. Imada, H. Higashimichi, H. Fujiwara, T. Saita, T. Miyamachi, A. Sekiyama, H. Sugawara, D. Kikuchi, H. Sato, A. Higashiya, M. Yabashi, K. Tamasaku, D. Miwa, T. Ishikawa, S. Suga, Phys. Rev. Lett. **98**, 156402-1-4 (2007)
27. J.J. Joyce, A.J. Arko, J. Lawrence, P.C. Canfield, Z. Fisk, R.J. Bartlett, J.D. Thompson, Phys. Rev. Lett. **68**, 236 (1992)
28. M. Garnier, K. Breuer, D. Purdie, M. Hengsberger, Y. Baer, B. Delley, Phys. Rev. Lett. **78**, 4127 (1997)
29. A. Fujimori, F. Minami, Phys. Rev. B **30**, 957 (1984)
30. A.E. Bocquet, T. Mizokawa, T. Saitoh, H. Namatame, A. Fujimori, Phys. Rev. **46**, 3771 (1992)
31. J. Zaanen, G.A. Sawatzky, J.W. Allen, Phys. Rev. Lett. **55**, 418 (1985)
32. A. Fujimori, E. Takayama-Muromachi, Y. Uchida, B. Okai, Phys. Rev. B **35**, 8814 (1987)
33. M.A. van Veenendaal, G.A. Sawatzky, Phys. Rev. Lett. **70**, 2459 (1993)
34. M.A. van Veenendaal, H. Eskes, G.A. Sawatzky, Phys. Rev. B **47**, 11462 (1993)
35. K. Okada, A. Kotani, Phys. Rev. B **52**, 4794 (1995)
36. F.C. Zhang, T.M. Rice, Phys. Rev. B **37**, 3759 (1988)
37. A. Georges, G. Kotliar, W. Krauth, M.J. Rozenberg, Rev. Mod. Phys. **68**, 13 (1996)

38. V.I. Anisimov, A.I. Poteryaev, M.A. Korotin, A.O. Anokhin, G. Kotliar, J. Phys.: Condens. Matter **9**, 7359 (1997)
39. S.-K. Mo, J.D. Denlinger, H.-D. Kim, J.-H. Park, J.W. Allen, A. Sekiyama, A. Yamasaki, K. Kadono, S. Suga, Y. Saitoh, T. Muro, P. Metcalf, G. Keller, K. Held, V. Eyert, V.I. Anisimov, D. Vollhardt, Phys. Rev. Lett. **90**, 186403 (2003)
40. A. Sekiyama, H. Fujiwara, S. Imada, S. Suga, H. Eisaki, S.I. Uchida, K. Takegahara, H. Harima, Y. Saitoh, I.A. Nekrasov, G. Keller, D.E. Kondakov, A.V. Kozhevnikov, Th Pruschke, K. Held, D. Vollhardt, V.I. Anisimov, Phys. Rev. Lett. **93**, 156402 (2004)
41. E. Pavarini, S. Biermann, A. Poteryaev, A. I. Lichtenstein, A. Georges, O.K Andersen, Phys. Rev. Lett. **92**, 176403 (2004)
42. O. Sakai, H. Harima, J. Phys. Soc. Jpn. **81**, 024717 (2012)
43. K. Horiba, M. Taguchi, A. Chainani, Y. Takata, E. Ikenaga, D. Miwa, Y. Nishino, K. Tamasaku, M. Awaji, A. Takeuchi, M. Yabashi, H. Namatame, M. Taniguchi, H. Kumigashira, M. Oshima, M. Lippmaa, M. Kawasaki, H. Koinuma, K. Kobayashi, T. Ishikawa, S. Shin, Phys. Rev. Lett. **93**, 236401 (2004)
44. O. Gunnarsson, O. Jepsen, Phys. Rev. B **38**, 3568 (1988)
45. N. Witkowski, F. Bertran, D. Malterre, Phys. Rev. B **56**, 15040 (1997)

Chapter 3
Instrumentation and Methodology

The progress of photoelectron spectroscopy has strongly been backed up by the development of synchrotron radiation (SR) from the late 1970s and, in particular, undulator radiation since the 1980s. In order to perform high resolution experiments, monochoromatization of not only the bending SR but also undulator SR was necessary by use of grating or crystal monochromators. The principle of both type of mochochromators is briefly described. Then typical examples of high performance SR beam lines in the region of VUV, soft X-rays and hard X-rays are described. Laboratory VUV rare gas light sources as well as laser sources for ultrahigh resolution are then introduced. X-ray tubes and phase plates for polarization switching are also described.

Photoelectron spectroscopy requires suitable high performance electron spectrometers or analyzers. Typical spectrometers as hemispherical analyzers, cylindrical mirror analyzers, 2D analyzers as well as time-of-flight analyzers are then described. Still best quality samples are necessary for reliable studies of materials. Such methods for preparing clean sample surfaces as (1) ion-sputtering followed by annealing, (2) in situ fracturing and/or cleavage, (3) in situ sample growth are described. In addition such environments surrounding the sample as low temperatures and ambient pressures are very briefly discussed.

Finally several methodologies used in the photoelectron spectroscopy nowadays are described in the order of (1) angle integrated PES, (2) resonance PES and constant initial state spectrum, (3) angle resolved PES, (4) PES in μm and nm regions and (5) momentum microscope.

3.1 Synchrotron Radiation and Undulator Radiation

The photoelectrons can be excited by photons which have enough energy higher than the work function (ϕ) of the individual materials [1–3]. In most cases of solids, the work function ϕ is $\sim$5 eV. Therefore photons with energy hν higher than this are usually required. For example, hν = 6 eV can be employed for the

S. Suga and A. Sekiyama, *Photoelectron Spectroscopy*,
Springer Series in Optical Sciences 176, DOI: 10.1007/978-3-642-37530-9_3,

measurement of photoelectrons to cover the region of ~ 1 eV from the Fermi level (E_F) in the conventional case of one photon excitation. This is not the case for two photon excitation discussed in Sect. 9.4. The wavelength λ (Å) is related to hν (eV) by the equation of λ (Å) = 12,398.1/hν (eV).

Traditionally X-ray tubes and He discharge light sources were often used for photoelectron spectroscopy. Most intensively employed were the hν of 1,253.6 eV from the Mg Kα, 1,486.6 eV from the Al Kα, 21.2 eV from the He I and 40.8 eV from the He II light sources. The full width at half maximum (FWHM) of the X-ray tubes was $\gtrsim 1$ eV which could be improved down to 0.4 or 0.3 eV by the use of crystal monochromators. On the other hand the FWHM of the He sources was of the order of a few meV [3].

The synchrotron radiation (SR) found in the 1940s was soon recognized to be a very useful light source to cover a wide hν range [2, 4–6] and powerful to excite photoelectrons. The characteristics of the SR emitted by relativistic electrons (or positrons) running through a bending magnet section with the velocity very close to the speed of light **c** in circular storage rings (Fig. 3.1) are summarized as follows: (1) wide hν spectral distribution from infrared to X-rays, which provides the tunability of monochromatized photons in a wide hν region. (2) linearly polarized in the orbital plane and elliptically polarized slightly above or below this plane. (3) divergence of the radiation is $1/\gamma$ radian [where the Lorentz factor γ is equal to $\sim$1957E, when the energy of the electrons accelerated in the storage ring is E(GeV)]. Therefore the divergence is of the order of 6.4×10^{-5} radian in the case of 8 GeV storage ring (SPring-8). (4) not a continuous wave (cw) light source but a pulsed light source with a quite accurate repetition. This pulse property

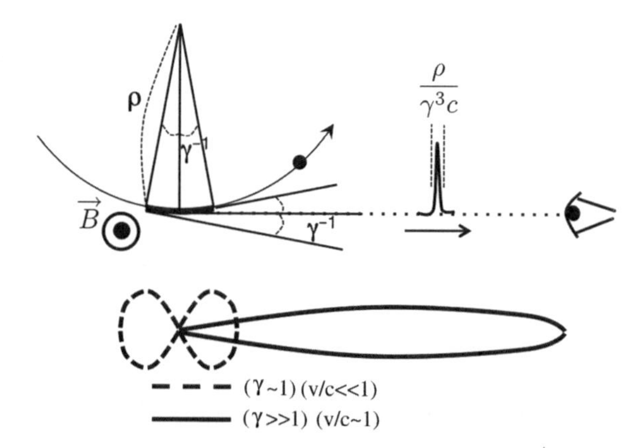

Fig. 3.1 (*Upper panel*) Schematic representations of the synchrotron radiation (*SR*) from a bending magnet with a very short pulse by one electron. (*Lower panel*) Instantaneous spatial distribution of the emitted power for the SR in comparison with that for a non-relativistic ($\gamma \sim 1$) radiation. The observer sees the electromagnetic wave emitted from an electron during it runs on the path indicated by the bold line in the upper panel with the length of $2\rho/\gamma$. The received pulse width is estimated as $(1 - v/c) \cdot 2\ \rho/(\gamma v) \sim [1/(2\gamma^2)] \cdot 2\rho/(\gamma c) \sim \rho/(\gamma^3 c)$ due to the relativistic effect, where v denotes the velocity of electron and $\gamma = 1/\sqrt{(1 - (v/c)^2)}$

originates from the fact that the electrons are bunched in the storage ring due to the RF (radio frequency) acceleration. (5) long 1/e decay time of the order a few hours to tens of hours even without a top-up injection, providing stable photon flux. (6) almost constant stored current and photon flux can be realized by the mode of top-up injection into the storage ring. (7) emitted from UHV environment inducing no light-source-induced contamination of the illuminated samples.

The quantitative theoretical derivation of the synchrotron radiation was given by considering the retardation potential (Lienard-Wiechert potential). The fundamental parameters of the storage rings are the curvature (radius) of the bending magnet ρ (m), the magnetic field B (Tesla) (of the order of sub to 1 Tesla in many cases) and the electron energy E (GeV). The electron rest mass energy m_oc^2 is 0.51 MeV and the Lorentz factor $\gamma = E/(m_oc^2) \sim 1957E$. The Lorentz force provides the practical formula of $\rho = E/(0.3B)$. The critical wavelength λ_c (Å) of the bending SR radiation was then given by

$$\lambda_c = 4\pi\rho/(3\gamma^3) = 18.6/\left(E^2B\right). \tag{3.1}$$

While the decrease in the photon flux on the lower hν side is gradual, the decrease on the higher hν side is rapid with a decrease by one order of magnitude at 0.1 λ_c. When higher hν is required higher energy storage rings with higher B in the bending magnet are required.

The higher energy storage ring, however, requires both higher B and higher RF power, because the energy loss of the electron per revolution is proportional to E^3B. The cost-effective way to generate higher hν photons in lower energy storage rings is to insert a high magnetic field device in the straight sections. Such a device is often called wiggler [7, 8]. Although the E is kept as it is, higher B provides a smaller λ_c according to the Eq. (3.1). Since such a high B insertion device has periodic magnetic structures with alternating up and down (or right and left) magnetic field directions with the period number N, the photon flux can be N times increased, as understood from Fig. 3.2. In this case the electron follows a "wiggling" orbit. This is the reason why such a device is called a wiggler and often employed in low energy storage rings to deliver higher hν photons. In both cases of SR radiation from a bending magnet and a wiggler, the radiation is not at all coherent. The maximum deviation angle ϕ_0 of the electron orbit in the wiggler from the central orbit can be expressed by K/γ where $K = eB_0\lambda_u/(2\pi m_0 c) = 93.4 \cdot B_0\lambda_u$ (Here B_o (Tesla) is the magnetic field of the wiggler and λ_u (m) is the length of a single period of the wiggler). When K is much larger than 1, ϕ_0 is larger than the natural divergence of $1/\gamma$. Then the spectral distribution of the wiggler is just the same as from a bending magnet. As for the characteristic features of the SR from the bending magnet, it should be recalled that the radiation is emitted into a wide angle covered by the bending arc, though the vertical divergence is still $1/\gamma$. Therefore, it is necessary to collect the radiation in a wide horizontal angle to get high photon flux. In the case of wiggler radiation, the total horizontal spread depends on ϕ_0 which can be a few times larger than $1/\gamma$. So the wiggler radiation must be more effectively collected than the radiation from a

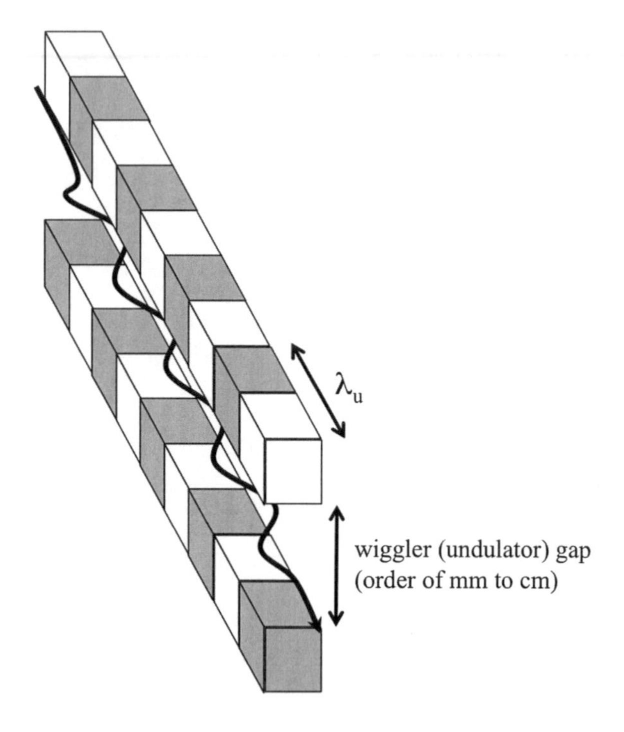

Fig. 3.2 A schematic representation of an insertion device called wiggler or undulator, where the magnetic field is alternatively switched between the *up* and *down* directions along its axis by a pair of magnet blocks (shown by *gray* and *white* color)

bending magnet in order to get higher photon flux. A sharp focus of high photon flux is better realized in the lower hν by undulator radiation as explained in the following.

If K is smaller than 1 in the case of an insertion device, ϕ_0 becomes smaller than $1/\gamma$. Then the emitted light intensity from one particular electron passing through this device is always non zero in the forward direction providing high photon flux [9, 10]. Many periods insertion devices called undulators are now installed in many SR facilities. The wavelength range of undulator radiation can be confined to several regions corresponding to the fundamental radiation and their higher order harmonics. Since the magnetic field required for this device can be lower than that in the bending sections, in-vacuum undulator technology with short period magnetic structures is feasible. For example, a period number N of up to 780 was achieved in the X-ray planar undulator of BL19LXU of SPring-8 [11]. Then the undulator radiation from this device becomes at least N times stronger compared with the SR radiation from a single period bending section. Thus undulator radiation has a very high brilliance represented by photon number/s/mrad^2/mm^2/1 % band width. There are two types of undulators: The planar undulator [11] with the alternating up and down (or right and left) magnetic fields and the helical undulator with a helical magnetic field [12, 13]. The undulator radiation is mostly used on the optical axis providing linearly (circularly) polarized light in the planar (helical) undulator. There can be other types of undulators made by combining planar and helical undulators. By defining the K values for the x (horizontal) and y (vertical) directions as K_x and K_y and the direction of the radiation from the undulator axis as θ, the wavelength of the n-th harmonics λ_n is represented as

$$\lambda_n = \lambda_u(1 + (K_x^2 + K_y^2)/2 + \gamma^2\theta^2)/(2n\gamma^2). \tag{3.2}$$

The undulator period length λ_u ranges usually from a few mm to a few tens cm. It is therefore possible to change the energy of the undulator radiation by changing the K value(s). In the case of permanent magnet undulators, the tuning is often done by varying the gap distance of the dipole magnet, where K increases with shortening the gap. In contrast to the case of the wigglers, the increase in the magnetic field results in the decrease of hν. It is also seen that the off-axis radiation has always lower hν. In the case of helical undulators, the radiation on the axis is mostly composed of the fundamental radiation ($n = 1$) because the velocity of the electrons for the observer on the axis is almost constant, whereas a finite acceptance angle near the axis and also a finite emittance of the electron beam provide some weak contribution of higher order harmonic radiation such as the third order harmonic. In the case of planar undulators the relative velocity of an electron to the observer changes with time and higher order harmonics are stronger than in the case of helical undulators, particularly in the off-axis direction. Therefore helical undulators cause only a low heat load off the axis, resulting in a much lower distortion of the optical components due to the heat load [14].

Now free electron lasers (FEL) are developed in several facilities [15–19], based on a linear accelerator with undulators. FEL can provide highly coherent pulsed laser light with femto second (fs) FWHM. Although the energy resolution of the photons is limited by the uncertainty principle expressed in the form of $\Delta t \cdot \Delta E \sim \hbar = 658$ (fs · meV), FEL will facilitate studies of the dynamics of various materials ranging from biomaterials to solids.

3.2 Principle of Grating and Crystal Monochromators

SR from bending magnets has a very wide energy spread and hence monochromators are required to obtain monochromatic radiation. Even in the case of undulator radiation, monochromators are required to perform high resolution experiments. The reflection grating monochromators are often used in the hν range from few eV up to 2,000 eV whereas crystal monochromators are often used above 1,500 eV [20–23]. In either case, the materials of the gratings or crystals should withstand the strong radiation without suffering from possible radiation damage. When the heat load onto these optical components is quite high, indirect cooling by water or liquid nitrogen is employed.

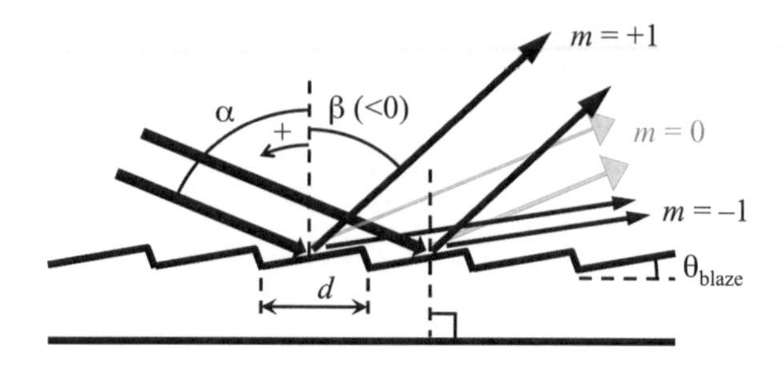

Fig. 3.3 Principle of grating monochromators

3.2.1 Grating Monochromators

The principle of diffraction by a grating is given by a simple diffraction Eq. (3.3) given later, where α is the incidence angle and β (<0) the exit angle measured from the macroscopic surface normal of the grating (see Fig. 3.3) [20]. In many cases, blazed or laminar gratings are employed for the monochromatization of the vacuum ultraviolet and soft X-rays. The definition of the blaze angle θ_{blaze} is also given in Fig. 3.3. If the period or the distance between neighboring grating grooves is d, the optical path difference between the two beams incident on neighboring facets is expressed as d(sin α + sin β) with a positive (negative) sign for α (β). When this quantity is equal to mλ (m is an integer and λ is the photon wavelength), interference between the two beams takes place and constructive monochromatic light is diffracted to the β direction. m can be either positive or negative and the diffraction under such conditions is called positive order light or negative order light. The light reflected for $\beta = -\alpha$ satisfies m = 0 and is called 0th order light which contains a wide hν range and is usually not used for spectroscopy. The surface of the grating is often coated with Au, Pt, or other high reflectivity materials (e.g. Al overcoated with thin MgF_2 against oxidation). The grating equation

$$\mathrm{m}\lambda = \mathrm{d}(\sin\alpha + \sin\beta) \tag{3.3}$$

can be transformed to $\mathrm{m}\lambda = 2\mathrm{d}\cos\theta \cdot \sin(\theta + \beta)$ if the deviation angle 2θ is defined as $2\theta = \alpha - \beta = \alpha + |\beta|$. Different λ can be obtained by changing α and/or β. For fixed photon incidence and diffracted directions, one can obtain different λ by rotating the grating. The blazed gratings can be made by such means as (1) original mechanical ruling, for example, by use of a ruling engine with a diamond cutter, (2) original ion beam etching after holographic exposure or (3) replica of the master grating. Therefore, microscopically speaking, the facet surface is rather irregular inducing scattered light and lowering the reflectivity. If the incidence and diffracted beam directions are symmetric along the facet normal, however, higher intensity is achieved. Using the blaze angle θ_{blaze}, this condition is expressed as α-$\theta_{\text{blaze}} = \theta_{\text{blaze}} - \beta$. By combining this equation with the diffraction equation, one obtains $\mathrm{m}\lambda_{\text{blaze}} = 2\mathrm{d}\sin\theta_{\text{blaze}} \cdot \cos(\alpha - \theta_{\text{blaze}})$. For normal incidence, $\alpha = 0$ and $\lambda_{\text{blaze}}(\mathrm{m} = 1) = \mathrm{d}\sin 2\theta_{\text{blaze}}$, which is often printed on the back side of the commercial gratings. The intensity of the monochromatized light intensity is somehow enhanced near λ_{blaze}. For practical use with $\alpha \neq 0$, however, one should consider the original equation to judge which λ is appropriately covered by the grating.

In the case of diffraction gratings, the *m*th order light with the wavelength of λ/m is also diffracted to the β direction and overlaps with the 1st order light with the wavelength λ. It is very important to suppress the higher order light to perform high quality spectroscopy. For this purpose, the total reflection by mirrors is often employed to suppress higher order light with hν higher than the cut-off energy. This technique is called order sorting. In the soft X-ray region with wavelength λ, the refractive index n(λ) of many materials used for coating diffraction gratings is slightly less than 1.0 except for the vicinity of core-absorption edge. For the incidence angle α onto the mirror, the total reflection takes place for α larger than $\alpha_c = \sin^{-1}$n. At this incidence angle, the photons with lower hν (or longer λ) is totally reflected and the higher hν photons are suppressed. To suppress the higher order light at low-hν (<100 eV) beam lines, thin-film filters (Mg, Al, Si etc.) can also be used [20].

From (3.3), one can obtain the wavelength resolution $\Delta\lambda$ of the monochromatized light as

$$\Delta\lambda = \left(\frac{d}{m}\cos\beta\right)\Delta\beta, \tag{3.4}$$

where $\Delta\beta$ stands for the acceptance angle of the monochromatized light at the exit slit as schematically shown in Fig. 3.4. When the distance between the centre of the grating and the exit slit, and its width are r′ and Δs_2, $\Delta\beta$ is expressed as Δs_2/r′. Therefore the resolving power defined as $\Delta\lambda/\lambda = \Delta(h\nu)/(h\nu)$ for a point light source is obtained as

$$\frac{\Delta(h\nu)}{h\nu} = \frac{h\nu}{hc}\left(\frac{d}{m}\cos\beta\right)\frac{\Delta s_2}{r'}, \tag{3.5}$$

yielding the energy resolution Δ(hν) roughly proportional to $(h\nu)^2$ when the same grating is used. On the other hand, the resolution is also limited by the monochromatized photon beam size at the exit slit, which is determined by the entrance

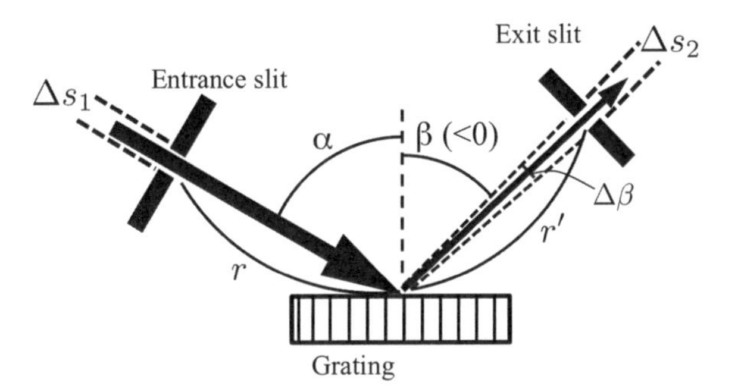

Fig. 3.4 Schematic layout of the entrance slit, grating and exit slit

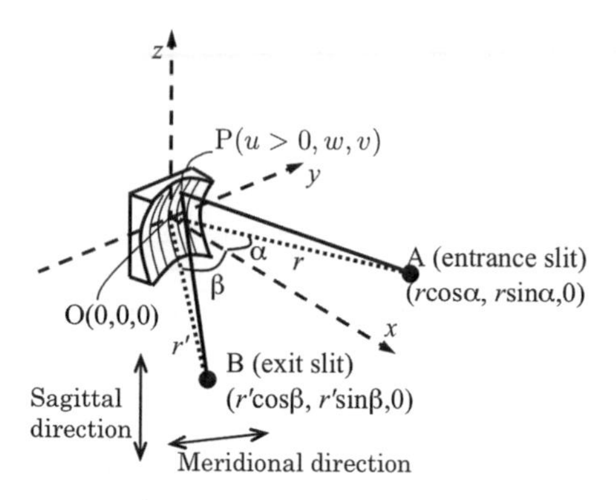

Fig. 3.5 Coordinate system for the optical path via the spherical grating with the radius of curvature R

slit width Δs_1 and the focusing conditions between the entrance and exit slits as discussed below.

The focusing conditions for a grating monochromator are obtained by considering the Fermat's principle. As shown in Fig. 3.5, the optical path, APB, from the entrance to the exit slit via the point $P(u, w, v)$ on the spherical grating with the radius of curvature R is expanded from the relation $(u-R)^2 + w^2 + v^2 = R^2$ as

$$\begin{aligned} APB &= r + r' - (\sin\alpha + \sin\beta)w \\ &+ \frac{1}{2}\left\{\left(\frac{\cos^2\alpha}{r} - \frac{\cos\alpha}{R}\right) + \left(\frac{\cos^2\beta}{r'} - \frac{\cos\beta}{R}\right)\right\}w^2 \quad \text{meridional focus} \\ &+ \frac{1}{2}\left\{\left(\frac{1}{r} - \frac{\cos\alpha}{R}\right) + \left(\frac{1}{r'} - \frac{\cos\beta}{R}\right)\right\}v^2 \quad \text{sagittal focus} \\ &+ \frac{1}{2}\left\{\frac{\sin\alpha}{r}\left(\frac{\cos^2\alpha}{r} - \frac{\cos\alpha}{R}\right) + \frac{\sin\beta}{r'}\left(\frac{\cos^2\beta}{r'} - \frac{\cos\beta}{R}\right)\right\}w^3 \quad \text{primary coma abbreation} \\ &+ \frac{1}{2}\left\{\frac{\sin\alpha}{r}\left(\frac{1}{r} - \frac{\cos\alpha}{R}\right) + \frac{\sin\beta}{r'}\left(\frac{1}{r'} - \frac{\cos\beta}{R}\right)\right\}wv^2 \quad \text{astigmatic coma abbreation} \\ &+ \ldots \text{higher-order abberations} \end{aligned} \tag{3.6}$$

It should be noted here that the coefficients for the terms with v, wv, w^2v and v^3 are analytically zero in (3.6). Since the optical path via the centre of the grating is $r + r'$, the first-order term with respect to w gives the grating Eq. (3.3). At the same time, to realize a high resolution, at least the meridional focusing condition

$$\frac{\cos^2\alpha}{r} + \frac{\cos^2\beta}{r'} = \frac{\cos\alpha + \cos\beta}{R}. \tag{3.7}$$

must be satisfied. In this way, the minimum condition for the coordination of the arrangements of entrance slit, grating and exit slit is determined. When the meridional focusing condition is held, the spot size of the monochromatized light at the exit slit is calculated from the emittance conservation law as

$$\frac{r' \cos \alpha}{r \cos \beta} \Delta s_1, \tag{3.8}$$

where M represents the magnification. When all abberations and slope errors for the optics are negligible and Δs_2 is set to $M\Delta s_1$, the energy resolution of the monochromatized light is limited as

$$\Delta(h\nu) = \frac{d}{mhc} \cos \beta \frac{M\Delta s_1}{r'} (h\nu)^2 \tag{3.9}$$

from (3.5) to (3.8). It should be noted that the resolution no longer becomes better even if Δs_2 is set as $< M\Delta s_1$.

Actually, the highest resolution of the grating monochromator beam lines is limited by the slope errors originating from the manufacturing precisions, heat load during the use on the beam line and by the abberations when all the optical components are perfectly aligned. Therefore the suppression of abberations is a key point to design the beam lines. One solution for the meridional focus (3.7) is $r = R\cos \alpha$ and $r' = R\cos \beta$ where the slits and the center of grating are located on a circle with a diameter R (therefore radius is R/2), which is called as a Rowland circle. It is known that the Rowland circle condition also yields the suppressions of the primary coma and several higher-order abberations. Therefore, a high-resolution monochromatization is expected for the Rowland-circle mount monochromators. The magnification M is 1 irrespective of the wavelength for the Rowland mount. On the other hand, the Rowland circle must be moved in space when the monochromatized wavelength is changed. Namely, simultaneous and precise multiple-axis motion and/or multiple-drives of optical components are required during the wavelength sweep. Such requirements may lead sometimes to the mechanical/optical instability during experiments. Non-Rowland mount monochromators with a simpler drive have thus been also developed. Another solution for the meridional focus in (3.7) is $r = \infty$ and $r' = R\cos^2\beta/(\cos \alpha + \cos \beta)$, which can be utilized when the incident light is parallel as in the case of SR. For $\cos \beta \sim 1$ (nearly normal diffraction), $r' = R/(1 + \cos \alpha)$. This is the condition for the Wadsworth mounting [20, 22].

Since the reflectivity of the grating surface becomes drastically decreased with increasing hν, grazing incidence monochromators are mostly used at high hν above ~40 eV, whereas normal incidence monochromators are often employed at lower hν. In order to compensate the changes of the focusing distance, diffraction angles and so on, especially for small samples in a huge UHV analyzer chamber, various techniques have been developed so far [20, 22]. Historically, such monochromators as Dragon-type and SX-700-type have been developed for the high-resolution grazing incidence monochromatization. Simultaneous motion (rotation and linear movement) of optical components is one of such techniques. However, high hν reproducibility became gradually required in accordance with the realization of high energy resolution. Reduction of the mechanical freedom is therefore quite important to realize high accuracy experiments. Soft X-ray monochromators based on the varied line spacing plane gratings are such examples [24, 25]. Without using the concept of the Rowland circle, good focusing and high resolution is realized by the varied line spacing plane grating (VLSPG) itself in a relatively wide soft X-ray hν region by only the grating rotation. In the low hν region below a few tens eV, various monochromators are used as explained in Refs. [20] and [22]. In addition to the Seya-Namioka monochromators, the modified Wadsworth monochromators, McPherson monochromators as the non-Rowland mount monochromators, and Eagle off-plane mounting monochomators with the Rowland mount are often used.

3.2.2 Crystal Monochromators

Above ~2,000 eV, Bragg diffraction by single crystals is used for getting monochromatic X-rays. If the white X-ray is incident onto the surface of a single crystal with the inter-plane distance d as shown in Fig. 3.6, the diffraction takes place for the condition $2d\sin\theta_B = m\lambda$, where θ_B is the Bragg angle defined as a supplementary angle of α defined in the case of diffraction grating. Here m is a positive integer. The crystal plane is defined by the Miller indices (hkl) and the inter plane distance d for a cubic crystal with the lattice constant a is given by $d = a/(h^2 + k^2 + l^2)^{1/2}$. In the nomenclature of the crystal monochromatization, the integer m is commonly included in the Miller indices as (mh mk ml) and thus d/m is redefined as d(mh mk ml). For example, d(111) with m = 4 is expressed as d(444). Then the Bragg's law is rewritten as

$$2d(hkl)\sin\theta_B = \lambda. \tag{3.10}$$

The most important requirement for crystal monochromators is the intensity of the monochromatized light (Bragg reflection intensity) and its wavelength (or hν) resolution. The single crystal should be as detect-free as possible without any distortion. Furthermore, the crystal is required to be robust against the heat load from SR (In this sense, artificial multilayer lattices are still behind the practical use.). A single-crystalline diamond (a = 3.58 Å) has an advantage in a high thermal conductivity, but the crystal quality is still poorer than that of Si. Therefore single-crystalline (of course undoped) Si (a = 5.43 Å) is very widely

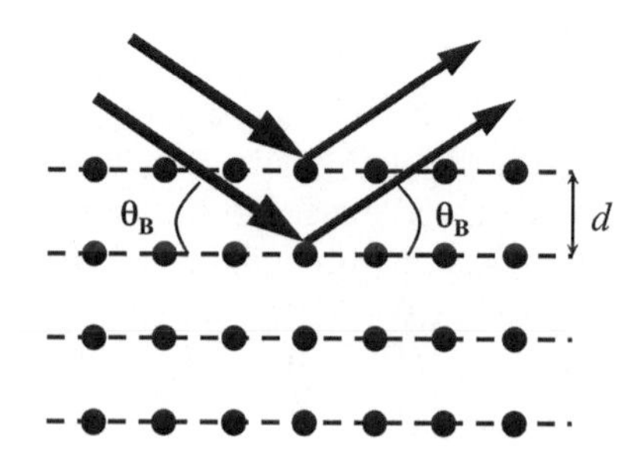

Fig. 3.6 Schematic description of crystal monochromators by using the Bragg reflection

used for crystal monochromators although its thermal conductivity is not so ideal for crystal monochromators.

Diffraction intensity of the crystal monochromators is proportional to a square of the absolute value of the structure factor of crystal F_{hkl}, namely, $|F_{hkl}|^2$. For the diamond crystal structure, it is well-known that

$$F_{hkl} = \begin{cases} 4(1 \pm \mathrm{i}) f_{hkl} & (\text{for odd } h, k, l) \\ 8 f_{hkl} & (\text{for even } h, k, l \text{ with a multiple of 4 for } h+k+l) \\ 0 & (\text{otherwise}) \end{cases}, \quad (3.11)$$

where i stands for the imaginary unit and f_{hkl} is the atomic factor which decreases monotonously with the interplane distance d(hkl). At hard X-ray beam lines, the Si(111) reflection is widely used for the monochromatization whereas the Si(222) reflection is forbidden and hence the second-order light is expected to be highly suppressed. On the other hand, the resolving power is given by

$$\frac{\Delta\lambda}{\lambda} \simeq \sqrt{\Delta\omega^2 + \Delta\theta^2} \cos\theta_B \equiv \Delta\Omega \cot\theta_B, \quad (3.12)$$

where $\Delta\omega$ (radian) stands for the width of the rocking curve of the diffraction, which depends on the individual materials and their quality whereas $\Delta\theta$ is the divergence angle of the incident beam. The rocking curve width ranges from about 1 to 10 s. 10 s correspond to $\sim 5 \times 10^{-5}$ radian and are comparable to $1/\gamma$ at 8 GeV, which is 6.2×10^{-5} radian. Therefore a proper combination of the light source and diffraction crystal is required in order to use the high resolution X-rays. According to the dynamical diffraction theory, $\Delta\omega$ is given by

$$\Delta\omega \simeq \frac{2r_e}{\pi V} \cdot \frac{\lambda^2}{\sin 2\theta_B} |F_{hkl}| \quad (3.13)$$

for the perfect single-crystalline monochromators, where r_e and V stand for the classical electron radius [$=2.82 \times 10^{-15}$ (m)] and unit cell volume of the crystal. When the parallel beam with $\Delta\theta = 0$ and the perfect crystal are used, the resolving power is thus evaluated as

$$\frac{\Delta\lambda}{\lambda} \simeq \frac{4r_e}{\pi V} \{d(hkl)\}^2 |F_{hkl}|. \quad (3.14)$$

The minimum resolving powers ($\Delta\lambda/\lambda$ or highest resolution) calculated by using (3.14) as well as the lowest photon energies covered by the diffraction with the Si crystal are summarized in Table 3.1. One can recognize that the high resolution is realized by using the diffraction with high Miller indices, high Bragg angle θ_B (although θ_B is automatically determined by the reflection plane) and small divergence angle $\Delta\theta$ irrespective of the spot size. This is in contrast to the conditions for the high-resolution spectroscopy by using grating monochromators, for which the focusing conditions of the grating must be carefully considered.

Table 3.1 Comparison of minimum $\Delta\lambda/\lambda$ (highest resolution) and lowest photon energy for several diffraction planes of Si

Diffraction plane	$d(hkl)$ (Å)	$h\nu_{min}$ (keV)	Minimum $\Delta\lambda/\lambda$
Si (111)	3.14	1.98	$\sim 10^{-4}$
Si (311)	1.64	3.78	$\sim 2 \times 10^{-5}$
Si (333)	1.05	5.93	$\sim 8 \times 10^{-6}$
Si (440)	0.96	6.46	$\sim 10^{-5}$
Si (444)	0.78	7.91	$\sim 4 \times 10^{-6}$
Si (555)	0.63	9.89	$\sim 10^{-6}$

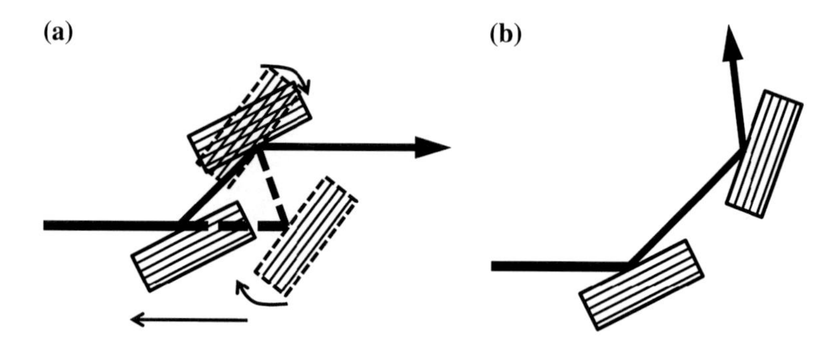

Fig. 3.7 Schematic layouts of double crystal monochromators. **a** So-called (+–) configuration usually used for hard X-ray beam lines to keep the optical path unchanged even after wavelength sweeping. **b** (++) configuration expected for realizing a high resolution by cancelling the divergence angle of the beam out

In fact, double crystal monochromators are usually used in the (+–) configuration as shown in Fig. 3.7(a) at hard X-ray beam lines to keep the same optical path irrespective of a target wavelength. One way to realize high resolution is to use the double crystal monochromators in the (++) configuration (Fig. 3.7(b)) by which the effects of the beam divergence are cancelled out, while the diffracted direction is deviated when the target wavelength is changed. Practically, a channel-cut crystal is used as a post monochromator downstream the primal double-crystal monochromator in the beam line.

When the Si (111) monochromators are used in the beam line, the suppression of higher-order light should be considered for the Si (333) reflection. One means of the suppression is to set the two crystals with a small angle offset (several second) from the parallel configuration. This is called as "de-tune", utilizing the fact that the centre angle of the Bragg reflection is slightly deviated at the higher-order reflection from the first-order one. One can drastically reduce the intensity of the higher order light by two successive diffractions. Another method is to use the focusing mirror after monochromatization where the grazing angle (supplemental angle of incidence angle) is set within the condition of total reflection for the target wavelength but out of total-reflection condition for the higher-order wavelength. These are based on the fact that the refractive index as a function of wavelength n(λ) is given by

$$n(\lambda) \simeq 1 - \delta\lambda^2 \tag{3.15}$$

in a hard X-ray region (except for the absorption edges) where δ depends on materials.

3.2.3 Focusing Mirrors

In order to focus the monochromatic radiation, reflection mirrors with surfaces of spherical, toroidal, cylindrical, ellipsoidal or parabolic shape are often employed for a wide hν region in addition to bent crystals for X-rays. As one example, the focusing by a toroidal mirror is discussed. Let us assume the curvatures (radii) of the toroidal mirror for the meridional and sagittal directions as R_m and R_s and the distance from the light source to the center of the mirror as r, where the incidence angle with respect to the surface normal of the mirror is α. By replacing R by R_m or R_s and set β as $-\alpha$ in the term of w^2 and v^2 in (3.6), the distances of the meridional and sagittal focusing r_m and r_s are given by

$$\frac{1}{r} + \frac{1}{r_m} = \frac{2}{R_m \cos\alpha} \quad \text{(meridional direction)} \tag{3.16}$$

$$\frac{1}{r} + \frac{1}{r_s} = \frac{2\cos\alpha}{R_s} \quad \text{(sagittal direction).} \tag{3.17}$$

The focused image by toroidal mirrors is, however, noticeably influenced by the aberrations, making a high quality point focus very difficult. Although the number of reflection surfaces is increased (which necessarily means an intensity reduction of photons), the combination of two mirrors for vertical and horizontal focusing is an alternative choice for focusing into a smaller area. For focusing into really small micro-nano regions, a combination of a bimorph X-ray mirror and a Bragg-Fresnel lens can be used in the hν range of 1.9–30 keV [26]. The so-called Kirkpatric-Baez configuration of two mirrors is also often employed [27]. Nowadays the use of ellipsoidal morrors with the Kirkpatric-Baez configuration in the total reflection condition allows the collimation down to ~25 nm [28]. On the other hand, Fresnel zone plates and Schwartzschild mirrors are used in the low hν region [22].

3.3 Examples of Light Sources

3.3.1 High Resolution Vacuum Ultraviolet Synchrotron Radiation Beam Lines

Since air and oxygen absorb photons with wavelengths below 2,000 Å, the region above h$\nu \gtrsim 6$ eV was called vacuum ultraviolet (VUV) [20]. We further use the term of soft X-ray to cover the λ region of 60 $\sim$ 6 Å (or hν in the region of ~200–2,000 eV) and the term of hard X-ray to mean the λ region below 6 Å (or hν above ~2,000 eV). Sometimes the term of extreme ultraviolet (EUV) is used for $\lambda < 1{,}000$ Å. However, the nomenclature of ultraviolet, VUV, soft X-ray and hard X-ray is not unique and depends on the individual users. So the hν value is used as much as possible in this book. The VUV and EUV handled in this book mean the hν region between 6 and 200 eV.

Although absorption takes place in a wide λ region even above 2,000 Å (or hν < 6 eV) in various solids, the photoelectron emission takes place only when hν is larger than the work function ϕ, which is often near 5 eV. Therefore photoelectron spectroscopy (PES) is useful in the regions of h$\nu \gtrsim 5$ eV except for multiphoton excitation PES. Since high energy-resolution is relatively easily realized in the hν region below 100 eV, detailed studies of electronic structures of materials were first performed in this hν region. Before 1990, a total energy resolution of 100 meV was rather typical in PES in the energy region with hν < 100 eV by using synchrotron radiation (SR). Such a resolution was then dramatically improved down to 30 meV and further down to a few or 1 meV. A lot of high precision studies of the band dispersions, Fermi surface topologies as well as band gap formations accompanying phase transitions were performed, demonstrating the potential of high resolution PES measurements. In this section high resolution vacuum ultraviolet beam lines in SR facilities are treated. For hν below 40 or 30 eV, the reflection efficiency of the gratings is still not so low compared with that in the soft X-ray region, where grazing incidence onto the grating is mandatory to obtain high enough reflectivity. Therefore normal incidence grating monochromators were often used below 30 eV for which the astigmatism by the spherical grating was not so serious in contrast to the case of grazing incidence monochromators.

Figure 3.8 schematically shows the optical system of a high resolution and low hν beam line with an off-plane Eagle mount monochromator using a grating with a radius of 3 m (beam line 9 (BL-9) of HiSOR, a 700 MeV electron storage ring, in Hiroshima University, Japan [20, 29]). The Eagle mount is rather close to the Rowland-type mount. For high resolution PES, not only high resolution of photons but also high photon flux are required. So the monochromators are in most cases installed on undulator beam lines. On this BL-9, a helical/linear polarization switchable undulator is installed [30]. Both the top and side views of this BL are schematically shown in Fig. 3.8. The fundamental order radiation on the axis is defined horizontally and vertically by a beam shaper with 4 movable shadowing plates. The radiation is vertically focused onto the horizontal entrance slit (S1) by a water-cooled toroidal mirror (M0) and a spherical mirror (M1). Then it falls onto a spherical grating (G) with a Rowland circle diameter of 3 m. The diffracted light is monochromatized by the horizontal exit slit (S2) and the hν scan is realized by rotating and linearly moving the grating [20, 29]. The focusing onto the sample is realized by a toroidal refocusing mirror (MF). In this configuration 4 reflections are employed for the optical system between the light source and the sample. The usable hν was reported to be 4–12 eV for a 600/mm groove grating and 10–40 eV

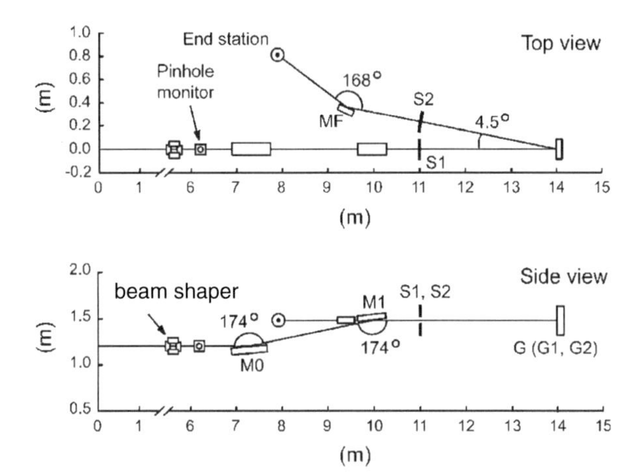

Fig. 3.8 Schematic optical layout of the 3 m off-plane Eagle mount monochromator at BL9 of HiSOR [29]

for a 1,200/mm groove grating [29]. The resolution of the monochromator was checked by absorption spectra of gases with known widths. With 10 μm slit, the resolution of 1.5 meV at 25 eV and 0.25 meV at 7 eV was achieved. For photoelectron spectroscopy, 100 μm slits were employed to get enough count rate providing low 10^{12} photons/s onto the sample. The total resolution of the photoelectron spectroscopy was estimated by measuring the Fermi edge of metals such as Au. Deconvoluting the Fermi–Dirac distribution function (FD), the total resolution of 4.5 and 7.5 meV was reported at $h\nu = 7.1$ and 22.6 eV, respectively.

Another example of BL7U of UVSOR (Okazaki, Japan, 750 MeV) [31] is equipped with a 3 m long APPLE-II undulator, which can provide horizontally and vertically polarized radiation besides left and right circularly polarized light (Fig. 3.9). The undulator is made of 38 periods with a unit length of 76 mm. High-energy resolution ($h\nu/\Delta h\nu \gtrsim 10^4$) and high photon flux above 1×10^{12} photons/s with good focusing ($<0.2 \times 0.2$ mm^2) onto the sample was targeted for this beam

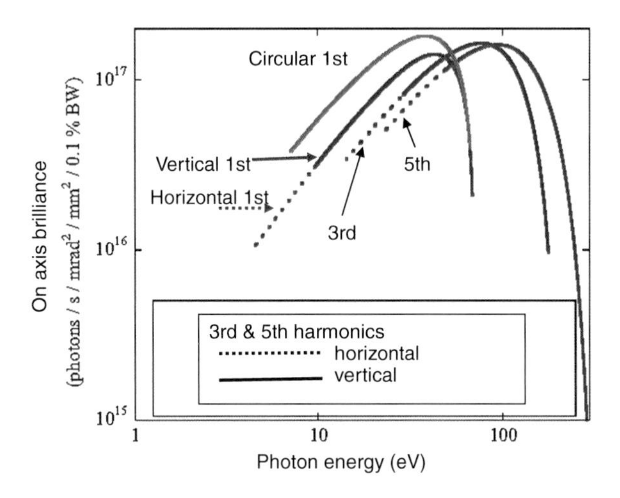

Fig. 3.9 UVSOR BL7U with a 3 m long APPLE-II type undulator. Brilliance spectra calculated for a stored current of 500 mA. In the low energy region are shown the fundamental (or 1st harmonic) radiation of the circularly polarized radiation (*top*) and the linearly polarized radiation (*lower*). In the higher hν region are shown the 3rd and 5th harmonics of the linearly polarized radiation [31]

line in the hν region between 6 and 40 eV. A Wadworth type (non-Rowland mount) monochromator without an entrance slit is employed in this beam line to realize high photon flux. The electron beam size 2σ at the center of the undulator is 0.08×1.1 mm^2 and this monochromator works as if a 80 μm wide horizontal entrance slit is employed. After passing through a pinhole (0.1, 0.2, 0.4, 1.0, 2.0, 4.0, or 8.0 mm in diameter), the light is deflected by two water cooled plane mirrors (M0 and M1 [31]) onto a spherical grating G with a radius of 10 m located at 22 m from the central source point. The incidence angles onto the M0 and M1 plane mirrors are set to 85 and 7.5°, respectively. Since the reflectivity of the 7.5° incidence M1 mirror becomes usually rather low in the high hν region, 7 plane mirrors with different surface coating (Au/Si, SiC, Si as well as 4 multilayer mirrors with Mg/SiC) are interchangeable in situ to cover the wide hν region [31]. The multilayer mirrors are used above hν = 29 eV. In order to realize high reflectivity in a relatively wide hν region, the number of the multilayer period is limited up to 6 or 7 and the period length is set to 16–22 nm. By using four different multilayer mirrors, the hν region from 29 to 43 eV can be covered with the reflectivity higher than 0.15. Three gratings (G1, G2, G3) with constant groove density of 3,600, 2,400 and 1,200 lines/mm optimized at hν = 32, 20 and 10 eV are interchangeable in situ. G3 is coated with SiC, whereas G2 and G1 are coated with Au. In addition to the rotational motion, the gratings are translationally movable to focus the monochromatic light onto the exit slit, which is located at 6.47 m from the grating center with an off-axis angle of 1°, similar to an off-plane Eagle type monochromator. A MgF_2 filter can be installed after the exit slit to cut the high hν higher order light above 10 eV from the G1. A toroidal focusing mirror is located at 2 m from this exit slit and focuses the light onto the sample at 1 m from this mirror. The beam size on the sample is 160×12 μm^2 with respect to the horizontal and vertical directions.

For the exit slit width of 100 μm, hν/Δhν was evaluated to be larger than 19,000 at ~24.4 eV using the 3,600 lines/mm grating and better than 9,000 at ~29 eV using the 2,400 lines/mm grating for a storage-ring coupling constant of 3 %. These results correspond to Δhν of 1.3 and 3.2 meV, respectively. For a reduced coupling constant of 1 % for the electron beam in the storage ring, the resolving power would be improved by ~40 %. For the exit slit width of 100 μm, a photon flux higher than 1×10^{12} photons/s can be realized in the regions of hν = 7–11 eV and 15–23 eV with hν/Δhν higher than 10^4. In this regard this beam line is very suitable to PES in the VUV region. By the PES measurement of the Fermi edge of Au at 15 K and hν = 10 eV with a monochromator exit slit width of 50 μm, an analyzer slit width of 0.2 mm and a pass energy of 2 eV, the system resolution (Gaussian FWHM) has been evaluated to be 2 meV besides the broadening by the FD function. If the photoelectron inelastic mean free path λ_{mp} was long enough, one can perform the three dimensional (3D) Fermiology. By changing both the sample tilt angle and hν, the 3D Fermi surface topology was partly probed in the case of paramagnetic CeSb [31].

3.3.2 High Resolution Soft X-ray Beam Lines

In contrast to laboratory-based X-rays and VUV light sources, the extensive use of soft X-rays became feasible by use of SR. Still rather long time passed to realize high resolution soft X-ray light sources because of the necessity of developing high performance undulators and monochromators. Since (1) the energy resolution of the X-ray from X-ray tubes was rather poor and (2) λ_{mp} of photoelectrons with kinetic energies realized by VUV light sources was often rather short, soft X-ray PES was desired for a long time. For example, λ_{mp} of photoelectrons was of the order of 3–5 Å for electron kinetic energies of 10–200 eV and PES for $h\nu <$ 200 eV was rather surface sensitive for many materials. In order to realize higher bulk sensitivity and higher energy resolution, the use of high brilliance soft X-ray SR above 500 eV was desired for decades. However, the energy resolution of grating monochromators deteriorates very rapidly with increasing $h\nu$ above a few hundred eV. In addition the tail of the 0th order light and output of higher order lights from the grating spoil the spectral quality.

It was proposed in the late 1990s to use the so-called helical undulator in a high energy storage ring as a soft X-ray light source because the higher order light off the undulator axis carries most of the heat load and it still provides low-heat-load fundamental order radiation on the axis defined by horizontal and vertical apertures. If the heat load on any optical component is too high, the thermally induced time-dependent and stored-current dependent slope error takes place and the energy shift with time ($h\nu$ drift) is serious, spoiling the high energy resolution and energy reproducibility of the monochromator. This effect was noticeable for the conventional planar undulators used up to several hundred eV with the decrease of the stored current in the storage ring. This problem was somehow reduced by use of the top-up injection operation of storage rings, which stabilized the thermal load onto optical elements. Still heat load problems appeared on changing $h\nu$ and undulator gaps, which induced a change of the heat load again. Even in this case, cooling of optical components for the fundamental radiation of the helical undulator is relatively easy because of the much lower heat load on the axis compared with that of the planar undulator.

For soft X-rays above a few hundred eV, the reflectivity of mirrors is much lower than in the low $h\nu$ region below a few tens eV. Therefore it was important to reduce the numbers of reflections as much as possible. In order to realize the reproducibility of $h\nu$, simple mechanical motion was preferred. The simplest, reproducible mechanical motion would be the axial rotation. After the invention of varied line spacing plane gratings (VLSPG), the design of the soft X-ray monochromators was drastically changed [24, 25]. The advantage of the VLSPG monochromator (VLSPGM) is that the surface slope error is generally very small in the case of a plane surface and focusing is automatically achieved by the initial designing of the ruling. Although the groove density was different up to several tens % from one end to the other end of the grating in the initial design to be employed for parallel incidence light [24], it could be reduced down to ±2 %

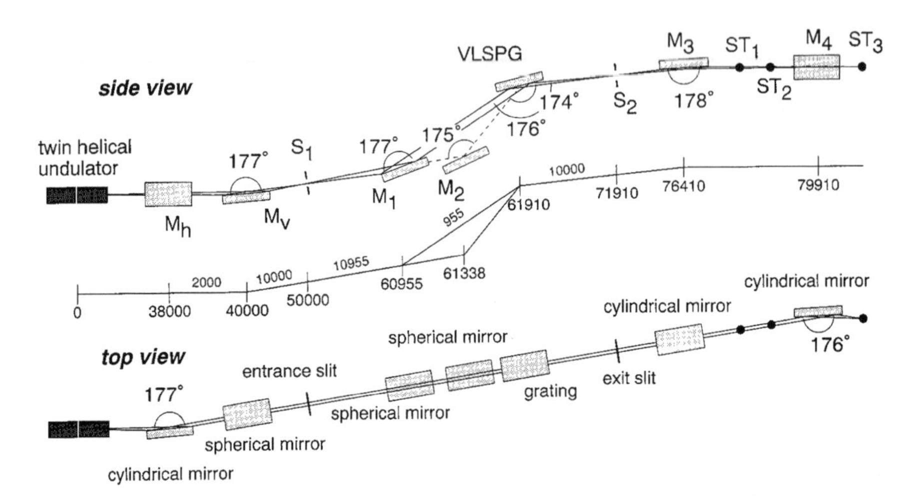

Fig. 3.10 Schematic layout of the varied line-spacing plane-grating soft X-ray monochormator (VLSPGM) at BL25SU of SPring-8. M_h and M_v stand for the horizontal- and vertical-focusing mirrors. S_1 (S_2) is the entrance (exit) slit, M1 (M2) is the spherical mirror to provide a converging beam onto the varied line-spacing plane-grating (*VLSPG*). Exit monochromatic light is further focused by the M3 or M4 mirror onto the sample in the experimental chambers ST_1, ST_2 or ST_3 [25]

across a ±100 mm ruling area for the converging incidence light in an improved design [25]. An example of an advanced soft X-ray beam line for PES is schematically shown in Fig. 3.10 [25] in the case of BL25SU of SPring-8 (8 GeV), which was the first soft X-ray beam line in SPring-8. The light source is a twin helical undulator [12], which can provide either a left- or right-handed circularly polarized light. Two helical undulators, each composed of 12 periods of 12 cm length, are aligned in tandem. Both undulators can be set to either right or left handed helicity, where either or both circularly polarized light(s) can be emitted on the same central axis. If both are set to opposite helicities, helicity modulation is possible by use of three kicker magnets. The circular polarization was checked to be beyond 99 % before the reflection by mirrors. By changing the undulator gap, the K value is tunable between 4.6 and 0.8 which corresponds to $h\nu$ from 0.22 and 3.0 keV. The x–y slits located upstream of the horizontal focus mirror M_h can remove more than 95 % of the total radiation power which is contained in the off-axis higher order lights. The central cone of the radiation of about ±50 μrad passes through these slits. The highest heat load from each undulator on the first mirror is 20 W. This highest heat load corresponds to $K = 0.8$ or $h\nu = 3$ keV and the heat load decreases for larger K or lower $h\nu$. The heat load is 10 W for $K = 2.0$ or 1 keV. This monochromator has a vertical dispersion and the monochromatic light is horizontally focused onto the sample owing to the M_h [25]. Although this monochromator design does not require any entrance slit as also confirmed later by experiments, a horizontal entrance slit S_1 is used not to spoil the monochromator performance in the case of a possible beam movement in the storage ring. The

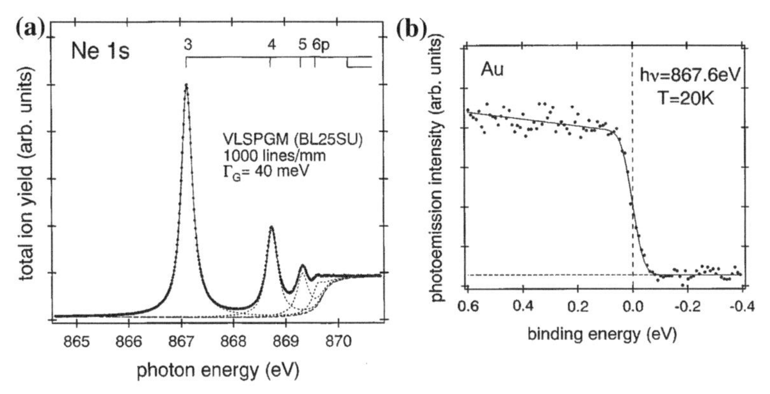

Fig. 3.11 **a** Soft X-ray photo-absorption spectrum of Ne gas to test the resolution of the varied spacing plane-grating monochromator at BL25SU of SPring-8 [25]. **b** Fermi edge of Au at 20 K excited by VLSPGM with the G_2 grating with the central groove density of 600 lines/mm

vertical focusing mirror Mv guides the undulator radiation onto the entrance horizontal slit S_1. In order to cover a wide energy region from 0.22 up to 2 keV by this VLSPGM, two deviation angles, 177 and 175° are employed with the mirrors M_1 and M_2, which are spherical to supply converging light onto the grating. Two VLSPG with a central groove density of 1,000 lines/mm(G_1) and 600 lines/mm(G_2) are mounted on the retractable rotation axis to cover the high and low $h\nu$ regions. The monochromatic light is focused onto the horizontal exit slit S_2 and then delivered to the experimental station via the focusing mirror M_3. ST_1 stands for the photoelectron station. The focus (FWHM) on the sample is less than 100 μm for both horizontal and vertical directions.

The energy resolution was evaluated by gas absorption. Figure 3.11(a) shows the Ne 1s core absorption spectrum by use of M_1 and G_1. By employing the minimum reported natural line width, the monochromator resolution was evaluated to be better than 40 meV at 867 eV. A resolution of more than 20,000 (16,000) resolution was achieved by use of the G_1 (G_2) grating in this system at $2\theta = 176°$ and an exit slit width of 10 (15) μm. A direct method to evaluate the monochromator resolution without the knowledge of the natural line width is to measure the Fermi edge by PES. Figure 3.11(b) shows the Fermi edge of an in situ deposited Au thin film cooled down to 20 K and excited by use of the G_2 grating. The total energy resolution of E_K was estimated to be 80 meV. From this result the resolution of this VLSPGM with G_2 and $2\theta = 176°$ was estimated to be $\sim$52 meV. By using the G_1 grating, a sharper Fermi edge and a higher resolution $\lesssim$50 meV was later realized in PES, though the photoemission count rate was much decreased. The advantage of this monochromator is the high $h\nu$ reproducibility within 10 meV, which is often required for reliable studies of photoelectron spectra. Such reproducibility is due to the simple one-axis rotation of the grating for the $h\nu$ scanning. For a resolution of 5,000 and a stored current of 100 mA, a photon flux beyond 10^{11} photons/s was realized with the maximum of about 4×10^{11} photons/s around hν = 800–1,100 eV. The ratio of the second order light intensity to the fundamental radiation was found to be below 10^{-3} for the G_2 grating above hν = 500 eV.

Another example of a high energy resolution soft X-ray beam line is the ADRESS beam line of the Swiss Light Source operated at 400 mA stored current (SLS, 2.4 GeV [32]). The undulator UE44 installed on this BL is composed of 75 periods of 44 mm long units occupying a total length of 3.4 m. This undulator is based on the APPLE II design and all four arrays with fixed-gap can be shifted in the longitudinal direction, providing both linear and circularly polarized light [32]. A plane grating monochromator (PGM) based on the SX-700 mount is installed on this beam line. Three gratings with constant groove densities (800, 2,000 and 4,200 lines/mm) are used in this PGM. A hν range between 300 and 1,600 eV is covered by this beam line. The highest flux is achieved near 1 keV and its value is 10^{13} photons/0.01 % (band width). This beam line is used for ARPES and resonance inelastic X-ray scattering (RIXS) in the soft X-ray region.

3.3.3 High Resolution Hard X-ray Beam Lines

In order to realize higher bulk sensitivity and a smaller influence of matrix element effects, hard X-ray photoelectron spectroscopy is attracting much attention nowadays. Since the photoionization cross section (PICS) decreases by more than two orders of magnitude between 1 and 10 keV as shown in Fig. 3.12 [33], high photon flux is required for PES in the hard X-ray region (Hereafter called HAXPES), necessitating the use of powerful undulators with a large number of periods. Still high resolution is required for the hard X-ray PES to compete with the VUV and soft X-ray PES (SXPES). Instead of the use of artificially manufactured gratings, single crystals are used in crystal monochromators for X-rays.

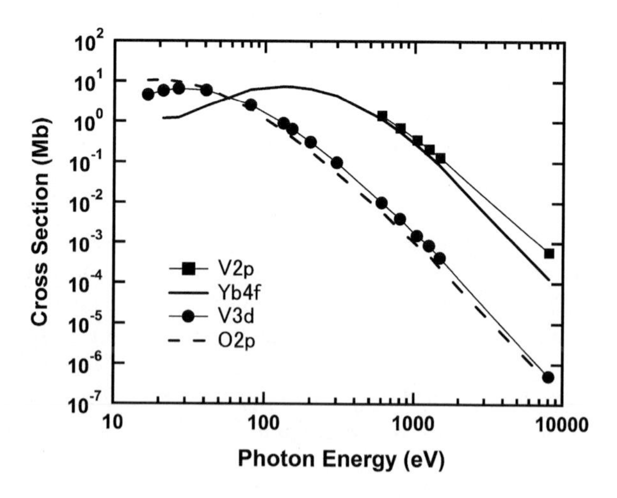

Fig. 3.12 hν dependence of photoionization cross sections (PICS) of typical p, d and f orbitals [33]

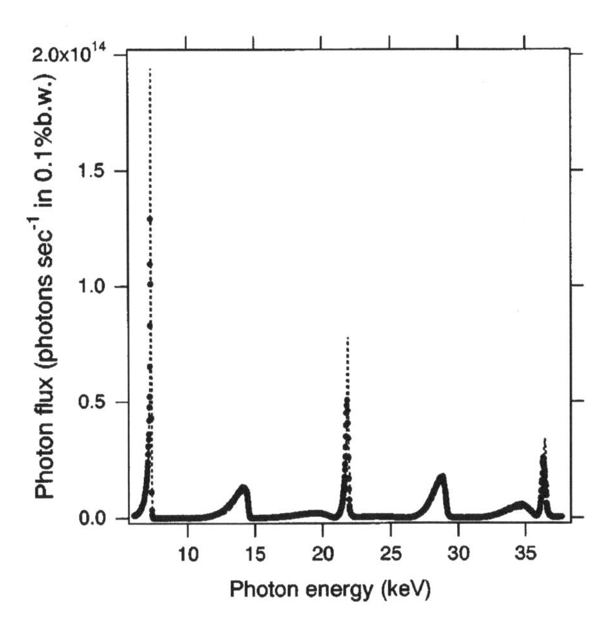

Fig. 3.13 Undulator radiation spectrum of the 25 m long 780 periods planar undulator at BL19LXU of SPring-8 [11]

Therefore a higher energy resolution than in the soft X-ray region can be expected in the hard X-ray region. Here a state of the art hard X-ray beam line, BL19LXU of SPring-8 [11], is described.

The advantage of the in-vacuum undulator is the feasibility to reduce the period length λ_u in (3.2) very short. Then it is easy to cover the hard X-ray region with the small value of λ_u/γ^2. The 25 m long in-vacuum planar undulator at BL19LXU is composed of 780 periods [11], each of which is composed of 32 mm-long permanent-magnet unit. The gap can be tuned between 12 and 50 mm and the K value at 12 mm gap is 1.76. The total power of the radiation at 12 mm gap is as high as 35 kW when the ring is operated at 8 GeV and 100 mA. The fundamental radiation covers $h\nu = 7.4$–18 keV. The typical spectral distribution at 12 mm gap and a square slit for accepting 3.9 μrad is shown in Fig. 3.13 [11], where rather narrow spectral width is observed whereas higher order light is noticeable, too. Still the available photon flux at the fundamental harmonic peak is quite high as 10^{14} photons/s in 0.1 % band width. Although the resolution down to 120 μeV is achieved at 14.4 keV [34], the photon flux is low 10^6 photons/sec, which is too low to do useful PES measurements.

For practical photoelectron spectroscopy, such an optical system as schematically shown in Fig. 3.14 is used. The undulator radiation is first monochromatized by the two Si(111) crystals. Then it is further monochromatized by a channel cut Si crystal with (440), (444) or (551) plane. In the case of (444) crystal, the resolution was set to $\sim$34 meV at 8 keV with the photon flux of the order of 10^9–10^{10} photons/s. Most parts of the optical path were pumped by diaphragm pumps to make free from the absorption of X-rays by atmosphere with using kapton films and/or Be windows. The monochromatic light was incident onto the sample in the

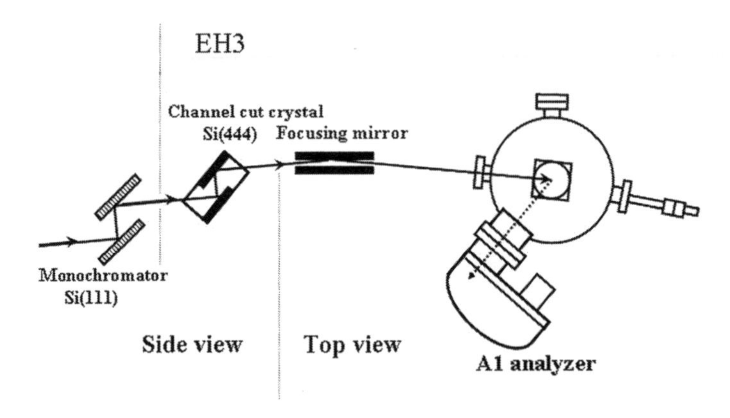

Fig. 3.14 An optical setup for the hard X-ray photoelectron spectroscopy (HAXPES) performed in the third or the first hutch of BL19LXU of SPring-8. The A1 analyzer is in reality rotated by 90° with the vertical dispersive plane

analyzer chamber through a 10 mmϕ Be window. The X-ray was horizontally focused down to 60 μm by the mirror while the vertical beam size was left as 0.6 mm in this case of locating the analyzer chamber in the third hutch of BL19LXU. In the case when the analyzer was placed in the first hutch of BL19LXU, vertical focusing down to 90 μm was also achieved by using bent mirrors in the optical hutch.

3.3.4 Laboratory Vacuum Ultraviolet Sources

PES in the low energy region was traditionally performed by use of the radiation from a gas discharge lamp [1] with or without a monochromator before the development of SR. Such a methodology below HeII ~40.8 eV has customarily been called ultraviolet photoelectron spectroscopy abbreviated as UPS. The objects to be studied by PES can be gases, liquids and solids, where low enough pressure is required in the chamber with the electron energy analyser. In this book, however, the electronic structures of solids and surfaces are mainly handled. In the angle integrated PES mode, the density of states (DOS) is mainly studied. This is applicable to polycrystal samples. In the case of single crystal surfaces, band dispersions and Fermi surface topologies can be studied by means of the angle resolved photoelectron spectroscopy (ARPES). When the angular resolution of the analyzer was not so high, low kinetic energy photoelectrons excited by low energy VUV light were suitable to ARPES. The symmetry of the electronic structures is also studied by use of the polarized light.

History of PES can be traced back in many literatures and its details are not repeated in this book. The advantage of UPS to the X-ray PES (XPS) was the higher energy resolution of the light source itself and higher counting rate of the

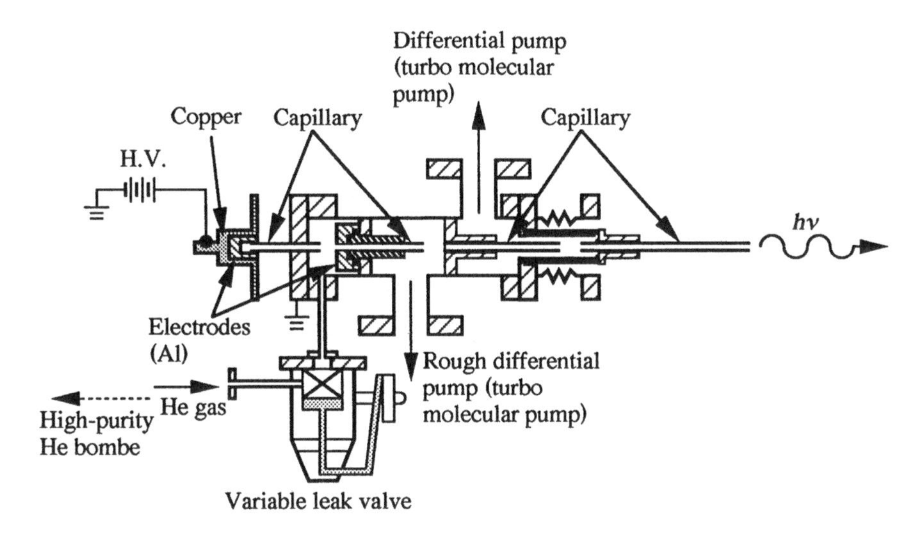

Fig. 3.15 Schematic picture of a conventional He discharge lamp

measurement due to higher PICS suitable to higher momentum resolution. The surface sensitivity also facilitated studies of surface electronic structures. The He discharge lamp was most conventionally used in the 20th century. HeI was useful to ionize most atoms and molecules. As shown in Fig. 3.15, one of the conventional He discharge lamp consisted of a narrow capillary to induce the discharge and one or two Pyrex capillaries to maintain the pressure difference between the discharge room ($<$ a few hundreds Pa) and the sample chamber to keep the sample surface as clean as possible. In the case of gas discharge lamp, the full width at half maximum (FWHM) of the resonance line was determined by the self-absorption by the gas itself and the Doppler effects. There was a tradeoff between the light intensity of He I and FWHM.

A much advanced way of producing high intensity radiation was to excite He atoms with microwave under the presence of magnetic field. This electron cyclotron resonance lamp enables one to remarkably reduce the He gas pressure. Then the self-absorption was much reduced and the FWHM of HeI lamp could be decreased to $\sim$1 meV and the intense He II radiation can be obtained as well. Since the discharge was induced only in the region of magnetic field, the plasma can be contained within a capillary wall. This type of source gains an intensity factor of 10^2.

Other rare gases can be used besides He in order to provide different $h\nu$. For example, Xe [35], Kr, Ar and Ne provide two strong lines at $h\nu = 8.44$ and 9.57 eV for Xe, 10.03 and 10.64 eV for Kr, 11.62 and 11.83 eV for Ar and 16.667 and 16.848 eV for Ne. It is very advantageous to use a LiF UHV compatible window [36, 37] to separate the analyzer chamber from the Xe, Kr and Ar light sources to shut out the contamination of the surface of samples (which are often cooled down to the liquid He temperature to realize the ultimate resolution) from the gas adsorption [36]. Figure 3.16 shows this microwave excited electron cyclotron resonance rare gas lamp [35–37]. Xe, Kr or Ar gas introduced into the ceramic tube with the inner diameter of 4 mm and the outer diameter of 6 mm with the length of $\sim$50 mm and pressure $\lesssim$1 Pa is excited by the microwave. Then the emitted light passes through an aperture (2 mmϕ) located at one end of the ceramic tube and is led into a differentially pumped chamber with better vacuum, where the emitted light is monochromatized. The small Doppler width and narrow FWHM (<600 μeV [36]) are very attractive characteristics of these light sources. The ceramic tube is cooled by the compressed air flow and the light exit aperture is cooled down by water. Still erosion of the plasma cavity wall (ceramic) is a serious problem for these light sources. One must replace the ceramic tube after the use of a few hundred to 1,000 h. But the replacement is quite simple and it takes only ten minutes for re-ignition of the lamp.

In order to perform high resolution and high quality PES, the removal of the higher hν components is required by means of high resolution grating monochromator [35] or exciton-absorption by ionic crystals [38]. For the latter purpose, thin crystals of sapphire, CaF_2 and slightly heated LiF are used for the Xe, Kr and Ar lamps, respectively [36, 37]. A toroidal mirror is used to focus the light with a diameter of 2 mm onto the sample with the comparable size by means of the 1:1 focus (Fig. 3.16(c)). The distances between the light exit aperture to the mirror center and the mirror center to the sample were set to 400 mm and 90° incidence angle is employed. The coating of the mirror is by Al which is over-coated with 25 nm thick MgF_2 film for anti-oxidation. In order to strengthen the LiF window, both sides are supported by lattices.

3.3.5 Laser Sources

The advantages of the laser light compared with SR in the low hν region are (1) higher energy resolution, (2) higher photon flux, (3) easy control of its polarization by use of phase plates and its slight disadvantages are the (4) limited hν tunability, (5) energy broadening by the space charge effect when too strong pulsed laser is focused into a too small region on the sample and by the uncertainty principle when the pulse width is too short. The high spatial and temporal coherence is a great advantage to focus the light into a very tiny region on samples. Then even a tiny new sample could be easily studied. High resolution time resolved PES is also facilitated by use of the pulsed laser, too. Powerful continuous wave (cw) lasers are quite rare above hν = 5 eV, which is a typical minimum energy required for single photon photoelectron excitation (called work function). The two-photon, three-photon and multi-photon excitations with lower hν lasers are possible in PES according to the nonlinear optical response in solids. The two- photon PES is briefly discussed later in Sect. 9.4. Several pulsed-lasers are available above this energy for single-photon PES. Higher order harmonics of lower energy pulsed lasers are often used in this region. However, high repetition rate and relatively

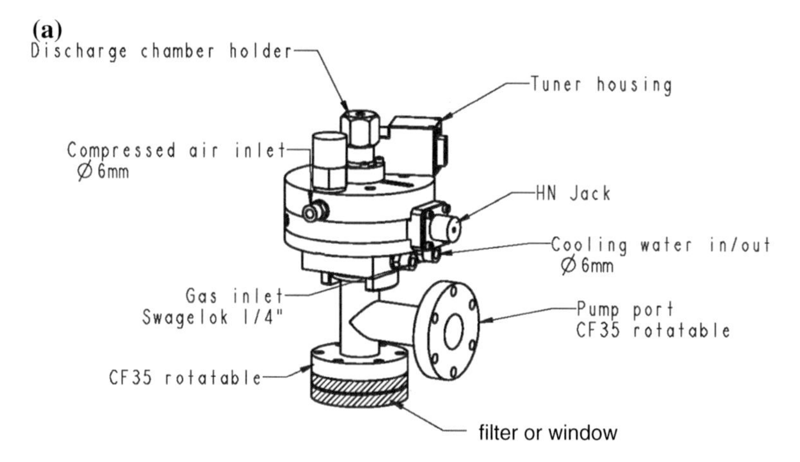

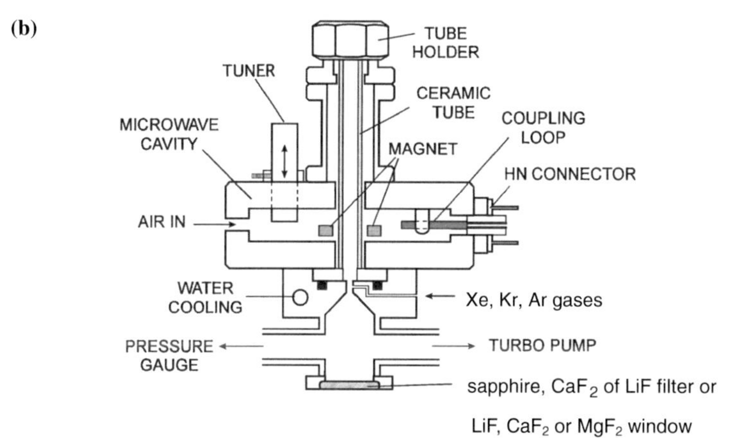

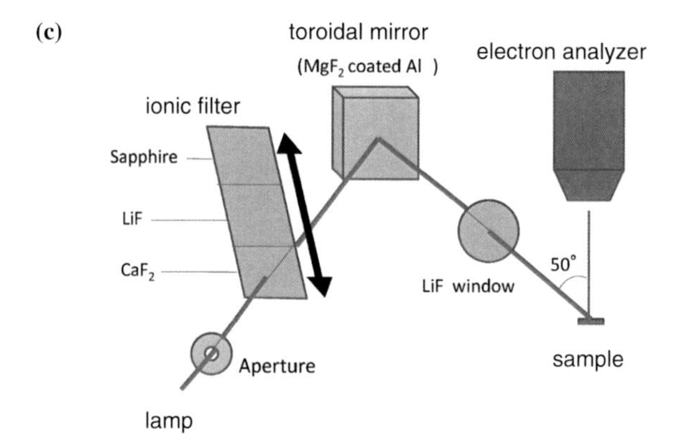

Fig. 3.16 Microwave excited electron cyclotron resonance lamp [35] manufactured by MB Scientific AB and monochromatization system by use of ionic filters [36, 37]. **a** Outer appearance and **b** cross section. **c** Schematic optical path for using the sharp lines from the Xe, Kr and Ar lamps into the analyzer chamber through a LiF UHV tight window

broad pulse width (τ) are necessary for high resolution PES, because of the reasons mentioned above. If each pulse width τ is too short, the uncertainty principle provides the broadening of the energy according to the following equation: $\tau \cdot \Delta E \sim \hbar$. For $\tau = 1$ ps, for example, the energy resolution ΔE is worse than 0.66 meV. Therefore quasi-cw laser with large enough τ is required for high count-rate, high resolution laser excited PES.

The total instrumental energy resolution ΔE_K of the photoelectron spectroscopy is given by the energy resolutions of the electron energy analyzer ΔE_A and the light source ΔE_L as follows:

$$\Delta E_K^2 = \Delta E_A^2 + \Delta E_L^2. \tag{3.18}$$

In addition, cooling of the samples is required to realize really high resolution below 1 meV, reducing the broadening by the Fermi–Dirac distribution function (FD). The high ΔE_A below 1 meV can be achieved by recent hemispherical analyzers such as VG Scienta R4000 and MB Scientific A1 analyzers if they were properly tuned. The residual magnetic field and exposure of non conductive materials near the sample must be minimized not to disturb the orbit of the photoelectrons and broaden ΔE_A. Electronic noise should also be minimized. In these analyzers with the radius of the electron orbit of 200 mm, the 250 μeV resolution is predicted for the slit width of 0.1 mm and the pass energy of 1 V. The realization of high ΔE_L (meaning small ΔE_L) is still under progress. The low hν might be, however, required for bulk sensitive measurements in the region of h$\nu < \sim 10$ eV and various developments are still progressing. Among them, a laser system operated at 6.994 eV turned out to be rather successful for some experiments [39].

The 3rd harmonic of the Nd:YVO_4 laser provides the 3.497 eV photons with the repetition of 80 MH_Z. This quasi cw laser light is impinged onto the KBe_2-BO_3F_2 (KBBF) non-linear optical crystals and the second harmonic at 6.994 eV is generated (Fig. 3.17). The width of this line is experimentally evaluated as 260 μeV. The photon numbers of the order of 10^{15} photons/s is reported to be generated at this hν, which are almost two orders of magnitude stronger than the microwave excited HeI light on the sample. In contrast to the large spot size of this He light, smaller spot is very useful for small samples so far as the space charge effect is suppressed not to influence the ultimate resolution.

For PES and ARPES measurements with energy resolution better than few meV, low temperature is the first necessary condition to overcome the thermal broadening. Here are explained more details of the quasi-cw laser PES system operated at h$\nu = 6.994$ eV [39, 40]. As shown in Fig. 3.17(a), h$\nu = 1.165$ eV photons from Nd:YVO_4 quasi-cw laser are amplified and then their second harmonics are generated. By mixing the latter with the former, the energy tripled (h$\nu = 3.497$ eV) quasi cw-laser is realized. This laser light is then incident onto an optically contacted two prisms (CaF_2) by means of a nonlinear crystal KBBF, generating its second harmonics (h$\nu = 6.994$ eV) [40]. This light can be easily focused by a CaF_2 lens and introduced into the sample-analyzer chamber through a UHV tight CaF_2 window as shown in Fig. 3.17(b). The laser light is linearly

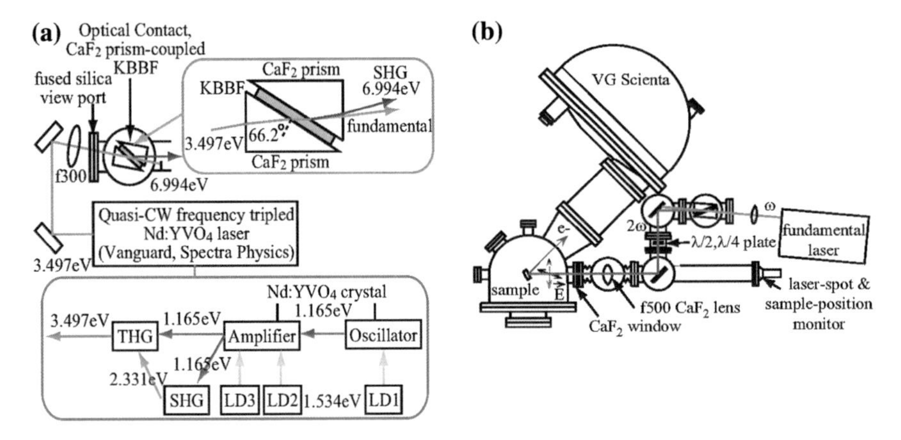

Fig. 3.17 Quasi-cw, high repetition laser source for extremely high resolution low hν PES. Details of the optical system (**a**) and the layout of the system (**b**) [39]

polarized and then one can easily change the polarization to the orthogonal linear polarization and circular polarization by use of a phase plate. Therefore this light source is rather convenient to the studies of linear and circular dichroisms of solid materials. Focusing down to 0.02 mm is not difficult and the photon flux can be $\sim 2 \times 10^{15}$ photons/s.

If strong laser light is focused into a very small area, many electrons are instantaneously emitted from a small region of the sample. Then the Coulomb interaction between excited electrons induces a noticeable broadening of the E_K of the photoelectrons which is called a space charge effect. The so far conventionally employed rare gas lamps and X-ray tubes have not induced such an effect within the resolution so far employed. This effect is sometimes influential even when strongly focused synchrotron radiation, which is intrinsically pulsed light, is used for the excitation of photoelectrons. Since pulsed lasers (including the free electron lasers (FEL)) are available for PES experiments, this phenomenon must be properly understood. A simulation of the space charge effect for PES was perfomed by considering the number of electrons per pulse, the pulse length, the size and shape of the electron source (excitation area), and the energy distribution of the emitted electrons (Baltzer unpublished). The electron interactions should be considered only when they are close to the surface of the photo-excited solid surfaces. Therefore spatially flat distribution of photoelectrons was assumed in the initial stage. When very strong and narrow pulsed light is used for the excitation, care should be taken to check the space charge broadening in order to realize a really high energy resolution. This can be done by measuring the PES spectrum and spectral broadening while changing the photon flux by use of neutral density filters for lasers.

3.3.6 Miscellaneous Subjects

In addition to SR, low hν gas discharge lamps, lasers and X-ray tubes are conventionally used for PES [1, 3]. X-rays are generated when high energy electrons are suddenly stopped by a target. The X-ray tube consists of an anode and a cathode, where the anode is usually grounded and the cathode is biased to a minus high voltage (−20,000 to −50,000 V). The cathode is made of a W filament and the anode of a target metal piece mounted on a water-cooled Cu block. The electrons thermally emitted from the filament are accelerated and hit the anode target. When high intensity is required, a rotating anode can be used. In all cases, the X-rays are emitted to all directions.

In addition to the characteristic lines, a continuum spectrum is emitted from the anode. The high energy cut-off of the continuous spectrum depends upon the acceleration voltage. However, only the characteristic spectrum such as the Al K_α and Mg K_α lines is useful for XPS. The characteristic X-ray is emitted only when the acceleration voltage exceeds a certain value characteristic to the individual anode material. They are called $K\alpha_1$, $K\alpha_2$, $K\beta_1$ and so on. $K\alpha$ is due to the transition from the 2p shell (L shell) to the 1s shell (K shell) and $K\beta$ is due to the transition from the 3p shell (M shell) to the K shell. The separation between $K\alpha_1$ and $K\alpha_2$ corresponds to the spin–orbit splitting of the 2p state. The hν of some characteristic X-ray lines is hν = 1,486.6 eV for Al $K\alpha$ and hν = 1,253.6 eV for Mg $K\alpha$ line. These lines have the FWHM of the order of 1 eV caused by their own natural width and unsolved spin–orbit splitting ($K_{\alpha 1}$ and $K_{\alpha 2}$, for example) and are not suitable for high resolution XPS with energy resolutions better than 100 meV. Even after monochromatization by crystals the resolution amounts down to only 0.4 or 0.3 eV.

In the case of conventional X-ray tubes and light sources like the He and rare gas discharge lamps, the emitted light is unpolarized. On the other hand, the SR from the bending section is well known to have a very high degree of linear polarization in the orbital plane. When the light is accepted through an aperture located above or below the median orbital plane of the storage ring, either the left or right ellipsoidally polarized light can be utilized, although the intensity is much reduced compared to the light emitted in the median plane. Although the degree of circular polarization of such light is not so high, still it is utilized for such experiments as magnetic circular dichroism in core absorption (XMCD).

In the SR X-ray region, the use of phase plates is also practical to change the linear polarization direction and/or convert the linearly polarized light into circularly polarized light. It is known for high quality Si and diamond crystals, that a phase shift is induced between the σ and π polarized light after transmission through the crystal [41–43], when the crystal is used near the Bragg diffraction condition. By utilizing this phenomenon, one can convert linearly polarized light from a planar undulator to circularly polarized light. By properly setting the offset angle from the Bragg angle, a phase plate can cover a rather wide hν region. If the crystal is vibrated around the Bragg angle of the order of tens seconds, one can moreover switch between right and left circularly polarized light with a frequency of 40–200 Hz [44]. The degree of circular polarization can also be very high. The $h\nu$ region of 5.6–13.7 keV is covered at BL39XU of SPring-8 [45]. The lock-in technique is applicable for XMCD in this case.

In the soft X-ray region, however, such transmission phase plates are not available. Therefore horizontally polarized radiation is mostly used. In order to reduce the heat load, helical undulators are mostly employed. The polarization of the helical undulator can be quite high in a wide $h\nu$ region [25]. The advantage of the helical undulator was already mentioned in Sect. 3.3.2 in this chapter. An additional advantage is that the PES results by circularly polarized light are not directly affected by the selection rules for the linearly polarized light associated with the crystal anisotropy effect. In the case of anisotropic materials, however, positive use of the horizontally and vertically polarized light is desirable. In order to reduce the heat load even in this case, a "Figure 8 undulator" was developed and practically used at BL27SU of SPring-8 [46]. It has the advantage that the on-axis power density is much lower than that of an ordinary planar undulator. Electrons inside this device move along a trajectory which looks like a figure-8 when projected on the plane perpendicular to the beam axis.

In PES by use of SR, focusing of light is often required to study a small area of the sample. For micro and nano materials, focusing elements such as zone plates [22, 47, 48] and Schwarzschild mirrors [22, 49] are employed as already explained in Sect. 3.2.3. In the near future, ellipsoidal Kirkpatric-Baez mirror will widely be used for micro- and nano-focusing.

3.4 Electron Spectrometers

The constraints on the vacuum in the analyzer chamber is briefly mentioned. In the first place, the discharge of the detector applied with high voltage should not take place. Therefore the pressure must usually be better than $\sim 1 \times 10^{-2}$ Pa in the analyzer chamber. In the case of solid state studies, the surface must be as clean as possible or the surface contamination (degradation) should be kept as low as possible. This usually requires the UHV condition around the sample. In the case of windowless rare gas light sources as e.g. He lamps, differential pumping between the light source and the sample is required to reduce the contamination of the clean sample surfaces by gases slightly contained in high purity He gas. Since higher resolution PES and ARPES measurements are becoming more and more required, good performance of the analyzer is also required in addition to both high resolution of the light source and the control of the sample T. Some discussions on the light sources were already given in Sect. 3.3 and some other discussions on the sample will be given later in Sect. 3.5.

3.4.1 Hemispherical Analyzers

Simple retarding type analyzers were used in the early stage of PES. These could collect a wide angle of photoelectrons but the energy spectra were only derived by differentiating the measured signals by the retardation voltage. Then deflection type analyzers were developed such as the parallel plate analyzer, cylindrical mirror analyzer (CMA) and hemispherical analyzer [1, 3]. However, the hemispherical analyzer is now most often used for high resolution measurements. Historically, the hemispherical deflector analyzer was used for the energy analysis of charged particles and then applied to UPS. CMA is superior to the hemispherical analyzer with respect to the acceptance angle and luminosity (the product of solid angle and sample area) by one order of magnitude for the same relative resolution when the single channel detection is employed for the angle-integrated PES. On the other hand the hemispherical analyzer has an image plane, where energy distribution and angular distribution can be detected simultaneously. These advantages compensate the lower luminosity. An input lens is often attached to the hemispherical analyzer providing more space around the sample. One important requirement for a hemispherical analyzer to realize high performance is how to terminate the spherical field at the entrance and exit slits. Although elaborate terminating electrodes have been used, it is possible to terminate the field with a simple plane electrode when the entrance direction is correctly set.

If the illuminated sample is located in one of the analyzer focal planes, the detector should be located in the other focal plane. Then the spectrum is measured by changing the inter-electrode voltage. However, this method has the drawback that the resolution varies over the spectrum. The use of multidetectors is also very difficult. It is therefore desirable to keep the analyzer pass energy E_p constant and sweep the voltage between the sample and the analyzer to get the energy spectrum with a constant energy resolution. Depending upon the kinetic energies of photoelectrons (E_K), either acceleration or deceleration of photoelectrons is required for a certain pass energy of the analyzer. An electron lens is mounted between the sample and the entrance of the analyzer to satisfy this requirement with a minimum variation in transmission efficiency. In the case of a hemispherical analyzer with the radius of the inner sphere R_1 and the radius of the outer sphere R_2 ($R_2 > R_1$) as shown in Fig. 3.18(a), the E_p of the electron passing on the spherical orbit with the radius of $R = (R_1 + R_2)/2$ is expressed as $E_p = eV_p/(R_2/R_1 - R_1/R_2)$, where V_p is the voltage between the outer and inner spheres. The energy resolution of the hemispherical analyzer ΔE_K is given by $\Delta E_p/E_p = w/(2R)$, where w stands for the entrance slit width of the hemisphere as schematically shown in Fig. 3.18(a). The actual energy resolution of the spectrometer for $\Delta E_p < \sim 10$ meV depends also on other extrinsic effects such as ground (earth) connection to the sample voltage, precision of the power supply as well as that of mechanical manufacturing. Residual (static) electric and magnetic fields around the sample induce the energy dependent image and angular distortions.

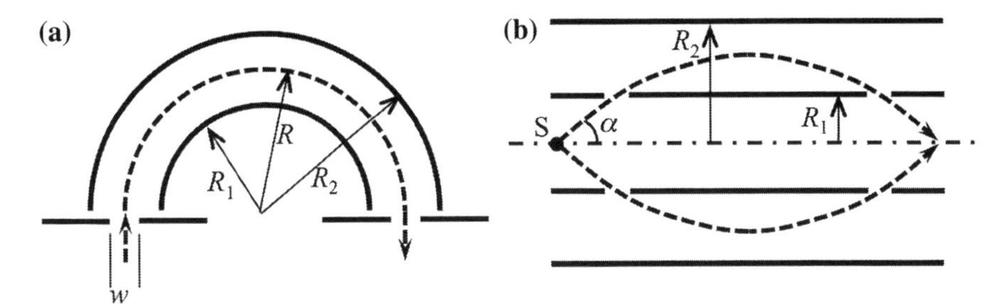

Fig. 3.18 Schematically shown cross-sectional view of electron spectrometers. **a** Hemispherical analyzer. It should be noted that the exit slit is often replaced by a two-dimensional position-sensitive detector. **b** Cylindrical mirror analyzer

In the case of hemispherical analysers a series of coaxial and rotationally symmetric electrodes are used in a lens, whose first element is on the same potential as the sample while the last element is on the same potential as the analyzer central potential. The sample is imaged onto the analyzer's entrance plane. By using only one intermediate element as in a three element lens, the magnification will be energy-dependent. By using 4 or more lens elements, however, the magnification can be kept constant over a wide energy range. Computer control of the voltages of the lens elements is required for measuring an energy spectrum. The emission angle dependence can be measured also as a dependence on the displacement over the entrance slit by a different voltage set [50]. In this case the simultaneous detection of the energy spectrum and angular distribution is feasible by use of a two-dimensional (2D) detector.

Electron counting is a popular technique for PES. While such detectors as channeltrons were used for single channel detection, MCPs (micro channel plates) combined with a proper readout are used for multichannel detection. Non linearities in the analyzer energy scale and the image distortion can be handled by the computer. Either CCD detection of a phosphorescence screen or a resistive anode behind the MCP is employed for 2D detection. New concepts for direct digital readout by one and two dimensional detectors are developed in order to increase the acceptable count rate before the saturation of the detection system. In most cases, it is advantageous to retard the electrons emerging from the analyzer before they impinge on the detector surface to reduce the secondary electrons from the analyzer walls. A mesh at the central trajectory potential is used to retard the electrons without disturbing the analyzer field.

In the deflector type analyzer, the energy resolution is inevitably restricted by the accuracy of the deflection voltage. Good magnetic shielding is also required. The static magnetic field should be kept below 10^{-7} T (100 nT, where a 1 eV electron has a radius of circular motion of 34 m). The earth magnetic field is $\sim 5 \times 10^{-5}$ T. The compensation of the earth magnetic field and residual magnetic field was formerly achieved by a 3-axes Helmholtz coil with feedback stabilization. But double μ-metal shielding is nowadays more popular. The fluctuating magnetic field has the origin in the transformer or motors nearby. Such fields are effectively damped by the eddy currents in the vacuum chamber and analyzer electrodes.

The electrode voltages should be very much stabilized to realize a resolution in the meV range. If the analyzer voltages for deflection and retardation fluctuate, the line position in the slit plane moves, deteriorating the energy resolution. Therefore high accuracy and high stability power supplies are required for these elements. The internal electrostatic field between metal surfaces is set up by the external voltage applied and their difference in contact potential. The contact potential of a metal is very sensitive to the surface condition and will change by a magnitude of up to 0.5 V with time and as a function of other conditions. The homogeneity of the work function over the electrode surfaces is quite important. If one single material such as Mo is used for all surfaces seen by the photoelectrons, this problem is mostly solved. Even when different materials are used for different electrodes for the reason of cost, weight, ease of manufacturing, suitable surface coating may provide a homogeneous work function. Colloidal graphite dissolved in isopropanol suspension is widely accepted for this purpose. The graphite coating is very stable against chemical attack, though it shows some sensitivity to UV radiation with respect to the work function. Nowadays various hemispherical analyzers are developed by several companies. A radius R of ~200 mm is quite often employed and a total energy resolution comparable or better than 1 meV is already achieved for real measurements by employing a pass energy of 1 eV and a slit width of 0.1 mm [39].

3.4.2 *Cylindrical Mirror Analyzers*

The simplest analyzer may be the plane mirror analyzer (PMA) made of two parallel plates 1 and 2 separated by the distance d. The plate 1 has two slits separated by L. The entrance and exit angle from this plate 1 is defined as α. The potentials of the plates 2 and 1 are set to V_2 and V_1. The electrons which have passed through the entrance slit follow a parabolic trajectory and pass through the exit slit if $V_1 - V_2 = 2E_p \cdot d/eL$ [3, 6], where E_p is the electron kinetic energy in the analyzer (pass energy). The plane mirror analyzer is very simple but has the disadvantage of low transmission, which is overcome by the cylindrical mirror analyzer (CMA) with an acceptance azimuthal angle up to 2π radian. The CMA was probably the most widely used analyzer type in the past for surface science. The deflection field is formed by two concentric cylinders with radii R_1 and R_2 ($R_1 < R_2$) as shown in Fig. 3.18(b), where the potentials applied are V_1 and V_2. The electrons enter the analyzing field with an angle α defined with respect to the cylinder axis. The electrons are deflected by the field between two cylinders. The full 2π azimuthal angle is available for detection. Among various selections of geometrical parameters for practical use, $\alpha = 42.3°$ is widely accepted to achieve second order focusing. In this case $E_p = 1.3e(V_1 - V_2)/\ln(R_2/R_1)$ [51] and $L = 6.1R_1$ [3]. The CMA is suited for high sensitivity at moderate resolution.

In some experiments double stage CMAs are applied. Cylindrical deflector analyzers (CDA) [3] can realize high resolution while the transmission is poor. If the cylinder spans $\pi/\sqrt{2} \sim 127°$, a good one dimensional focus is realized. The electron pass energy is evaluated by $E_p = e(V_1 - V_2)/[2 \ln(R_2/R_1)]$ [6].

3.4.3 Two-Dimensional Analyzers

In accord with the progress of angle resolved photoelectron spectroscopy (ARPES), two dimensional (2D) detectors were developed to simultaneously detect the photoelectrons emitted to different directions in order to save total measuring time. Since PES is a surface sensitive method, the reduction of the measuring time can somehow overcome the problem of surface contamination with time. The first practical 2D analyzer is schematically shown in Fig. 3.19(a) [52]. This analyzer was equipped with an ellipsoidal mirror and spherical grids with the sample at one focusing point and the entrance to the detector at the other focusing point of the ellipsoid. The electrons with an energy lower than a certain energy E_1 are reflected by the mirror and focused into the aperture located at the other focusing point. Then the electrons with an energy higher than a certain energy E_2 can pass through the retardation grid and are multiplied by a microchannel plate or a CEMA and hit the phosphorescence screen. Since the angle onto the detector is not equal to the emission angle, the image on the screen is inevitably distorted with respect to the emission angle from the sample. Therefore a numerical transformation by a computer was required to restore the original angular distribution pattern. A wide collection angle was also difficult to realize in this design and the manufacturing of an accurate ellipsoidal surface was also a hard job. An angular acceptance of $\sim 85°$ with a resolution of $\sim 2°$ and an energy resolution <100 meV were reported [52].

Later a 2D display type analyzer with a simpler design was developed as shown in Fig. 3.19(b), which had the advantage of undistorted images [53–55]). The SR introduced from a small aperture through the electrodes functioning as obstacle-rings (J) hit the sample and excite photoelectrons which pass through the small concentric hemispherical grids around the sample and then enter the space surrounded by a larger hemispherical grid (G), obstacle rings (J), the guard rings (U), and the outer sphere (D). The obstacle rings provide a spherical Coulomb field inside the analyzer and function as a low pass filter. Aperture A has a diameter of 1 mm and is located at a mirror symmetry position of the light focus on the sample with respect to the center of the grid G. The distance between the sample and aperture A is 100 mm in Fig. 3.19(b). Retardation grids functioning as a high pass filter are located after the aperture A. The detector part is composed of the microchannel plate (MCP) and a phosphorescence screen. The shape of the inner surface of each obstacle ring is part of the sphere with a specific angle and radius. The potential on each obstacle ring was set to a proper value to realize a spherical electric field. The acceptance cone could be as large as $\pm 50°$. The advantage of

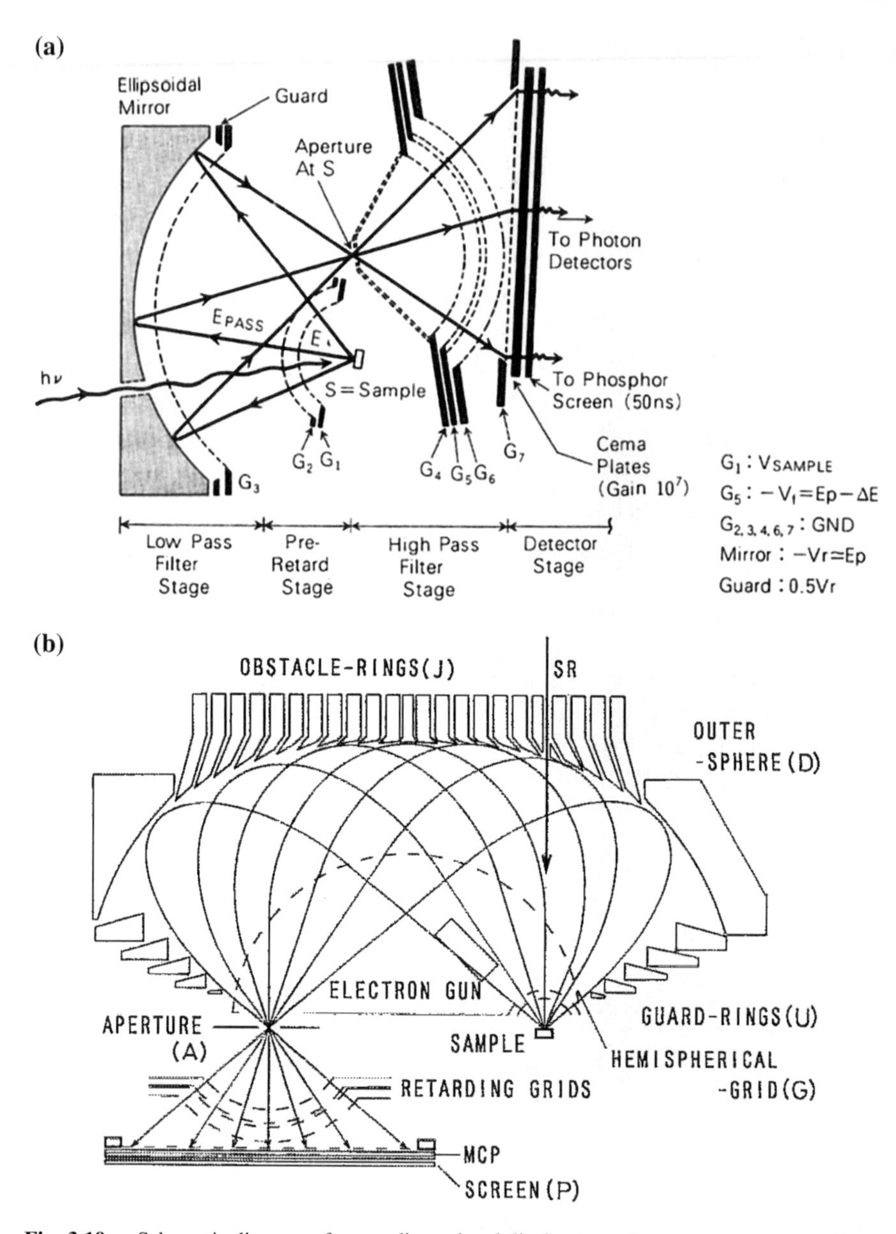

Fig. 3.19 **a** Schematic diagram of a two-dimensional display type electron spectrometer with an ellipsoidal mirror [52]. The ellipsoidal mirror was used as a low pass filter and a 4-fold spherical grid worked as a high pass filter. Around the sample was a spherical grid pre-retard stage. CEMA means two continuous electron multiplier arrays. **b** Two dimensional display type electron spectrometer without distortion of the angular distribution [53, 54]

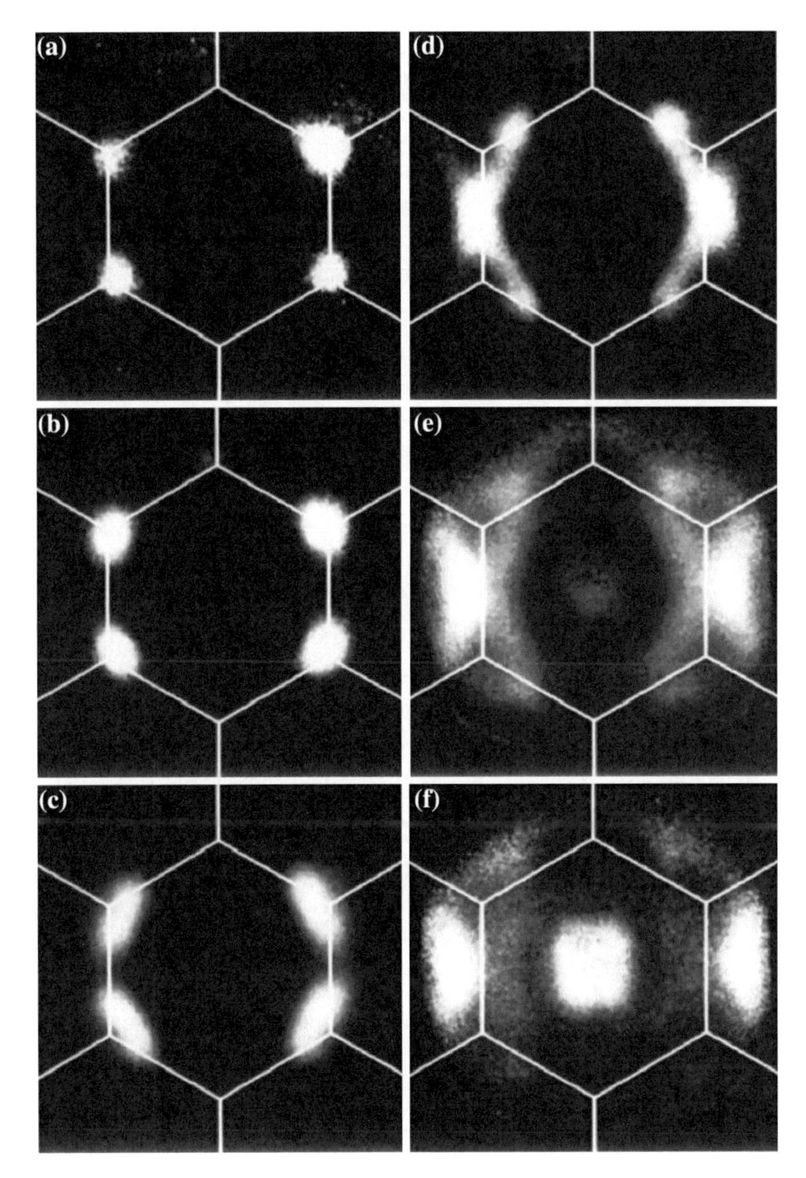

Fig. 3.20 Two dimensional angle resolved photoelectron spectra of Kish graphite at hν = 54 eV. The incident SR was horizontally polarized. The energy resolution (FWHM) was set to about 0.5 eV. The E_B was −0.4, 0.7, 1.7, 2.7, 3.7, and 4.7 eV for (**a**)–(**f**). The hexagonal white lines show the Brillouin zone boundaries as a guide to the eye [55]

this analyzer was that the emission angle was exactly the same as the detection angle, meaning the angular patterns were measured without any distortion and the results recorded on the screen did not require any distortion correction by computers. A typical results for Kish graphite is reproduced in Fig. 3.20, where the change of the two dimensional angular distribution of photoelectrons from the states with different E_B is clearly resolved for incident light with a horizontal polarization. The nonequivalent intensities around the six K points are well explained by the dipole selection rule [55]. The energy resolution can be set to 1 % of the pass energy. Since the full azimuthal angle of 360° is covered by this analyzer, any artificial symmetrization is not necessary for further data analysis. The abundant information obtained by this analyzer is convenient for the application to photoelectron diffraction and holography as explained in Chap. 12.

3.4.4 Time of Flight Analyzers

Photoelectron spectra can also be measured by means of time of flight (TOF) analysis. When the photoelectrons are excited by pulsed light and emitted from the sample, the dispersion of the velocities (energies) is induced in a field-free drift region between the particle source and the detector. Although the duty cycle is rather low, the whole range of velocities can be detected as a function of time. TOF analyzers match well with SR because of the well defined pulsed time structure and accurate repetition as an alternative to deflection type analyzers. Relatively large opening angle is easily realized because no deflection field is required.

The time resolution is limited by (1) the pulse width of the excitation light, (2) the drift length, (3) the size of the light focus on the sample in addition to the intrinsic time resolution of the detector. The typical pulse width of synchrotron radiation ranges from ns down to 0.1 ns. In the case of free electron lasers, it is tens fs and the shortest pulse of laser can be a few fs. The final energy resolution $\Delta E/E$ can be estimated as [56]:

$$\Delta E/E = [(2\Delta t/t)^2 + (2\Delta l/l)^2]^{1/2},$$

where Δt stands for the total time resolution including the pulse width and Δl represents the uncertainty of the flight path to the detector. Δt is determined by the timing capabilities of the electron multiplier detector and the associated electronics, together with the synchrotron light pulse width. The contribution of the detector and electronics can be <70 ps whereas the light pulse width may be larger. If Δt is $\sim$300 ps, the first term gives a contribution of 40 meV at 10 eV for a typical drift tube length of $\sim$30 cm.

Δl results from the finite source size as well as the finite collection angle. The angular acceptance of $\pm 3°$ provides the energy uncertainty of 0.25 %. However, a much larger contribution comes from the finite source size. For 1 mm source size the contribution is about 0.7 %. TOF PES has so far been mostly applied to gas phase samples due to its high count rate. An interesting device is the double TOF spectrometer, which can be used for coincidence detection of photoelectrons and photoions.

3.5 Sample Preparation and Characterization

3.5.1 Ion Sputtering, Scraping, Fracturing and Cleavage

Preparation of clean surfaces is often mandatory for PES of solids because of its surface sensitivity, particularly in the conventional E_K range from ~ 10 to 200 eV [1–3, 57] except for some special cases. Even though the bulk electronic structures are the target of bulk sensitive PES studies, measurements of samples with clean surfaces are usually required to guarantee the reproducibility of the results. In the case of surfaces not clean enough from the beginning or gradually degraded within the measuring time, it is very difficult to obtain reliable results. Therefore UHV conditions as well as stable surface conditions are usually necessary for PES except for ambient pressure PES explained later or studies of chemical reactions on the surface. For low T measurements this requirement is very stringent because of the noticeable adsorption of gases onto the clean sample surfaces which distorts the valence band as well as core level spectral shapes. On the other hand the surface oxidation tends to be relatively slowed down at low T. In PES below $h\nu = 1$ keV, a pressure better than low 10^{-8} Pa is usually required. In the case of much more bulk sensitive hard X-ray PES, the required vacuum condition may be somehow relaxed since the surface spectral weight is relatively reduced compared with soft X-ray PES.

For single element metals and alloy metals, ion sputtering cleaning with rare gases (such as Ne and Ar), followed by annealing is often employed to obtain clean crystalline surfaces. The samples can be mounted on Mo, Ta, or other high melting point metallic substrates, which are strong enough against sputtering around 1 keV, via wires or nails made of Ta and W. The materials should be selected to have only low outgassing at elevated temperatures for annealing. In the case of compounds, however, great care should be taken to make the chemical composition remain the same as the starting material upon sputtering and annealing. The surface quality (chemical composition, crystal structure and surface roughness) after these treatments should be checked by low energy electron diffraction (LEED) and Auger electron spectroscopy (AES).

Filing (or scraping) by use of a (diamond) file was often used in the past to get clean surfaces of fragile compounds for PES and IPES, because the scraping could be repeated many times during the measurement under similar conditions. However, the damage or strain in the surface region is unexpectedly serious in many cases and the electronic structures in the shallow region accessible by PES are noticeably modified from those in the undamaged clean surface [58]. In the case of filing, the relative weight of the surface with respect to the bulk becomes inevitably increased. Therefore either fracturing or cleaving is generally recommended for PES measurements of single crystal samples except for some special cases. It should be stressed that the momentum resolved measurements are only feasible in the case of cleaved surfaces in photoelectron spectroscopy besides successfully in-situ grown or sputter-annealed surfaces.

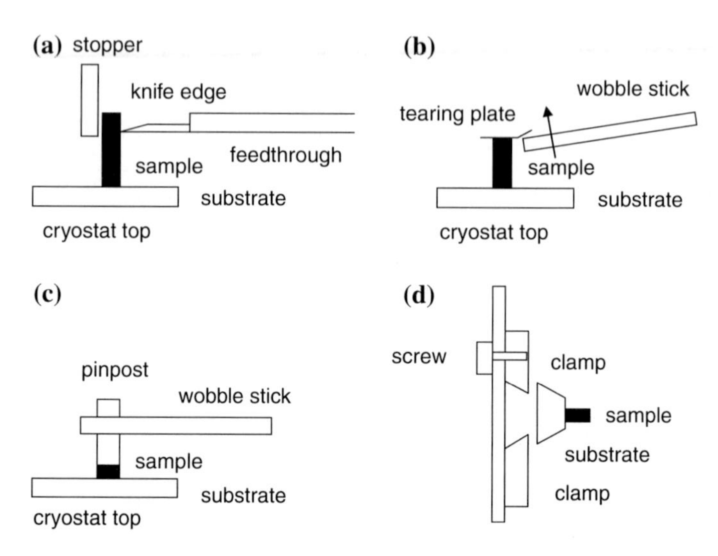

Fig. 3.21 Schematic view of knife-edge (**a**), tearing-plate (**b**) and pinpost fracturing (*cleavage*) (**c**) for obtaining clean sample surfaces. Cleavage is possible on some samples by these methods. How to clamp the sample substrate at the top of the cryostat is illustrated in (**d**)

Fracturing or cleavage can be performed by use of either the knife edge (Fig. 3.21(a)), tearing plate (Fig. 3.21(b)) or pinpost (Fig. 3.21(c)). Linear-forward motion, swing up motion or horizontal swing motion is employed in these cases for fracturing or cleavage. In some cases, a shallow groove is in advance engraved into the side surface of the sample in the air to facilitate fracturing later in vacuum. In the case of a low temperature measurement, the sample substrate with the clean sample must be firmly fixed to the top of the He cryostat by screws to realize a good thermal contact (Fig. 3.21(d)) as explained later in more detail.

Cleavage is, however, often more easily realized by the methods shown in Fig. 3.21(b), (c) than Fig. 3.21(a). One wobble stick is enough in these cases. In these methods, the cutoff sample is always left in the UHV chamber. Sometimes, its volume or thickness exceeds that of the measured sample. A narrow pinpost fixed with a rigid tearing plate is useful to cleave a very tiny sample. Namely such a pinpost-tearing plate with a small amount of glue can be supported by a micrometer driven lift and accurately fixed onto a very tiny sample (≤ 1 mm) already fixed to a substrate. Then fast hardening of the glue can be done by heating on a hot plate. The organic epoxy between the pinpost and the tiny sample never spreads over the sample substrate. If the tearing plate size is unified and could be grabbed by a mouth-opening rotatable feed through during the cleavage, this part of the sample can be easily recycled for the next experiments.

In these cases, the samples were mostly glued to a metal substrate by use of conductive materials or further adding organic insulating epoxy, which usually withstands stronger applied force. The contact surfaces should be roughened in advance by scraping to provide more gluing surface area. Organic epoxies such as the thermally hardening Torr Seal prepared by mixing two materials just before the use is employed in our experiment to provide enough bonding strength. Since a sufficient electrical conductivity to the sample is necessary to perform PES without charging up, we often put conductive epoxy in the central part of the contact, then surround it by the organic epoxy before putting the sample on top if the sample area is wide enough ($\geq$2 mm). Then the glued sample is heated up on a hot-plate. If the sample size is much smaller ($\leq$1 mm), the sample is first glued to the substrate by the organic epoxy and heated up and dried. Then a conductive material (for example, colloidal graphite), is painted and dried to electrically connect the substrate and the sample and also protect the charging up effects on the epoxy. After hardening, the outgassing from the glues become low enough to do PES measurements in low 10^{-8} Pa.

It is empirically known that ARPES measurements to probe the momentum dependence of the electronic structure are only successful in the case of specular surfaces of single crystals. Generally speaking, fractured surfaces with many small cleavage planes are not suitable for ARPES. If the photon beam size is small enough and one can reliably select such a small specular region, ARPES is still possible [59]. Note that grain boundaries are sometimes exposed to the vacuum during fracturing. Since such off-stoichiometric sample surfaces provide different spectra than those of the bulk, a careful check, for example, by means of Auger electron spectroscopy, will be required for higher reliability. It is well known that spectra on fractured surfaces of single crystals without grain boundaries are much more reliable than those on filed/scraped surfaces of polycrystals.

Usually a few sample holders with samples are stored in the sample bank in the entry air lock system, whose vacuum is typically 10^{-5} Pa or better evacuated by a turbomolecular pump. One of the samples is then selected and transferred into the sample preparation chamber with a vacuum of 10^{-7}–10^{-8} Pa, where such a treatment as fracturing, cleavage or ion-sputtering can be performed under proper conditions. Sample banks may also be mounted in the sample preparation chamber. In some cases, fracturing or cleavage at low temperatures is indispensable, for example, to reduce the thermal diffusion of impurities towards the surface. In such a case the final cleaning treatment should be done in situ on the sample mounted to the variable temperature He cryostat. For example, we use a trapezoidally shaped sample carrier, which is inserted into two trapezoidal clamp nails fixed to the cold stage of the He cryostat (Fig. 3.21(d)). One of the nails (upper one in this drawing) can be loosened or tightened by a screw driven from outside by a rotatable feedthrough. A mechanism to stop possible falling and azimuthal rotation of this clamp nail should be properly designed for a practical use. The fracturing or cleavage by a knife edge, tearing plate or pinpost can be performed under this condition at low temperatures.

On the other hand, high T measurements at several hundred K above room temperature require careful outgassing of the local surroundings over a long time before cleaning the sample. Usually glue can not be employed because of the high outgassing rate at high T. Then the samples should be mechanically supported by wires or clamps and be locally heated up in order not to induce the outgassing from the local environment.

The orientation of single crystal samples is important for ARPES. One can use X-ray Laue diffraction for aligning the sample in advance. If the polar and azimuthal angles of the sample are adjustable in situ on the manipulator, the orientation of the sample is also possible by use of LEED after the preparation of a clean surface of the single crystal in UHV. If the ARPES measurements can cover rather wide angles by rotating the sample, one can determine the sample orientation and high symmetry lines in the BZ later even by the ARPES measurements themselves. One can also use photoelectron diffraction (PED) techniques handled in Chap. 12 to judge the sample orientation.

3.5.2 In Situ Sample Growth and Surface Analysis

In order to get clean sample surfaces, it is also possible to grow crystals in a sample preparation chamber connected to the analyzer chamber via a UHV gate valve. In addition to classical methods of evaporation of starting materials by crucible heating, electron beam sputtering is used for growing metals such as Fe, Co, Ni and similar materials. In the case of compounds and alloys, molecular beam epitaxy (MBE) and pulsed laser deposition (PLD) can be utilized. By proper selection of the substrate and growth conditions including the substrate temperature and thermal treatment, single crystalline samples can be grown in situ. Surface structural analysis can be performed not only by LEED but also by reflection high energy electron diffraction (RHEED). RHEED is applicable during the sample growth and used to judge not only the layer by layer growth but also how many layers are piled up onto the substrate. Growth of superlattices is also possible by combining PLD and/or MBE and RHEED. In the case of surface sensitive PES, the presence of the surface reconstruction and relaxation can strongly modify the electronic structure. Therefore surface characterization is a prerequisite to discuss the obtained experimental PES results.

Nowadays both SR and laser light for PES can be focused down to a small spot with a dimension below 100 μm. During the sample deposition on the other hand, a well defined chemical and/or geometrical variation can be accomplished in the lateral direction of the sample by manipulation of the shutter opening time and sample substrate movement. For example, wedge type thin films with monotonously increasing thickness towards one direction and double wedge thin films with increasing thicknesses towards two (orthogonal) directions can be grown by controlling the shutter for evaporation or substrate movement. The chemical composition of the sample can be likewise controlled. Then the measurement of the thickness dependence or composition dependence of the electronic structure is feasible across the sample surface. Such combinatorial samples are now available for high lateral resolution PES measurements. It is also possible to prepare such combinatorial samples in other vacuum chambers and protect their surfaces by in situ evaporated thin non-reactive materials. After transportation to a preparation chamber for PES in air, for example, such thin over-coated materials can be removed by heating or weak sputtering. Then photoemission can be performed on the uncoated samples.

3.5.3 Samples at Low Temperatures or at Ambient Pressure

The sample T from around the room temperature down to a few-20 K is often required to study various SCES materials by use of He cryostat. When the sample sizes are larger than sub mm, one can use a closed cycle He cryostat with some care to reduce the vibration of the top of the cryostat. It is sometimes very important that cleavage, fracturing and scraping is possible in situ at any temperature as explained above. The vacuum of the analyzer chamber is desirably kept in 10^{-8} Pa or low 10^{-7} Pa region after proper baking by tandem turbomolecular pumps or an ion pump to protect adsorption of residual gases onto the sample surfaces during the measurement at low temperatures.

In accordance with the gradual improvement in the energy resolution of the PES instruments and light sources [39, 60], the Fermi-Dirac distribution function (FD) became to play a major role in the spectral broadening. Namely the broadening up to $\sim 4k_BT$ is inevitable at the Fermi level. Since the total resolution resulting from the light source and the PES instrument became better than 1 meV at low energies, the realization of low T is now essential, for example, to study the opening of a small gap across the phase transition temperature. The width of the FD of $\sim 4k_BT$ corresponding to 1 meV means a sample T of ~ 3 K. So far a sample T down to 2.9 K was reported by use of a He flow cryostat together with another He cryostat by extensive use of multiple-thermal-shielding [39]. A typical design is shown in Fig. 3.22, where the sample is cooled by thermal conductance from the He flow cryostat. The heat from the environment should be minimized by reducing the opening of the thermal radiation shield. The sample transfer and screw tightening can be done after retracting the He flow cryostat from the shielding attached to the closed cycle He cryostat.

Ambient pressure PES (AP-PES) is thought to be very powerful for the in situ study of environmental and materials sciences, energy related sciences and so on by use of SR with high brilliance [61]. Studies of the oxidization of metal surfaces and catalyst properties under reaction conditions of vapor/solid and vapor/liquid interfaces are typical examples of its application. Then AP-PES is recognized to be a unique surface science technique. Chemical kinetics leading to thermodynamic equilibrium may slow down by orders of magnitude at reduced temperatures. So it is desirable to be able to perform AP-PES in a wide temperature range from tens of K up to high temperatures beyond 1,000 K. Although as high an ambient pressure as possible around the sample in the analyzer chamber is desired, UHV conditions are still required for both the light source and the electron energy analysis except for some special cases. UHV conditions for X-ray light sources can be satisfied by

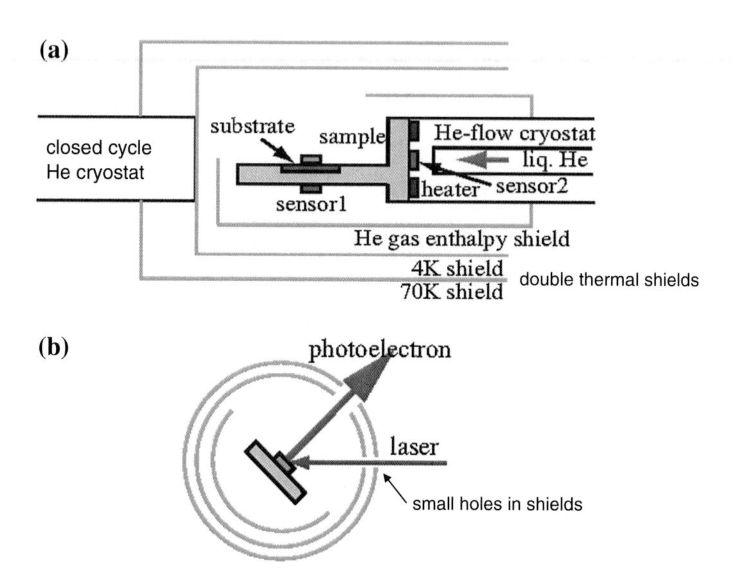

Fig. 3.22 An example of thermal radiation shielding for low temperature PES, where the openings for the light and the photoelectrons path are kept in three shielding stages [39] coaxial along the axis (**a**). A cross section perpendicular to the axis is shown in (**b**)

use of a thin SiN or Al window. The required vacuum for the electron detector can be realized by multiple stage differential pumping of the space between the detector and the sample by use of a series of apertures. An example of such systems is illustrated in Fig. 3.23 [62]. The radius of the first aperture closest to the sample is typically set to 0.1–0.5 mm. The size of this aperture, which dominates in the conductance between the analysis chamber and the analyzer lens column, is decisive for realizing the highest possible ambient pressure within the normal operation conditions of the electron detector. Then the distance between the sample and this first aperture determines the angular acceptance. A shorter distance is favored for higher counting rates. In this case, the scattering or attenuation of photoelectrons by molecules on the way from the sample to the first aperture is much reduced. On the other hand, the photoelectron mean free path in the gas is inversely proportional to the gas pressure. In practice, the optimum distance is set to a few times the aperture diameter.

Since SR, in particular, the undulator radiation should be focused for AP-PES in order to realize a high efficiency despite the use of the small aperture, damages of the sample surfaces might be induced. Charging up of the surfaces of insulators and semiconductors as well as chemical decomposition including the reduction of oxide surfaces often induce artifacts to the observed spectra. In the case of samples in a reactive gas environment, ionized gas molecules and adsorbates as well as secondary electrons may induce reactions and the surface may be modified during the measurement. Very careful examination of the results is always required in AP-PES in these cases.

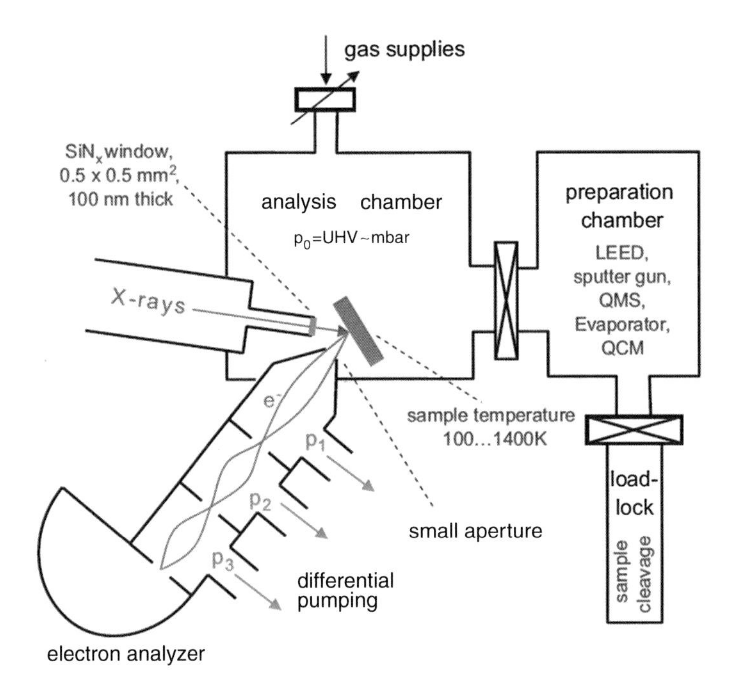

Fig. 3.23 Schematic view of a AP-PES system with an X-ray source equipped with a SiN_x window, analysis chamber with a sample at variable temperature, differential pumping system and electron energy analyzer. The pressure in the analysis chamber can be changed from UHV up to mbar [62]

3.6 Methodology

3.6.1 Angle Integrated Photoelectron Spectroscopy

In the photoelectron excitation process, an electron in the occupied state is excited by photons to an unoccupied state and escapes into the vacuum. Although occupied valence band states are probed even at low photon energies ($h\nu$), deep core level states can only be probed by employing higher $h\nu$. Let us define the energies of the Fermi level, the vacuum level and the photoelectron kinetic energy in the vacuum by E_F, E_V and E_K. E_F is located in the band gap in the case of semiconductors and insulators while the exact position depends upon doping. In the case of metals, E_F is located in the conduction band. The excited photoelectron can escape into the vacuum when its energy is above E_V with the kinetic energy E_K calibrated to E_V. In systems with negligible electron correlations, the binding energy E_B of the photoelectron in solid is evaluated as $E_B = h\nu - E_K - \phi$ with respect to E_F according to Koopmans' theorem, where ϕ represents the work function of the sample surface. Since E_K is measured by an electron energy analyzer and the position of E_F is observed as a sharp cut off in the photoelectron spectrum in the case of metals, E_B is directly measured by the PES experiment. When the spectral intensity is changing drastically near E_F in a specific material, the Fermi edge of, for example, Au with common ground should also be measured and the position of E_F be determined. This method is also applied to narrow gap semiconductors as far as the charging up effect is negligible. The work function ϕ is evaluated from the values of hν and E_K for the electrons from E_F. The vacuum level cut-off due to zero E_K is observed at larger E_B. The photoemission intensity plotted as a function of E_B is called energy distribution curve (often abbreviated as EDC) [1–4]. In the angle integrated photoemission spectrum, this is understood to be reflecting the joint density of states (JDOS) between the occupied and unoccupied states. In fact, the EDC for the valence band excitations obtained by relatively low hν is modified from the JDOS by the interband transition matrix element effects. When hν becomes large enough (>100 eV), the EDC of the valence band mostly reflects the density of states (DOS) of the occupied states, modified by the PICS. The photoelectrons following the before mentioned energy conservation law are called sometimes primary photoelectrons, which provide the most useful information on the electronic structure of the materials. However, part of the photoelectrons strongly interacts with other electrons on their way through the solid, producing many lower energy electrons, inducing a background to the EDC. The background increases with E_B and finally disappears at the E_B corresponding to E_V. Such electrons are called secondary electrons. Therefore subtraction of this background is necessary to discuss the intrinsic spectral shapes of valence and core levels. Because of such electron–electron interaction and the interaction with phonons, the photoelectron inelastic mean free path is relatively short as discussed already in Chap. 2 and in more detail in Sect. 4.1 later.

3.6.2 Resonance Photoemission and Constant Initial State Spectrum

Resonance photoemission (RPES) [3] was first clearly observed for Ni excited at various hν through the 3p → 3d absorption threshold [63]. When hν was increased across this threshold, the intensity of the 6 eV satellite was dramatically enhanced. Before going into a detailed discussion of RPES, the Auger process is briefly explained. When a core electron was excited by photons, either radiative or non-radiative transition can take place from a shallower core or valence/conduction electron level sometimes with additional excitations. The radiative decay is called the fluorescence. In the latter case, a shallower core electron or a valence/conduction electron can be excited to the empty state following the non-radiative decay, satisfying the energy conservation law. This process is called an Auger electron excitation process. Even when the core electron is photoexcited to the core exciton state, Auger process is feasible. In RPES, a dramatic enhancement of a certain electronic state is observed with increasing hν across some core excitation threshold (sometimes a slight decrease is also observed just below this energy due to Fano resonance as explained later in Sect. 5.1). On the further increase of hν, the intensity of this peak was reduced but the Auger peak (in the case of Ni $M_{2,3}VV$, where one 3d electron relaxes into the 3p core hole and another 3d electron is excited into the empty state) branched off from this satellite and moved towards larger E_B linearly with hν. The direct photoemission could be described as $|3d^9> + h\nu \rightarrow |3d^8[4s]> + \varepsilon l$, where the 3d hole is screened by the 4 s electron and εl represents the photoelectron. Under the resonance excitation, the core excitation is followed by the direct recombination of the core hole as $|3p^6 3d^9> + h\nu \rightarrow |3p^5 3d^{10}> \rightarrow |3p^6 3d^8[4s]> + \varepsilon l$. Since this process has the same initial and final states as the direct photoemission, the two processes interfere quantum–mechanically with each other providing a resonance minimum and maximum. In the case of Ni, the origin of the 6 eV satellite was interpreted as a two hole bound state [64]. In the direct recombination process in transition metal (TM) and/or rare earth (RE) compounds, the photoexcited electron and the electrons occupied from the beginning can not be discriminated in the RPES.

It is known that RPES is very prominent in the case of rather localized states [65]. An example is shown in Fig. 3.24 for the valence fluctuating compound Sm_3Se_4 [66, 67]. The hν covers the 4d → 4f core excitation region. The multiplet structures of the EDC in the E_B region of 7–12 eV are due to the $|4f^4>$ final states from the Sm^{3+}

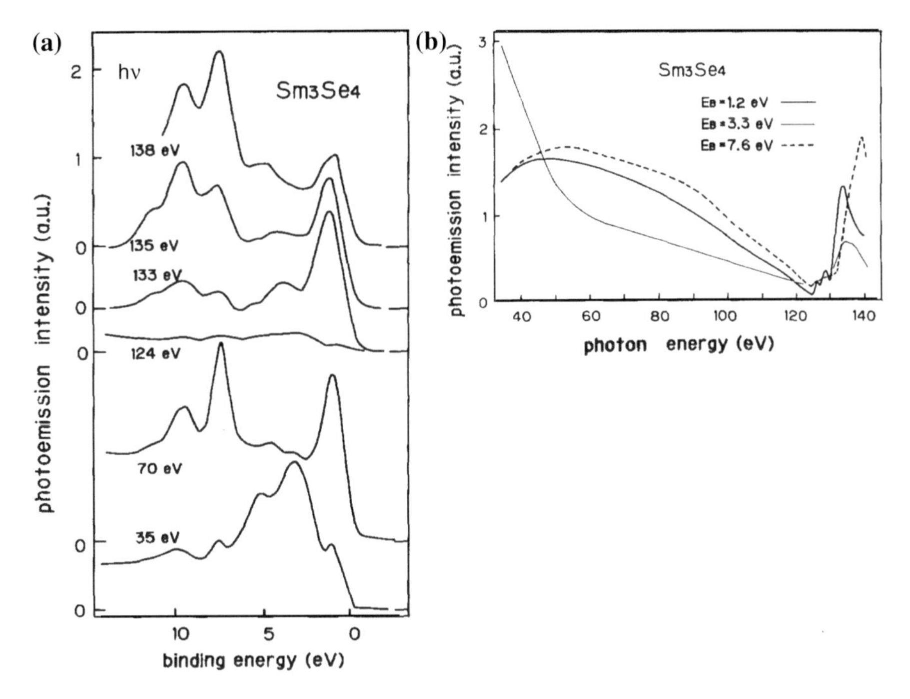

Fig. 3.24 Resonance photoemission of Sm_3Se_4 [66]. hν dependence of the spectra (**a**) and the constant initial state spectra (*CIS*) at different E_B (**b**)

initial states with a $|4f^5>$ configuration. The structures closer to E_F are due to the $|4f^5>$ final states from the Sm^{2+} initial states with a $|4f^6>$ configuration. It was known that the PICS of the 4f states decreased sharply below $h\nu = 50$–70 eV, whereas that of the Se 4p states increased below $h\nu = 50$ eV [33]. Thus the strong contribution for $h\nu = 35$ eV centered around $E_B = 3$ eV was ascribed to the Se 4p contribution. When the photoemission intensity at a certain E_B was plotted as a function of $h\nu$ after normalization by the photon flux, the so-called constant initial state spectrum (CIS) is obtained. It was noticed in the CIS in Fig. 3.24(b) that the $|4f^5>$ and $|4f^4>$ components were dramatically enhanced above $h\nu = 125$–130 eV in a qualitative coincidence with the 4d $\rightarrow$ 4f excitation threshold as a result of RPES. Namely 4f photoemission was resonantly enhanced when an additional 4f electron was excited from the core levels. One recognized, however, the slight difference between the CIS spectra for the $|4f^5>$ and $|4f^4>$ final states. This was corresponding to the difference of the 4d $\rightarrow$ 4f absorption for the $|4f^6>$ and $|4f^5>$ initial states. RPES technique is quite powerful to extract the electronic structures related to low concentration atoms, the states with low electron occupation and with low PICS under the off-resonance condition.

The Ce 4f states in alloys are such examples [68, 69]. In this case extraction of the electronic states related to the 4f states is very important from the viewpoint of Kondo physics and valence fluctuation. The resonance enhancement of the 4f states is expected for the resonance excitation from the 3d and 4d core states. For simplicity, the 3d–4f RPES is explained below. The direct photoemission of the 4f state is generally described as $|3d^{10}4f^1> + h\nu \rightarrow |3d^{10}4f^0\varepsilon l>$, where the 3d state is included for completeness. In parallel to this process, the strong 3d $\rightarrow$ 4f core absorption takes place when $h\nu$ is tuned to this threshold. The process is described as $|3d^{10}4f^1> + h\nu \rightarrow |3d^9 4f^2>$. If the 4f electron relaxes down nonradiatively to the 3d core level, the energy can be transferred to the other 4f electron, which is excited far above the Fermi level as a result of this direct recombination process as described by $|3d^9 4f^2> \rightarrow |3d^{10}4f^0\varepsilon l>$. If one compares the direct photoemission process with the indirect process, it is recognized that both the initial state and the final state are the same. Then a quantum-mechanical interference is expected to take place, resulting in a resonance enhancement of the f photoemission intensity in the case of constructive interference.

If the PICSs of the other orbitals are not much different between the two $h\nu$ for on- and off-resonance excitations, the subtraction of the two PES spectra after normalization by the photon flux can provide the PES spectrum of the particular orbital which is resonantly enhanced. In this way the electronic structures of the 4f and 5f states in rare earth lanthanide and actinide compounds as well as the 3d and 4d states in transition metal compounds are revealed. If the core absorption is composed of multiplet structures, the resonantly enhanced PES spectral shapes change slightly depending on the excitation $h\nu$. If the PES spectral shape variation with $h\nu$ in the multiplet absorption region is noticeably large, the unique evaluation of the spectral shape of the particular electronic structure by subtraction will not be straightforward. According to the Auger nature of the core hole decay, even the 5d states are slightly resonance enhanced in lanthanide compounds, because the energy can be transferred from the decaying 4f electron to the 5d electron as explained later in Chap. 7.

In many cases of RPES, Auger structures are observed at larger E_B for excitation far above the threshold energy. If the two (three) principal quantum numbers of the three electronic states involved in the Auger process are the same, the process is called Coster-Kronig (super Coster-Kronig) decay. In these cases a rather effective Auger decay is expected providing strong Auger structures. When the hν is decreased down to the threshold region, the Auger features are gradually enhanced in proportion to the absorption intensity which represents the numbers of core holes. These features approach the main PES peak(s) located at fixed E_B and then the resonance enhancement of the main peak(s) takes place at the original and rather constant E_B, where the two processes are no more discriminated. The resonance photoemission process is, however, different from the simple overlap between the Auger features and the main PES structures as judged from the detailed CIS studies.

The technique for CIS was already developed in early 1970s [70] and a complementary technique called constant final state spectroscopy (CFS) was also started to be used around that time [71, 72]. In contrast to CIS obtained at constant E_B, CFS is performed at constant E_K. In this sense, CFS can be considered as partial (photoelectron) yield spectra. Both CIS and CFS are measured as a function of hν by use of SR. It is empirically known that the total (photoelectron) yield spectrum represents the core absorption spectrum. The CFS and CIS spectra are slightly different from the core absorption spectrum. Namely, CFS and CIS spectra change their shapes for different E_K and E_B. It is known in the case of insulators and semiconductors that the direct recombination process via the core exciton can provide additional structure(s) in the CFS spectra below the interband excitation threshold. In this case, the E_K becomes slightly larger than the E_K expected for the Auger electron [71]. If empty surface states exist in the bulk band gap of semiconductors, the partial yield spectrum, for example at $E_K = 4$ eV, shows a strong enhancement below the hν corresponding to the core excitation to the bulk conduction band minimum. Such empty surface states were observed by CFS studies in GaAs (110), Ge(111) as well as GaSb, InSb and InAs [72]. Since occupied surface states are observable by PES, we can now obtain useful information on both empty and occupied surface states.

3.6.3 Angle Resolved Photoelectron Spectroscopy

While information on the wave vector **k** of the photoelectrons is integrated and hence momentum is smeared out in the EDC obtained by angle integrated photoelectron spectroscopy, the wave vector **k** of electrons can be determined and evaluated in the angle resolved photoelectron spectroscopy (ARPES). Then the mutual relation between the binding energy E_B and the wave vector **k** provides the so-called band dispersions $E_B(\mathbf{k})$. By tracing the crossing point of the dispersion across the Fermi energy E_F, the Fermi surface topology can experimentally be mapped as well and the electron-like or the hole-like character of the Fermi surface can be judged from the whole dispersion behavior. This methodology called Fermiology has progressed intensively through the study of high temperature Cu superconductors [73].

Although the ARPES was explained already in Chap. 2, it will be simply explained here again. This technique has traditionally been developed for low photon energy ($h\nu$) excitations, where the finite value of the photon wave vector **q** can be neglected in comparison with the electron wave vector scaled by the Brillouin wave number π/a, where a is the lattice constant. The relation between the binding energy (E_B) of the peak and the wave vector along the surface $k_{//}$ can be evaluated according to (2.19). However, the wave vector perpendicular to the surface ($k_\perp$) is not conserved on passing through the surface. Therefore the evaluation of the band dispersion is only possible in this case by using only one specific $h\nu$ when the $k_\perp$ dependence of the E_B is negligible. Such a condition is satisfied in the cases of two dimensional (2D) and one dimensional (1D) systems, where the electron transfer perpendicular to the 2D layer or 1D chain is negligibly small. If the $k_\perp$ dependence of the electron energy is not negligible, the evaluation $E_B(k_i)$ is not straight forward.

In the three dimensional (3D) systems with the noticeable $k_\perp$ dependence of the electron energy, a different approach is employed. Namely normal emission spectra are measured as a function of $h\nu$. In the case of the soft X-ray angle resolved photoelectron spectroscopy (SXARPES), however, the photon wave vector **q** cannot be generally neglected in comparison with the Brillouin wave vector, because the photon wavelength λ is $\sim$12 Å, for example, in the case of $h\nu = 1{,}000$ eV and $|\mathbf{q}| = 2\pi/\lambda$. If the incidence angle of the photon is θ from the surface normal, the $\perp$ and $\|$ component of **q** are expressed as $q_\perp = (2\pi/\lambda)\cos\theta$ and $q_\| = (2\pi/\lambda)\sin\theta$. These quantities are just transferred to the photoelectron on the photoelectron excitation according to the momentum conservation rule. The SXARPES can provide the information on the 3D band dispersions in the probing depth region larger than the low energy ARPES.

The angular resolution $\Delta\theta$ of the electron analyzer is given by the design of the input lens and the applied voltages to the lens elements together with the entrance slit width. For a state of the art analyzer employing a two dimensional detector, the angular resolution is also influenced by the lateral resolution of the micro channel plate. The resolution of the wave vector of the photoelectrons is proportional to $(E_K)^{1/2}$ according to (2.21) (note that the second term of (2.21) is predominant while the first term is relatively negligible). This means that a smaller $\Delta\theta$ is required for high energy ARPES to realize the equivalent wave vector resolution as at low E_K. If the same $\Delta\theta$ is used at $E_K = 700$ eV, for example, the momentum resolution becomes 3.7 times worse than at 50 eV. For $\Delta\theta$ of 0.1° and a lattice constant of 3 Å, the resolution of the wave vector is better than 3 % of the Brillouin wave number. Therefore there is almost no serious difficulty for ARPES

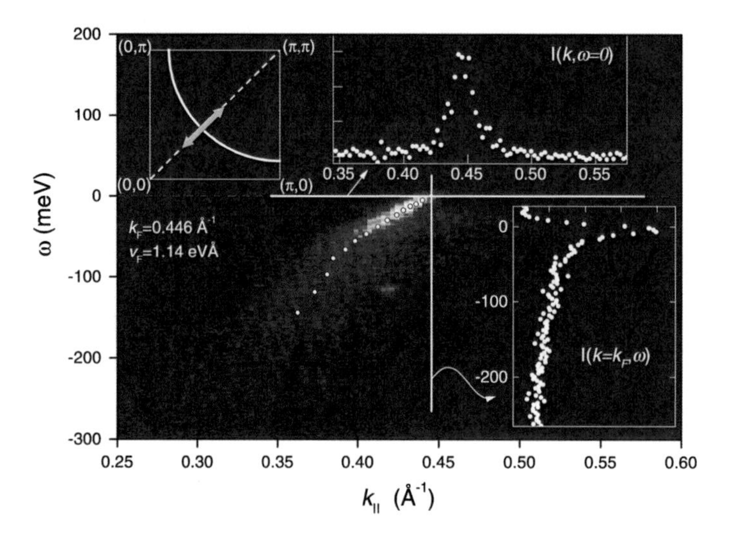

Fig. 3.25 ARPES spectrum of $Bi_2Sr_2CaCu_2O_{8+\delta}$ along the (0, 0)−(π/a, π/a) direction. The main panel (*bottom*) shows the E_B–$k_{||}$ dependence of the photoemission intensity. This k-space cut is along the *arrow line* in the *upper left inset*. Dispersion of a quasiparticle is clearly seen in this main panel. The energy distribution curve (*EDC*) at the wave number $k = k_F$ is shown in the *right inset*. The momentum distribution curve (*MDC*) at $E_B = 0$ eV is shown in the *upper right inset* [75]

in the soft X-ray region [74]. The advantage of the SXARPES is its bulk sensitivity and the potential to evaluate the $k_\perp$ dependence of the bulk band dispersion.

Since the wave vector is well defined in addition to the energy in ARPES, the so called momentum distribution curves (MDCs) can be obtained in addition to the EDCs. MDCs sometimes help a delicate evaluation of the Fermi wave vector and details of the band dispersions [75] (Fig. 3.25). Details of ARPES will be given later in Chaps. 6 and 7.

3.6.4 Photoelectron Spectroscopy in the μm and nm Regions

In parallel with the development of high resolutions in energy and momentum in PES and ARPES, high spatial resolution PES also attracts wide attention. The spatial resolution of conventional imaging XPS was >10 μm until recently. In the case of PES, sub μm and nm resolved chemical and electronic information is thought to be quite useful under some specific occasions. High spatial resolution imaging XPS with a lateral resolution better than several hundred nm has been desired for a long time [76–78]. The advent of high-brightness SR or undulator sources has stimulated the development of high spatial resolution PES. There are two different approaches for this purpose. One approach is to focus the excitation photon beam into an extremely small spot (sub μm) by photon optics and scan either the sample area or the photon focus area (scanning photoelectron microscope: SPEM). In this case, a standard analyzer can be employed for photoelectron detection. Fresnel zone plates (FZP) or special mirrors can be used to focus the beam down to a few nm. This method was mostly employed in SR facilities, because the excellent photon beam quality facilitated such fine focusing. Since the wavelength of the high hν photons in the soft and hard X-ray regions was rather short, the diffraction limit did not restrict the possible spatial resolution. The other approach was the photoelectron emission microscopy (generally abbreviated as PEEM). In this case one can magnify the images of the illuminated surface region by electron optics.

SPEM were installed, for example, at beam line 7.0.1 of ALS, Berkley (5 cm period planar undulator, SGM with hν = 100–800 eV), ESCA microscopy branch beam line at ELETTRA (2 GeV) with a 81 periods undulator (with a 56 mm period, SGM with hν in the range of 100–1,500 eV) and U5-SGM undulator (total length of 3.9 m) beam line with hν = 60–1,500 eV at SRRC, Taiwan. A typical setup in SRRC is shown in Fig. 3.26 [77], where the undulator radiation was monochromatized by a spherical grating monochromator (SGM). FZP and an order sorting aperture (OSA) were used to focus the monochromatic photon beam onto the sample with eliminating the unwanted higher order diffraction light. Coarse and fine scan of the sample was done in the x and y directions. The emitted photoelectrons are analyzed by a hemispherical electron analyzer with input lens. Three operation modes are available: (1) conventional EDC mode for fixed hν and sample position, (2) the constant final state mode with fixed E_K and hν scanning, which can only be realized by simultaneous movement of the FZP/OSA and (3) the

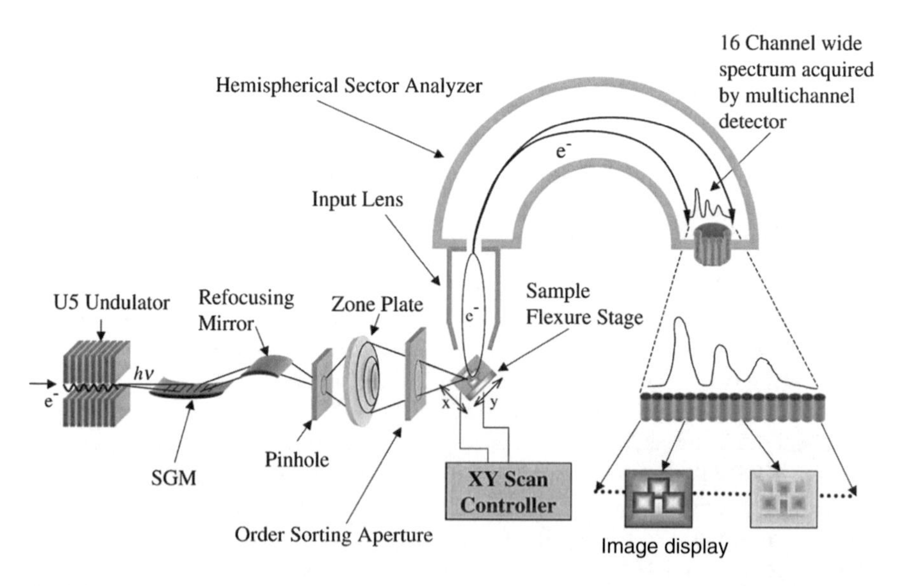

Fig. 3.26 Schematic diagram of the U5-SPEM system at SRRC, Taiwan for scanning photoelectron microscopy (SPEM) [77]

spectroscopy mode with fixed hν and scanning of the sample position. In this mode (3) the data acquisition must be synchronized with the sample scanning. If a multichannel detector is mounted in the focal plane of the analyzer, images with different E_K are simultaneously obtained. By properly setting the pass energy of the analyzer, one can get a 2D map of the different elements or different chemical states. Around hν = 430 eV, a total energy resolution of 0.25 eV and a spatial resolution better than 0.3 μm were realized [77]. In the case of SPEM in ELETTRA, FZP can focus the monochromatic light down to less than 90 nm with the flux on the sample being 10^9–10^{10} photons/s while the accuracy of the sample positioning by a piezo motor is 10 nm. A spatial resolution down to 0.15 μm is realized [76]. Recently, a nanoscale 3D spatial-resolved photoelectron spectroscopy (3D nano-ESCA) was realized by means of a rather bulk-sensitive SX excitation (hν = 870 eV), where FZP is combined with SPEM with a wide-acceptable-angle "angular" (ARPES) mode at SPring-8 BL07LSU [79]. Here the angle dependence in ARPES corresponds to the probing depth-dependence in PES for flat samples where the spectra with a grazing (nearly normal) incidence angle mainly reflect the electronic structure of the topmost surface (deeper region). The spatial resolution of 70 nm was achieved.

A typical PEEM instrument is schematically shown in Fig. 3.27 [80]. It is composed of the sample stage, objective lens, contrast aperture, stigmator-deflector, field aperture, projective lens, micro channel (or multichannel) plate, screen and a CCD camera. Secondary electrons are extracted by the high potential of the objective lens. It is empirically known that the amount of the secondary electrons is proportional to the core absorption intensity. If circularly polarized light is used near the core absorption threshold, core absorption magnetic circular dichroism (MCD) is reflected in the secondary electron intensity. Therefore small magnetic domains can be imaged by PEEM. Many such experiments were so far made for studying nanomagnetism [81, 82]. A resolution better than 0.1 μm is expected in PEEM instruments. Surface studies by low-energy electron microscopy (LEEM) and PEEM emerged in 1980s to directly image surfaces of solids without scanning [83, 84]. The specimen works as the cathode and the resolution limit can reach to 2 nm in LEEM and 5–10 nm in low hν PEEM. Compared to several other methods such as scanning tunneling microscopy (STM) and scanning electron microscopy (SEM), LEEM and PEEM have many advantages as fast parallel imaging and a large field of view.

PEEM with high time resolution is a powerful technique to study the dynamics of nanoscale electronic and magnetic systems by ultrafast imaging. Combined with the excellent time structure of SR and/or pulsed lasers, time resolved PEEM can supply such useful information as remagnetization dynamics of micro- and nano-magnetic systems and ultrafast electron dynamics [85–88]. A promising technique of real time imaging and stroboscopic imaging with high time resolution can be realized by combining the contrast mechanism in PEEM with such pulsed excitation sources. Laser with fs pulses and SR with a width down to a few ps can be used for excitation. Such examples as the stroboscopic XMCD-PEEM imaging of precession magnetic excitations by means of the magnetic field pulse pumping and

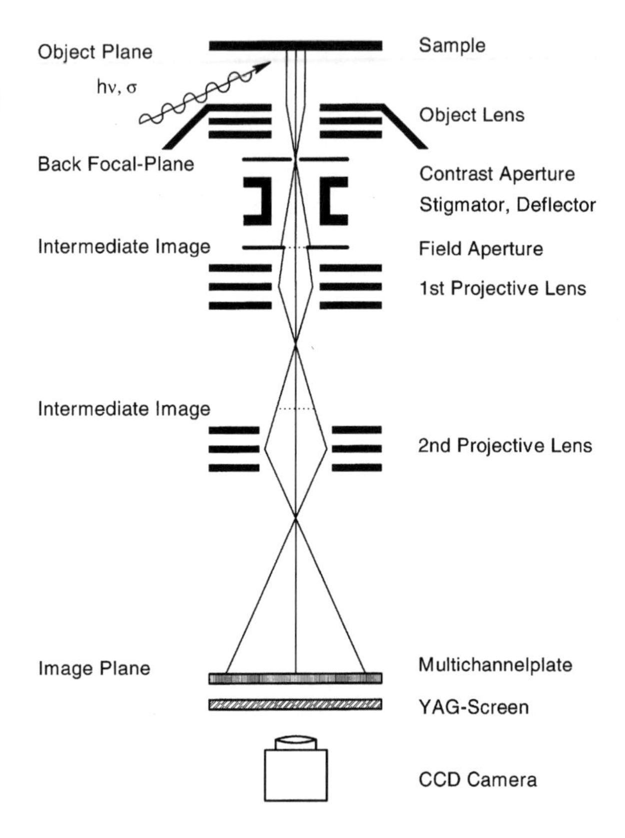

Fig. 3.27 Schematic diagram of a photoelectron emission microscope (PEEM) [80]

X-ray probing have been reported. Jitter between the pump and probe pulses is inevitable but a time resolution down to 15 ps has been obtained at BESSY. Time-resolved stroboscopic PEEM with a time resolution of a few fs and a spatial resolution of 20 nm was successfully performed by use of 2-photon photoelectron excitation by means of an intense fs laser [89]. Physics of hot-electron dynamics in such systems as nanostructures and low-dimensional materials will be probed by this new technique.

An instrument called NanoESCA is used for imaging photoelectron spectroscopy [78, 90]. Based on a tandem and antisymmetric arrangement of two hemispherical analyzers as shown in Fig. 3.28, the main spherical aberration of the analyzer can be eliminated. In addition, the time structure of the electron signal is not broadened due to this tandem arrangement unlike in the case of a single hemisphere. Thus energy- and time-resolved measurements are feasible. This instrument facilitates such measurement modes as (1) non-energy-resolved PEEM imaging, (2) area-selected spectroscopy and (3) energy-filtered photoelectron imaging for relatively high E_K up to few keV. The PEEM column is used as the entrance lens of the analyzer. Area-selection is done by use of the continuously

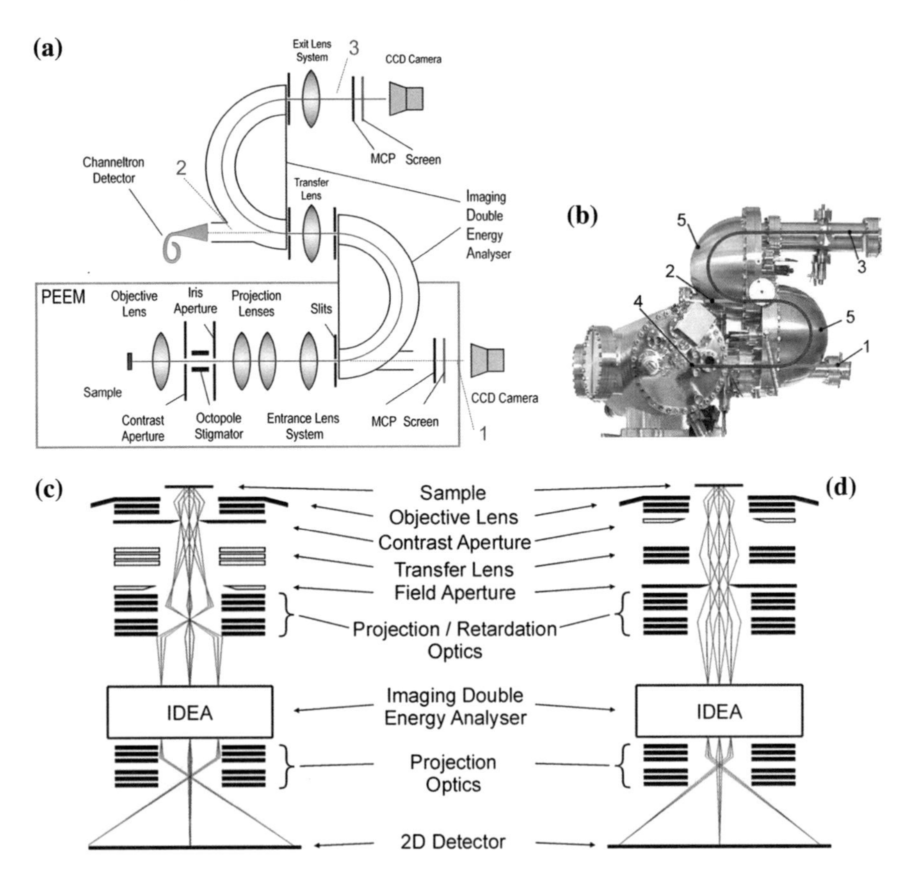

Fig. 3.28 **a** Schematic layout of the NanoESCA instrument with aberration correction based on imaging double energy analyzer (IDEA) [78]. The three paths of the electrons are indicated: *1* PEEM-mode, *2* selected area spectroscopy, *3* energy-filtered ESCA imaging. Energy resolution of 190 meV and a lateral resolution of 120 nm are reported. **b** photograph of this NanoESCA instrument. **c**, **d** two typical operation modes of this set up. Namely real-space imaging mode in **c**, where this system is called imaging microscope and **d** angular resolved k-space imaging. Then this instrument is called a momentum microscope [90]. The important design concept of this instrument is the ability to switch between the spatial and momentum imaging [90, 91]

adjustable iris aperture in the first intermediate image plane (Field aperture in Fig. 3.28(c)). The projection lenses are used either to project the first intermediate image onto the first screen or to adapt the electron beam to the pass energy of the first hemispherical analyzer. The information obtained by PEEM imaging (1) is very useful to select the area for (2). Between the two hemispherical analyzers, a transfer lens is installed to image the electrons leaving the first hemisphere onto the entrance of the second hemisphere. So it is called an imaging double energy analyzer (IDEA). Without deflection field of the second hemisphere, the electrons are focused into the channeltron detector in the (2) mode. Good lateral resolution of the microscope is realized by the immersion objective lens, where the sample is located near the ground potential and a high voltage up to 16 keV is applied to the extractor anode. For non-energy-filtered threshold emission, the theoretical resolution limit is about 10 nm. A low pass energy is favored to have good spatial and energy resolution with high transmittance. Energy resolution of 190 meV and lateral resolution of 120 nm were so far achieved in the spectroscopic imaging [78] by use of SR at UE–52/SGM undulator micro-spot BL in BESSY II. By use of a laboratory X-ray source (focused and monochromatized Al Kα source), a lateral resolution of 610 nm was still achieved (final goal is <150 nm) [90].

3.6.5 Momentum Microscope

By the use of the similar concept as the NanoESCA explained in Sect. 3.6.4 and shown in Fig. 3.28(a) and (c), it is also shown that the 2D momentum resolved ARPES can be performed at a constant E_K [91] with use of a different mode as described in Fig. 3.28(d). Then this instrument is called photoelectron momentum microscope. Namely, 2D momentum resolved photoelectron spectra are measured by an imaging energy filter. Although the achievement of the ultimate energy resolution is not easy due to the limitation of the complex lens, this system is very useful to simultaneously measure the 2D angular distribution of the photoelectrons at any fixed E_K. So it can in some sense compete with the 2D analyzer explained in Sect. 3.4.3.

The ARPES by use of this momentum microscope is steadily progressing for example on Cu(111), Cu(001) [92], clarifying very complex dispersions of band at various E_B and facilitating the discussion on many-body self-energy effects on the Shockley surface state and so on. Direct k-space imaging by this instrument on clean and Bi-covered Cu (111) surfaces also facilitated such a discussion as the bulk excited electrons are subsequently diffracted by an atomic 2D surface grating [93, 94]. The high potential of this instrument can also be combined with the spin analyser as later explained in Sect. 11.2.

References

1. M. Cardona, L. Ley (eds.), *Photoemission in Solids I* (Springer, Berlin, 1978)
2. R.Z. Bachrach (ed.), *Synchrotron Radiation Research, Advances in Surface and Interface Science*, vols. 1 and 2 (Prenum Press, New York, 1992)
3. S. Hüfner, *Photoelectron Spectroscopy, Principles and Application* (Springer, Berlin, 2003)
4. C. Kunz, K. Codling, *Synchrotron Radiation, Techniques and Applications* (Springer, Berlin, 1979)
5. S. Winich, S. Doniach, *Synchrotron Radiation Research* (Plenum, New York, 1980)
6. E.E. Koch, *Handbook of Synchrotron Radiation* (North-Holland, Amsterdam, 1983)
7. B.M. Kincaid, J. Appl. Phys. **48**, 2684 (1977)

8. H. Winick, G. Brown, Nucl. Instrum. Meth. **195**, 347 (1982)
9. G.P. Fomenko, Y.G. Yushkov, Soviet Phys. Tech. Phys.—USSR **14**, 1243 (1970)
10. H. Kitamura, Jpn. J. Appl. Phys. **19**, L185 (1980)
11. T. Hara, M. Yabashi, T. Tanaka, T. Bizen, S. Goto, X.M. Maréchal, T. Seike, K. Tamasaku, T. Ishikawa, H. Kitamura, Rev. Sci. Instrum. **73**, 1125 (2002)
12. T. Hara, T. Tanaka, T. Tanabe, X.M. Maréchal, K. Kumagai, H. Kitamura, J. Synchrotron Rad. **5**, 426 (1998)
13. H. Kitamura, J. Synchrotron Rad. **5**, 184 (1998)
14. Y. Saitoh, T. Nakatani, T. Matsushita, T. Miyahara, M. Fujisawa, K. Soda, T. Muro, S. Ueda, H. Harada, A. Sekiyama, S. Imada, H. Daimon, S. Suga, J. Synchrotron Rad. **5**, 542 (1998)
15. J.M.J. Madey, H.A. Schwettman, W.M. Fairbank, IEEE Transactions Nucl. Sci. **NS20**, 980 (1973)
16. M. Altarelli, Crystallogr. Rep. **55**, 1145 (2010)
17. Y. Ding, A. Brachmann, F.-J. Decker, D. Dowell, P. Emma, J. Frisch, S. Gilevich, G. Hays, P.H. Hering, Z. Huang, R. Iverson, H. Loos, A. Miahnahri, H.-D. Nuhn, D. Ratner, J. Turner, J. Welch, W. White, J. Wu, Phys. Rev. Lett. **102**, 254801 (2009)
18. T. Shintake, H. Tanaka, T. Hara, T. Tanaka, K. Togawa, M. Yabashi, Y. Otake, Y. Asano, T. Fukui, T. Hasegawa, A. Higashiya, N. Hosoda, T. Inagaki, S. Inoue, Y. Kim, M. Kitamura, N. Kumagai, H. Maesaka, S. Matsui, M. Nagasono, T. Ohshima, T. Sakurai, K. Tamasaku, Y. Tanaka, T. Tanikawa, T. Togashi, S. Wu, H. Kitamura, T. Ishikawa, T. Asaka, T. Bizen, S. Goto, T. Hirono, M. Ishii, H. Kimura, T. Kobayashi, T. Masuda, T. Matsushita, X.M. Maréchal, H. Ohashi, T. Ohata, K. Shirasawa, T. Takagi, S. Takahashi, M. Takeuchi, R. Tanaka, A. Yamashita, K. Yanagida, C. Zhang, Phys. Rev. ST Accel. Beams **12**, 070701 (2009)
19. T. Tanaka, T. Shintake (eds.), *SCSS X-FEL Conceptual Design Report* (RIKEN Harima Institute, Hyogo, 2005)
20. J.A. Samson, *Techniques of Vacuum Ultraviolet Spectroscopy* (Wiley, New York, 1967)
21. J.A. Samson, D.L. Ederer, *Vacuum Ultraviolet Spectroscopy* (Academic Press, London, 2000)
22. G. Margaritondo, *Introduction to Synchrotron Radiation* (Oxford University Press, New York, 1988)
23. P. Pianetta, J. Arthur, S. Brennan (eds.) *Synchrotron Radiation Instrumentation*. AIP Conference Proceedings vol 521, 1999
24. M. Fujisawa, A. Harasawa, A. Agui, M. Watanabe, A. Kakizaki, S. Shin, T. Ishii, T. Kita, T. Harada, Y. Saitoh, S. Suga, Rev. Sci. Instrum. **67**, 345 (1996)
25. Y. Saitoh, H. Kimura, Y. Suzuki, T. Nakatani, T. Matsushita, T. Muro, T. Miyahara, M. Fujisawa, K. Soda, S. Ueda, H. Harada, M. Kotsugi, A. Sekiyama, S. Suga, Rev. Sci. Instrum. **71**, 3254 (2000)
26. A. Erko, F. Schäfers, A. Firsov, W.B. Peatman, W. Eberhardt, R. Signorato, Spectrochim. Acta **B59**, 1543 (2004)
27. P. Kirkpatric, A.V. Baez, J. Opt. Soc. Am. **38**, 766 (1948)
28. H. Mimura, H. Yumoto, S. Matsuyama, Y. Sano, K. Yamaura, Y. Mori, M. Yabashi, Y. Nishino, K. Tamasaku, T. Ishikawa, K. Yamauchi, Appl. Phys. Lett. **90**, 051903 (2007)
29. T. Matsui, H. Sato, K. Shimada, M. Arita, S. Senba, H. Yoshida, K. Shirasawa, M. Morita, A. Hiraya, H. Namatame, M. Taniguchi, Nuclear Instrum. Meth. A **467–468**, 537 (2001)
30. A. Hiraya, K. Yoshida, S. Yagi, M. Taniguchi, S. Kimura, H. Hama, T. Takayama, D. Amano, J. Synchrotron Rad. **5**, 445 (1998)
31. S. Kimura, T. Ito, M. Sakai, E. Nakamura, N. Kondo, T. Horigome, K. Hayashi, M. Hosaka, M. Katoh, T. Goto, T. Ejima, K. Soda, Rev. Sci. Instrum. **81**, 053104 (2010)
32. V.N. Strocov, T. Schmitt, U. Flechsig, T. Schmidt, A. Imhof, Q. Chen, J. Raabe, R. Betemps, D. Zimoch, J. Krempasky, X. Wang, M. Grioni, A. Piazzalunga, L. Patthey, J. Synchrotron Rad. **17**, 631 (2010)
33. J.J. Yeh, I. Lindau, At. Data Nucl. Data Tables **32**, 1 (1985)
34. M. Yabashi, K. Tamasaku, S. Kikuta, T. Ishikawa, Rev. Sci. Instrum. **72**, 4080 (2001)

35. S. Souma, A.T. Sato, T. Takahashi, P. Baltzer, Rev. Sci. Instrum. **78**, 123104 (2007)
36. G. Funabashi, H. Fujiwara, A. Sekiyama, M. Hasumoto, T. Itoh, S. Kimura, P. Baltzer, S. Suga, Jpn. J. Appl. Phys. **47**, 2265 (2008)
37. S. Suga, A. Sekiyama, G. Funabashi, J. Yamaguchi, M. Kimura, M. Tsujibayashi, T. Uyama, H. Sugiyama, Y. Tomida, G. Kuwahara, S. Kitayama, K. Fukushima, K. Kimura, T. Yokoi, K. Murakami, H. Fujiwara, Rev. Sci. Instrum. **81**, 105111 (2010)
38. H. Sumi, Y. Toyozawa, J. Phys. Soc. Jpn. **31**, 342 (1971)
39. T. Kiss, T. Shimojima, K. Ishizaka, A. Chainani, T. Togashi, T. Kanai, X.-Y. Wang, C.-T. Chen, S. Watanabe, S. Shin, Rev. Sci. Instrum. **79**, 023106 (2008)
40. T. Togashi, T. Kanai, T. Sekikawa, S. Watanabe, C.T. Chen, C.Q. Zhang, Z.Y. Xu, J.Y. Wang, Opt. Lett. **28**, 254 (2003)
41. K. Hirano, K. Izumi, T. Ishikawa, S. Annaka, S. Kikuta, Jpn. J. Appl. Phys. **30**, L407 (1991)
42. K. Hirano, T. Ishikawa, S. Kikuta, Nucl. Instrum. Meth. **A336**, 343 (1993)
43. K. Hirano, T. Ishikawa, S. Kikuta, Rev. Sci. Instrum. **66**, 1604 (1995)
44. M. Suzuki, N. Kawamura, M. Mizumaki, A. Urata, H. Maruyama, S. Goto, T. Ishikawa, Jpn. J. Appl. Phys. **37**, L1488 (1998)
45. H. Maruyama, M. Suzuki, N. Kawamura, M. Ito, E. Arakawa, J. Kokubun, K. Hirano, K. Horie, S. Uemura, K. Hagiwara, M. Mizumaki, S. Goto, H. Kitamura, K. Namikawa, T. Ishikawa, J. Synchrotron Rad. **6**, 1133 (1999)
46. T. Tanaka, H. Kitamura, Nucl. Instrum. Meth. **A364**, 368 (1995)
47. C.-H. Ko, J. Kirz, H. Ade, E. Johnson, S. Hulbert, E. Anderson, Rev. Sci. Instrum. **66**, 1416 (1995)
48. A.V. Baez, J. Opt. Soc. Am. **51**, 405 (1961)
49. I.A. Artioukov, K.M. Krymski, Opt. Eng. **39**, 2163 (2000)
50. N. Mårtensson, P. Baltzer, P.A. Brühwiler, J.-O. Forsell, A. Nilsson, A. Stenborg, B. Wannberg, J. Electron Spectrosc. Rel. Phenom. **70**, 117 (1994)
51. H. Hafner, J.A. Simpson, C.E. Kuyatt, Rev. Sci. Instrum. **39**, 33 (1968)
52. D.E. Eastman, J.J. Donelon, N.C. Hien, F.J. Himpsel, Nucl. Instrum. Meth. **172**, 327 (1980)
53. H. Nishimoto, H. Daimon, S. Suga, Y. Tezuka, S. Ino, I. Kato, F. Zenitani, H. Soezima, Rev. Sci. Instrum. **64**, 2857 (1993)
54. H. Daimon, T. Nakatani, S. Imada, S. Suga, Y. Kagoshima, T. Miyahara, Rev. Sci. Instrum. **66**, 1510 (1995)
55. H. Nishimoto, T. Nakatani, T. Matsushita, S. Imada, S. Suga, J. Phys. Condens. Matter. **8**, 2715 (1996)
56. O. Hemmers, S.B. Whitfield, P. Glans, H. Wang, D.W. Lindle, R. Wehlitz, I.A. Sellin, Rev. Sci. Instrum. **69**, 3809 (1998)
57. L. Ley, M. Cardona (eds.), *Photoemission in Solids II* (Springer, Berlin, 1979)
58. Y. Sekiyama, M. Fujita, S. Tsunekawa, A. Kasai, S. Shigemoto, D.T. Imada, T. Adroja, F. Yoshino, T. Iga, T. Takabatake, S. Nanba, J. Suga, Electron Spectr. Relat. Phenom. **144**, 655 (2005)
59. T. Muro, Y. Kato, T. Matsushita, T. Kinoshita, Y. Watanabe, H. Okazaki, T. Yokoya, A. Sekiyama, S. Suga, J. Synchrotron Rad. **18**, 879 (2011)
60. T. Kiss, F. Kanetaka, T. Yokoya, T. Shimojima, K. Kanai, S. Shin, Y. Onuki, T. Togashi, C. Zhang, C.T. Chen, S. Watanabe, Phys. Rev. Lett. **94**, 057001 (2005)
61. M.E. Grass, P.G. Karlsson, F. Aksoy, M. Lundqvist, B. Wannberg, B.S. Mun, Z. Hussain, Z. Liu, Rev. Sci. Instrum. **81**, 053106 (2010)
62. D.F. Ogletree, H. Bluhm, E.D. Hebenstreit, M. Salmeron, Nucl. Instrum. Meth. A **601**, 151 (2009)
63. C. Guillot, Y. Ballu, J. Paigné, J. Lecante, K.P. Jain, P. Thiry, R. Pinchaux, Y. Pétroff, L.M. Falicov, Phys. Rev. Lett. **39**, 1632 (1977)
64. T. Jo, A. Kotani, J.-C. Parlebas, J. Kanamori, J. Phys. Soc. Jpn. **52**, 2581 (1983)
65. J.W. Allen, Resonant photoemission of solids with strongly correlated electrons, in *Synchorotron Radiation Research: Advances in Surface and Interface Science, 1 Techniques*, ed. by R.Z. Bachrach (Plenum Press, New York, 1992)

66. S. Suga, M. Yamamoto, M. Taniguchi, M. Fujisawa, A. Ochiai, T. Suzuki, T. Kasuya, J. Magn. Magn. Matt. **52**, 293 (1985)
67. S. Suga, S. Imada, T. Jo, M. Taniguchi, A. Fujimori, S.-J. Oh, A. Kakizaki, T. Ishii, T. Miyahara, T. Kasuya, A. Ochiai, T. Suzuki, Phys. Rev. **B51**, 2061 (1995)
68. S. Suga, S. Imada, H. Yamada, Y. Saitoh, T. Nanba, S. Kunii, Phys. Rev. **B52**, 1584 (1995)
69. J.W. Allen, S.J. Oh, O. Gunnarsson, K. Schönhammer, M.B. Maple, M.S. Torikachvili, I. Lindau, Adv. Phys. **35**, 275 (1986)
70. G.J. Lapeyre, J. Anderson, P.L. Gobby, J.A. Knapp, Phys. Rev. Lett. **33**, 1290 (1974)
71. G.J. Lapeyre, A.D. Baer, J. Hermanson, J. Anderson, J.A. Knapp, P.L. Gobby, Solid State Commun. **15**, 1601 (1974)
72. D.E. Eastman, J.L. Freeouf, Phys. Rev. Lett. **33**, 1601 (1974)
73. A. Damascelli, Z. Hussain, Z.-X. Shen, Rev. Mod. Phys. **75**, 473 (2003)
74. M. Yano, A. Sekiyama, H. Fujiwara, T. Saita, S. Imada, Y. Onuki, S. Suga, Phys. Rev. Lett. **98**, 036405 (2007)
75. T. Valla, A.V. Fedorov, P.D. Johnson, B.O. Wells, S.L. Hulbert, Q. Li, G.D. Gu, N. Koshizuka, Science **285**, 2110 (1999)
76. L. Casalis, W. Jark, M. Kiskinova, D. Lonza, P. Melpignano, D. Morris, R. Rosei, A. Savoia, A. Abrami, C. Fava, P. Furlan, R. Pugliese, D. Vivoda, G. Sandrin, F.-Q. Wei, S. Contarini, L. DeAngelis, C. Gariazzo, P. Nataletti, G.R. Morrison, Rev. Sci. Instrum. **66**, 4870 (1995)
77. I.-H. Hong, T.-H. Lee, G.-C. Yin, D.-H. Wei, J.-M. Juang, T.-E. Dann, R. Klauser, T.J. Chuang, C.T. Chen, K.-L. Tsang, Nucl. Instrum. Meth. A **467–468**, 905 (2001)
78. M. Escher, N. Weber, M. Merkel, C. Ziethen, P. Bernhard, G. Schönhense, S. Schmidt, F. Forster, F. Reinert, B. Krömker, D. Funnemann, J. Phys. Condens. Matter **17**, S1329 (2005)
79. K. Horiba, Y. Nakamura, N. Nagamura, S. Toyoda, H. Kumigashira, M. Oshima, K. Amemiya, Y. Senba, H. Ohashi, Rev. Sci. Instrum. **82**, 113701 (2011)
80. W. Swiech, G.H. Fecher, C.H. Ziethen, O. Schmidt, G. Schönhense, K. Grzelakowski, C.M. Schneider, R. Frömter, H.P. Oepen, J. Kirschner, J. Electron. Spectrosc. Rel. Phenom. **84**, 171 (1997)
81. W. Kuch, J. Gilles, S.S. Kang, S. Imada, S. Suga, J. Kirschner, Phys. Rev. **B62**, 3824 (2000)
82. W. Kuch, J. Gilles, S.S. Kang, F. Offi, J. Kirschner, S. Imada, S. Suga, J. Appl. Phys. **87**, 5747 (2000)
83. E. Bauer, W. Telieps, Scanning Microsc **1**, 99–108 (1987)
84. E. Bauer, M. Mundschau, W. Swiech, W. Telieps, Ultramicroscopy **31**, 49 (1989)
85. C.M. Schneider, A. Kuksov, A. Krasyuk, A. Oelsner, D. Neeb, S.A. Nepijko, G. Schönhense, I. Mönch, R. Kartofen, J. Morais, C.D. Nadai, N.B. Brookes, Appl. Phys. Lett. **85**, 2562 (2004)
86. G. Schönhense, H.J. Elmers, Surf. Interface Anal. **38**, 1578 (2006)
87. A. Krasyuk, A. Oelsner, S.A. Nepijko, A. Kuksov, C.M. Schneider, G. Schönhense, Appl. Phys. A **76**, 863 (2003)
88. T. Ohkochi, H. Fujiwara, M. Kotsugi, A. Tsukamoto, K. Arai, S. Isogami, A. Sekiyama, J. Yamaguchi, K. Fukushima, R. Adam, C.M. Schneider, T. Nakamura, K. Kodama, M. Tsunoda, T. Kinoshita, S. Suga, Jpn. J. Appl. Phys. **51**, 073001 (2012)
89. O. Schmidt, M. Bauer, C. Wiemann, R. Porath, M. Scharte, O. Andreyev, G. Schönhense, M. Aeschlimann, Appl. Phys. B **74**, 223 (2002)
90. M. Escher, K. Winkler, O. Renault, N. Barrett, J. Electron. Spectrosc. Rel. Phenom. **178–179**, 303 (2010)
91. B. Krömker, M. Escher, D. Funnemann, D. Hartung, H. Engelhard, J. Kirschner, Rev. Sci. Instrum. **79**, 053702 (2008)
92. A. Winkelman, C. Tusche, A.A. Ünal, M. Ellguth, J. Henk, J. Kirschner, New J. Phys. **14**, 043009 (2012)
93. A. Winkelmann, A.A. Ünal, C. Tusche, M. Ellguth, C.-T. Chiang, J. Kirschner, New J. Phys. **14**, 083027 (2012)
94. A. Winkelmann, M. Ellguth, C. Tusche, A.A. Ünal, J. Henk, J. Kirschner, Phys. Rev. B **86**, 085427 (2012)

Chapter 4
Bulk and Surface Sensitivity of Photoelectron Spectroscopy

Due to the relatively short inelastic mean free path (λ_{mp}) of photoelectrons in solids in a wide E_k region, photoelectron spectroscopy is rather surface sensitive. The surface electronic structure is known to be noticeably different from the bulk electronic structure in the case of strongly correlated electron systems. Even in the case of much less correlated electron systems, any surface reconstruction and/or relaxation provide the different electronic structure on the surface in comparison with that in the bulk. Therefore probing of the bulk electronic structure, which is not much influenced by the surface and sub-surface electronic structure in a few atomic layers from the surface is required in many fields of materials sciences. The observed photoelectron intensity is thought to be proportional to the probabilities of photoelectron excitation in solids $P(E_K, h\nu)$, electron transport in solids $T(E_K, h\nu)$ and its escape from the surface $D(E_K)$. The λ_{mp} is thought to be governed by the second process among these processes. Then the concept of the λ_{mp} is first discussed together with some concrete examples experimentally evaluated. Then the method to deconvolute the surface, subsurface and bulk spectral weights from the experimental spectra measured at several different $h\nu$ and/or different emission angles is introduced in this chapter.

4.1 Concept of Inelastic Mean Free Path

As briefly introduced in Sect. 2.1, the probability $Ip(E_K, h\nu)$ of the photoelectrons to be observed is proportional to $P(E_K, h\nu) \cdot T(E_K, h\nu) \cdot D(E_K)$ in the framework of the three-step model for photoelectron emission. Here $P(E_K, h\nu)$, $T(E_K, h\nu)$ and $D(E_K)$ depend actually on the "photoelectron kinetic energy inside the solid" E_k rather than the photoelectron kinetic energy in the vacuum E_K. The E_k has the relation of $E_k = E_K + V_0$ in the nearly free-electron final state model, where $V_0 > 0$ is the energy difference between the vacuum level E_v and the bottom energy E_0 of the nearly free-electron-like final-state band. On the other hand, E_k is rigorously expressed as $h\nu$ added to the energy in the initial (just before photoexcitation) state,

S. Suga and A. Sekiyama, *Photoelectron Spectroscopy*,
Springer Series in Optical Sciences 176, DOI: 10.1007/978-3-642-37530-9_4,

which is practically unknown. Thus, the value of E_k is often substituted by the photoelectron kinetic energy measured from E_F, namely, $E_K + \phi$ (work function) in the discussion of λ_{mp}. Hereafter the kinetic energy inside the solid, E_k, is used for the parameter of $P(E_k, h\nu)$, $T(E_k, h\nu)$ and $D(E_k)$ in this chapter. Matrix element effects including the polarization dependence strongly influence $P(E_k, h\nu)$, which is the most important quantity to be tackled by PES. The excitation of photoelectrons takes place down to the region characterized by $1/\alpha(h\nu)$ from the surface in solids, where $\alpha(h\nu)$ is the absorption coefficient of the individual materials. In many cases, photoelectrons are created in the region down to several hundred or even several thousand Å from the surface for normal incidence photons employed for PES. Before discussing the complex behavior of $T(E_k, h\nu)$ in more detail, let us briefly consider $D(E_k)$. $D(E_k)$ corresponds to the escape probability of the photoelectrons from the surface into the vacuum. The escaping electrons should have enough kinetic energy as discussed in Sect. 2.1, where the component of the kinetic energy perpendicular to the surface must be sufficient to overcome the surface potential barrier. Such a condition is written as $(\hbar^2/2\,m)k_\perp^2 \geq E_V - E_0 = V_0$. Since $k_\perp$ is expressed as $k{\cdot}\cos\theta$, the concept of the escape-cone is often applied. Other electrons outside this escape cone are reflected back into the bulk. No essential effect is, however, introduced in the PES spectra by this term.

$T(E_k, h\nu)$ is the characteristic parameter related to the inelastic mean free path of the photoelectrons (λ_{mp}) (Effects of elastic scattering is only lightly touched in the text not only in this chapter but also in Chaps. 6 and 9). The photoelectron with energy E_k created at depth z below the sample surface has a certain probability to reach the surface without inelastic scattering. The purpose of PES is to detect such primary photoelectrons and discuss the electronic structures of materials through $P(E_k, h\nu)$. On their way to the surface, however, some electrons may experience electron–electron, electron–phonon or electron–impurity interactions. The secondary electron background increasing with the increase in the binding energy (E_B) results from such inelastic scattering processes. The impurity scattering is, however, not playing a major role except for discussing the lifetime broadening (see Chap. 9). Inelastic scattering with phonons may cause energy losses of the order of several tens meV and is less influential than the electron–electron interaction, by which a noticeable portion of the original electron energy E_k is transferred to other electrons. Only when the electron–electron interaction becomes much reduced making the contribution to λ_{mp} longer than several tens Å, the electron–phonon interaction plays an important role in λ_{mp}. As far as the electron–electron interaction is considered, there may be no or very few electrons to interact with the photoelectron with E_k in the case of insulators and then λ_{mp} increases suddenly with decreasing E_k below the band gap energy of the material. Thus λ_{mp} is material dependent, though it generally has a minimum in the energy region of 15–200 eV. The minimum value is less than 10 Å and often close to 2–5 Å [1–4]. It is experimentally as well as theoretically known that λ_{mp} increases gradually with E_k above a few hundred eV. The E_k for the minimum λ_{mp} value might be corresponding to the interband transitions and the excitation energy of plasmons.

If the electronic structure in the surface region is the same as in the bulk, the concept of λ_{mp} is not so important. In the case of materials with a depth dependent electronic structure near the surface region, however, one cannot obtain correct information on the bulk electronic structure without properly treating the effects of λ_{mp}. Systems with surface structural relaxation or depth dependent doping, shallow superlattice systems, systems with a reactive surface and so on belong to this case. Among many solids, strongly correlated electron systems with the electron correlation energy U comparable to or larger than the band width W belong also to this category, because the electrons are less itinerant at the surface due to the large U/W compared to those in the bulk where U/W is much reduced. Therefore the mutual relation between λ_{mp} and the characteristic thickness s or depth z must be carefully checked in these cases.

By using a model dielectric function, λ_{mp} was calculated as a function of the electron energy above the Fermi level E_F for Cu, Ag, Au and Al [2]. The imaginary part of the electron self-energy is proportional to $Im(-1/\varepsilon_L)$, where ε_L is the Lindhard dielectric function. This treatment is expected to give reasonable results for free-electron like materials. The calculation was based on a model dielectric function $\varepsilon(q, \omega)$, which is obtained from a modification of the statistical approximation for a free electron gas [3]. For zero momentum transfer, $\varepsilon(0, \omega)$ was determined from the experimentally obtained optical dielectric function or electron energy loss spectra to which plasmons contribute predominantly. Such an assumption was thought to fail for energies near E_F in the case of transition metals because of the presence of unfilled d bands but it was thought to be applicable to noble metals for example. The results [2] are compared with preceding calculations [3, 4] and some experimental results [5–8] in Fig. 4.1(a). While many experimental evaluations have been made based on the overlayer type experiments (open squares in Fig. 4.1(a)), the full circles in Fig. 4.1(a) are from the analysis of photoemission line broadening [5, 6]. It was recognized that the two preceding theories shown by dotted and dashed lines gave either too long λ_{mp} at low electron energies [3] or too short λ_{mp} at high electron energies [4] compared with the calculated results shown by the solid line [2]. In this treatment [2], a free-electron-like band structure was assumed above E_F. It was, however, conjectured in [2] that the employed approximation would not introduce any serious error as long as the electron energy was greater than 10 eV above E_F even in the case of transition metals. It was thought based on qualitative arguments that the theory in [2] was more accurate at high energies than at low energies. An experimental evaluation with higher accuracy was anyway necessary to check the validity of the theoretical treatments.

Since λ_{mp} is a key decisive material-parameter in XPS (or ESCA) and Auger electron spectroscopy for chemical and surface analysis, the calculation of λ_{mp} has very enthusiastically been performed so far by S. Tanuma, C. J. Powell and D. R. Penn for more than 2 decades. The calculation of λ_{mp} for photoelectron energies between 200 and 2,000 eV was reported for 27 elemental solids and 4 compounds, namely LiF, SiO_2, ZnS and Al_2O_3 [9]. Later a calculation was performed for 27 elements over the photoelectron energy range of 50–2,000 eV

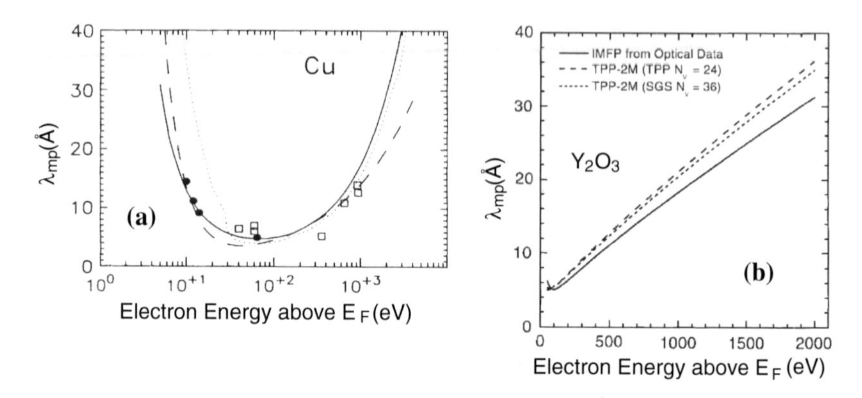

Fig. 4.1 Inelastic mean free path (λ_{mp}) of photoelectrons versus the energy above the Fermi energy. **a** Calculated results for Cu. The solid curve represents the calculated results by the treatment in Ref. [2]. Theoretical result by Tung et al. [3] is shown by the dotted curve and that by Seah and Dench by the dashed curve [4]. Experimental results [5, 6] are added by solid circles and that from overlayer type experiments by open squares [7, 8]. **b** λ_{mp} for Y_2O_3. Comparison of the calculated results from optical data and TPP-2 M for different N_v [13]

with 4 parameters related to atomic weight, bulk density, number of valence electrons per atom and band gap energy [10]. The calculation based on equations called TPP-2 was performed for 15 inorganic materials [11]. The λ_{mp} (in Å) is expressed there as

$$E/\{E_p^2[\beta \ln(\gamma E) - (C/E) + (D/E^2)]\} \tag{4.1}$$

Here E is the electron energy (in eV), $E_p = 28.8\ (N_v\rho/M)^{1/2}$ is the free-electron plasmon energy (eV), where ρ is the density (g cm^{-3}), N_v the number of valence electrons per atom or molecule and M the atomic or molecular weight. The terms β, γ, C and D are parameters. Here β is empirically expressed as

$$\beta = -0.0216 + 0.944/\left(E_p^2 + E_g^2\right)^{1/2} + 7.39 \times 10^{-4}\rho. \tag{4.2}$$

Eg is the band gap energy. Later, λ_{mp} of 14 organic compounds was simulated by using a modified empirical expression for one parameter (β) in TPP-2, called TPP-2M method [12], where

$$\beta = -0.10 + 0.944/\left(E_p^2 + E_g^2\right)^{1/2} + 0.069\rho^{0.1} \tag{4.3}$$

was used. Results of such calculations for a certain material, Y_2O_3, are shown in Fig. 4.1(b) in comparison with λ_{mp} obtained from optical data [13]. Further calculations were performed up to higher energies with checking their reliability [13, 14] as partly shown already in Fig. 2.2. From these results, it is confirmed that HAXPES with E_k above several thousand eV can generally reveal bulk electronic structures as discussed in detail in Chap. 8. Still the behavior of λ_{mp} in the low energy region below 50 eV is very much material dependent and one cannot too much rely on the theoretical predictions as later seen in Chap. 9.

The values of λ_{mp} can experimentally be examined by transmission of photoelectrons through overlayer materials [15, 16]. For the overlayer thickness z, the PES intensity I from the substrate through the overlayer is attenuated as $I = I_0 \cdot \exp(-z/L)$, where L is the effective attenuation length (EAL) in the overlayer material, which corresponds roughly to λ_{mp} if the elastic scattering effects are neglected [16]. In the E_k region where EAL is very short, inhomogeneous thickness distribution and/or defects in the overlayer film as well as the possible diffusion of the substrate atoms into the overlayer induce a serious ambiguity in the evaluation of EAL, although the evaluation is rather reliable in the case of large EAL. So evaluation of short EAL by means of the overlayer method should be made very carefully. The behavior of EAL and λ_{mp} in high E_k region above several hundred eV and with the polar angle $\lesssim 60°$ are more or less convergent in terms of a monotonous increase with E_k.

4.2 How to Separate the Bulk and Surface Contributions in the Spectra

If the surface core level(s) is (are) clearly observable at different E_B in the vicinity of the bulk core level(s) in PES, it is feasible to reliably evaluate the EAL without resorting to the overlayer method as demonstrated in the case of Yb [17]. In the case of an in situ evaporated Yb thin film on clean W, one can clearly see both bulk and surface Yb $4f_{7/2}$ and $4f_{5/2}$ spin-orbit doublets. Due to the small surface core level shift [18], the bulk spectral weight is $\sim\exp(-s/L(E_k))$ and the surface spectral weight becomes $\sim(1 - \exp(-s/L(E_k)))$ assuming a proper surface thickness s, where EAL is represented by L. Since the bulk and surface Yb 4f peaks are still well separated in E_B, one can easily fit the bulk and surface spectral shape and evaluate the relative spectral weight and then $L(E_k)$ for a certain s as shown in Fig. 4.2. Even for a much higher s, the result does not change so much. While EAL increases with the decrease in the electron energy below $\sim$15 eV (or E_k below $\sim$10 eV) in Yb, it remains less than what was experimentally obtained above 1 keV. These L are found to be much shorter (a factor of 4) than the theoretically predicted values at $E_k < \sim 10$ eV, revealing that the low energy PES is not so bulk sensitive as predicted by theory in the case of Yb.

It is known that the surface electronic structure is often noticeably different from that in the bulk when surface relaxation and/or surface reconstruction takes place. It has also been recognized that the surface electronic structure of strongly correlated electron systems (SCES) is often much different from that in the bulk. So reliable experimental evaluation of EAL or λ_{mp} in the region of their small values is always desirable. Since the primary photoelectrons not only from the bulk but also from the surface region are contained in the detected photoelectrons

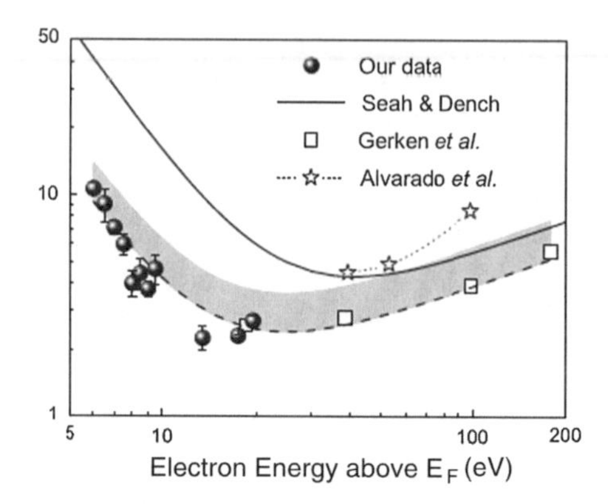

Fig. 4.2 Effective attenuation length (EAL) evaluated by PES from the relative intensity of the bulk and surface Yb 4f peaks in Yb thin film (solid symbols with error bars) for s = 2.74 Å corresponding to 1 atomic layer in comparison with other evaluations. The energy scale is from E_F. The upper limit of the shaded area corresponds to s = 4.1 Å (1.5 atomic layer) [17]. Additional contributions other than the Yb 4f states, for example, those from the Yb 5d states are properly subtracted before the evaluation of the relative intensity of the bulk and surface Yb 4f states

at the same E_K, separation of them into the surface and bulk components is necessary to reliably estimate their contributions. However, this is not a straight forward task.

The short λ_{mp} inevitably requires clean surfaces for PES studies. UHV in 10^{-8} Pa region is usually required for conventional surface PES measurements. Even when one tries to probe the bulk electronic structure, any change of the surface electronic structure induces ambiguity of the analysis. Therefore UHV condition is required except for hard X-ray PES, where the contribution from the surface is negligibly small.

If the angle of the electron emission direction is θ with respect to the sample surface normal, the probability for a photoelectron created at a depth $z \sim z + dz$ below the sample surface to reach the surface is estimated by

$$\propto \exp(-z/(\lambda_{mp} \cos\theta))dz. \tag{4.4}$$

For normal emission with $\theta = 0°$ the probability of the electron to escape from the surface without inelastic scattering is 1/e for $z = \lambda_{mp}$. With increasing z, the probability decreases monotonously. If there were a surface layer with thickness s, in which the electronic structure is different from that in the bulk, the measurement of photoelectrons with the mean free path λ_{mp} provides the bulk spectral weight of $\exp(-s/\lambda_{mp})$ and the surface spectral weight of $(1 - \exp(-s/\lambda_{mp}))$. In the case of angle integrated PES, the deconvolution of the surface and bulk components is possible under some assumption as explained above. However, a similar treatment in low energy angle resolved photoelectron spectroscopy (ARPES) for valence bands is not simple because the spectra are measured as a function of θ and/or ϕ.

For example, the effect of the λ_{mp} was rarely discussed in the case of ARPES of high Tc cuprates, though there are many ARPES results reported below 100 eV. Inelastic scattering effects themselves can be a main subject in these systems in low energy ARPES as briefly explained in Chap. 9. Generally speaking, comparison of high and low energy ARPES is quite important to see the effect of the surface and to derive the information on bulk band dispersions. Such examples are discussed in Chaps. 7 and 8.

In the case of angle integrated PES, the hν dependence of the spectra enables one to make the deconvolution into bulk and surface components by assuming the surface layer thickness s and the values of λ_{mp} at two different photoelectron kinetic energies E_k. Such an example is given in Sect. 7.1.3 for $Sr_{1-x}Ca_xVO_3$, where the values λ_{mp} at two kinetic energies were obtained from theoretical calculations. λ_{mp} is assumed to be isotropic in most theoretical works.

When not only the surface but also the subsurface electronic structure is different from that in the bulk as in the case of $YbInCu_4$ [19], a slightly modified formalism is required. Let us assume for simplicity that the subsurface region between the surface and the bulk is homogeneous and has a characteristic electronic structure. In this approximation, the thicknesses of the surface and the subsurface are defined as s and ss. Since this system has different valences in the surface, subsurface and bulk, they are defined as V_s, V_{ss} and V_b. These are the parameters to be clarified from PES experiments. As already discussed the spectral weight of the bulk, subsurface and surface in the PES is then given by

$$\begin{aligned} &W_b = \exp(-(s+ss)/\lambda),\ W_{ss} = \exp(-s/\lambda) - \exp(-(s+ss)/\lambda) \text{ and} \\ &W_s = 1 - \exp(-s/\lambda), \end{aligned} \tag{4.5}$$

respectively (λ_{mp} is abbreviated as λ in this equation) [19]. The valence evaluated from the Yb 3d PES spectra corresponds to the weighted sum of

$$W_s * V_s + W_{ss} * V_{ss} + W_b * V_b, \tag{4.6}$$

since the contributions from the bulk, sub-surface and surface are contained in the Yb 3d spectra. On the other hand, the surface contribution can rather reliably deconvoluted out in the experimentally evaluated valence from the Yb 4f SXPES. Then the derived valence value after excluding the surface contribution can be expressed by

$$w_{ss} * V_{ss} + w_b * V_b \tag{4.7}$$

where w_{ss} and w_b stand for

$$1 - \exp(-ss/\lambda) \quad \text{and} \quad \exp(-ss/\lambda). \tag{4.8}$$

For the purpose of analysis, theoretically predicted λ_{mp} values are employed. Here the surface valence V_s can be safely assumed to be 2.0 in the case of $YbInCu_4$, because the surface Yb $4f^{13}$ component in the valence band was observed at around 1 eV higher E_B than the bulk component, suggesting the stability of the surface divalent component. Then the remaining four unknown parameters, s, ss, V_{ss} and V_b can be uniquely determined from PES experiments performed at four rather different hν [19]. The results are shown later in Sect. 8.4. For more complex systems such as superlattice structures with more unknown parameters, more numbers of PES experiments at more hν energies will facilitate the analysis.

Another method to separate the bulk and the surface PES spectra is to use the emission angle (θ) dependence of the spectra, which is correlated with the change of the probing depth. According to (4.4), the photoemission intensity detected in a small acceptance angle can be expressed as $dI \propto \exp[-z/(\lambda_{mp} \cdot \cos\theta)]dz$. If we measure PES spectra at different θ values and properly normalize them, we can separate the contributions from the surface and the bulk [20]. This method is, however, only applicable to samples with very flat specular surfaces.

In order to separate the contributions from the surface, bulk and sometimes subsurface, it is thus recommended to perform the PES measurement at several rather different hν ranging from the soft X-ray to hard X-ray. In addition, careful check of the bulk sensitivity is required in the case of PES below $E_k < 15$ eV.

References

1. S. Hüfner, *Photoelectron Spectroscopy* (Springer, Berlin, Heidelberg, 2003)
2. D.R. Penn, Phys. Rev. **B35**, 482 (1987)
3. C.J. Tung, J.C. Ashley, R.H. Ritchie, Surf. Sci. **81**, 427 (1979)
4. M.P. Seah, W.A. Dench, Surf. Interface Anal. **1**, 2 (1979)
5. J.A. Knapp, F.J. Himpsel, D.E. Eastmann, Phys. Rev. **B19**, 4952 (1979)
6. F.J. Himpsel, W.E. Eberhardt, Solid State Commun. **31**, 747 (1979)
7. M.P. Seah, Surf. Sci. **32**, 703 (1972)
8. M.P. Seah, J. Phys. **F3**, 1538 (1973)
9. S. Tanuma, C.J. Powell, D.R. Penn, Surf. Interface Anal. **11**, 577 (1988)
10. S. Tanuma, C.J. Powell, D.R. Penn, Surf. Interface Anal. **17**, 911 (1991)
11. S. Tanuma, C.J. Powell, D.R. Penn, Surf. Interface Anal. **20**, 77 (1993)
12. S. Tanuma, C.J. Powell, D.R. Penn, Surf. Interface Anal. **21**, 165 (1993)
13. S. Tanuma, C.J. Powell, D.R. Penn, Surf. Interface Anal. **35**, 268 (2003)
14. S. Tanuma, C.J. Powell, D.R. Penn, Surf. Interface Anal. **43**, 689 (2011)
15. A. Jablonski, S. Tougaard, J. Vac. Sci. Technol. **A8**, 106 (1990)
16. A. Jablonski, C.J. Powell, J. Electron Spectrosc. Rel. Phenom. **100**, 137 (1999)
17. F. Offi, S. Iacobucci, L. Petaccia, S. Gorovikov, P. Vilmercati, A. Rizzo, A. Ruocco, A. Goldoni, G. Stefani, G. Panaccione, J. Phys. Condens. Matter **22**, 305002 (2010)
18. The width of the surface core levels and the magnitude of the surface core level shift seem to depend on the substrate [17]
19. S. Suga, A. Sekiyama, S. Imada, J. Yamaguchi, A. Shigemoto, A. Irizawa, K. Yoshimura, M. Yabashi, K. Tamasaku, A. Higashiya, T. Ishikawa, J. Phys. Soc. Jpn. **78**, 074704 (2009)
20. A. Sekiyama, S. Kasai, M. Tsunekawa, Y. Ishida, M. Sing, A. Irizawa, A. Yamasaki, S. Imada, T. Muro, Y. Saitoh, Y. Onuki, T. Kimura, Y. Tokura, S. Suga, Phys. Rev. B**70**, 060506 (R) (2004)

Chapter 5
Examples of Angle Integrated Photoelectron Spectroscopy

The photoelectrons are emitted into the vacuum with the information on energy, momentum (wave vector) as well as spin. The counting rate in the angle integrated photoelectron spectroscopy is orders of magnitude higher than that in the angle resolved photoelectron spectroscopy (ARPES). However, momentum information is smeared out or averaged in the angle integrated photoelectron spectroscopy. Then the information on density of states (DOS) of valence bands can be relatively easily obtained. By use of the constant initial state spectroscopy (CIS) and resonance photoelectron spectroscopy (RPES) one can even evaluate the contribution of a particular orbital buried in the valence bands. A lot of angle integrated experiments were so far performed on various intriguing phenomena in many materials. Some typical results on valence bands are given in Sect. 5.1. Then studies of core level spectra with satellite structures and/or well- and poorly-screened peaks are introduced in Sect. 5.2. Finally multiplet structures prominent in materials with incomplete outer shell due to the Coulomb and exchange interactions are briefly discussed in Sect. 5.3. Some theoretical treatments are also briefly introduced time to time.

5.1 Valence Band Spectra

Very many results of valence band spectra of semiconductors and metals by angle integrated photoelectron spectroscopy were so far reported in literature [1–5]. Gross features of the valence bands studied by angle integrated measurements on polycrystalline solids and non-specular surfaces of single crystals were compared with the density of states (DOS) obtained by various band calculations. The electron correlation or on-site repulsive Coulomb interaction is quite weak in conventional semiconductors as Si, Ge, GaAs and so on, where the Koopmans' theorem can be safely applied and the PES results can be relatively easily compared with the DOS. Here is shown some examples of layered semiconductors. The PES spectra of black phosphorus (P) at $h\nu = 110$ and 1,253 eV are shown in Fig. 5.1(a) [6] and (b) in comparison with the result of self-consistent pseudo

S. Suga and A. Sekiyama, *Photoelectron Spectroscopy*,
Springer Series in Optical Sciences 176, DOI: 10.1007/978-3-642-37530-9_5,

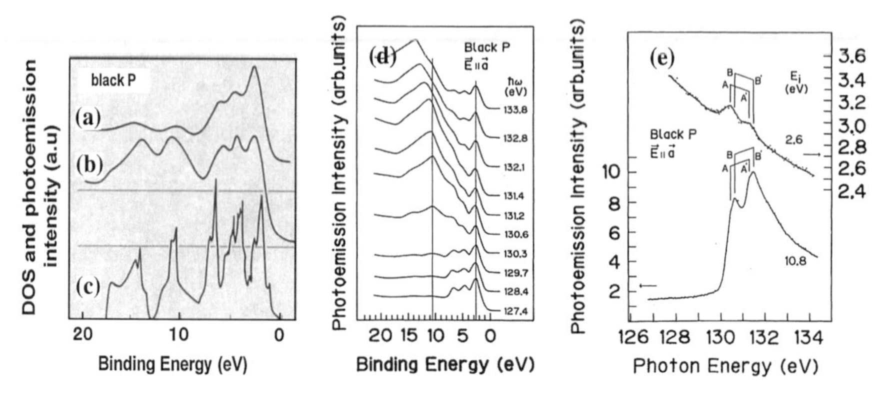

Fig. 5.1 Photoemission spectra at hν = 110 eV (**a**) and XPS (**b**) compared with the DOS (**c**) [7]. hν dependence of the spectra around the P 2p core excitation threshold is shown in (**d**). CIS spectra at E_B = 2.6 and 10.8 eV are shown in (**e**) [10]

potential band calculations in Fig. 5.1(c) [7]. The weak shoulder at ~1.2 eV and three structures at ~2.7, ~4.7 and ~6.6 eV were assigned to the P 3p orbitals, while the peaks at ~10.8 and ~15.4 eV were attributed to the P 3s orbitals. The relative increase of the intensity of the 4.7 eV peak with increasing hν above 40 eV may suggest a significant admixture of P 3s as suggested by tight binding band-structure calculations [8] in agreement with the behavior of the PICS [9]. By measuring the hν dependence of the spectra around the P 2p core threshold, resonance photoemission behavior was observed [10] as reproduced in Fig. 5.1(d) and (e). The splitting A–A′ and B–B′ in the constant initial state (CIS) spectrum (Fig. 5.1(e)) corresponds to the core exciton absorption structures associated with the P $2p_{3/2}$–$2p_{1/2}$ spin–orbit splitting. A and B were associated with different regions of the conduction band. The CIS spectrum for E_B = 10.8 eV is rather similar to the core absorption spectrum and was interpreted as due to the core-exciton-induced resonance Auger process [11, 12]. In the CIS spectrum for E_B = 2.6 eV, structure corresponding to the A–A′ exciton is prominent in contrast to the core absorption spectrum. Then the resonance behavior could not be explained by the contribution of the $L_{2,3}VV$ Auger tail. A Fano type resonance interference behavior was recognized for this CIS for 2.6 eV. This behavior was interpreted as due to the interference between the direct photoelectron excitation of the 3p valence electron and the core exciton excitation from the 2p core level followed by the non-radiative direct recombination of this core exciton exciting one 3p valence hole. However, the magnitude of interference was smaller than in the case of strongly correlated electron systems, reflecting the itinerant character of the P 3p states [11].

Chalcogenide amorphous semiconductors attracted much attention because of their unique optical responses such as photo-structural change and photo-doping. In addition to XPS and UPS [13–15], valence band PES was measured in the Se 3d core exciton excitation region in $GeSe_2$ [16]. Many structures were observed in the valence

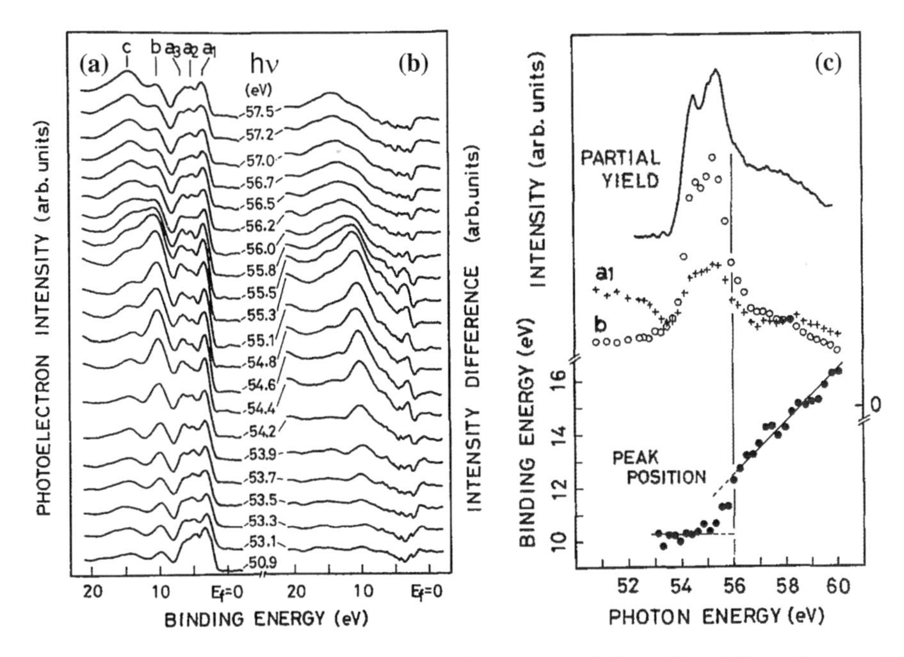

Fig. 5.2 Valence band photoemission spectra of amorphous $GeSe_2$ [16] at different hν around the Se 3d core threshold (**a**). The difference spectra (**b**) are shown by subtracting the spectrum at hν = 50.9 eV. (**c**) The *upper part* shows the partial yield spectrum representing the core absorption spectrum and CIS spectra for the structures a_1 and b in (**a**). The *lower part* shows the peak positions of the Auger structure as a function of hν

band region as shown in Fig. 5.2. The peak a_1 at ~3 eV was ascribed to the Se lone-pair 4p electrons. a_2 and a_3 at ~5 and ~7 eV were assigned to the bonding electron states between Ge and Se. The structures b at $E_B = 10$ eV and c at $E_B \sim 14$ eV correspond to the Ge 4s and Se 4s states. The CIS spectrum for the a_1 structure clearly showed the interference minimum and maximum in the Se 3d core exciton region (Fig. 5.2(c)) as in the case of black P. The CIS spectral shape for the Ge 4s structure b in the Se 3d core exciton excitation region was, however, very similar to that of the partial photoelectron yield behavior reflecting the Se 3d level associated absorption spectrum. This CIS behavior was most likely ascribed to the accidental overlap of the core exciton Auger process as discussed in detail in Ref. [16].

The metal to insulator transition (MIT) and accompanying change of the electronic structure attract wide attention in solid state physics. VO_2 is a d^1 system with a clear first order MIT near $T_c = 340$ K. In the metallic (M) phase above T_c it has a TiO_2 type rutile structure whereas it has a monoclinic structure in the insulator (I) phase below T_c. The PES measured with modest resolution on a cleaved surface are shown in Fig. 5.3 [17] with the dashed and solid curves for the M and I phases at 375 and 298 K. The dotted sharper curves near the Fermi level represent the spectra deconvoluted with the instrumental resolution of 0.25 eV. In the M phase two structures in $E_B = 0{-}2$ eV are observed, which could definitely

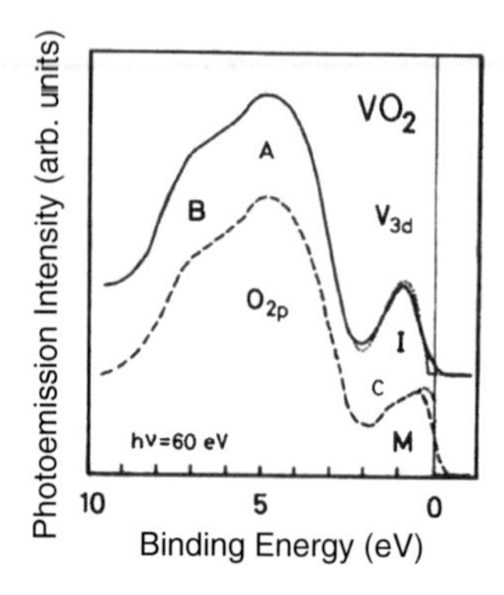

Fig. 5.3 Photoemission spectra of a cleaved single crystal VO_2 measured at $h\nu = 60$ eV in the metal phase (375 K) and insulator phase (298 K) [17]

be ascribed to the coherent and incoherent peaks. In the I phase a peak at $\sim$0.9 eV was seen together with the band gap opening. More bulk sensitive and systematic measurements were later performed at higher $h\nu$ and their interpretations are given in Sect. 8.5.

Changes of electronic structures accompanying phase transitions beyond MIT were also interesting subjects of PES. One example is given for 1T-TaS_2 in Fig. 5.4. The spectral changes in the valence band are strongly correlated with the changes in the core level spectra. This material shows a MIT together with a charge density wave (CDW) phase transition. Figure 5.4 shows the angle integrated PES spectra in the $1T_1$ incommensurate, $1T_2$ quasi-commensurate and $1T_3$ commensurate CDW phases [18]. The Ta 4f core level spectra measured at $h\nu = 66$ eV in Fig. 5.4(a) are compared with the valence band spectra measured at $h\nu = 21$ eV (Fig. 5.4(b)). The broad Ta $4f_{7/2}$ and $4f_{5/2}$ peaks in $1T_1$ incommensurate phase are split into a doublet in the $1T_2$ quasi-commensurate phase, where the FWHM of each peak is $\sim$0.5 eV. The FWHM decreases further down to 0.25 eV in the $1T_3$ commensurate CDW phase with a splitting of $\sim$0.73 eV between α_1 and α_2 as well as between β_1 and β_2. From these Ta 4f core level splitting, one could discuss the inequivalent potentials at different Ta sites and check the validity of the proposed model. From the 0.73 eV α_1–α_2 splitting of the Ta $4f_{7/2}$ components, a CDW amplitude of 0.05 electrons was derived [18]. The valence band spectra in Fig. 5.4(b) clearly showed the MIT from the $1T_2$ phase to the $1T_3$ phase.

Fe_2O_3 is an antiferromagnetic insulator and crystallizes in the α-corundum structure with an Fe atom octahedrally coordinated by six O atoms [19]. This material called hematite belongs to the charge-transfer-type insulators like NiO rather than the Mott–Hubbard-type insulators. Figure 5.5(a) summarizes PES spectra measured for a wide range of $h\nu$ with modest resolution. Broad spectral shapes are observed with some features at $\sim$2.5, 5 and 7 eV. According to the ligand field theory the former two structures were assigned to the $|d^4\rangle$ final state multiplet (^5E_g and $^5T_{2g}$) excited from the high spin $|d^5\rangle$ ground state and the third one to the O 2p band. However, this simple interpretation was not fully acceptable for the following reason. The PICS of the O 2p states relative to that of the Fe 3d states is known to increase towards lower $h\nu$ in this $h\nu$ region [9]. The Fe 3d states

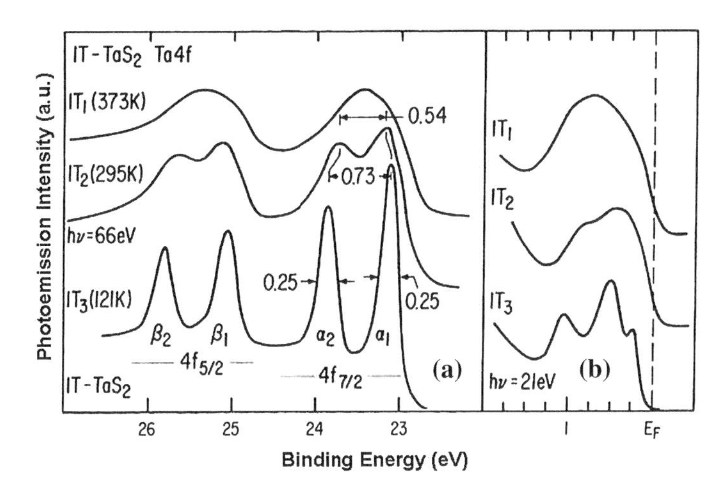

Fig. 5.4 Temperature dependence of the valence band PES spectra (**b**) of 1T-TaS_2 measured at $h\nu = 21$ eV compared with the Ta 4f core level PES (**a**) measured at $h\nu = 66$ eV in different charge density wave (CDW) phases with a total energy resolution of 0.15 eV [18]

were expected to be much weakened below $h\nu = 40$ eV. However, one noticed that the observed spectral shape did not change much with decreasing $h\nu$, suggesting strong hybridization between the Fe 3d and O 2p states. The spectrum at $h\nu = 21.2$ eV was mostly ascribed to the O 2p states, where the higher energy part at ~8 eV was somehow suppressed. The structure at ~4 eV was ascribed to the O 2p nonbonding state and the structure located around ~8 eV was ascribed to the Fe 4sp-O 2p bonding states [19].

The resonance PES (RPES) with changing $h\nu$ was performed as shown in Fig. 5.5(b). The Fe 3p–3d resonance excitation was employed in this case. For the excitation above the Fe 3p core threshold, the $M_{2,3}M_{4,5}M_{4,5}$ Auger emission appears at a constant E_K. The expected energy is shown by the vertical bars in Fig. 5.5(b). This Auger feature does not overlap with the valence band structures in the region of $E_B = 2$–8 eV but the resonance behavior of all valence band structures is clearly seen as shown by the CIS spectra in Fig. 5.5(c). Even the structure near 7 eV shows a resonance behavior with a minimum and a maximum below and above the threshold, suggesting a noticeable contribution of the Fe 3d component. Such a behavior could not be explained by a simple overlap of the Auger emission tail but had to be attributed to the resonance of the Fe 3d component. The Fe 3d component could be empirically extracted by taking a difference spectrum between the two spectra measured at the resonance maximum and minimum as shown at the bottom in Fig. 5.5(b).

In order to interpret the Fe 3d spectrum, a configuration interaction cluster calculation was applied to the FeO_6^{9-} cluster as shown in Fig. 5.5(d). The ground state was assumed to be a linear combination of the $|d^5\rangle$ and $|d^6\underline{L}\rangle$ states and the final states were represented by the mixing between $|d^4\rangle$, $|d^5\underline{L}\rangle$ and $|d^6\underline{L}^2\rangle$ states.

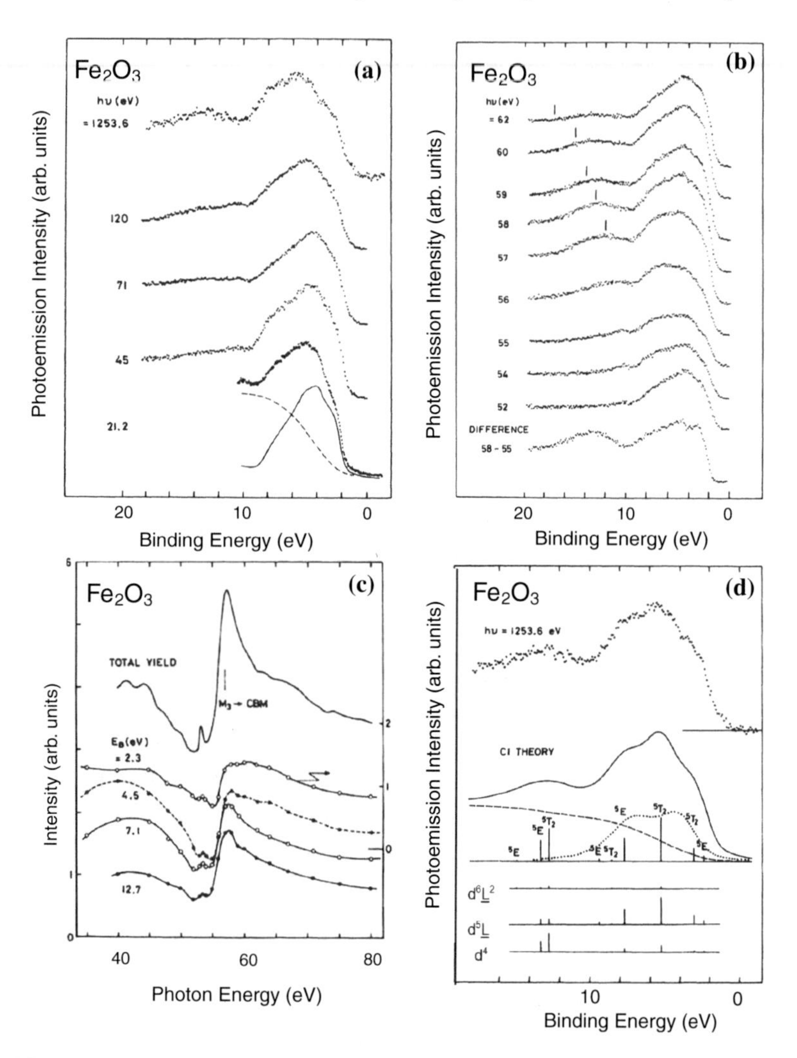

Fig. 5.5 PES of Fe_2O_3 scraped in situ by a diamond file. **a** Spectra in a wide hν region. **b** Resonance PES near the Fe 3p excitation threshold. **c** CIS spectra at several E_B. **d** Results of configuration interaction cluster calculation for the valence bands in Fe_2O_3 [19]

The correlation energy U was explicitly taken into account in the calculation. The decomposition of the whole spectrum into the three different configurations is given at the bottom. Very similar results were obtained for MnO, which is intermediate between the charge-transfer-type and Mott–Hubbard-type insulators [20] and wustite Fe_xO [21] with $0.85 \leq x \leq 0.95$, which is classified to be a charge-transfer insulator.

Next example is on alkali-metal dosing into a low dimensional Mott insulator TiOCl [22], which is a prototypical Mott insulator with $3d^1$ configuration in a good contrast to cuprates with a single hole in the 3d shell. In 2D TiOCl, Ti–O bilayers are stacked along the c-axis and weakly coupled by the van der Waals force. Ti ions are centered in a distorted octahedron of O and Cl ligands and the degeneracy of the 3d t_{2g} orbital is lifted, providing a 0.3 eV splitting for the lowest d-orbital. In situ cleaved clean surfaces were dosed by alkali-metal (Na or K) vapor. Al Kα and HeIα lines were used for excitation. Sample temperature (T) was kept at 360 K to minimize the charging up. With increasing the dosing, additional structures in the Ti 2p core-level spectra (Fig. 5.1 of [22]) with the same spin–orbit splitting as for the Ti^{3+} 2p states emerged on the smaller E_B side, which are ascribed to the Ti^{2+} 2p states, induced by the electron doping concomitant with the intercalation of alkali metal atoms into the van der Waals gaps. The degree of the electron doping can be evaluated as $x = (I(2+))/(I(2+)+I(3+))$, where I represents the integrated intensity. Then valence band spectra were measured at $h\nu = 21.2$ eV as shown in Fig. 5.6(a). The whole spectrum jumped by 0.6 eV toward larger E_B already for $x = 0.05$. Although the appearance of a QP peak at E_F was expected with increasing x, no such feature was observed. Instead a broad hump evolved at E_B near the original LHB. The doping induced peak weight was found to increase with x while the weight of the original LHB located at higher E_B decreased with x. After subtracting the background and fitting the spectra by two Gaussians, the intensity change of the LHB was approximated by $1 - x$. Since the total intensity increases as $1 + x$, the spectral intensity of the new peak was approximated by 2x as shown in Fig. 5.6(b) for the case of K dosed TiOCl. Insulating behavior upon doping was also reported in Na intercalated TiOCl. The spectral weight distribution for the doped and undoped cases is schematically shown in Fig. 5.6(c), where AB stands for the alloy band and J is the Hund's rule coupling and δ is the small difference in orbital energies. In this picture the spectral weight transfer occurs from the LHB to the AB. By considering the energy lowering of electrostatic potential $\Delta \sim 2$ eV on the Ti site with neighboring alkali-metal ion, U-3 J $\sim$ 2.5–3.5 eV, and the $\sim$1 eV AB-LHB separation, the chemical potential jump by $\sim$0.5 eV with x can be reasonably understood. Both single particle and many body aspects are reconciled within a Hubbard model description in this system [22].

Quasi-one-dimensional (1D) metallic systems are also intriguing in angle-integrated PES in addition to ARPES as discussed in Sect. 6.4. In these systems, an anomalous intensity decay of the spectral weight towards E_F represented by a power-law $|E_B - E_F|^{\alpha}$, is theoretically expected at zero temperature due to a Tomonaga-Luttinger liquid [23]. The angle-integrated spectral functions were examined for quasi-1D inorganic [24–26] and organic [27, 28] conductors in the 1990s. All valence-band spectra of these compounds showed a suppression of intensity in the vicinity of E_F in consistence with the power-law behaviour based on the Tomonaga-Luttinger model. However, the value of the power-law α was comparable or larger than 1, being far beyond the upper limit of the predicted value 1/8 from the single-band Hubbard model. Note that the upper limit increases when long-range interactions are taken into account in the theoretical models [28]. In order to explain such an experimentally observed large α, several possible origins as the long-range electron–electron interactions, electron–phonon interactions, and extrinsic effects like a surface relaxation, were discussed. Meanwhile

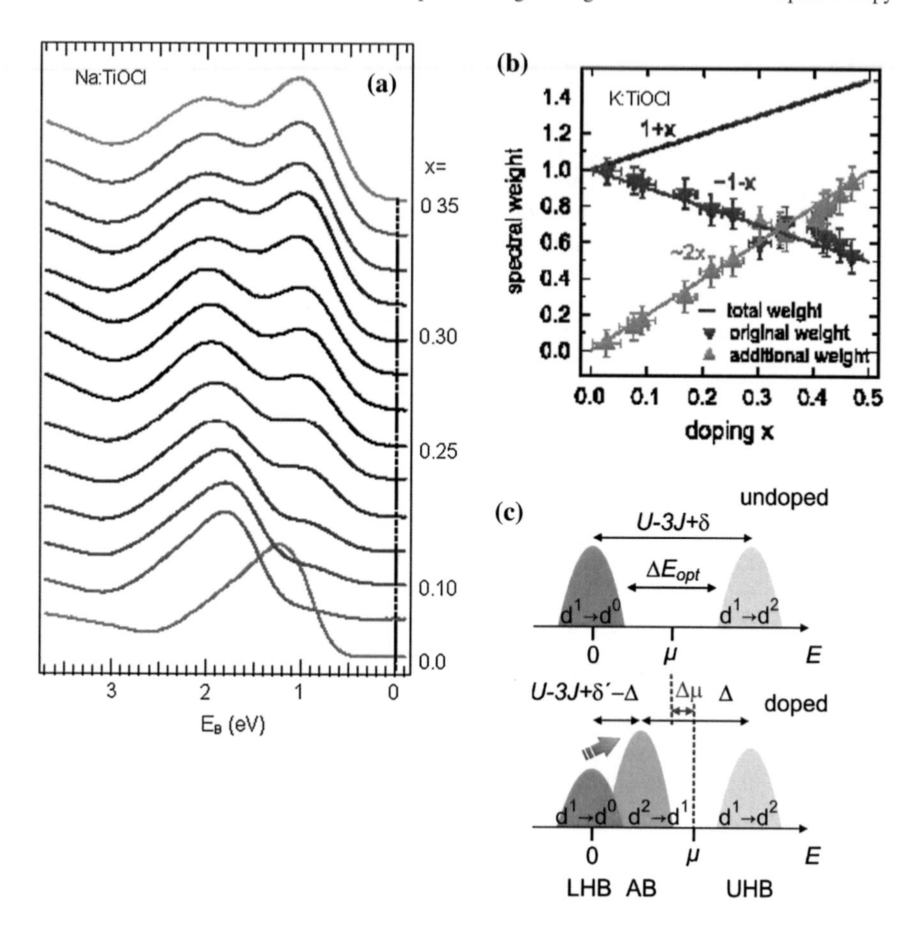

Fig. 5.6 **a** Evolution of the Ti 3d valence band near E_F with electron doping x. **b** Quantitative change of the spectral weight with doping. **c** Model spectral weight for undoped and doped TiOCl [22]

a high-resolution angle-integrated PES study of metallic single-walled carbon nanotubes was reported showing the Tomonaga-Luttinger liquid behaviour with $\alpha = 0.46 \pm 0.10$ near E_F ($|E_B - E_F|^{0.46\pm0.10}$) and the spectral weight at E_F with the similar power-law of temperature ($T^{0.48}$), which are highly consistent with the theoretical study for carbon nanotubes [29]. Now it seems to be empirically established that the spectral weight in the vicinity of E_F is really suppressed at low temperatures for the metallic quasi-1D systems.

5.2 Core Level Spectra

Core level spectroscopy was widely applied for chemical analysis. In the case of atoms and molecules, the so-called equivalent-core approximation was most frequently used for the analysis of the core-level binding energies. In this model, the effect of photoionization of a core-level with much less spatial extent than that of the valence electrons was approximated by the hypothetical addition of a proton to the nucleus. Then this model is called the (Z + 1) approximation [4]. The E_B of a particular state in molecules and solids changes depending on the chemical environment. Many examples are found in literature [4, 30]. One of the examples was the surface core level shift as shown in Fig. 5.7 [31] for the (110) cleaved surface of GaAs. When the proper hν with the highest surface sensitivity was employed for the measurement of each core level, additional structures were observed in addition to the conventional spin–orbit split As 3d and Ga 3d core level doublets. Among the two corresponding spectra in Fig. 5.7, the upper spectra (a) and (b) show a higher surface sensitivity, with a noticeable bump on the smaller (larger) E_B side for the As (Ga) 3d core level. By a proper deconvolution, the surface spectra (S) were separated from the bulk spectra (B) as shown by the dashed curves. The S components were found to decrease in intensity for smaller E_k of $\sim$10 eV (Fig. 5.7 (c), (d)) compared with the results for the kinetic energies E_k of $\sim$40 eV (Fig. 5.7(a), (b)) in agreement with the increase of the λ_{mp}, which was estimated as 19 Å at E_k $\sim$9 eV compared to 5.9 Å at $\sim$40 eV. The opposite shifts of the As and Ga core levels on the surface were interpreted as due to the initial-state charge transfer accompanying the surface relaxation, where the surface As atoms moved outwards and the surface Ga atoms moved inwards with a $\sim$25° bond-angle rotation and a charge transfer from Ga to As.

In the case of core level PES in transition metal compounds, the presence of satellite structures is widely known. One of the examples is shown for Fe_2O_3 in Fig. 5.8, where Fe 3p, Fe 3 s and Fe 2p core level spectra excited by Mg Kα line are compared with the valence band PES on a relative E_B scale. The Fe 3p, $2p_{3/2}$ and $2p_{1/2}$ core levels are accompanied by a satellite structure at higher E_B [19]. Meanwhile, two satellite structures, one strong and the other rather weak, were observed in the case of the Fe 3 s core level PES. The splitting of the satellite from the main peak is different in the case of the Fe 3 s core level compared with those of Fe 3p and 2p core levels. In contrast to these Fe core levels, no satellite structure is observed for the O core levels. The origin of the satellites is not due to energy losses by plasmons but due to screening involving the Fe 3d orbitals at the core–hole site. According to the similarity of the satellites between the core level and valence band, the main line of the core PES was interpreted as well-screened $|\underline{c}3d^6\underline{L}\rangle$ final state where the core hole $\underline{c}$ is screened by the charge transfer from the ligand L to the 3d state, whereas the satellites are poorly screened $|\underline{c}d^5\rangle$ final states. This new interpretation is in contrast to the traditional one which assumed a shake-up transition from the ligand to the 3d state [32].

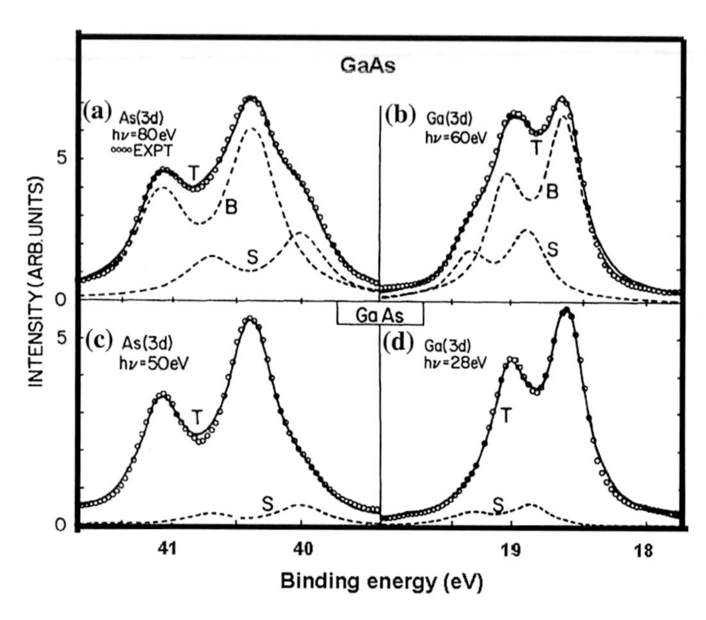

Fig. 5.7 PES of the As 3d and Ga 3d core levels of GaAs (110) measured at two different hν. Surface core level shifts were clearly observed [31]

The Fe 3 s core level in Fe_2O_3 is not simply split into the ^7S and ^5S components separated by the intra-atomic spin exchange interaction with the magnetic $3d^5$ shell [1, 2, 4] but shows rather complicated structures as generally observed for 3s core levels as demonstrated in Fig. 5.9 for the case of transition metal trifluorides and difluorides [33] measured by XPS. In fact, not only the intra-atomic exchange splitting but also the charge transfer and the final-state core–hole screening are responsible for the observed structures. Cluster model calculations taking these effects into account became available for core levels [34–36].

Due to the element selectivity, core level XPS is very useful to study the electronic structure of low concentration atoms. One example is shown for intercalation compounds M_xTiS_2 with M = Mn, Fe, Co and Ni. The transition metal atoms (M) intercalated into the van der Waals gaps of the layered transition metal dichalcogenides can noticeably modify the physical properties of the host materials beyond the rigid-band model, where the guest 3d atoms are considered to donate electrons to the partially filled d_{z^2} band of the host. In the case of M_xTiS_2, $\sqrt{3} \times \sqrt{3}$ and 2 × 2 bulk superlattices are formed for x = 1/3 and 1/4, respectively. Since the number of M atoms on the cleaved surface becomes just half of that in the bulk, the low hν PES with high surface sensitivity is not an appropriate method to study the electronic structure modified by the intercalation. Therefore more bulk sensitive XPS was required for this study. Figure 5.10(a) summarizes the M 2p core level PES of the guest transition metal atoms, where the $2p_{3/2}$ and $2p_{1/2}$ spin–orbit split main peaks are prominently observed together with weaker satellites on the larger energy side slightly overlapping with Auger features such as

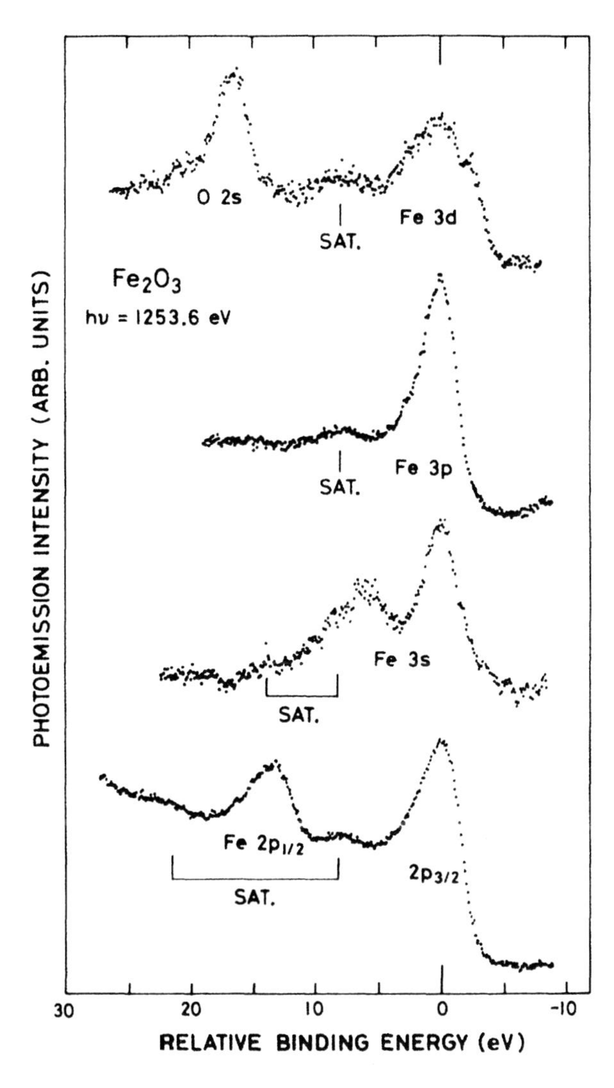

Fig. 5.8 Core level spectra of Fe_2O_3 in comparison with the valence band spectrum for hν = 1,253.6 eV [19]

the Ti $2p_{3/2}3d3d$ in some cases [37]. The location of the satellites is much different from that of the bulk plasmon satellite. The main peaks and the satellites were ascribed to the $|\underline{c}d^{n+1}\underline{L}\rangle$ and $|\underline{c}d^{n}\rangle$ final states, where n is the nominal d electron number in the ground state. By comparing these spectra with those of corresponding mono- and di-sulfides, one can judge whether the observed spectra are resulting from the high-spin divalent state. Among these results in Fig. 5.10(a), only the spectrum of $Co_{1/4}TiS_2$ showed resemblance to that of the low spin pyrite CoS_2, whereas other spectra of M_xTiS_2 with M = Mn, Fe and Ni showed the features of high spin valence states.

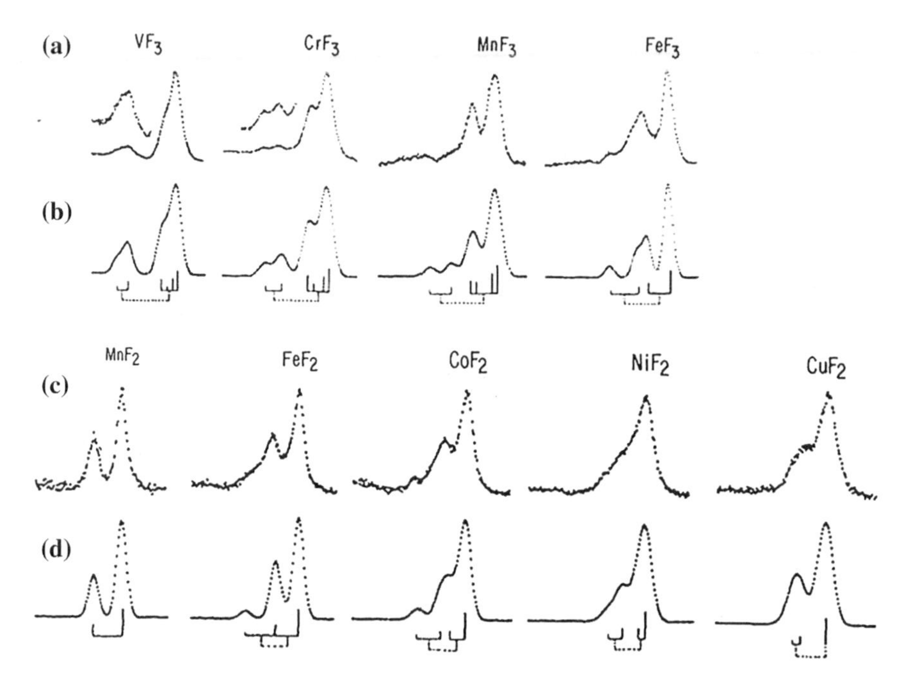

Fig. 5.9 Transition metal 3s core level XPS spectra of transition metal trifluorides (**a**) and transition metal difluorides (**c**). Measurements were performed on powder samples pressed onto In coated substrates. Moreover, sample surfaces were charge neutralized with an electron flood gun tuned to minimize the observed line widths. So one should not too much rely on the detailed spectral shapes of these spectra. (**b**) and (**d**) are the results of model calculations [33]

In contrast to the core level XPS of the guest M atoms, S 2p core level spectra showed only the spin–orbit split doublet component without satellite structures. The Ti 2p core level spectra showed pronounced broadening, asymmetry and energy shift with intercalation as displayed in Fig. 5.10(b). The shift is opposite to the expectation of band filling effect. It is difficult to ascribe the increased width to the initial state effect due to an overlap of the $3d^0$ and $3d^1$ components, because the Ti 3d states in M_xTiS_2 are known to behave itinerantly. It was proposed that the Ti 2p core hole potential pulled down a localized screening level of Ti 3d character below E_F, which could be either occupied by charge transfer from the conduction band or unoccupied in the final state, providing the well screened and poorly screened peaks [38] as schematically illustrated in Fig. 5.10(c). Figure 5.10(b) shows the well screened component at lower (=smaller) E_B with asymmetry and the poorly screened component at higher E_B without asymmetry. The larger width or shorter lifetime of the poorly screened peak for $x = 1/3$ compared with $x = 1/4$ can naturally be understood as due to the increase of the number of Ti 3d conduction electrons supplied from the guest atom. The splitting between the well-screened and poorly screened Ti 2p peaks was evaluated as $\sim$1.5 eV. The contribution of the M 3d states to the valence bands can be evaluated by

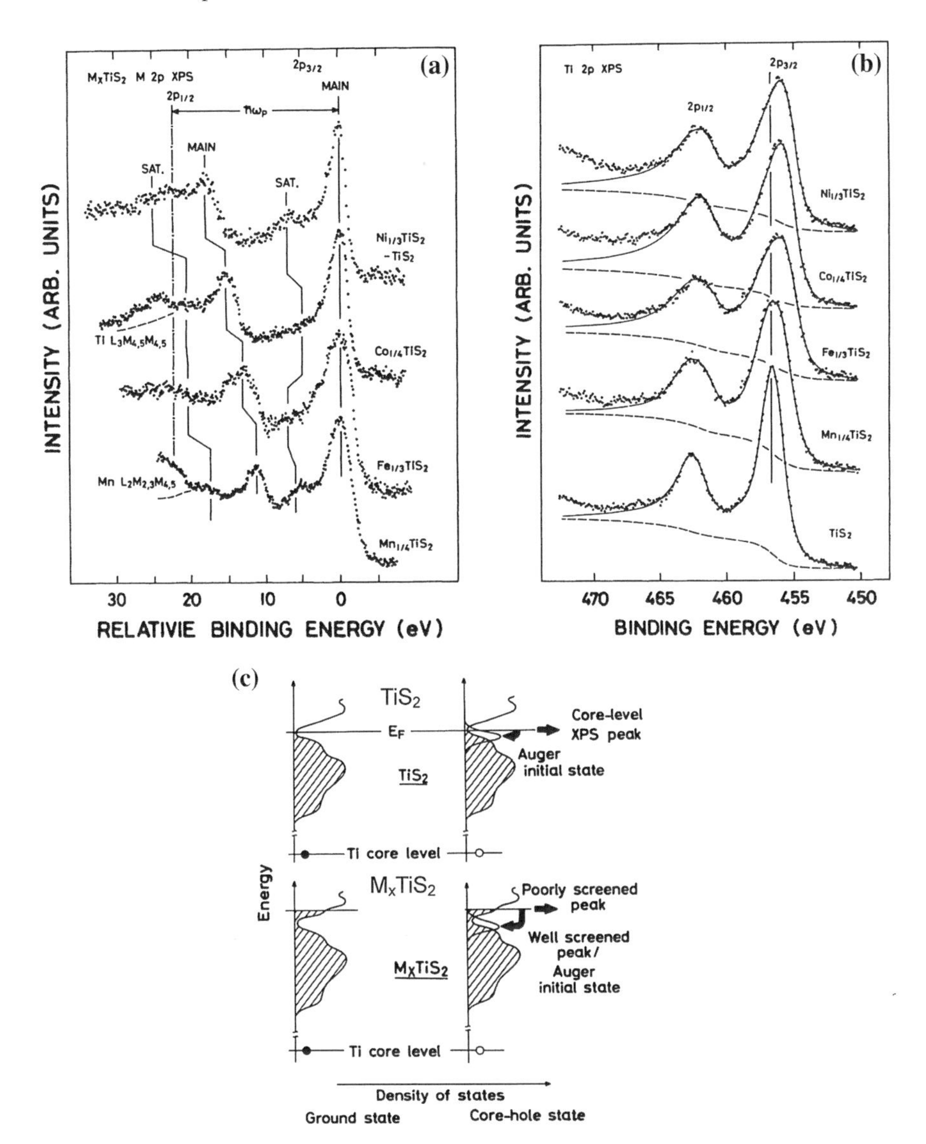

Fig. 5.10 **a** 2p core level XPS spectra of guest 3d transition metal atoms in M_xTiS_2. The energy scale was aligned at the $2p_{3/2}$ main peak and is given in relative E_B. $\hbar\omega_p$ roughly shows the bulk plasmon energies [37]. **b** Host Ti 2p core level XPS spectra of M_xTiS_2 compared with that in TiS_2. *Dots* represent the experimental results and the solid lines represent the fitted theoretical curves. **c** Screening process of the Ti core holes in M_xTiS_2 and TiS_2

subtracting the TiS_2 valence band spectrum of from that of the M_xTiS_2. Configuration interaction calculations on the MS_6^{10-} cluster model were performed to explain the M 3d states [37].

5.3 Multiplet Structures

Multiplet structures in PES were most clearly observed in rare earth (RE) systems with localized and open 4f shells. In many cases of RE compounds except for some Ce, Sm, Eu, and Yb compounds, the valence of the rare earth atom is very close to trivalence, where 4f electrons are rather localized and atomic-like. With the increase of the atomic number, the extension of 4f wave functions decreases as known lanthanide contraction. Their 4f PES spectra can therefore be predicted well by atomic multiplet calculations as reported in many studies [39, 40]. The 4f charge remains inside the mean extension of the rare earth 5s, 5p, 5d and 6s shells [40]. When the E_B of the 4f states is very small, however, they can in principle participate in the formation of bonds. Then the key issue in the studies of lanthanide compounds is to know the hybridization behavior. The RE 5d states form bands in solids and they determine the atomic radii which remain almost constant for most trivalent lanthanides. If the hybridization between the 4f states and the conduction band states is noticeable, the valence fluctuation (VF) can take place, where the multiplets resulting from different initial configurations ($|f^n\rangle$ and $|f^{n+1}\rangle$ or $|f^{n-1}\rangle$) can be observed by PES. Such examples were most clearly seen in some Sm systems. Temperature and composition dependent valence mixing of Sm in cation- and anion-substituted SmS was studied by XPS [41] as reproduced in Fig. 5.11(a–e). SmS was understood as predominantly divalent under atmospheric pressure. On the other hand SmAs was regarded a trivalence dominant compound. In $SmAs_{0.18}S_{0.82}$, however, both $|4f^5\rangle$ and $|4f^4\rangle$ final state multiplets were clearly observed. A similar situation was seen in $Sm_{0.81}Y_{0.19}S$, too. These mixed valence features and their temperature (T) dependence were found to be strongly related to the lattice constant behavior [41]. As discussed later in Chaps. 8 and 9, however, the separation of the bulk and surface contributions are desirable for detailed discussions.

Multiplet structures were also observed in PES of core levels of an evaporated Ce film with open outer shells [42]. Detailed studies of many intermetallic Ce compounds were so far performed. Typical results of Ce 3d core level XPS of various $CePd_x$ are summarized in Fig. 5.11(f) [43]. The hybridized $|3d^9 4f^0\rangle$ and $|3d^9 4f^2\rangle$ components are recognized in addition to the main $|3d^9 4f^1\rangle$ features. The relative intensities and energy positions depend upon such parameters as the magnitude of hybridization, f level energy, correlation energy, Coulomb interaction between the core hole and the outer electrons, exchange interaction, and so on [44–47]. In the case of the La metal and La compounds, there is no 4f electron in the ground state. Still the $3d_{5/2}$ core level XPS shows a doublet structure. This could only be explained by considering the lowering of the 4f level below E_F due to the Coulomb interaction with the core hole [38], where this f level was either occupied (well screened) by the conduction electron or remained empty (poorly screened). In the case of Ce systems, two types of core excited final states ($|4f^1\rangle$ and $|4f^2\rangle$) are expected starting from the $|4f^1\rangle$ ground state. If valence fluctuation is taking place in the ground state between the $|4f^0\rangle$ and $|4f^1\rangle$ configurations, the final states associated with both the $|4f^0\rangle$ ground state and the $|4f^1\rangle$ ground state were

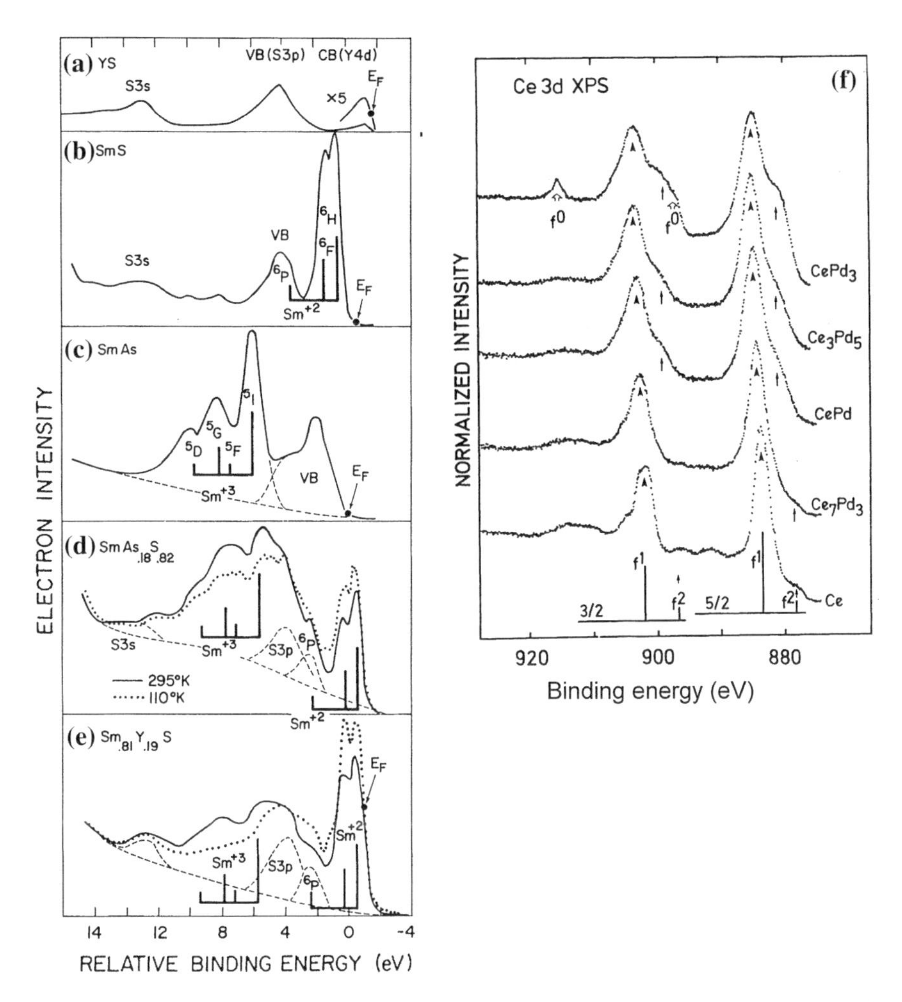

Fig. 5.11 **a–e** XPS spectra of anion or cation substituted SmS measured on cleaved surfaces with a resolution of 0.6 eV [41]. Solid line spectra were obtained at 295 K and the dotted line spectra in **d** and **e** were measured at 110 K. Energies are given in relative E_B. The results of the atomic multiplet calculations are displayed by the histogram. **f** Ce 3d XPS spectra of $CePd_x$ with the surfaces cleaned by Al_2O_3 file scraping [43]

expected to coexist providing three peaks corresponding to the $|4f^0\rangle$, $|4f^1\rangle$ and $|4f^2\rangle$ configurations as shown in the figure. It was noticed that an analysis within a single impurity Anderson model (SIAM) or configuration interaction cluster model calculation is effective for discussing bulk electronic structures, which could only be probed by high $h\nu$ photons providing high enough E_k to photoelectrons from deep core levels. This notice is quite important for systems with localized surface electronic structures and more itinerant bulk electronic structures.

The effect of hybridization is less important in heavy rare earth compounds except for some Sm, Eu and Yb systems. Therefore the 3d core level XPS was well explained by ionic model calculations (SIAM with negligible hybridization). In the case of the 4d core level XPS of lanthanide systems, however, equivalent calculations provided a less satisfactory reproduction of the observed spectra. In the case of Dy metal and Dy_2O_3, for example, the 4d core level XPS was considered to reflect the multiplet structure with a $|4d^9 4f^9\rangle$ configuration, where the spectral broadening was found to be very prominent for Dy_2O_3. The calculated result by SIAM convoluted with constant width turned out to be much different from the experimental results in the sense that the high E_B structures are much more suppressed in the experimental results. This discrepancy was understood as due to the multiplet term dependence of the lifetime of the 4d core level XPS final states induced by the 4d4f4f ($N_{4,5}N_{6,7}N_{6,7}$) super-Coster-Kronig decay, which could be numerically calculated [48]. Multiplet structures of the core level XPS are due to Coulomb and exchange interactions between the core hole and valence electrons. This rather general concept can also be applied to transition metal compounds.

References

1. C.R. Brundle, A.D. Baker (eds.), *Electron Spectroscopy: Theory Techniques and Applications I* (Academic, New York, 1977)
2. M. Cardona, L. Ley (eds.), *Photoemission in Solids I, Topics in Applied Physics*, vol. 26 (Springer, Berlin, New York, 1978)
3. L. Ley, M. Cardona (eds.), *Photoemission in Solids II, Topics in Applied Physics*, vol. 27 (Springer, Berlin, New York, 1978)
4. J.W. Allen, Resonant photoemission of solids with strongly correlated electrons, synchrotron radiation research, in *Advances in Surface and Interface Science, 1 Techniques*, ed. by R.Z. Bachrach (Plenum Press, New York, 1992)
5. S. Hüfner, *Photoelectron Spectroscopy* (Springer, Berlin, 2003)
6. M. Taniguchi, S. Suga, M. Seki, H. Sakamoto, H. Kanzaki, Y. Akahama, S. Terada, S. Endo, S. Narita, Solid State Commun. **45**, 59 (1983)
7. H. Asahina, K. Shindo, A. Morita, J. Phys. Soc. Jpn. **51**, 1193 (1982)
8. Y. Takao, H. Asahina, A. Morita, J. Phys. Soc. Jpn. **50**, 3362 (1981)
9. J.J. Yeh, I. Lindau, Atomic Data Nucl. Data Tables **32**, 1 (1985)
10. M. Taniguch, S. Suga, M. Seki, H. Sakamoto, H. Kanzaki, Y. Akahama, S. Endo, S. Terada, S. Narita, Solid State Commun. **49**, 867 (1984)
11. M. Taniguchi, J. Ghijsen, R.L. Johnson, S. Suga, Y. Akahama, S. Endo, Phys. Rev. **B39**, 11160 (1989)
12. T. Nakano, A. Kotani, J. Phys. Soc. Jpn. **55**, 2867 (1986)
13. T. Takahashi, Y. Harada, J. Non-Cryst. Solids **35, 36**, 1041 (1980)
14. S. Hino, T. Takahashi, Y. Harada, Solid State Commun. **35**, 379 (1980)
15. T. Ueno, A. Odajima, Jpn. J. Appl. Phys. **21**, 1382 (1982)
16. K. Inoue, M. Kobayashi, K. Murase, M. Taniguchi, S. Suga, Solid State Commun. **54**, 193 (1985)
17. S. Shin, S. Suga, M. Taniguchi, M. Fujisawa, H. Kanzaki, A. Fujimori, H. Daimon, Y. Ueda, K. Kosuge, S. Kachi, Phys. Rev. **B41**, 4993 (1990)

18. R.A. Pollak, D.E. Eastman, F.J. Himpsel, P. Heinmann, B. Reihl, Phys. Rev. B **24**, 7435 (1981)
19. A. Fujimori, M. Saeki, N. Kimizuka, M. Taniguchi, S. Suga, Phys. Rev. B **34**, 7318 (1986)
20. A. Fujimori, N. Kimizuka, T. Akahane, T. Chiba, S. Kimura, F. Minami, K. Siratori, M. Taniguchi, S. Ogawa, S. Suga, Phys. Rev. **B42**, 7580–7586 (1990)
21. A. Fujimori, N. Kimizuka, M. Taniguchi, S. Suga, Phys. Rev. B **36**, 6691 (1987)
22. M. Sing, S. Glawion, M. Schlachter, M.R. Scholz, K. Goß, J. Heidler, G. Berner, R. Claessen, Phys. Rev. Lett. **106**, 056403 (2011)
23. J. Voit, J. Phys. Condens. Matter **5**, 8305 (1993)
24. B. Dardel, D. Malterre, M. Grioni, P. Wibel, Y. Baer, F. Lévy, Phys. Rev. Lett. **67**, 3144 (1991)
25. C. Coluzza, H. Berger, P. Alméras, F. Gozzo, G. Margaritondo, G. Indlekofer, L. Forro, Y. Hwu, Phys. Rev. B **47**, 6625 (1993)
26. M. Nakamura, A. Sekiyama, H. Namatame, A. Fujimori, H. Yoshihara, T. Otani, M. Misu, M. Takano, Phys. Rev. B **49**, 16191 (1994)
27. B. Dardel, D. Malterre, M. Grioni, P. Weibel, Y. Baer, J. Voit, D. Jérôme, Europhys. Lett. **24**, 687 (1993)
28. A. Sekiyama, A. Fujimori, S. Aonuma, H. Sawa, R. Kato, Phys. Rev. B **51**, 13899 (1995)
29. H. Ishii, H. Kataura, H. Shiozawa, H. Yoshioka, H. Otsubo, Y. Takayama, T. Miyahara, S. Suzuki, Y. Achiba, M. Nakatake, T. Narimura, M. Higashiguchi, K. Shimada, H. Namatame, M. Taniguchi, Nature **426**, 540 (2003)
30. K.A. Gschneider, L. Eyring, S. Hüfner (eds.), *Handbook on the Physics and Chemistry of Rare Earths*, vol. 10 (Elsevier Science, New York, 1987)
31. D.E. Eastman, T.-C. Chiang, P. Heimann, F.J. Himpsel, Phys. Rev. Lett. **45**, 656 (1980)
32. T.A. Carlson, J.C. Carver, L.K. Saethre, F.G. Santibánez, G. Santibánez, G.A. Vernon, J. Electron Spectrosc. Relat. Phenom. **5**, 247 (1974)
33. B.W. Veal, A.P. Paulikas, Phys. Rev. Lett. **51**, 1995 (1983)
34. K. Okada, A. Kotani, Phys. Rev. B **52**, 4794 (1995)
35. K. Okada, A. Kotani, J. Electron Spectrosc. Rel. Phenom. **71**, R1 (1995)
36. P. Krüger, J.C. Parlebas, A. Kotani, Phys. Rev. B **59**, 15093 (1999)
37. A. Fujimori, S. Suga, H. Negishi, M. Inoue, Phys. Rev. B **38**, 3676 (1988)
38. A. Kotani, Y. Toyozawa, J. Phys. Soc. Jpn. **37**, 912 (1974)
39. M. Campagna, G.K. Wertheim, Y. Baer, Photoemission in solids II, in *Topics in Applied Physics*, vol. 27, ed. by L. Ley, M. Cardona (Springer, Heidelberg, New York, 1978)
40. Y. Baer, W.D. Schneider, See Fig. 8, for example, in Chap. 62, in *Handbook on the Physics and Chemistry of Rare Earths*, vol. 10, ed. by K.A. Gschneider Jr, L. Eyring, S. Hüfner (Elsevier, NY, 1987)
41. R.A. Pollak, F. Holtzberg, J.L. Freeouf, D.E. Eastman, Phys. Rev. Lett. **33**, 820 (1974)
42. I. Nagakura, T. Ishii, T. Sagawa, J. Phys. Soc. Jpn. **33**, 754 (1972)
43. J.C. Fuggle, F.U. Hillebrecht, Z. Zolnierek, R. Lásser, Ch. Freiburg, O. Gunnarsson, K. Schönhammer, Phys. Rev. **B27**, 7330 (1983)
44. O. Gunnarson, K. Schönhammer, Phys. Rev. B **28**, 4315 (1983)
45. A. Kotani, Inner shell photoelectron process in solids, in *Handbook on Synchrotron Radiation*, vol. 2, Chap. 2 ed. by G.V. Marr (Elsevier, NY, 1987).
46. A. Kotani, T. Jo, J.C. Parlebas, Adv. Phys. **37**, 37 (1988)
47. A. Kotani, J. Electron Spectrosc. Rel. Phenom. **100**, 75 (1999)
48. H. Ogasawara, A. Kotani, B.T. Thole, Phys. Rev. B **50**, 12332 (1994)

Chapter 6
Angle Resolved Photoelectron Spectroscopy in the hν Region of ~15 to 200 eV

The angle resolved photoelectron spectroscopy (ARPES) is performed in most cases on single crystal sample surfaces. Momentum information becomes available in ARPES, facilitating the experimental evaluation of band dispersions. By fully utilizing the energy distribution curves (EDCs) and momentum distribution curves (MDCs), the Fermi surface (FS) topology can be quantitatively estimated. By use of the polarized light excitation, the ARPES spectra can provide useful information on the parity of the initial states. In the low hν excitation in $\sim$15–200 eV, however, the λ_{mp} is rather short for the electrons from the region near E_F and the k_z (z $\perp$ surface) dependence of the band dispersions and FS topology is rather difficult to be accurately evaluated in most cases. Then 2D and 1D dispersions were mostly studied by ARPES up to now. Examples of layer materials, rare earth compounds, 1D materials, topological insulators, superconductors and quantum well states (QWS) are discussed in this order.

6.1 General

The clarification of "bulk" Fermi surfaces (FSs) is essential to understand the physical properties of a huge amount of new functional materials such as superconducting transition metal oxides, heavy fermion systems, and organic conductors. Quantum oscillation measurements by virtue of the de Haas-van Alphen and/or Shubnikov-de Haas effects are known as useful techniques to detect bulk FSs. However, the character of the FSs whether they were electron- or hole-like can not be directly judged by these measurements alone. In addition, low temperatures and almost defect-free single crystals are required. So these techniques are not easily applicable to doped or partially substituted systems such as the high-temperature superconductors $La_{2-x}Sr_xCuO_4$, $Bi_2Sr_2CaCu_2O_{8+\delta}$, or $Sr_{2-x}Ca_xRuO_4$. On the other hand, angle-resolved photoemission (ARPES) was known as a useful tool for probing FSs as well as QP dispersions [1–3].

S. Suga and A. Sekiyama, *Photoelectron Spectroscopy*,
Springer Series in Optical Sciences 176, DOI: 10.1007/978-3-642-37530-9_6,

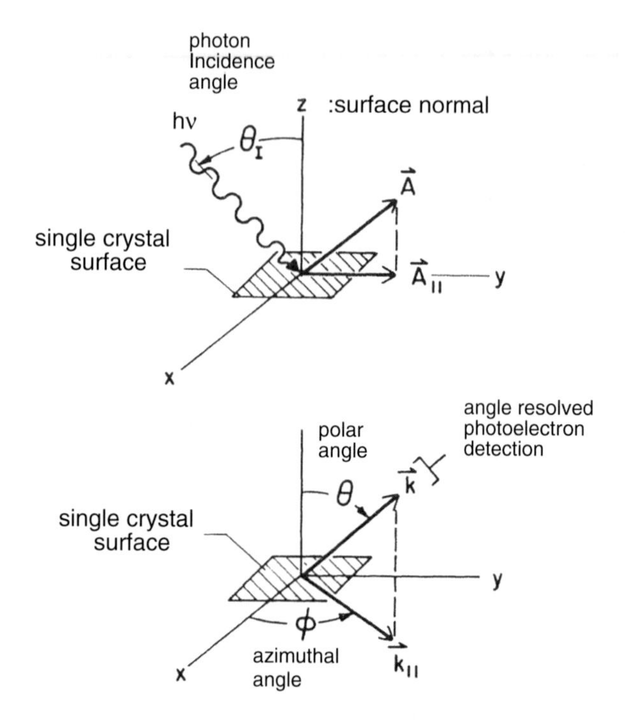

Fig. 6.1 Schematic drawing of the geometry of the polarization dependent angle resolved photoelectron spectroscopy [4]. The hatched region shows the crystal surface within the x–y plane, where the y axis is along the surface component of the vector potential **A** of the incident light

It is nowadays well known that the ARPES for single crystals depend strongly on the polarization of the incident excitation light. Fermi surface mapping may be facilitated owing to the matrix element effects and/or dipole selection rules by use of polarized lights. Dipole selection rules for optical transitions are considered first in regard to the symmetry of the electronic states probed by linearly polarized synchrotron radiation. The parity of photoelectrons propagating in the mirror symmetry plane is considered according to the scheme illustrated in Fig. 6.1 [4]. The z axis is the surface normal and the y axis is taken along the surface component of the vector potential **A**. The incidence angle of photons with respect to the surface normal is defined as θ_1.The direction of photoelectrons to be detected is defined by the polar angle θ with respect to the surface normal and the azimuthal angle ϕ defined with respect to the normal to the plane of incidence [4]. The photoelectron excitation matrix element is given by

$$\langle\psi_f|\mathbf{A}\cdot\mathbf{P}|\psi_i\rangle,$$

where ψ_f and ψ_i are the final and initial state wave functions and the dipole operator **A**·**P** is the scalar product of the vector potential **A** and the momentum operator **P**. If the electrons emitted in a mirror plane are detected, the detected final state wave function has even parity with respect to this plane. Hereafter $A_{||}$ denotes the direction of the surface component of the polarization vector **A**. If the light polarization vector **A** is parallel to the collection mirror plane, **A**·**P** is of even parity. If **A** is perpendicular to the collection mirror plane, on the other hand, **A**·**P**

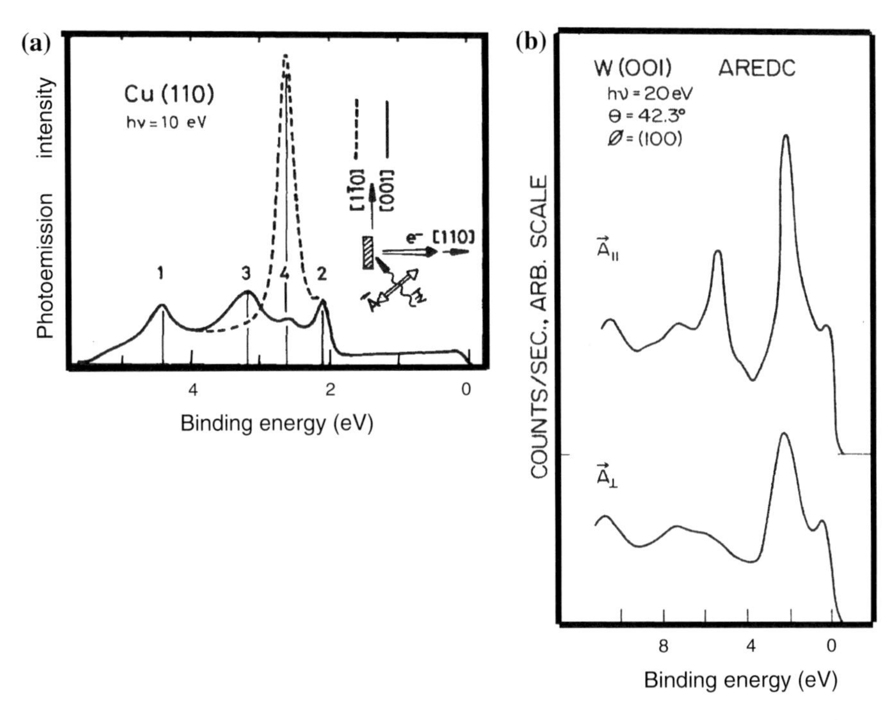

Fig. 6.2 **a** ARPES of Cu (110) measured at hν = 10 eV for p-polarized light with **A** parallel to [1$\bar{1}$0] and [001] [6]. **b** ARPES of W (001) measured at hν = 20 eV for both θ_I and $\theta = 42.3°$ and an azimuthal angle ϕ corresponding to the (100) direction in the (100) mirror plane. The results are for **A** parallel and perpendicular to the mirror plane. A clear polarization dependence is seen at $E_B \sim 5.4$ eV [8]

is of odd parity. Therefore a non vanishing matrix element is obtained for the even wave function ψ_i by excitation with $A_{||}$ in the mirror plane in the case of p-polarized light. In contrast if $A_{||}$of the p-polarized light is perpendicular to the collection mirror plane, both even and odd ψ_i states are detected. These parity rules are applicable to both surface and bulk electronic states [4]. If the spin–orbit coupling is considered, states with different parity can be mixed resulting in less prominent polarization effects [1].

In solids, group theoretical symmetry selection rules hold [5]. Such arguments were given for fcc, bcc and hcp structures. Figure 6.2(a) shows the results for normal emission from Cu(110) excited by p-polarized light with the polarization always in the plane of light incidence and the normally emitted photoelectrons [6]. By azimuthally rotating the sample by 90°, the surface parallel component of **A** could be set parallel to [001] or [110]. In this normal emission configuration the states along the Σ line are probed. Σ1, Σ2, Σ3 and Σ4 states associated with the high symmetry point states result from the Cu 3d states. The electron emission from Σ4 is allowed for **A** parallel to [110] and that from Σ3 for **A** parallel to [001] to the Σ1 symmetry final state. This correspondence can be well understood by considering the $x^2 - y^2$ character of the Σ4 state and the yz + zx character of the Σ3 state. Extensive studies were also made for Cu(100) and Cu (111) surfaces in the hν region from 5 to 35 eV [7].

Polarization dependent ARPES was also performed on other classes of materials. Figure 6.2(b) shows the results on W(001) at hν = 20 eV where the emission direction is in the (100) mirror plane [8]. It is known for k_i and k_f off the mirror plane that Bloch functions have no definite parity. In such a case, the emission intensity will not vanish for any direction of **A** in contrast to the present case. In the results in Fig. 6.2(b) on the clean W(001), a strong polarization dependence was observed at $E_B \sim 5.4$ eV. In order to identify the initial and final bands, the results were compared with the results of band-structure calculations [9], where only $k_{||}$ was thought to be conserved in the photoemission process. The $k_{||}$ was estimated as 1.1 $Å^{-1}$ for the structure at $E_B \sim 5.4$ eV for hν = 20 eV. Since the photon wave vector **q** is almost negligible in this case one should consider the nearly vertical transition along this $k_{||}$ ($k_\perp$ near the Δ point) in the BZ [8]. The parity of each state was obtained in the BZ region with negligible spin–orbit effects from the nonrelativistic band-structure calculation [10]. Both initial and final states were found to have even parity in consistence with the experimental observation. Similar measurements were performed at hν = 16 eV for the emission in the (100) mirror plane and at hν = 13 eV in the (110) mirror plane. A strong structure was observed at $E_B \sim 3.4$ eV for $A_\perp$ in both cases. Since the final state corresponding to this E_B could have only even parity, the idea of an odd parity initial state near the N point of $k_\perp$ was derived from the hν = 16 eV result [8].

The measurements were also performed on a W(001) surface saturated with H_2. Many of the chemisorption induced peaks exhibited polarization dependence similar to clean W. Although symmetry arguments were applicable to both localized and band like states, there was ample evidence that the hydrogen states formed bands with 2D character [11]. The polarization dependence was then discussed by considering the proposed hydrogen bands and their symmetries. The polarization measurements were also applied to Ni and CO chemisorbed Ni and so on. The experimental results could be used to check the validity of theoretical models.

6.2 Layered Materials

Many ARPES studies were so far reported for various quasi-2D layered materials, where the dispersion along $k_\perp$ could be neglected. ARPES was commonly applied to evaluate the shape of Fermi surfaces (FSs) from the maximum PES intensity at E_F in EDC or the discontinuity in the integrated MDCs for determining k_F parallel to the surface. However, these criteria induce large uncertainties, in particular for narrow band systems. So a reliable method called a ΔT method for the determination of k_F from high resolution ARPES at different temperatures (T) was proposed [12] and the accuracy of this method was discussed on the ARPES data of the quasi 2D system 1T-$TiTe_2$. This material does not show a charge density wave (CDW) and exhibits Ti 3d conduction band emissions well separated from other bands. The angular precision was better than 0.1° and the energy resolution was set to 30 meV at $h\nu = 21.2$ eV. The accuracy of E_F was ±1 meV. Spectra measured at various T (30, 100, 190 K) were analyzed.

The photoemission intensity $I(k, \omega)$ can be expressed within the sudden approximation as $I(k, \omega) = I_0(k, \omega) \cdot f(\omega) \cdot A(k, \omega)$, where $f(\omega)$ is the Fermi–Dirac distribution function (FD) and $A(k, \omega)$ the one-particle spectral function (imaginary part of the Green's retarded function). The prefactor $I_0(k, \omega)$ involves the transition matrix element and is therefore **k** dependent. This **k** dependence influences the reliability of the k_F determination. The $k_{F\parallel}$ could be identified by the peak of $A(k, \omega)$ at $E_B = 0$ eV. Since the photocurrent corresponds to $f(\omega) \cdot A(k, \omega)$, it is not trivial to determine the peak of $A(k, \omega)$. The difference of the FD for the temperatures T_1 and T_2 should be an odd function of ω and $A(\mathbf{k}_F, \omega)$ is an even function of ω. Therefore their product becomes odd and its integral over a symmetric energy window should vanish [12]. In practice, the integration could be replaced by the finite energy resolution defined by the symmetric analyzer function w, which vanishes outside $[-\epsilon, \epsilon]$. Then the difference of the intensities at $E_F = 0$ is expressed as

$$\Delta I(k_\parallel) = I_0(k_\parallel) \int_{-\epsilon}^{\epsilon} A(k, \omega)[f_{T1}(\omega) - f_{T2}(\omega)]w(\omega)d\omega. \tag{6.1}$$

$\mathbf{k}_\parallel = \mathbf{k}_{F\parallel}$ is then derived from $\Delta I(\mathbf{k}_\parallel) = 0$. This should be valid for all A which satisfy in $[-\epsilon, \epsilon]$ that $A(\mathbf{k}_F, \omega) = A(\mathbf{k}_F, -\omega)$ and $A(\mathbf{k}_F, \omega) \neq A(\mathbf{k}_F, -\omega)$ for $\mathbf{k} \neq \mathbf{k}_F$. At general **k** different from $\mathbf{k}_F$, $\Delta I \neq 0$ would be obtained. These conditions are fulfilled in a variety of spectral functions such as Lorentzians, Gaussians, Voigt profiles and so on [12]. **k** dependence of I_0 might lower the accuracy of this method. This possibility could directly be excluded when the experimental $k_\parallel$ resolution windows lies well within the regime where point symmetry of $\Delta I(\mathbf{k}_\parallel)$ around $\Delta I = 0$ is observed in the experimental data. A simulation by the ΔT method for both a Gaussian profile and the Luttinger model showed that the Fermi vector $k_\parallel$ was exactly determined from $\Delta I = 0$.

ARPES for 1T-$TiTe_2$ at $T = 30$ K along the ΓM direction is shown in Fig. 6.3(a). The k_F values so far reported were between 14° (0.51 $Å^{-1}$) and 14.75° (0.53 $Å^{-1}$) [13–15]. An excellent fit could already be achieved for all emission angles by employing simple Voigt profiles times the FD function convoluted by the spectrometer resolution as shown in Fig. 6.3(a), where the band crossing of the peak positions was found at 16.7° (0.60 $Å^{-1}$) (Fig. 6.3(b)). By using the Luttinger profile, however, the band crossing was still observed at 14.75° as in Fig. 6.3(b). The ΔT method was applied to the ARPES results measured at 100 and 190 K shown in Fig. 6.3(c). Intensities at E_F and intensities integrated over the whole spectrum are shown in Fig. 6.3(d) and (f). According to the ΔT method, k_F is given by the intersection of the curves and zero crossing in Fig. 6.3(d) and (e), which was clearly identified as 16.6 ± 0.1° (0.598 ± 0.004 $Å^{-1}$). On the other hand, the k_F

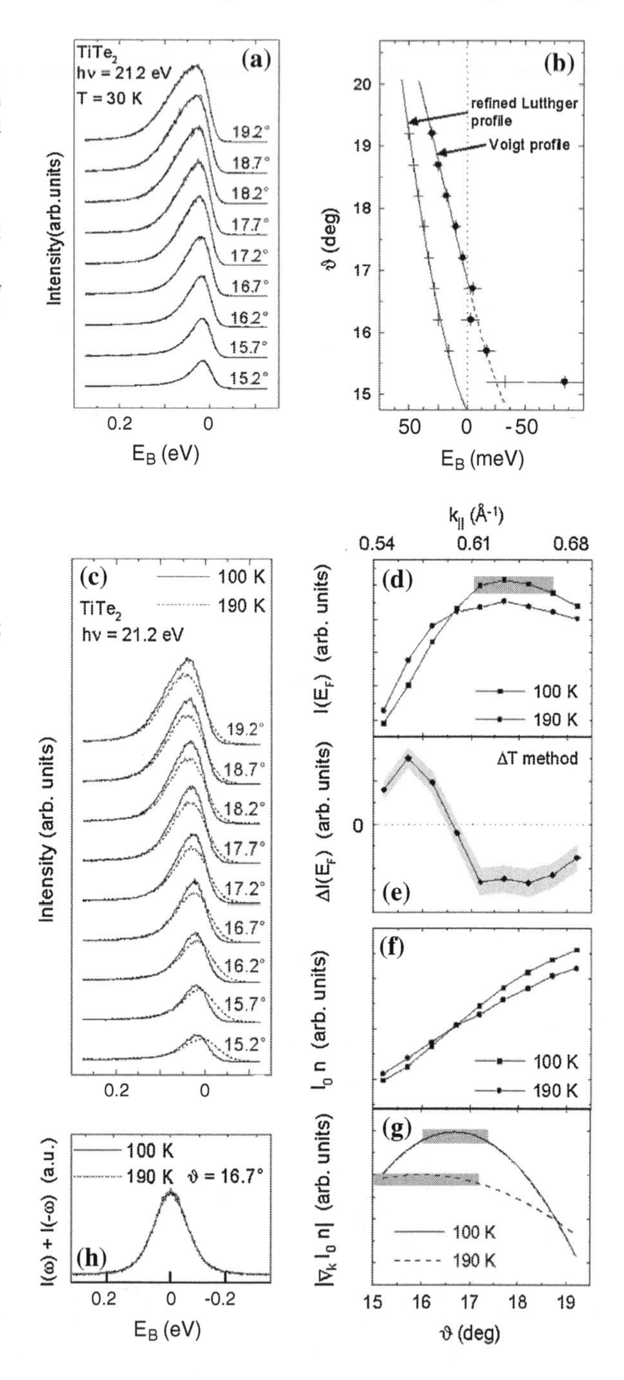

Fig. 6.3 ARPES of 1T-$TiTe_2$ at $h\nu = 21.2$ eV along the ΓM direction. Spectra are fit by Voigt profiles in (**a**) and the band dispersion obtained by fitting with Voigt and refined Luttinger profiles is shown in (**b**) [12]. ARPES at T = 100 and 190 K are shown in (**c**). **d** Intensity at E_F as a function of the emission angle. **e** The intersection corresponds to k_F by the ΔT method. **f** Intensities integrated over the whole spectrum at two different T. **g** k derivative of the integrated intensity. Grey bars show the uncertainties for the determination of k_F (95 % of the maximum value). **h** Symmetrized spectra at $\vartheta = 16.7°$ showing no difference between 100 and 190 K

derived from the maximum intensity at E_F was around 17.8°. The maximum gradient in Fig. 6.3(g) is marginally changing in the range of interest making a quantitative evaluation difficult. Thus a reliable and simple method to determine Fermi vectors by ARPES became feasible by means of the ΔT method, though it is not practically applicable to a system with a phase transition at a certain T.

As shown by Eq. (2.36) in Chap. 2, the QP spectral function $A(\mathbf{k}, \omega)$ is given by

$$A(\mathbf{k}, \omega) \propto \frac{\mathrm{Im}\,\Sigma(\mathbf{k}, \omega)}{[\omega - \varepsilon_{\mathbf{k}} - \mathrm{Re}\,\Sigma(\mathbf{k}, \omega)]^2 + [\mathrm{Im}\,\Sigma(\mathbf{k}, \omega)]^2} \quad (6.2)$$

where $\Sigma(\mathbf{k}, \omega)$ is the QP self energy reflecting all the many body interactions such as (1) electron–electron scattering, (2) electron–phonon scattering and (3) impurity (defect) scattering. The lifetime of the photoexcited valence hole or QP is given by Im $\Sigma(\mathbf{k}, \omega)$, to which the processes (1), (2) and (3) contribute. These terms are represented by Γ_{e-e}, Γ_{e-ph} and Γ_{e-imp}. Although the QP concept is restricted to T $\sim$ 0 K and in a narrow region around E_F, it is somehow useful also at finite T and in the energies away from E_F except for possible non-Fermi-liquid behavior in some organic 1D conductors [16] and in the normal state of high temperature superconductors close to T_C [17]. The (3) process is elastic and Γ_{e-imp} is proportional to the impurity concentration but independent of energy and T to a first approximation. In contrast, (1) $\Gamma_{e-e}(\omega, T) = 2\beta\left[(\pi k_B T)^2 + \omega^2\right]$ in 3D systems, while the quadratic terms are modified in 2D systems by a logarithmic factor [18]. Since β is often too small, however, the contribution of Γ_{e-e} was not accurately known until recently. On the other hand, (2) $\Gamma_{e-ph}(\omega, T)$ increases almost linearly with T at high T with a slope of $2\pi\lambda_{e-ph}k_B$, where λ_{e-ph} is the electron–phonon coupling constant. For most metals, the cut off energy of phonons lies in the range of 20–100 meV and λ_{e-ph} is between 0.1 and 1.5. If the mechanisms (1)–(3) are independent, the total scattering rate is given by $\Gamma_{tot} = \Gamma_{e-e} + \Gamma_{e-ph} + \Gamma_{e-imp}$. All T dependence in Γ_{tot} is thought to be resulting from λ_{e-ph}, while the energy dependence is divided into two distinct regions. In a narrow ω region smaller than the phonon cut-off energy, the λ_{e-ph} term is the only significant term. Any measurable change outside this region is most likely resulting from the Γ_{e-e} contribution. The real part of the QP self energy is also strongly influenced by the electron–phonon coupling effects, which can change the QP mass in a narrow region around E_F.

Before moving to layered Mo dichalcogenide compounds, the self-energy of the surface QP on Mo (110) is briefly discussed [19]. ARPES was measured at $h\nu \sim$ 15 eV along the $\overline{\Gamma}$–$\overline{N}$ symmetry line of the surface BZ (SBZ). Information on $A(\mathbf{k}, \omega)$ can be obtained by energy distribution curves, EDCs or momentum distribution curves, MDCs. The QP dispersion was estimated from the MDCs and the QP width was evaluated from the EDCs [19]. Figure 6.4(a–c) shows the results. The ImΣ and ReΣ at T = 70 K are plotted in Fig. 6.4(d). The peak width increases rapidly within 40 meV from E_F, followed by a slower increase at higher E_B. The sharp increase near E_F reflects the electron–phonon scattering. The result that the

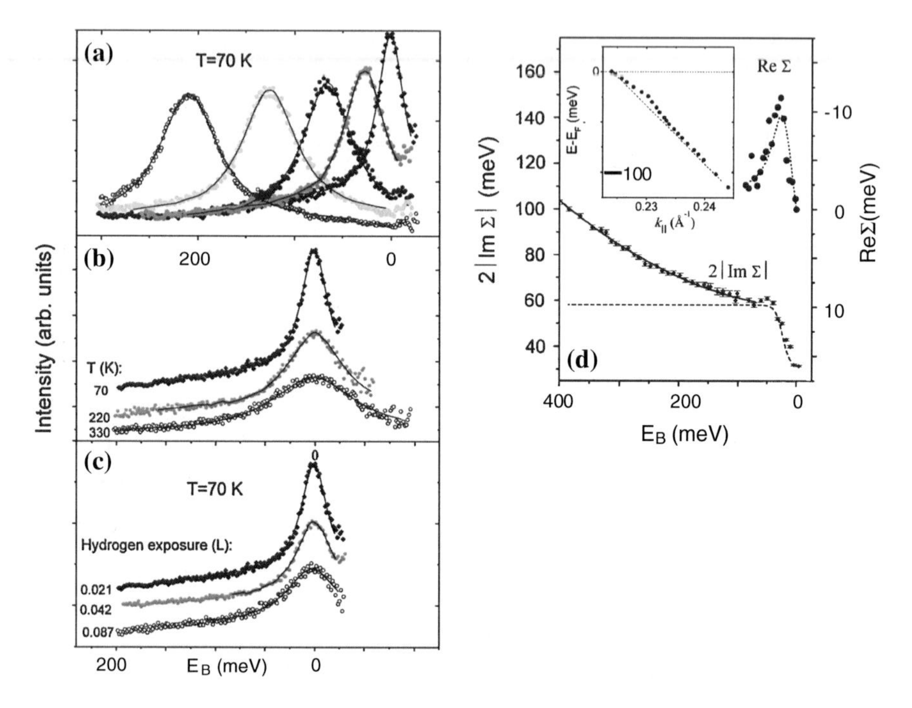

Fig. 6.4 θ dependence of the surface QP peak on the Mo (110) at hν = 15.16 eV along $\bar{\Gamma}$–$\bar{\text{N}}$ in the surface BZ [19]. Spectra are divided by the FD function (**a**). **b** T dependence and (**c**) hydrogen exposure dependence. **d** E_B dependence of the photohole self energy at T = 70 K

QP width always has a constant term (26 meV) indicates the contribution of the impurity scattering. If the calculated electron–phonon scattering contribution (dashed line in Fig. 6.4(d)) is subtracted from the total width, a monotonic increase in E_B is left above $E_B \sim 80$ meV as well fitted with a parabola. This component was thought to be resulting from the electron–electron scattering term. If a linear contribution was subtracted from ReΣ, the electron–phonon contribution to ReΣ could be evaluated.

Layered Mo dichalcogenides (i.e., MoS_2, $MoSe_2$, and $MoTe_2$) attract wide attention from the standpoint of application, for example, as electrodes in high efficiency photoelectrochemical cells in the case of α-$MoTe_2$. In this material Te-Mo-Te sheets are stacked as shown in Fig. 6.5(a) and its BZ is illustrated in Fig. 6.5(b). This material is a semiconductor with an indirect gap of 1.0 eV and a direct gap of 1.1 eV. ARPES was measured on a layered α-$MoTe_2$ in the hν region from 11.0 to 29 eV for different θ in order to study its precise band structure beyond the analysis based on the assumption of flat bands (with no $k_\perp$ dependence) [20]. Normal emission spectra in the form of EDC with $k_{||} = 0$ for hν = 11–29 eV are shown in Fig. 6.5(c). The E_B of the peak closest to E_F shows E_B minima at hν ~ 15.75 and 26 eV, corresponding to high symmetry points in the BZ, namely the Γ point in this case. The inner potential was then evaluated as

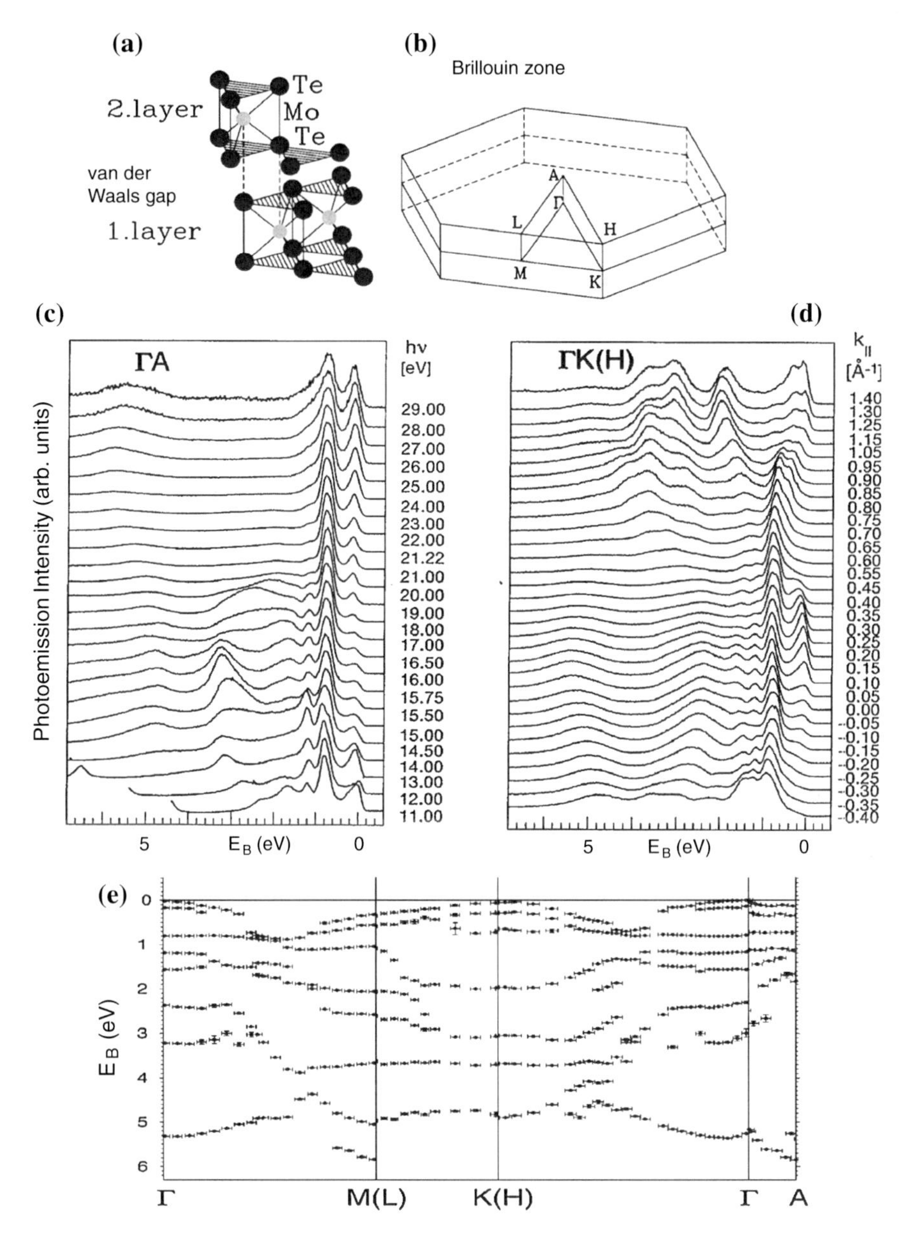

Fig. 6.5 Crystal structure (**a**) and the first BZ (**b**) of α-$MoTe_2$. (**c**) ARPES measured at hν = 11−29 eV with the total energy resolution of ∼60 meV. (**d**) CFS along the ΓK(H) direction [20]. **e** Band dispersions along the high symmetry lines of α-$MoTe_2$

$V_0 = 14 \pm 1$ eV. Then other ARPES spectra were measured in the constant-final-state (CFS) mode [20] for a certain E_K while scanning the hν. In this case, both $k_{||}$ and $k_{\perp}$ are thought to be kept constant during the hν scan. A new spectrum at a different $k_{||}$ could be measured by choosing E_K and θ for which $k_\perp$ remains constant. CFS ARPES spectra along the ΓK(H) direction are shown in Fig. 6.5(d). Since $k_\perp$ was taken into account, the results of the band dispersions shown in Fig. 6.5(e) along several high symmetry lines in the BZ are rather reliable. It is generally accepted that the VB consists of Mo 4d and Te p/d states. The two topmost emission features at the Γ point can be understood to be resulting from the mixing between Mo 4d and Te 5p orbitals. On the half way to the M and H point, these bands disperse noticeably to higher E_B and merge with the third band as recognized in Fig. 6.5(d) and (e). This third band located near $E_B = 1$ eV was ascribed to a state with Mo $4d_{z^2}$ character.

6.3 Rare Earth Compounds

Low hν ARPES was also widely applied to study the band dispersions in various rare earth (RE) compounds. Since the localized RE lanthanide 4f states become often more localized near the surface, one should always pay attention to the surface sensitivity of the low hν ARPES as later discussed in Chap. 7. In addition, the dispersions related to the RE 4f states are usually much smaller than those of the d states. So much higher energy resolution is required to study the 4f band dispersion for RE compounds. The first example is LaSb, which has no 4f electrons and is regarded as a nonmagnetic reference to CeSb and USb. ARPES was performed at $h\nu = 21.2$ eV at 30 K with an energy resolution of 50 meV [21]. Noticeable dispersions were observed down to $E_B \sim 3$ eV. The results in a narrow E_B region near E_F are reproduced in Fig. 6.6(a) in comparison with the results obtained for the paramagnetic (PM) and antiferroparamagnetic (AFPM) phases of CeSb with one 4f electron shown in Fig. 6.6(b) and (c) [22], where θ is the polar angle from the surface normal of the (001) plane. In the results for LaSb in Fig. 6.6(a), three electron FSs and two hole FSs were clarified. The experimental results were compared with the 1D DOS along $k_\perp$ calculated by the LMTO method with local-density approximation (LDA) for the ΓXXX plane [21]. The 1D DOS was calculated at each $k_{||}$. The calculation predicted two hole pockets with dominant Sb $5p_{3/2}$ contribution at the Γ point and an electron pocket with La 5d character at the X point, making LaSb a semimetal. A qualitative agreement was seen between the calculated and ARPES results. The present ARPES results were also compared with the results of dHvA measurements. Similarity to and differences from the results of calculation and dHvA measurements were discussed [21].

Ce monopnictides show very complicated magnetic phase transitions with temperature T and/or magnetic field B. Anisotropic p-f mixing with the crystal field effect on the Ce 4f level plays an essential role in such phase transitions. CeSb shows a first order transition at $T_N = 16$ K from a PM to an AFPM phase, which is further followed by several different AFPM phases and then changed into an antiferromagnetic (AFM) phase at 6 K. The ARPES results in the PM and AFPM phases at T = 30 (PM) and 10 K (AFPM) show slight differences as recognized in

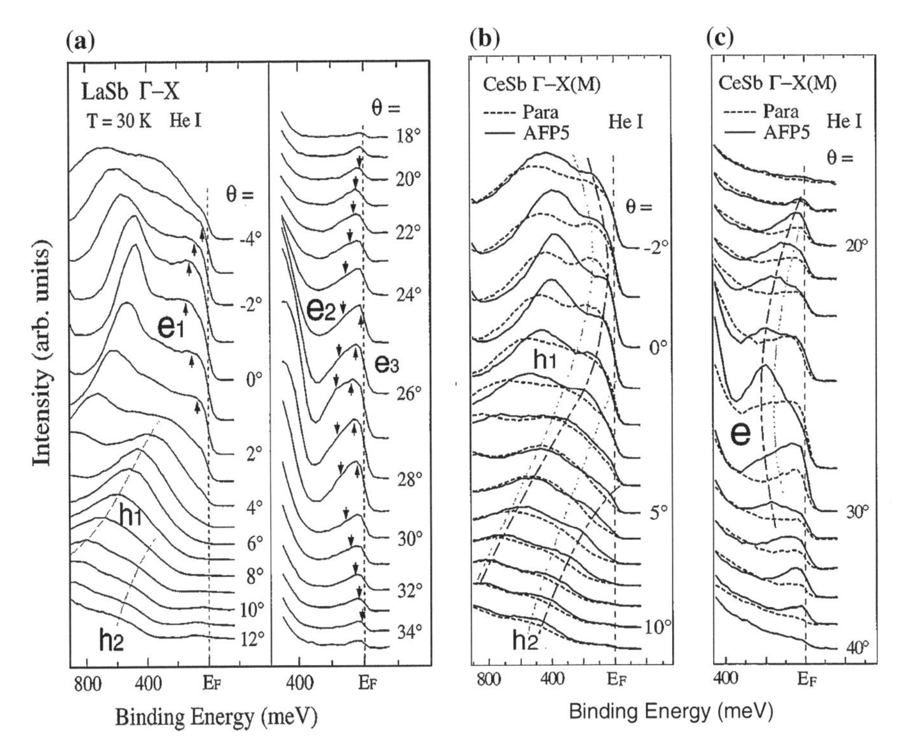

Fig. 6.6 ARPES at hν = 21.2 eV and 30 K on cleaved surfaces of (**a**) LaSb [21]. **b** ARPES of CeSb at hν = 21.2 eV measured at 30 K (paramagnetic phase) and (**b**, **c**) ARPES of CeSb at hν = 21.2 eV measured at 30 K (paramagnetic phase) and 10 K (antiferroparamagnetic phase) [22]

Fig. 6.6(b) and (c). However, the AFPM magnetic ordering seems to have not induced symmetry lowering. Very recently, however, extremely low hν ARPES [23] was performed on CeSb at T = 30, 15 and 5 K, where clear differences were observed between PM, AFPM and AFM phases [24], suggesting a close correlation between the near E_F band dispersions and the magnetic ordering. The bulk sensitivity and high resolution are thus the key issues of ARPES for strongly correlated electron systems.

Extensive ARPES studies were also performed for XRu_2Si_2 (X = La, Ce, Th, U) isostructural layered compounds. They have different nominal f occupations of $|4f^0\rangle$, $|4f^1\rangle$, $|5f^0\rangle$ and $|5f^2\rangle$. The 4d and 5d edge RPES measurements were also performed to distinguish the f-character in the complex valence bands. The Fermi-level intensity mapping of FS contours was performed in the hν region between 14 and 200 eV [25]. $CeRu_2Si_2$ is a $|4f^1\rangle$ heavy fermion system with a large specific heat coefficient of $\gamma = 350$ mJ/mol·K^2 with a Kondo temperature $T_K \sim 20$ K, which is discussed in Sect. 7.1.a in more detail. URu_2Si_2 is a nominally $|5f^2\rangle$ system with a moderately large $\gamma = 180$ mJ/mol·K^2, whereas $LaRu_2Si_2$ and

$ThRu_2Si_2$ are $|f^0\rangle$ systems with a nominal valence of 3+ and 4+, respectively. In addition, URu_2Si_2 is in a "hidden" order below 17.5 K, whose origin is still unclear for more than a quarter century, and superconducting below 1.2 K. Therefore this material attracts wide interest among the U compounds. The ARPES experiments on these materials were performed at T $\approx$ 25 K in the hν region of 14–35 eV with an energy resolution ΔE of 50–100 meV and an angular resolution $\Delta\theta$ of 2°. ARPES in the region of hν = 80–200 eV was performed with ΔE of 60 meV and $\Delta\theta$ of 0.7°, where the lanthanide samples were cleaved and measured at T $\leq$ 150 K, while high resolution ARPES on URu_2Si_2 was performed at T $\geq$ 20 K with ΔE of 20 meV and $\Delta\theta$ of 0.36°.

ARPES were measured at a fixed hν by rotating either the analyzer or the sample. Then the information along the spherical arc as shown in Fig. 6.7(a) could be obtained. In this case, $k_\perp$ was not constant in contrast to the case explained in Sect. 6.2. Therefore one should pay attention to the deviation from the high symmetry points for relatively low hν. If the deviation from the high symmetry point was noticeable, the point above or below the high symmetry point was expressed by adding a quotation mark like "Γ". $k_\perp$ could be estimated by determining V_0 from the hν dependence of the normal emission spectra. In the case of URu_2Si_2, V_0 was evaluated to be 12 eV. The direct evaluation of the sizes of the FSs is thus influenced by the effect of $k_\perp$ in the case of low hν ARPES. This effect might be somehow alleviated by the k_z broadening resulting from the finite λ_{mp}, which blurs the three dimensionality of the FS topologies.

ARPES for $LaRu_2Si_2$ was performed at hν = 122 and 152 eV, which correspond to the Z and Γ points at normal emission. 120° azimuthal angles (ϕ) and 30° polar angles (θ) were covered by the measurements and then a two-fold symmetrization was made to show the FS topology. It was found that 3 main pieces of FSs were repeated in multiple BZs. Namely small circular FSs centered at Γ and Z points and a large FS centered at Z. Although not clear for the small FSs, three dimensionality was seen for the large FS. Namely, the large FS contour around the Z-point at hν = 122 eV normal emission ($\theta = 0^\circ$) shifted to Z-point centers in the outer BZs ($\theta \neq 0^\circ$) at hν = 152 eV, in consistence with the stacking of body centered tetragonal (bct) BZs.

In the case of $CeRu_2Si_2$, ARPES was performed at hν = 154, 122 and 91 eV. The 4f states were thought to contribute also to the ARPES results. The band dispersions for $E_B > 0.5$ eV were rather similar to those in $LaRu_2Si_2$. The major difference from $LaRu_2Si_2$ was the noticeable spectral weight at E_F over large regions of the FS at hν = 154 eV. This intensity was thought to be necessarily originating from the Ce 4f spectral weight. ARPES at hν = 91 eV under off-resonance condition could reveal the near E_F d-band dispersions. As recognized in Fig. 6.7(a), (b) and (d), the large FSs centered on the second BZs Z-points extending into the central BZ are dim but they were still recognized to be similar to $LaRu_2Si_2$. Noticeable intensity in the form of pairs of straight segments was observed on both sides of the X-points (Fig. 6.7(b)). In addition, bright intensity was observed at the central Γ point and intense closed contours were seen in the second BZ near "Γ" in Fig. 6.7(b).

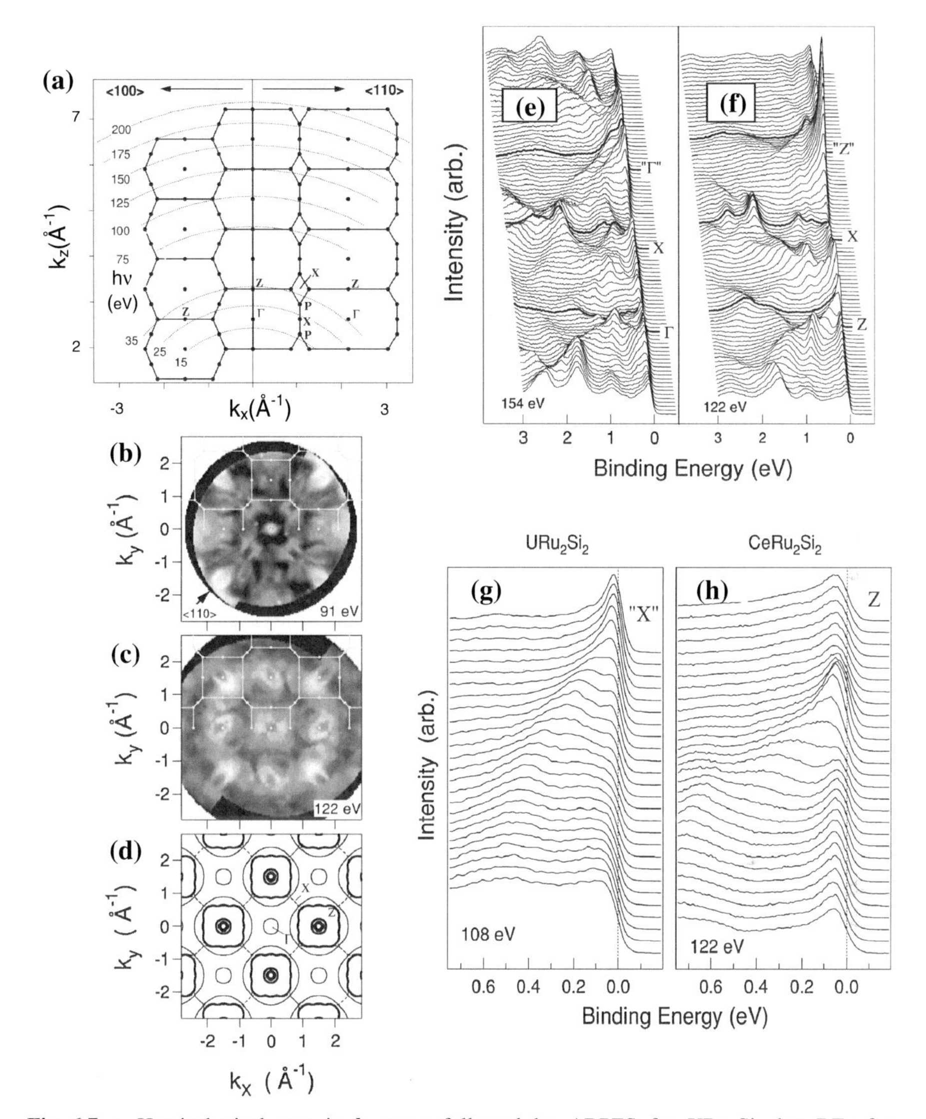

Fig. 6.7 **a** Hemispherical arcs in **k** space followed by ARPES for XRu_2Si_2 bct BZs for $h\nu = 15-35$ eV and 75–200 eV. Cross sections along the [100] and [110] directions are illustrated. **b** and **c** show the off- and on-resonance intensity maps of $CeRu_2Si_2$ at E_F at $h\nu = 91$ and 122 eV. **d** Theoretical prediction of contours of hole and electron FSs in $CeRu_2Si_2$ drawn by bold and thin lines. **e** and **f** show the valence band spectra of $CeRu_2Si_2$ under the off- and on-resonance excitation at $h\nu = 154$ and 122 eV along the $\langle 110 \rangle$ direction displaying large spectral weight near E_F. **g** and **h** show on-resonance ARPES spectra near the d-band E_F-crossing in URu_2Si_2 compared with those of $CeRu_2Si_2$ [25]

Compared with the LDA calculation in Fig. 6.7(d), qualitative agreement was recognized in Fig. 6.7(b) for the small closed FS contours at the Γ and Z points. As for the large Z-point FS contour, however, the interpretation as an electron FS theoretically predicted in Fig. 6.7(d) was doubted. On the contrary, it was suggested to be a large hole FS appeared in the ARPES in Fig. 6.7(b). Such an interpretation based on the similarity between La and Ce compounds was supported by the Z–X–Z dispersions clearly showing the single hole band on each side of the X point as was appropriate for the single large hole FS of the La compound (see Fig. 9(d) of [25]). It was surprising that a FS, essentially the same as that of the La compound, was observed in contrast to dHvA results which were measured at $T < 3$ K in the Fermi liquid regime. The authors of [25] claimed that this might be because the ARPES was measured at $T \gg T_K$, where T was 6–7 times higher than $T_K \sim 20$ K.

In order to directly study the contribution of the Ce 4f states, the 4d → 4f resonance ARPES was then performed at $h\nu = 122$ eV corresponding to the Z point at normal emission (Fig. 6.7(c), (f)), where no 4f weight was observed on the large hole surface around the Z point. On the other hand, the 4f spectral weight was observed on the elliptical contours centered at the Γ and Z points (Fig. 6.7(c)). These 4f-character ellipses have a size and shape that closely match the strongest d-character contours seen in Fig. 6.7(b) under the off-resonance condition, with close correspondence to the locations of the d-band FSs except for the large hole pocket seen in Fig. 6.7(b). In resonance ARPES along the Z–X–“Z” in Fig. 6.7(f), 4f weight is especially enhanced near the normal emission Z point with a clear correspondence to the location of the E_F crossing of a hole-like d-band as explained before. In addition, more prominent enhancement of the 4f weight was observed on one side of the “Z” point in the 2nd BZ (Fig. 6.7(f)) as also recognized in Fig. 6.7(c). The correspondence of this strong enhancement to a distinct elliptical FS contour feature in the FS map in Fig. 6.7(c) strongly suggests the relation between the d-band E_F crossing and the 4f spectral weight. While the E_F peak exhibits dramatic intensity variation as a function of θ and ϕ, no distinct dispersion of spectral weight away from E_F was observed except for the notable E_F crossing of a distinct d-band hole pocket centered on the Z point as mentioned before. This location represents a fairly isolated region with 4f hybridization to a d-band. Hence the observed unique dispersion was speculated to be reflecting the mixing between the f- and d-character states.

Similar ARPES studies were made for $ThRu_2Si_2$ and URu_2Si_2. An isolated d-band hole pocket was observed at the “X” point in URu_2Si_2, providing a good opportunity to study the d-f mixing by means of resonance ARPES. U 5f states were enhanced under the resonance ARPES at $h\nu = 112$ eV, where the dominant 5f weight was found to be confined to the interior of the off-resonance d-character hole pockets. The dispersion of the f states was thought to occur through hybridization with dispersing s-p-d states. In the vicinity of the d-band E_F crossing, a two band mixing model was employed to discuss the dispersing behaviors. In this model, f-weight is expected on the unoccupied side of the d-band dispersion or at the interior of the d-band hole pocket. The high resolution resonance ARPES of URu_2Si_2 at $h\nu = 108$ eV near the “X” point (Fig. 6.7(g)) was compared with those of $CeRu_2Si_2$ at hν = 122 eV near the Z point (Fig. 6.7(h)). While the URu_2Si_2 spectra in Fig. 6.7(g) show a complete lack of the 5f-weight outside the hole-pocket, $CeRu_2Si_2$ displays a different behavior. Namely, 4f-weight is observable both interior and exterior to the Z hole pocket, and it is slightly suppressed near the E_F crossing of the d-band. Such a difference could be explained by considering the difference of the renormalized 4f level energy ε_f' and the d-f hybridization V'. It is known that the hybridization is much weaker in $CeRu_2Si_2$ than in URu_2Si_2.

6.4 One Dimensional Materials

The electronic structures of quasi-1D materials attract wide interest since the quantum phenomenon called spin-charge separation was discovered in the theoretical framework of the 1D Hubbard model [26]. It was shown that the low energy excitations in a 1D system are not due to QPs with charge e and spin 1/2 in a Fermi liquid but decoupled excitations of spin and charge called spinons and holons as illustrated in Fig. 6.8(a). Separation of the spin and charge degrees of freedom for a single electron corresponds to a decay of a hole into a holon and a spinon. Experimental studies of this phenomenon are significant since the holons and spinons can be considered to be new elementary particles in solids.

ARPES is a powerful approach to study such a behavior of electrons in 1D systems. ARPES on $SrCuO_2$ (Fig. 6.8(b)), which is a 1D AFM charge-transfer-type insulator with an optical gap of 1.8 eV was first performed at hν = 22.4 eV [27, 28]. The two branches expected between $k_{||}$ = 0 to $\pi/(2c)$ were, however, not clearly resolved at this hν possibly due to the strong overlap with the small E_B tail of the O states located near E_B = 2.3 eV. Then ARPES was performed at 7 different hν between 85 and 200 eV (or at 7 different $k_\perp$ values from −0.3 to 0.85 $Å^{-1}$) covering the wide $k_{||}$ region from around $-\pi/(3c)$ to beyond $4\pi/c$ (or −0.3–4.0 A^{-1}) [28, 29]. Since the contribution of the O states in the main valence band measured in the region hν = 85–200 eV was suppressed compared with the results at hν = 22.4 eV, the states with Cu 3d character were clearly observed as shown in Fig. 6.8(c) for $k_{||}$ = 0.0–0.6π/c at $k_\perp$ = 0.7 $Å^{-1}$. Two dispersing branches are resolved and explained by the spin-charge separation model calculated by considering the nearest neighbor hopping of t = 0.65 eV and the exchange coupling of J = 0.27 eV in Fig. 6.8(d). From the negligible $k_\perp$ dependence, the 1D nature of the chain electronic structures was also confirmed. In this experiment it was observed that the two-peak structures were prominent only in a certain region of $k_\perp$ (0.6–0.85 $Å^{-1}$). It was also noticed from a comparison with calculated spectral functions on a realistic Cu–O ring that the spectral weight between $\pi/(2c)$ and π/c has mostly O 2p character whereas the weight between Γ and $\pi/(2c)$ has both Cu 3d and O 2p character. Then the spectral weight in the former region might be suppressed as experimentally observed [29].

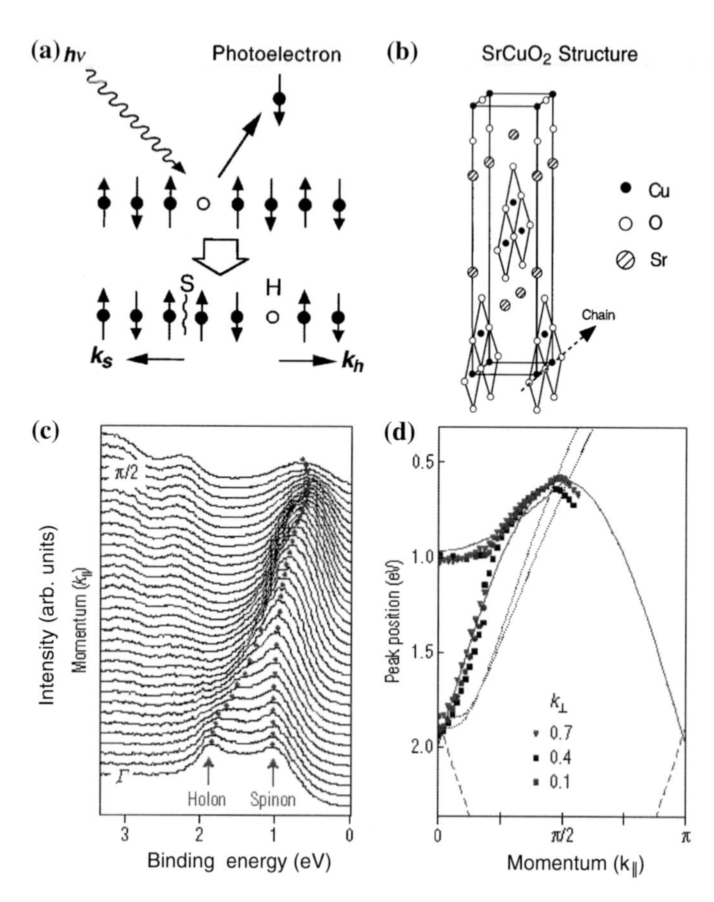

Fig. 6.8 Schematic explanation of the spinon (S) and holon (H) excitations in the case of a 1D AFM, which propagate independently along the chain (**a**). **b** Arrangement of Cu, O and Sr atoms in a $SrCuO_2$ unit cell [28]. **c** $k_{||}$ dependence of ARPES spectra on $SrCuO_2$ measured at $h\nu = 85$ eV for $k_{||}$ between Γ and $0.6\pi/c$ at $k_\perp = 0.7$ Å^{-1} [29]. The energy resolution at $h\nu = 85$ eV was set to 60 meV and the angular resolution to 0.25° (**d**) Experiments were performed for 7 $h\nu$ between 85 and 200 eV. Dispersions of the observed two branches are plotted and compared with theoretical dispersions (Figs. (**c**) and (**d**) are reprinted by permission from Macmillan Publishers Ltd.)

As another example of a quasi 1D system, here are introduced the ARPES results of the organic conductor TTF-TCNQ [30]. Planar TTF and TCNQ molecules form separate linear stacks along the **b** direction and the π molecular orbitals overlap with those of neighboring molecules stacked above and below as illustrated in the lower panel of Fig. 6.9(a). The charge transfer from TTF to TCNQ is ~0.59 electrons per molecule. The conductivity along **b** is 3 orders of magnitude larger than perpendicular to it. Therefore this system is thought to be a quasi 1D metal. A Peierls instability takes place at 54 K. The importance of U was experimentally confirmed and various properties were understood by the single-

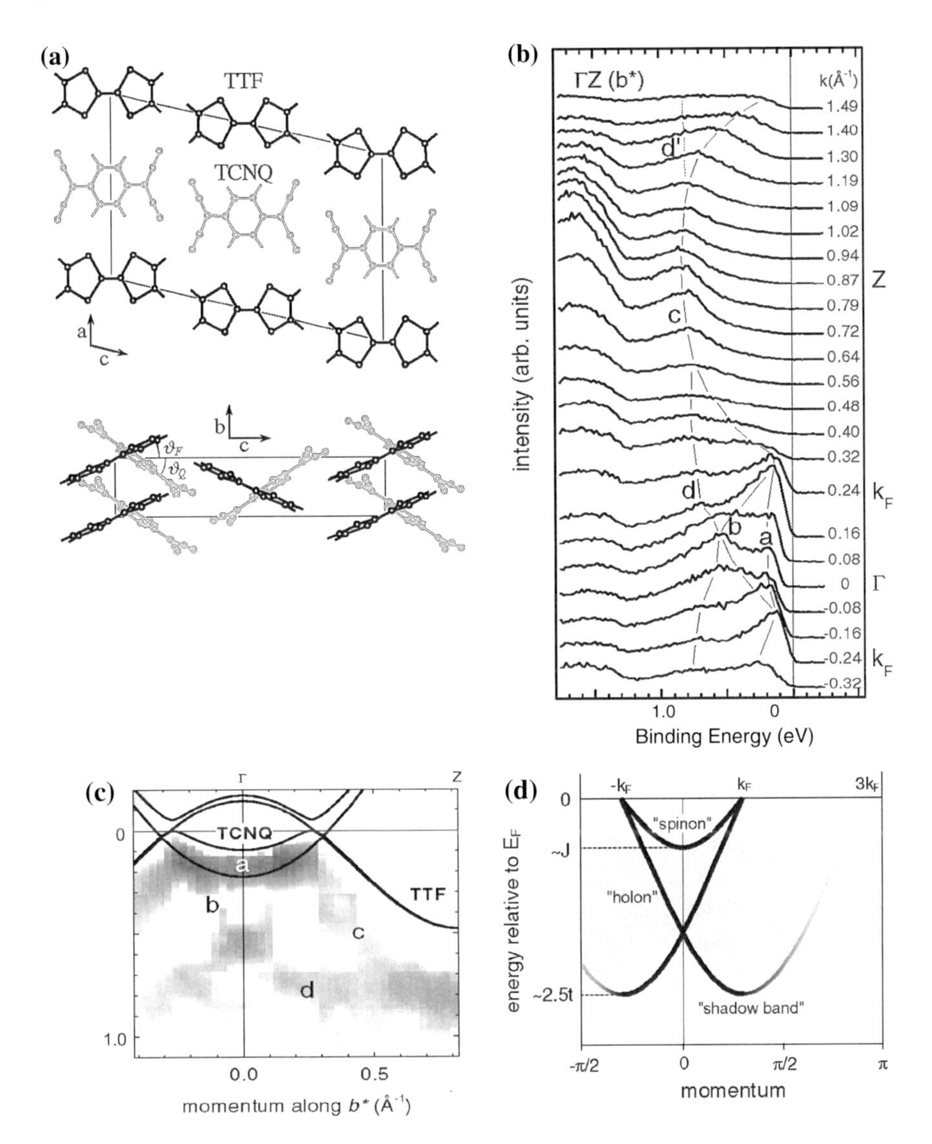

Fig. 6.9 **a** Monoclinic crystal structure of TTF-TCNQ shown in the a-c plane and in the b-c plane. **b** ARPES at hν = 25 eV and T = 61 K for **k** along the ΓZ direction [30]. **c** Experimentally obtained dispersions (gray scale) compared with the conduction band structures obtained by density functional band theory. **d** Schematic electron removal spectrum of the 1D Hubbard model at finite doping (band filling n < 2/3)

band 1D Hubbard model [31], where U/4t $\sim$ 1 and 4t $\sim$ 0.5 eV were often employed. Surfaces parallel to the **ab** plane were obtained by in situ cleavage. The ARPES spectra along the **b** axis in Fig. 6.9(b) were measured at hν = 25 eV and T = 61 K with an energy resolution of 60 meV and a momentum resolution of 0.07 Å^{-1}, where a periodic shift of spectra in E_B was recognized in contrast to the dispersionless behavior perpendicular to the **b** axis. Four structures were recognized to show dispersions. A gray scale plot of the negative second energy derivative in the $(E, k_{||})$ plane is shown in Fig. 6.9(c), in which the results of density functional band-structure calculations are also drawn. A qualitative correspondence of the experimentally observed structures a, b and c to the results of the band-structure calculation could be inferred, where the peaks a and b are attributed to the TCNQ and c to the TTF stacks. The observed band widths, however, exceed those from the calculation by a factor of ~ 2 and more importantly there is no counterpart for structure d in the calculations. Various scenarios were considered but they could not explain the observed results. Finally the ARPES dispersions were compared with the electron removal spectra of the 1D Hubbard model at finite doping depicted in Fig. 6.9(d). One then found a good agreement between the TCNQ related dispersions in Fig. 6.9(c) and the Hubbard model spectrum in Fig. 6.9(d). Experimental features (a) and (b) in Fig. 6.9(c) closely resemble the spinon and holon branches in Fig. 6.9(d). In addition, the feature d can be accounted for as the shadow band of the Hubbard model within a certain $k_{||}$ range, although its upward dispersion beyond k_F and its $3k_F$ crossing were not observed possibly because of the loss of spectral weight at large k and interference with the TTF band [30]. Within the 1D Hubbard model the spin-charge separation was experimentally observed in TTF-TCNQ.

This Hubbard model interpretation was further supported by the T dependence of the ARPES spectra measured at k_F. Namely, the spectral weight showed a dramatic shift of intensity from the features a, b to the feature d with increasing T from 60 K to 260 K. Such a weight transfer over an energy scale far beyond k_BT cannot be due to the electron–phonon interaction or the effect of Peierls fluctuations. For the t–J model at quarter filling, it was predicted that a redistribution of the spectral weight takes place on a scale given by t already for $k_BT \ll t$ [32] in agreement with these results. At small E_B, it was found that the intensity nearly linearly decreased. This result was not reproduced by the simple Hubbard model. Close to E_F, the density of states of an interacting 1D metal is expected to follow a $|E - E_F|^\alpha$ power law behavior as described in Sect. 5.1. Hubbard model provides $\alpha \leq 1/8$ if only on-site interactions were considered [33, 34], whereas up to $\alpha \sim 1$ is possible for an extended Hubbard model including nearest- and next-nearest-neighbor interactions [35].

6.5 Topological Insulators

As another class of low dimensional systems, topological insulators are discussed. Over the last three decades, the quantum Hall effect was intensively studied and the concept of topological order was proposed [36–38]. In recent years, the realization of topological insulating electronic phases due to spin–orbit interactions was extensively discussed [39–49]. Topological insulators have a bulk band gap like an ordinary insulator but have protected spin-polarized conducting states on their edge in the case of 2D systems or surface in the case of 3D systems. These states arise from the strong spin–orbit coupling that inverts the energy gap, i.e., the states expected above the gap appear below the gap in topological insulators, and are protected by time reversal symmetry. The 2D topological insulator is a quantum spin Hall insulator, which is closely related to the integer quantum Hall state. The HgTe quantum well is an example of such 2D topological insulators [50]. Regarding the 3D systems, such materials as $Bi_{1-x}Sb_x$, Bi_2Se_3, Bi_2Te_3, Sb_2Te_3 were recognized to be 3D topological insulators, where Bi is a semimetal and has strong spin–orbit interaction. Topological insulators attract wide attention not only from the fundamental scientific point of view but also from the view point of application as spintronics materials.

Bulk $Bi_{1-x}Sb_x$ single crystals were predicted to be candidates for topological insulators and to show 3D Dirac particle behavior [51]. In spite of many experiments on transport and magnetic properties since the 1960s, however, no evidence of the topological Hall states or bulk Dirac particles was found until recently. Then ARPES was applied to observe Dirac particles in the bulk of $Bi_{0.9}Sb_{0.1}$ [46]. The valence band (VB) top and conduction band (CB) bottoms are located at the T point and three equivalent L points in the bulk BZ, respectively. VB and CB at L are derived from antisymmetric and symmetric p-type orbitals (abbreviated as L_a and L_s), respectively. At $x \sim 0.04$ the gap between L_a and L_s closes and a massless 3D Dirac point is realized. For further increased x, this gap opens again but with inverted symmetry ordering as schematically shown in Fig. 6.10(a). For $x > 0.07$, there is no overlap between the VB at T and the CB at L and the material becomes an inverted-band insulator. The material becomes further a direct-gap insulator for $x > 0.08$, where its low energy physics is dominated by the spin–orbit coupled Dirac particles at L [44]. Band-inverted $Bi_{1-x}Sb_x$ families were thought to be possible 3D bulk realizations of the quantum spin Hall phase where the 1D edge states were expected to take the form of 2D surface states.

High **k** resolution ARPES was performed on $Bi_{0.9}Sb_{0.1}$ at $T = 15$ K and $h\nu$ between 17 and 55 eV with an energy resolution between 9 and 40 meV and a **k** resolution better than 1.5 % of the surface BZ (SZB). ARPES spectra along k_y, which corresponds to the $\bar{\Gamma}-\bar{K}$ direction of the SBZ as indicated in Fig. 6.10(a), are shown in Fig. 6.10(b), where a Λ shaped dispersion is clearly seen with its tip at $E_B \sim 50$ meV. Then the dependence of the spectra on k_z was measured with changing $h\nu$ at a certain (k_x, k_y) with (0.8, 0.0 Å^{-1}) and the surface character of the state near E_F was confirmed. At some $h\nu$ the Λ shaped bands were observed to have noticeable overlap with the bulk valence band at the L point near E_F. The gapless character of the surface states in bulk insulating $Bi_{0.9}Sb_{0.1}$ was thus confirmed. In the case of the present (111) surface of $Bi_{0.9}Sb_{0.1}$, four time-reversal-invariant momenta are located at $\bar{\Gamma}$ and three $\bar{M}$ points. Due to the three-fold crystal symmetry and the observed mirror symmetry of the surface FS across $k_x = 0$, the three $\bar{M}$ points were thought to be equivalent. From the time-reversal invariance, $E(k_y) = E(-k_y)$ mirror symmetry was also expected. As shown in Fig. 6.10(c) it was found in the dispersions along the $\bar{\Gamma}-\bar{M}$ direction that two

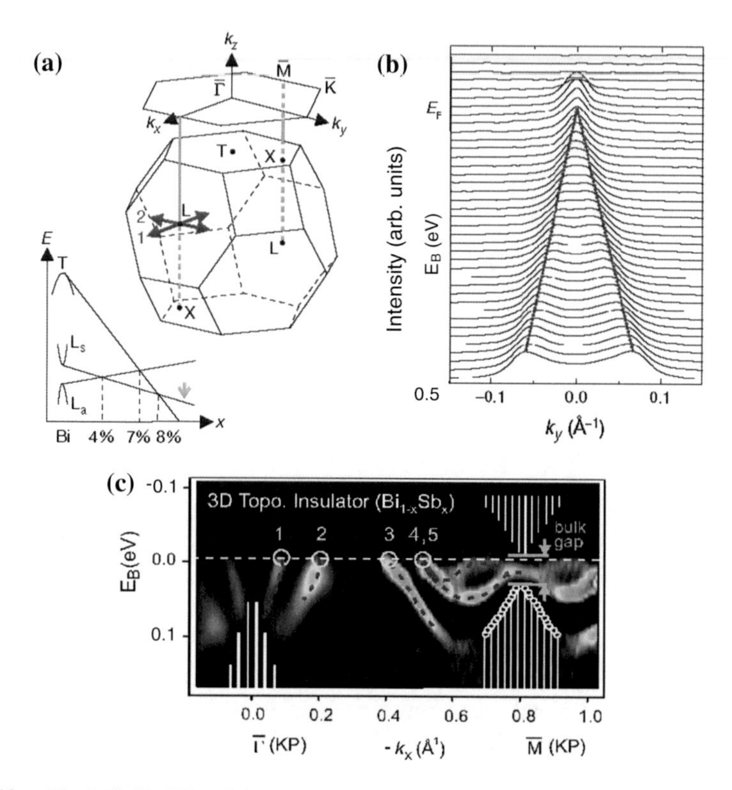

Fig. 6.10 **a** The bulk 3D BZ and the 2D BZ of $Bi_{1-x}Sb_x$ projected onto the (111) surface, where $\bar{\Gamma}$, $\bar{M}$ and $\bar{K}$ are high symmetry points in the SBZ. Evolution of the bulk VB and CB energies with x is schematically shown. The L band inversion takes place at x ~ 0.04, where a 3D gapless Dirac point is realized. **b** MDC of $Bi_{0.9}Sb_{0.1}$ at $h\nu = 29$ eV resolving the dispersion near the L point in the 3D bulk BZ (selected L_z is ~2.9 Å^{-1} in the third BZ). **c** Surface band dispersions of $Bi_{0.9}Sb_{0.1}$ along the $\bar{\Gamma}$–$\bar{M}$ direction. Projection of the bulk bands associated with the L point is shown by vertical hutches [46, 49]. The E_F crossings of the surface state are denoted by the yellow circles. The crossing near $-k_x \approx 0.5$ Å^{-1} is counted twice considering its double degeneracy. The red lines are guides to the eye. There are five crossings between $\bar{\Gamma}$ and $\bar{M}$, indicating that these surface states are topologically non-trivial (Figures are reprinted by permission from Macmillan Publishers Ltd.)

surface bands emerged from the bulk band continuum near $\bar{\Gamma}$ forming a central electron pocket and an adjacent hole lobe. It was established that these two bands result from the spin-splitting of a surface state and are singly degenerate. On the other hand, the surface band crossing E_F at $-k_x \approx 0.5$ Å^{-1} and forming a narrow electron pocket around $\bar{M}$ is clearly doubly degenerate. This splitting becomes zero at $\bar{M}$ in accordance with the Kramers' theorem. In contrast to semimetallic Bi, where only a single surface band was observed around $\bar{M}$ forming an electron pocket, the surface states near the $\bar{M}$ point fall completely inside the bulk energy gap in $Bi_{0.9}Sb_{0.1}$. Then the point of the surface Kramers doublet could be defined in the bulk insulator gap around $E_B \sim 15 \pm 5$ meV at the $\bar{M}$ point [46].

The gapless surface states thus observed were topologically non-trivial and supported the interpretation that $Bi_{0.9}Sb_{0.1}$ is in the insulating band-inverted regime containing an odd number of bulk (gapped) Dirac points and topologically analogous to an integer-quantum-spin Hall insulator [46, 49].

The undoped Bi_2Se_3 is a semiconductor with a rhombohedral structure, whose unit cell contains five atoms and is composed of quintuple layers ordered in the Se(1)-Bi-Se(2)-Bi-Se(1) sequence. The intrinsic gap is experimentally evaluated as $\sim$350 meV. In contrast to bulk band inversions at three equivalent L points in $Bi_{1-x}Sb_x$ responsible for its characteristic topological order, only one band is expected to be inverted similarly to the 2D quantum spin Hall insulator phase. ARPES was performed on an in situ cleaved (111) plane of Bi_2Se_3 at hν between 17 and 45 eV with an energy resolution of 15 meV and a momentum resolution of 1.5 % of the SBZ. A clear V-shaped band pair was observed near $\overline{\Gamma}$ to approach E_F as shown in Fig. 6.11(a) [47]. It was found that the V bands cross E_F at 0.09 Å^{-1} along the $\overline{\Gamma} - \bar{M}$ and at 0.10 Å^{-1} along $\overline{\Gamma} - \bar{K}$ with nearly equal band velocities of 5×10^5 m/s. Since a theoretical study showed that no surface band crosses E_F when the spin–orbit interaction is not included, this V-shaped band must be originating from the strong spin–orbit coupling of this system and therefore realizing a topological surface state. In addition to this V-shaped band, a continuum feature was recognized inside the V-shaped feature. This feature is due to the bulk conduction band, indicating that the bulk electronic state became metallic in the sample shown in Fig. 6.11(a) caused by the electron-doping. Regarding the FS map, none of the three time-reversal invariant momenta located at $\bar{M}$ is enclosed by a FS sheet in contrast to $Bi_{1-x}Sb_x$. On the other hand, a ring-like feature corresponding to the outer V-shaped surface band surrounds the conduction band continuum around $\overline{\Gamma}$. In Bi_2Se_3 a singly degenerate surface FS encloses only one Kramers point of the SBZ. Spin of this V-shaped surface band was later measured by spin polarized ARPES and the validity of the present interpretation as the topological insulator with a single Dirac cone was confirmed [52].

First principles electronic structure calculations were performed on layered stoichiometric crystals of Sb_2Te_3, Sb_2Se_3, Bi_2Te_3 and Bi_2Se_3 and predicted that they are topological insulators except for Sb_2Se_3 [48]. These topological insulators have robust and simple surface states consisting of a single Dirac cone at the $\overline{\Gamma}$ point like the result shown above. The ARPES studies were also performed on Bi_2Te_3 [53, 54] and Sb_2Te_3 [54]. Hole doping effects on electronic structures were studied by ARPES and transport experiments in $(Bi_{1-\delta}Sn_\delta)_2Te_3$ for $\delta = 0, 0.0027$, 0.0067 and 0.09 [53]. Bi_2Te_3 has an indirect gap of about 150 meV induced by the strong spin-orbit interaction and the surface electronic structure was characterized by a single topological spin-Dirac cone. The self-doped Sb_2Te_3 exhibited also similar topological properties [54].

Tl based ternary compounds were proposed to be new members of 3D topological insulators. In the case of the ternary chalcogenide $TlBiSe_2$ the topological surface state was also observed on the (111) surface [55, 56]. Since each Tl or Bi layer is sandwiched between up and down Se layers, stronger coupling was

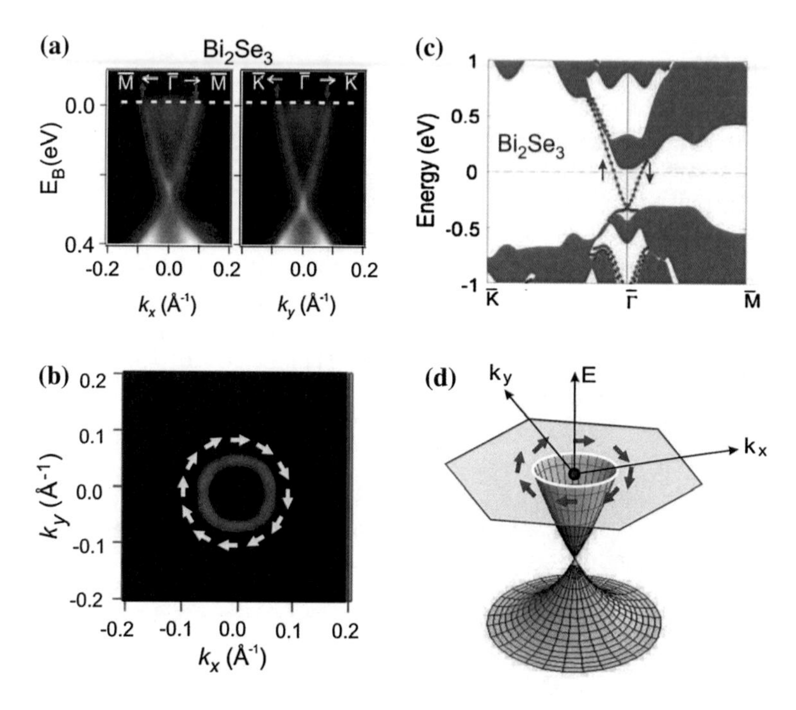

Fig. 6.11 **a** ARPES of Bi_2Se_3 measured at hν = 22 eV along the $\bar{\Gamma}$–$\bar{M}$ and $\bar{\Gamma}$–$\bar{K}$ directions [47]. **b** The surface FS exhibits a chiral left-handed spin texture as observed by spin-polarized ARPES [49, 52]. **c** Surface electronic structure calculated by the local-density approximation. The shaded regions represent the bulk states. **d** Illustration of the spin-polarized surface state dispersion (Figures are reprinted by permission from Macmillan Publishers Ltd.)

expected between neighboring atomic layers, providing 3D character in comparison with the layered binary chalcogenides. Like the case of Bi_2Se_3, $\bar{\Gamma}$ and $\bar{M}$ are the time-reversal invariant momenta or the surface Kramers points. ARPES was measured at T = 10 K and hν between 22 and 98 eV. Both p- and s-polarization light were employed for excitation. The energy and angular resolutions were set to 20 meV and ±0.1°, respectively. k_z was evaluated for V_0 = 11.7 eV. The band dispersions and the MDC along the $\bar{\Gamma}$–$\bar{M}$ and $\bar{\Gamma}$–$\bar{K}$ directions for p- and s-polarized light excitation at hν = 58 eV are shown in Fig. 6.12. An X-shaped massless dispersion is clarified. From the negligible dependence of the observed dispersion on hν, the 2D nature of the dispersion was confirmed. These surface state bands crossing E_F at $k_{\|} \sim 0.1$ Å^{-1} intersect at $E_B \sim 390$ meV corresponding to the Dirac point. The bulk VB maximum is at $E_B \sim 540$ meV and located far away from the $\bar{\Gamma}$ point along the $\bar{\Gamma}$ – $\bar{M}$ direction. It is much deeper and weaker along the $\bar{\Gamma}$–$\bar{K}$ direction as indicated in Fig. 6.12((a) ii and (b) ii), showing a strong anisotropy. Although the bulk CB is rather weak at hν = 58 eV, it becomes enhanced at low hν in the region of 22 and 30 eV. The minimum of the bulk CB within the cone was inferred to be located at $E_B \sim 200$ meV at hν = 22 eV.

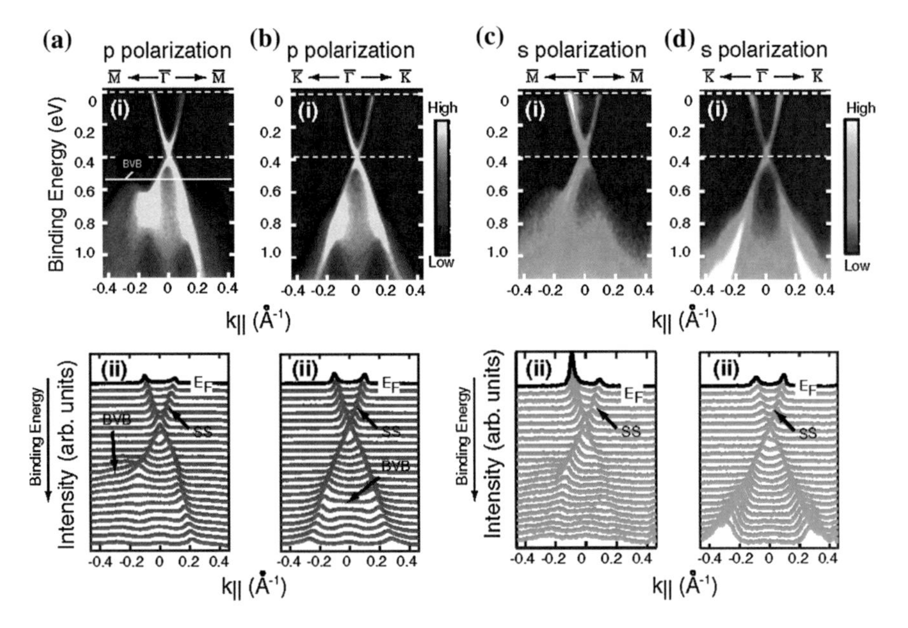

Fig. 6.12 ARPES of $TlBiSe_2$ (111) measured at hν = 58 eV with (**a**, **b**) p- and (**c**, **d**) s-polarized light along the $\bar{\Gamma}-\bar{M}$ (**a**, **c**) and $\bar{\Gamma}-\bar{K}$ (**b**, **d**) directions [56]

Though the Dirac cone is not exactly linear, it is more linear than in Bi_2Se_3, particularly in the local energy gap window between $E_B \sim 540$ meV and $\sim$200–240 meV. So the topological spin-polarized transport with high mobility and long spin lifetime are expected for $TlBiSe_2$ [56].

6.6 Superconductors

After high temperature superconductors were found, many ARPES studies were performed and a huge amount of knowledge was accumulated. In other words, methodology and instrumentation for ARPES were accelerated and improved dramatically in order to study the mechanism and physics of high Tc superconductivity. The ARPES experiments were mostly performed in the hν region between $\sim$15 and 200 eV. (Extensive ARPES results by high hν > 400 eV and extremely low hν < 10 eV ARPES are separately reviewed in Chaps. 7 and 9 in this book). Many review articles were so far published on this subject. Among them the review paper by Damascelli et al. [57] covered the major achievements until 2002 on various materials such as $YBa_2Cu_3O_{7-\delta}$ (YBCO), $Bi_2Sr_2CaCu_2O_{8+\delta}$ (Bi2212), $Bi_2Sr_2CuO_{6+\delta}$ (Bi2201), $Bi_2Sr_2Ca_2Cu_3O_{10+\delta}$ (Bi2223), $La_{2-x}Sr_xCuO_4$ (LSCO) and $Nd_{2-x}Ce_xCuO_4$ (NCCO). Here in this book mainly introduced are the results obtained afterwards. Rather recent results on pseudogap, superconducting gap and Fermi arcs in high T_c cuprates up to mid 2011 were reviewed in Ref [58].

mainly on LSCO, Bi2212, Bi2201 and Bi2223. Underdoped samples showed a momentum dependence of the energy gap deviating from a simple d-wave form below T_c, suggesting the coexistence of multiple energy scales in the superconducting (SC) state [58]. As two distinct energy scales, the gap sizes near the node (represented by Δ_0) and that in the antinodal region (Δ^*) were considered. It was found that Δ_0 is strongly material dependent reflecting the magnitude of $T_{c,max}$ and that Δ^* at the same doping level is approximately material independent. In contrast to the gap size near the node which almost follows the BCS-like T dependence, the antinodal gap does not close at T_c. The effective SC gap Δ_{SC} defined at the end point of the Fermi arc was found to be proportional to T_c in various materials [58].

ARPES results of superconductors not reviewed in these two references are here introduced. The two dimensional layer material 2H-$NbSe_2$ was known to show the incommensurate charge density wave (CDW) below ~35 K and phonon mediated superconductivity below $T_c = 7.2$ K. ARPES was first performed on a cleaved (0001) surface of a single crystal sample at 10 K above T_c with use of HeII line [59]. A hole-like small FS derived from the Se 4p states was observed around the Γ(A) point in addition to the Nb 4d derived large hexagonal FS around the Γ(A) and K(H) points in consistence with band-structure calculations. The detailed T dependence of the SC gap was measured at these momentum points by use of HeI with a total resolution of 2.5 meV at 10 and 5.3 K, which are above and below $T_c = 7.2$ K. Although no clear gap opening was observed for the Se 4p derived band, the Nb 4d derived FS shows a shift of the leading edge and the opening of the superconducting gap with a coherent peak. The magnitude of the Nb 4d derived gaps was estimated to be 1.0 ± 0.1 meV for the inner FS around the Γ(A) point and 0.9 ± 0.1 meV for the outer FS around the K(H) point. The high quality results of very high resolution ARPES could be compared with such experiments as the magnetic field dependent specific heat, quantum oscillation dHvA measurements and STS, opening high potential application to a wide field.

The next example is MgB_2, which clearly showed the multiple SC gaps below $T_c = 39$ K. Angle integrated PES was first performed at $h\nu = 21.2$ eV by use of He Iα radiation with a total resolution (analyzer and light source) of 3.8 meV [60]. While the normal Fermi edge structure was observed above T_c, SC gap opening and a prominent peak at 7 meV were observed at 5.4 K. A very weak shoulder feature was also recognized at $E_B \sim 3.5$ meV. The first ARPES measurement was performed by use of SR at $h\nu = 28$ eV along the $\Gamma-$M and $\Gamma-$K directions [61] with the linearly polarized light perpendicular to these directions. Although the k_z dependence was measured with changing $h\nu$ from 17 to 28 eV, the PICS decreased with the decrease in $h\nu$ and k_z could not be experimentally determined. The SC gap was not identified because of the insufficient resolution of $\leq$40 meV. Then ARPES was performed at 40.814 eV by use of a high flux RF discharge lamp with a toroidal grating monochromator providing He IIα radiation [62]. The origin of the possible multiple gaps and their relation to the high T_c were interesting subjects. First a wide energy scan was done with the resolution of ~100 meV and $\Delta\theta \sim \pm 0.1°$($\Delta k \sim 0.010$ Å^{-1}). Then high resolution measurements for the SC

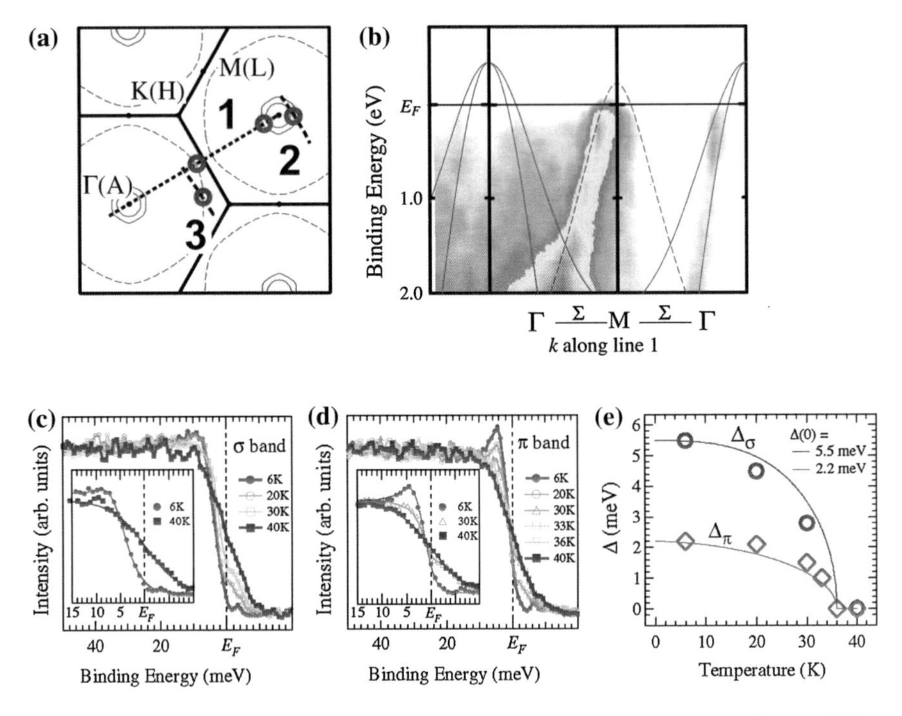

Fig. 6.13 **a** Thin solid and dashed lines show the FSs at $k_z = 0$ of MgB_2 [62]. ARPES at $h\nu = 40.8$ eV was performed along the directions 1, 2 and 3 at $T < T_c$. **b** The result along the Γ(A)–M(L) direction indicated by the line 1 is shown. **c** and **d** T dependent ARPES spectra for the σ and π bands. **e** T dependence of the SC gap Δ of the σ and π bands

gap on the σ and π bands were performed with a resolution of 5.0 and 3.9 meV. MgB_2 consists of honeycomb B layers and the bands at E_F results mainly from B orbitals. Band-structure calculations predicted two 2D cylindrical FSs around the $\Gamma-A$ line corresponding to the σ-bonding states with B $p_{x,y}$ character and two 3D tubular FS derived from the bonding and antibonding π orbitals with B p_z character. Figure 6.13(a) shows the calculated FS at $k_z = 0$, where the two small closed FSs shown by full lines around the Γ(A) point represent the 2D cylindrical FS derived from the B $p_{x,y}$ σ orbitals and the large FS shown by the dashed line represents one of the 3D FSs derived from the B p_z π orbitals [63]. ARPES spectra were measured along the **k** direction indicated by the dashed lines 1, 2 and 3 in Fig. 6.13(a). Along the Γ(A)-M(L)-Γ(A) line was clearly observed a feature dispersing towards E_F in the first BZ as shown in Fig. 6.13(b), which was ascribed to the π orbital. The absence of the counterpart in the second BZ in Fig. 6.13(b) may be due to matrix element effects. Likewise a band dispersing to E_F was observed along the lines 2 and 3 (Fig. 1(c), (d) in [62]), where the dispersive band, only weakly observed near the Γ point in the second BZ in Fig. 6.13(b) was strongly observed along the line 2 corresponding to the σ band. Thus both the σ and π bands were experimentally confirmed.

The detailed T dependence of the spectra near E_F was measured for the σ band and π band with an energy resolution of 5.0 and 3.9 meV, respectively. In order to overcome the low count rate, the ARPES results along the lines 2 and 3 containing k_F were independently summed up to provide angle integrated spectra with better statistics (Fig. 6.13(c) and (d)) at different T, where a noticeably different T dependence of the σ and π bands was observed. The two SC gaps are more clearly shown in Fig. 6.13(e) after the analysis by considering the QP lifetime broadening [62]. Both gaps were found to close at the same temperature ($\sim$35 K). In this study $2\Delta/k_BT_c$ was evaluated as 3.54 for the σ band and 1.42 for the π band from the lowest T gaps of 5.5 meV and 2.2 meV. The comparable magnitude of the gap for the σ band between this experiment and angle integrated PES results [60] suggest that the anisotropy of the SC gap of the σ band is very small. The gap anisotropy is less than 20 % for the π band. The importance of the **k** dependent electron–phonon coupling as the origin of multiple-gap high Tc superconductivity in this materials was suggested.

Here some results are briefly added on Bi2212 systems. Unusual pseudogap behavior and Fermi arcs in underdoped Bi2212 attracted wide attention for a long time. Two main questions were whether the pseudogap was a precursor to SC or arose from another order competing with SC. Therefore its anisotropy in a wide **k** space as well as the resulting Fermi arcs dependent on T were studied by ARPES in underdoped Bi2212 [64]. In the SC phase, sharp spectral peaks were observed outside the SC gap, which vanish only at the nodal points ($\phi = 45°$) as shown in Fig. 6.14(a). On the contrary, the peak in the pseudogap phase is much broader and the pseudogap only observed in the cuts 1–7 at 140 K in Fig. 6.14(b). The spectra peaked at E_F were observed for the cuts 8–15 showing the Fermi arc of gapless excitations. The gap size estimated are shown as a function of ϕ in Fig. 6.14(d), where a sharp opening of the pseudogap is observed at the end of the Fermi arc in contrast to a d-wave shaped SC gap. In the case of a very underdoped sample with $T_c = 25$ K, a much shorter Fermi arc was observed at T = 55 K (Fig. 6.14(c)). Further it was found that the momentum dependence of the pseudogap $\Delta(\phi)$ at different T and different doping level (x) could be well scaled by a reduced temperature t = T/T*(x) after normalization by the value at the antinode ($\Delta(\phi = 0^{\circ})$). Here T*(x) is the temperature below which the pseudogap first developed at a given hole doping x. $\Delta(\phi)/\Delta(0)$ was found to depend only on t = T/T*(x) for t = 0.9 or 0.45 [64]. T_c did not have a meaningful correspondence to the pseudogap behavior. It was also found that the length of the Fermi arc was a linear function of t and it extended rather straight to the point node as a function of t. Above t = 1 or T > T*, the arc of course covered the full length of the FS. In other words, the low temperature limit of the momentum dependence of the pseudogap is identical to that of the d-wave superconductor.

The FS topologies of underdoped Bi2212 sample with $T_c = 65$ K and Ca doped sample with $T_c = 45$ K were separately measured by ARPES and analyzed [65], whereby results rather contrasting to [64] were obtained. Determination of the FS crossing was made by analyzing the raw data on Fermi arcs taking the

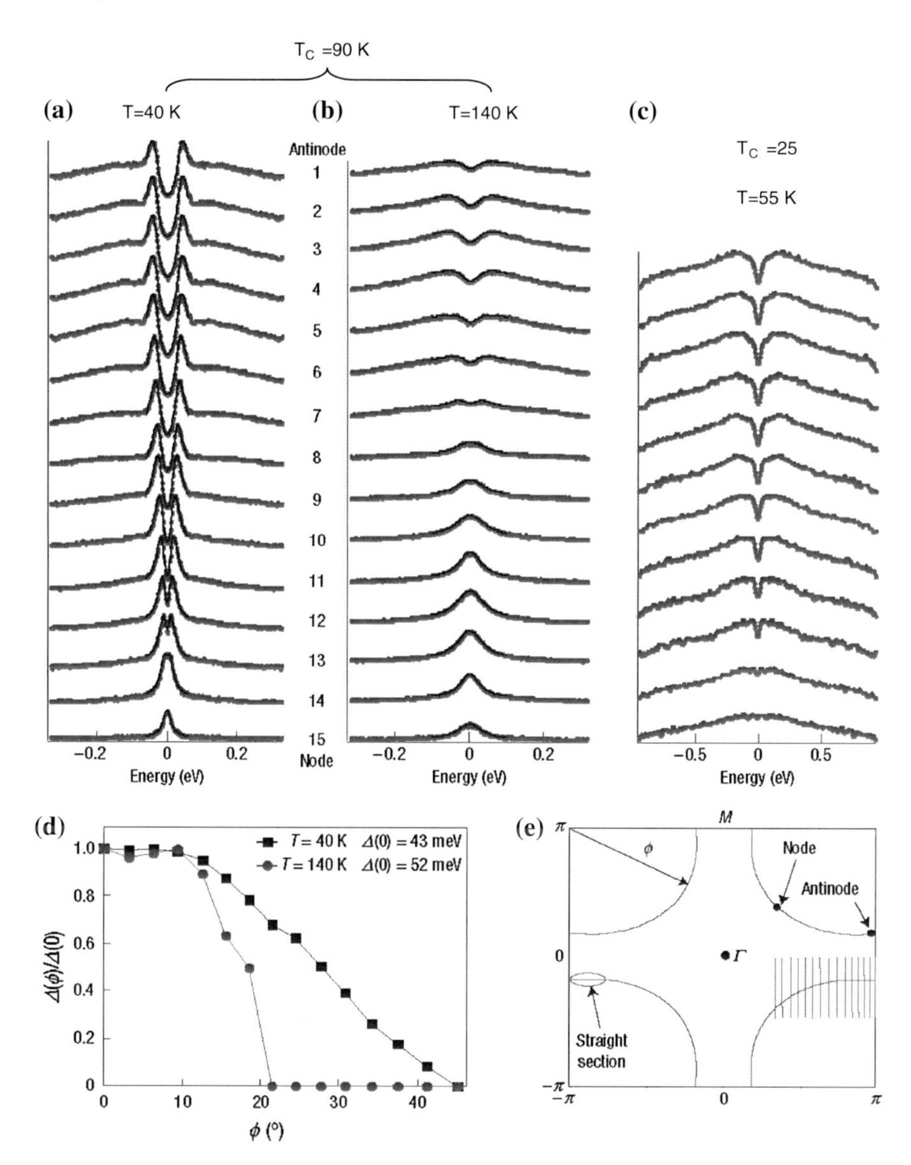

Fig. 6.14 **a** and **b** Symmetrized EDCs for underdoped Bi2212 with $T_c = 90$ K measured from the antinodal (*top*) region to the nodal (*bottom*) region indicated in (**e**). **a** T = 40 K in SC phase, **b** T = 140 K in the pseudogap phase. **c** Similar results for a very underdoped sample with $T_c = 25$ K measured at T = 55 K in the pseudogap phase and k corresponding to 4–15 in (**a**) or (**b**). (**d**) ϕ dependence of the gap from (**a**) and (**b**) [64] (Figures are reprinted by permission from Macmillan Publishers Ltd.)

experimental resolution and FD function into account. The hole FS pocket thus determined is clearly asymmetric with respect to the magnetic zone boundary as shown in Fig. 6.15(a). The deviation from the FS derived from the local density

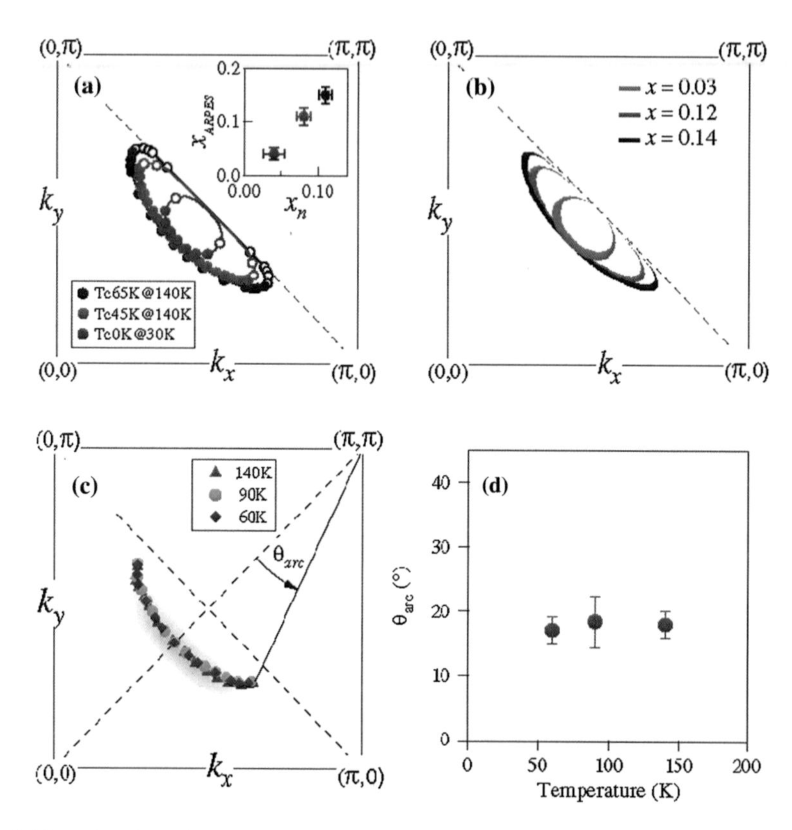

Fig. 6.15 **a** Pseudopocket FS determined by ARPES for three different doping levels, $T_c = 65$ K Bi2212, $T_c = 45$ K Ca doped sample and $T_c = 0$ K oxygen-deficient non superconductor sample [65]. **b** FS pocket derived from Yang-Rice-Zhang ansatz [66]. **c** FS crossings of the $T_c = 45$ K Ca doped sample at three different T. The length of the Fermi arc was found to be almost constant against T as shown in (**d**)

approximation (LDA) was found to become more noticeable with reduced doping level. Then a further question was whether the FS pocket area was scaled with the doping level or the observed arc length was proportional to T/T*. It is clear from Fig. 6.15(a) that the area of the pocket represented by χ_{ARPES} in the inset is roughly proportional to the doping level χ_n. The FS pocket shapes are also consistent with those derived theoretically on the doped resonant valence bond (RVB) spin liquid concept [66] in Fig. 6.15(b). The length of the directly observed Fermi arc in Fig. 6.15(c) is almost constant against T as shown in Fig. 6.15(d). Several theories on the pseudogap proposed the preformed singlet pairs above T_C in the antinodal region of the BZ. One such model [66] identified the formation of resonating pairs of spin-singlets along the Cu–O bonds of the square lattice as the configuration with the lowest energy. It was confirmed that a symmetric gap existed in the antinodal region almost parallel to the $(\pi/a, 0)$-$(\pi/a, \pi/a)$ line at $T = 140$ K for Bi2212 with $T_c = 65$ K, indicative of incoherent preformed pairs

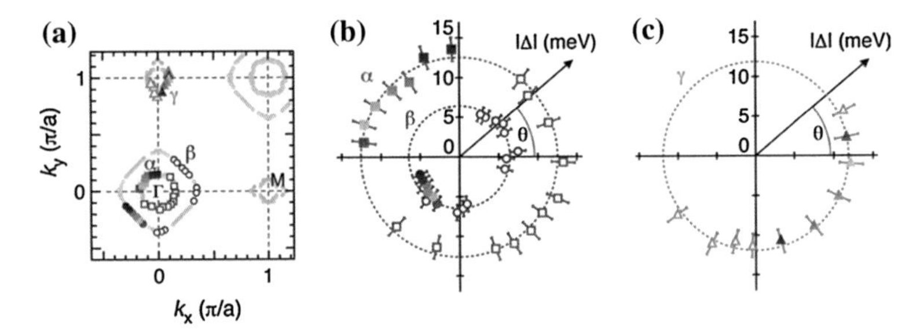

Fig. 6.16 ARPES results on $Ba_{0.6}K_{0.4}Fe_2As_2$ ($T_c = 37$ K). **a** FSs. **b** SC gap behavior in the k_x–k_y plane for the α and β sheets. **c** Same but for the γ sheet [67]

of the electrons in singlet states (here a denotes the lattice constant with the body-centered tetragonal symmetry). The particle-hole symmetry in E_B observed here in this region was in marked contrast to the particle-hole symmetry breaking predicted in the presence of the density wave order. The size and shape of the hole pockets were well reproduced by phenomenological models of the pseudogap phase as spin liquid [66].

Superconductivity was also discovered in Fe-As compounds, for which conflicting results were reported ranging from one gap to two gaps, from line nodes to nodeless gap functions. High resolution ARPES was performed on $Ba_{0.6}K_{0.4}$-Fe_2As_2 ($T_c = 37$ K) at hν = 21.2 eV with an energy resolution of 2–4 meV and a momentum resolution of 0.007 $Å^{-1}$ at T = 7–50 K [67]. ARPES at T = 50 K in the normal phase shows three FSs, an inner hole-like FS pocket (α sheet), an outer hole-like FS pocket (β sheet) around the Γ point and an electron-like FS (γ sheet) around the M(π/a,0) as reproduced in Fig. 6.16(a). The observed FS topology is consistent with the LDA band-structure calculation at $k_z = 2\pi/c$ in the body-centered tetragonal symmetry. At T = 15 K, the leading edge of all these three FSs show the opening of the SC gap. The gap on the β sheet is smaller than those on the α and γ sheets as shown in Fig. 6.16(b) and (c). Namely, the k-averaged gap is 12 meV for the α and γ sheets, while it is 6 meV for the β sheet. These values are smaller than those for electron doped $NdFeAsO_{1-x}F_x$, and $2\Delta/k_BT_c$ is 7.5, 3.7 and 7.5 for the α, β and γ sheets. The sharper QP peak in the β sheet suggested its longer lifetime or reduced scattering rate compared with the α and γ sheets. The multi-orbital nature of the SC states, nodeless SC gaps and nearly isotropic SC gap behavior were thus experimentally confirmed.

In order to understand the pairing mechanism in FeAs based superconductors, the study of the electronic structure of the parent compound $BaFe_2As_2$ (BFA) was thought to be important and ARPES was made at hν = 30–175 eV with a total energy resolution of 10 meV (hν = 30 eV) to 20 meV (hν = 175 eV) and an angular resolution of 0.2°–0.3°. $BaFe_2As_2$ is a metal with antiferromagnetic (AFM) ordering, which was supposed to occur by a nesting of hole pockets at the BZ center and an electron pocket at the zone corner (X point in a primitive tetragonal BZ). Measurements were mostly done around the zone center at

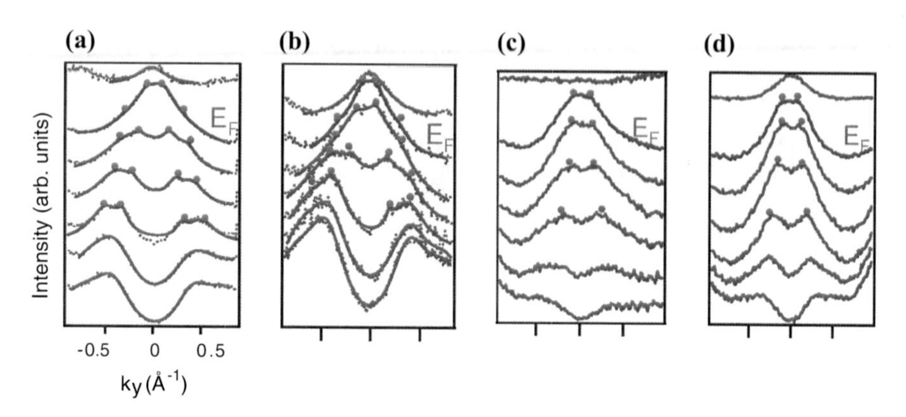

Fig. 6.17 ARPES MDCs at $E_B = 30$–150 meV. **a** $h\nu = 50$ eV for BFA100 K at T = 20 K. **b** the same as (**a**) but at T = 300 K. **c** $h\nu = 75$ K for BFA65 K at T = 20 K. **d** same as (**c**) but at T = 300 K [68]

$\Gamma(0, 0, 0)$ and $Z(2\pi/c(0, 0, 1))$ as well as around the zone corner $X(\pi/a(1, 1, 0))$. $BaFe_2As_2$ single crystals grown by Sn flux method contained Sn with ~ 1.6 atomic %. The structural transition temperature from the tetragonal PM phase to the orthorhombic AFM phase was reduced by the Sn contamination. Although Sn free polycrystals show a structural transition temperature close to 140 K, here employed two categories of crystals grown out of Sn flux showed either the transition temperature of ≈ 65 K (later called BFA65K) or ≈ 100 K (BFA100K) [68]. ARPES was performed for probing the electronic structure near Γ and X with k_x direction set parallel to the Γ−M direction and k_y set parallel to the Γ−N direction. Then both horizontal and vertical polarization light (parallel and perpendicular) with respect to the scattering plane (defined by the photon incidence and electron emission) was used for excitation [68]. The parity of the initial state could be discussed as explained in Sect. 6.1. For the measurement around the X point, the sample was rotated by 45° around the ΓZ axis. ARPES was compared with the results of band-structure calculations for tetragonal PM $BaFe_2As_2$ performed within the DFT in the generalized gradient approximation using WIEN2K code.

As for the FS of BFA65K around the Γ point, it was clarified to be a hole pocket FS and both the Fermi wave vector k_F and the Fermi velocity did not change between T = 20 and 300 K (Fig. 6.17(c), (d)). On the other hand, two hole pockets were distinguished for BFA100K as shown in Fig. 6.17(a), (b) [68] in agreement with previous ARPES [69, 70]. k_F and Fermi velocity of these two bands were estimated as $k_F = 0.10 \pm 0.01$ Å^{-1}, $v_F = 0.69 \pm 0.06$ eV·Å and 0.27 ± 0.01 Å^{-1}, 0.75 ± 0.05 eV·Å along the k_y direction. The spectra in BFA65K were noticeably broader than those in BFA100K possibly because of the higher Sn content and more impurity scattering. Then ARPES was measured around the X point with k parallel to Γ−X and also along the Γ−N direction parallel to k_y. The results were discussed in comparison with the results of band-structure calculations including the matrix element effects. Differences of the spectra between PM and AFM metal phases were much smaller than the prediction by DFT band structure calculation.

Most experimental ARPES data on Fe pnictide superconductors so far showed that the SC gap opens on the entire FS, suggesting the s$\pm$ wave gap in contrast to the d-wave SC gap in high T_c cuprate superconductors. However, recent experiments by other means such as thermal conductivity and NMR studies (Refs. 3, 4, 5 in Ref. [71]) for $BaFe_2(As_{1-x}P_x)_2$ revealed signatures of a line node SC gap. The substitution of P for As was thought to suppress magnetic order but without changing the number of Fe 3d electrons. A maximum of $T_c \sim 30$ K was expected for $x \sim 0.3$. Strong three dimensionality was also proposed by band-structure calculations for the $BaFe_2As_2$ system [72]. A shrinkage of the electron-like FSs on approaching the optimal composition from $x = 1$ ($BaFe_2P_2$) was revealed by dHvA measurements [73]. In addition, significant mass renormalization was observed. So $h\nu$ dependence of the ARPES was measured for $BaFe_2(As_{1-x}P_x)_2$ with $x = 0.38$ ($T_c = 28$ K) in order to study the effects of FS nesting and dimensionality, since the c-axis length is reduced and the interlayer hopping was expected to increase with P substitution [71]. ARPES was performed at $T = 10$ K with total energy and momentum resolutions of ~ 15 meV and $0.02\pi/a$ at $h\nu$ above 40 eV. FS mapping around the Z, X and Γ points in the k_x–k_y plane is shown in Fig. 6.18(a), (b), where $V_0 = 13.5$ eV is taken into account. At least two hole FSs are observed around the center of the BZ (Z and Γ). An electron FS is also observed around X. The diameter of the outer hole FS is much reduced near the Γ point compared with those near the Z point, suggesting its strong 3D character. Corresponding band dispersions are reproduced in Fig. 6.18(c–f). Intensity mapping in the $k_{||}$–k_Z plane around the Z–Γ–Z and X–X-X axes are shown in Fig. 6.18(g–i), in comparison with the results of band-structure calculations in Fig. 6.18(j), (k) for $x = 0.4$ and 0.6. A clear 3D character could be traced for some bands in these figures in Fig. 6.18(g), (h). The right-left intensity asymmetry in these results relative to $k_{||} = 0$ might be due to matrix element effects. Figure 6.18(i) shows the symmetrized experimental results. In the band calculation given in Fig. 6.18(j), (k), three hole FSs named α, β and γ are predicted around the Γ–Z axis and two electron FSs named δ and ε are predicted around the X–X axis. The experimentally observed warping of the electron FS along the X–X axis is qualitatively consistent with the prediction of the band calculation. From this, it was concluded that the prominent inner electron FS along the X–X direction had the $d_{xz/yz}$ character. As for the hole FSs, only two branches were experimentally revealed in contrast to three hole FSs theoretically predicted. The inner hole FS around the Γ point cannot be the α FS with d_{xy} character but was ascribed to the γ FS with $d_{xz/yz}$ character. The nearly 2D hole FS was assigned to the β FS with $d_{xz/yz}$ character. The three dimensionality of the γ hole FS showed poor nesting with the electron FSs. Although β hole FS with $d_{xz/yz}$ orbital character and δ electron FS with d_{xy} character had nearly the same size with cylindrical shapes, the difference in the orbital character weakens the FS nesting. Only partial nesting among the

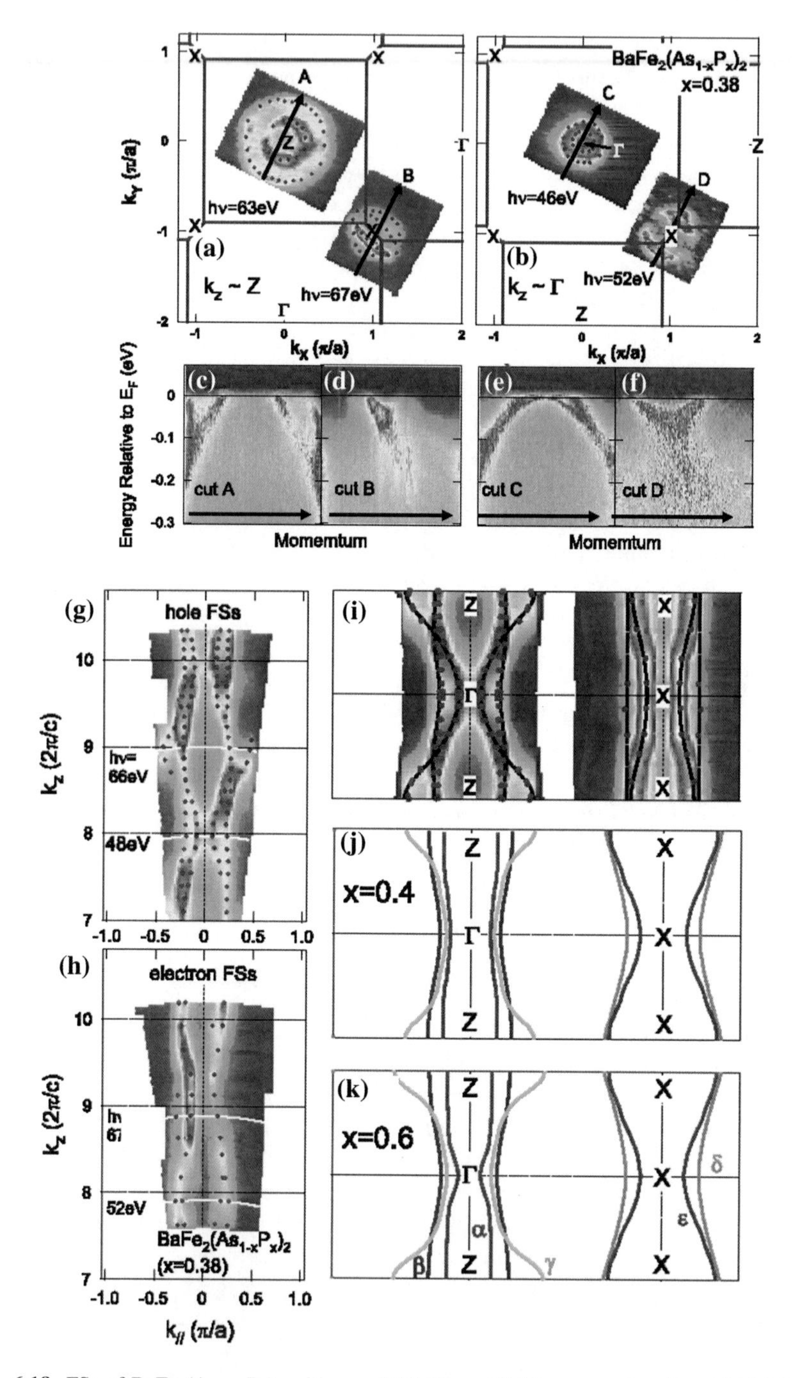

Fig. 6.18 FSs of $BaFe_2(As_{1-x}P_x)_2$ with x = 0.38 (T_C = 28 K) measured at T = 10 K [71]. **a**, **b** FSs in the k_x–k_y plane through the Z and Γ points. **c**–**f** Band dispersions. (**g**, **h**) FS mapping along the $k_{||}$–k_z plane, where **g** and **h** shows the hole and electron FSs around the center and corner of the BZ, respectively. (**i**) Symmetrized results of (**g**): *left* and (**h**):*right*. **j**, **k** Band structure calculation for x = 0.4 and 0.6

outer electron δ FSs with d_{xy} orbital character and/or within the 3D hole FS might lead to the nodal superconductivity [71].

6.7 Quantum Well States

In accordance with the development of the crystal growth techniques, superlattices and quantum well states (QWS) attracted materials scientists since around 1985. Tunneling, transport and photoelectron spectroscopy were applied to such materials. The first PES experiment was performed on GaAlAs/GaAs/GaAlAs QWS or superlattice [74]. Soon later ARPES was performed at $h\nu = 10$ eV for QWS in Ag overlayers epitaxially grown on a Au (111) substrate (hereafter expressed as Ag/Au (111)) [75], for which the coverage dependence was studied up to 40 monolayers (ML) of Ag. In contrast, no such features were resolved for the complementary system Au on Ag (111). Further the spin polarized photoelectron spectroscopy (SP-PES) technique (described in Chap. 11) was applied to study the spin polarization of QWS. Since the oscillatory exchange coupling in magnetic multilayers was a subject of general interest, intensive studies were performed for such systems as Fe/Cr and Co/Cu. It was demonstrated that the coupling between adjacent ferromagnetic layers depended strongly on the thickness of the spacer layer. In the case of Cu thin films up to 16 ML deposited on fcc Co (001) substrates, QWS were observed with some spin polarization [76]. The measurement was performed at $h\nu = 24$ eV and the QWS crossing E_F showed always minority spin dominated polarization even for thicker Cu films up to 8 ML. This spin polarization was understood as reflecting the hybridization at the interface between Co and Cu.

SP-ARPES studies were also made for the Cu thin overlayers on a Co(100) substrate at $h\nu = 17$ eV [77]. On the clean Co, a peak was observed at $E_B \sim 0.6$ eV. The Cu 3d-derived structure developed at $E_B \sim 2.5$ eV. The strongly delocalized Cu sp states were sensitively affected by the Cu film thickness. New structures appeared between E_F and $E_B = 2$ eV or between the Co 3d and Cu 3d states with increasing Cu thickness up to $\sim$10 ML. The E_B of these new structures observed in the normal emission configuration changed with the Cu thickness but stayed constant with changing $h\nu$. The observed new structures were well explained by considering the discretization of energy and $k_\perp$ due to the finite Cu film thickness [78] in the framework of QWS. Although the Cu 3d bands of the overlayer did not show any recognizable spin polarization, the QWS clearly showed spin polarization. Spin polarized QWS were clearly resolved between 6 and 9 ML coverage. A pair of minority spin features and a majority spin feature were found to shift towards E_F with increasing coverage. The possible origin of the spin polarization of these states was discussed [78].

Layer-by-layer resolved QWS were observed in ultrathin Ag islands of different thicknesses on highly oriented pyrolytic graphite (HOPG) [79]. Ag was deposited onto a clean HOPG at 50 K and then annealed at room temperature for 1 min. Ag

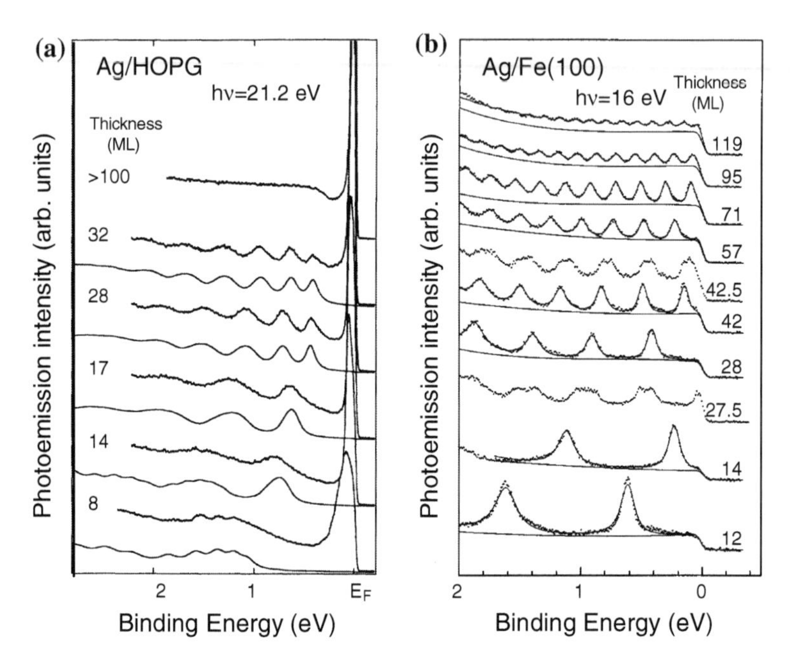

Fig. 6.19 **a** Normal emission spectra of ultrathin Ag islands grown on HOPG measured at hν = 21.2 eV [79]. Calculated spectra are also shown by *smooth curves* below each experimental spectrum. **b** Normal emission spectra (*dots*) of Ag/Fe (100) with various Ag thickness measured at hν = 16 eV and 100 K [81]. Background in the form of a slightly curved step function as well as the fits are added

(111) surfaces were exposed to the vacuum. Since the photoemission spectral shape from the HOPG was visible, Ag islands with (111) surface were thought to be formed under the employed experimental conditions. Measurement was made at hν = 21.2 eV, where strong Ag 4d states were observed in the E_B region between 4 and 7 eV. Then the sp band appeared as flat and structureless up to E_F. If magnified, however, oscillatory structures were clearly observed demonstrating the QWS as shown in Fig. 6.19(a), though the intense structure at E_F was identified as the well known surface state of Ag (111). The spectral width dependence on island thickness and on E_B was well described within the phase accumulation model with considering the QW barriers [79].

There are several review articles on photoelectron spectroscopy of QWS [80, 81], where various results and detailed discussions are provided in addition to many references. So only one more example is explained here. In the case of Ag on Fe (100), for example, growth of uniform films in the thickness range of 1 to 100 ML was known for which the Fabry–Perot formula could describe the results over the entire two orders of magnitude different Ag film thickness range [81, 82]. The normal emission spectra measured at hν = 16 eV on Ag/Fe (100) with the Ag thickness of 12–119 ML were reported (Fig. 6.19(b)). The multiple structures are

QWS derived from the Ag sp valence electrons. Among the results shown there, the 27.5 ML and 42.5 ML results showed double peak structures resulting from the mixture of 27 and 28 ML films as well as 42 and 43 ML films. Except for these two cases uniform films were grown on Fe (100) substrates. Ag films were prepared by depositing Ag onto a Fe (100) whisker kept at 100 K followed by annealing at 300 K. The nearly free-electron-like Ag sp electrons were confined in the film, resulting in the QWS. Extensive ARPES on QWS was also performed on such examples as Ag/Au(111), Ag/Cu(111) and Ag/Ni(111) systems. A detailed study has demonstrated that it was possible to discuss the lifetime broadening and the interfacial reflectivity and phase shift as well. Quantification of the ARPES results was achieved via a simple phase analysis based on the Bohr-Sommerfeld quantization rule. By taking the explicit form of wave functions into account, a detailed understanding of simple QWS including the effects of lattice mismatch was developed, providing a useful basis for studying the properties of multilayers. Examples of multilayer systems including coupled quantum wells and superlattices were also reported [81].

References

1. G.W. Gobeli, F.G. Allen, E.O. Kane, Phys. Rev. Lett. **12**, 94 (1964)
2. H. Becker, E. Dietz, U. Gerhardt, H. Angemüller, Phys. Rev. **B12**, 2084 (1975)
3. E. Dietz, H. Becker, U. Gerhardt, Phys. Rev. Lett. **36**, 1397 (1976)
4. E.W. Plummer, W. Eberhardt, Phys. Rev. **B20**, 1444 (1979)
5. W. Eberhardt, F.J. Himpsel, Phys. Rev. **B21**, 5572 (1980)
6. E. Dietz, F.J. Himpsel, Solid State Commun. **30**, 235 (1979)
7. J.A. Knapp, F.J. Himpsel, D.E. Eastman, Phys. Rev. **B19**, 4952 (1979)
8. J. Anderson, G.J. Lapeyre, R.J. Smith, Phys. Rev. **B17**, 2436 (1978)
9. N.E. Christensen, B. Feuerbacher, Phys. Rev. **B10**, 2349 (1974)
10. I. Petroff, C.R. Viswanathan, Phys. Rev. **B4**, 799 (1971)
11. B. Feuerbacher, R.F. Willis, Phys. Rev. Lett. **36**, 1339 (1976)
12. L. Kipp, K. Roßnagel, C. Solterbeck, T. Strasser, W. Schattke, M. Skibowski, Phys. Rev. Lett. **83**, 5551 (1999)
13. S. Harm, R. Dürig, R. Manzke, M. Skibowski, R. Claessen, J.W. Allen, J. Electron Spectrosc. Rel. Phenom. **68**, 111 (1994)
14. R. Claessen, R.O. Anderson, J.W. Allen, C.G. Olson, C. Janowitz, W.P. Ellis, S. Harm, M. Kalning, R. Manzke, M. Skibowski, Phys. Rev. Lett. **69**, 808 (1992)
15. R. Claessen, R.O. Anderson, G.-H. Gweon, J.W. Allen, W.P. Ellis, C. Janowitz, C.G. Olson, Z.X. Shen, V. Eyert, M. Skibowski, K. Friemelt, E. Bucher, S. Hüfner, Phys. Rev. **B54**, 2453 (1996)
16. B. Dardel, D. Malterre, M. Grioni, P. Weibel, Y. Baer, J. Voit, D. Jérôme, Europhys. Lett. **24**, 687 (1993)
17. C.G. Olson, R. Liu, D.W. Lynch, R.S. List, A.J. Arko, B.W. Veal, Y.C. Chang, P.Z. Jiang, A.P. Paulikas, Phys. Rev. **B42**, 381 (1990)
18. C. Hodges, H. Smith, J.W. Wilkins, Phys. Rev. **B4**, 302 (1971)
19. T. Valla, A.V. Fedorov, P.D. Johnson, S.L. Hulbert, Phys. Rev. Lett. **83**, 2085 (1999)
20. Th Böker, A. Müller, J. Augustin, C. Janowitz, R. Manzke, Phys. Rev. **B60**, 4675 (1999)

21. H. Kumigashira, H.-D. Kim, T. Ito, A. Ashihara, T. Takahashi, T. Suzuki, M. Nishimura, O. Sakai, Y. Kaneta, H. Harima, Phys. Rev. **B58**, 7675 (1998)
22. H. Kumigashira, H.-D. Kim, A. Ashihara, A. Chainani, T. Yokoya, T. Takahashi, A. Uesawa, T. Suzuki, Phys. Rev. **B56**, 13654 (1997)
23. This technique is discussed later in Sect. 9.3 as ELEARPES
24. A. Takayama, S. Souma, T. Sato, T. Arakawa, T. Takahashi, J. Phys. Soc. Jpn. **78**, 073702 (2009)
25. J.D. Denlinger, G.-H. Gweon, J.W. Allen, C.G. Olson, M.B. Maple, J.L. Sarrao, P.E. Armstrong, Z. Fisk, H. Yamagami, J. Electron Spectrosc, Rel. Phenom. **117–118**, 347 (2001)
26. E.H. Lieb, F.Y. Wu, Phys. Rev. Lett. **20**, 1445 (1968)
27. C. Kim, A.Y. Matsuura, Z.-X. Shen, N. Motoyama, H. Eisaki, S. Uchida, T. Tohyama, S. Maekawa, Phys. Rev. Lett. **77**, 4054 (1996)
28. C. Kim, Z.-X. Shen, N. Motoyama, H. Eisaki, S. Uchida, T. Tohyama, S. Maekawa, Phys. Rev. **B56**, 15589 (1997)
29. B.J. Kim, H. Koh, E. Rotenberg, S.-J. Oh, H. Eisaki, N. Motoyama, S. Uchida, T. Tohyama, S. Maekawa, Z.-X. Shen, C. Kim, Nat. Phys. **2**, 397 (2006)
30. R. Claessen, M. Sing, U. Schwingenschlögl, P. Blaha, M. Dressel, C.S. Jacobsen, Phys. Rev. Lett. **88**, 096402 (2002)
31. S. Kagoshima, H. Nagasawa, T. Sambongi, *One Dimensional Conductors* (Springer, Berlin, 1987), pp. 1–3
32. K. Penc, K. Hallberg, F. Mila, H. Shiba, Phys. Rev. **B55**, 15475 (1997)
33. J. Voit, J. Phys. Condens. Matter **5**, 8305 (1993)
34. J. Voit, Rep. Prog. Phys. **58**, 977 (1995)
35. A.K. Zhuravlev, M.I. Katsnelson, Phys. Rev. **B64**, 033102 (2001)
36. D.J. Thouless, M. Kohmoto, M.P. Nightingale, M. den Nijs, Phys. Rev. Lett. **49**, 405 (1982)
37. X.G. Wen, Adv. Phys. **44**, 405 (1995)
38. S. Murakami, N. Nagaosa, S.-C. Zhang, Phys. Rev. Lett. **93**, 156804 (2004)
39. C.L. Kane, E.J. Mele, Phys. Rev. Lett. **95**, 146802 (2005). ibid. 95, 226801 (2005)
40. L. Fu, C.L. Kane, E.J. Mele, Phys. Rev. Lett. **98**, 106803 (2007)
41. J.E. Moore, L. Balents, Phys. Rev. B **75**, 121306 (2007)
42. R. Roy, Phys. Rev. B **79**, 195322 (2009)
43. X.-L. Qi, Y.-S. Wu, S.-C. Zhang, Phys. Rev. **B74**, 085308 (2006)
44. L. Fu, C.L. Kane, Phys. Rev. B **76**, 045302 (2007)
45. M. König, S. Wiedmann, C. Brüne, A. Roth, H. Buhmann, L.W. Molenkamp, X.L. Qi, S.C. Zhang, Science **318**, 766 (2007)
46. D. Hsieh, D. Qian, L. Wray, Y. Xia, Y.S. Hor, R.J. Cava, M.Z. Hasan, Nature **452**, 970 (2008)
47. Y. Xia, D. Qian, D. Hsieh, L. Wray, A. Pal, H. Lin, A. Bansil, D. Grauer, Y.S. Hor, R.J. Cava, M.Z. Hasan, Nat. Phys. **5**, 398 (2009)
48. H. Zhang, C.X. Liu, X.L. Qi, X. Dai, Z. Fang, S.C. Zhang, Nat. Phys. **5**, 438 (2009)
49. M.Z. Hasan, C.L. Kane, Rev. Mod. Phys. **82**, 3045 (2010)
50. B.A. Bernevig, T.L. Hughes, S.-C. Zhang, Science **314**, 1757 (2006)
51. H. Fukuyama, R. Kubo, J. Phys. Soc. Jpn. **28**, 570 (1970)
52. D. Hsieh, Y. Xia, D. Qian, L. Wray, J.H. Dil, F. Meier, J. Osterwalder, L. Patthey, J.G. Checkelsky, N.P. Ong, A.V. Fedorov, H. Lin, A. Bansil, D. Grauer, Y.S. Hor, R.J. Cava, M.Z. Hasan, Nature **460**, 1101 (2009)
53. Y.L. Chen, J.G. Analytis, J.-H. Chu, Z.K. Liu, S.-K. Mo, X.L. Qi, H.J. Zhang, D.H. Lu, X. Dai, Z. Fang, S.C. Zhang, I.R. Fisher, Z. Hussain, Z.-X. Shen, Science **325**, 178 (2009)
54. D. Hsieh, Y. Xia, D. Qian, L. Wray, F. Meier, J.H. Dil, J. Osterwalder, L. Patthey, A.V. Fedorov, H. Lin, A. Bansil, D. Grauer, Y.S. Hor, R.J. Cava, M.Z. Hasan, Phys. Rev. Lett. **103**, 146401 (2009)
55. T. Sato, K. Segawa, H. Guo, K. Sugawara, S. Souma, T. Takahashi, Y. Ando, Phys. Rev. Lett. **105**, 136802 (2010)

56. K. Kuroda, M. Ye, A. Kimura, S.V. Eremeev, E.E. Krasovskii, E.V. Chulkov, Y. Ueda, K. Miyamoto, T. Okuda, K. Shimada, H. Namatame, M. Taniguchi, Phys. Rev. Lett. **105**, 146801 (2010)
57. A. Damascelli, Z. Hussain, Z.-X. Shen, Rev. Mod. Phys. **75**, 473 (2003)
58. T. Yoshida, M. Hashimoto, I.M. Vishik, Z.-X. Shen, A. Fujimori, J. Phys. Soc. Jpn. **81**, 011006 (2012)
59. T. Yokoya, T. Kiss, A. Chainani, S. Shin, M. Nohara, H. Takaagi, Science **294**, 2518 (2001)
60. S. Tsuda, T. Yokoya, T. Kiss, Y. Takano, K. Togano, H. Kito, H. Ihara, S. Shin, Phys. Rev. Lett. **87**, 177006 (2001)
61. H. Uchiyama, K.M. Shen, S. Lee, A. Damascelli, D.H. Lu, D.L. Feng, Z.-X. Shen, S. Tajima, Phys. Rev. Lett. **88**, 157002 (2002)
62. S. Tsuda, T. Yokoya, Y. Takano, H. Kito, A. Matsushita, F. Yin, J. Itoh, H. Harima, S. Shin, Phys. Rev. Lett. **91**, 127001 (2003)
63. H. Harima, Physica C **378–381**, 18 (2002)
64. A. Kanigel, M.R. Norman, M. Randeria, U. Chatterjee, S. Souma, A. Kaminski, H.M. Fretwell, S. Rosenkranz, M. Shi, T. Sato, T. Takahashi, Z.Z. Li, H. Raffy, K. Kadowaki, D. Hinks, L. Ozyuzer, J.C. Campzano, Nat. Phys. **2**, 447 (2006)
65. H.-B. Yang, J.D. Rameau, Z.-H. Pan, G.D. Gu, P.D. Johnson, H. Claus, D.G. Hinks, T.E. Kidd, Phys. Rev. Lett. **107**, 047003 (2011)
66. K.Y. Yang, T.M. Rice, F.C. Zhang, Phys. Rev. **B73**, 174501 (2006)
67. H. Ding, P. Richard, K. Nakayama, K. Sugawara, T. Arakane, Y. Sekiba, A. Takayama, S. Souma, T. Sato, T. Takahashi, Z. Wang, X. Dai, Z. Fang, G.F. Chen, J.L. Luo, N.L. Wang, Europhys. Lett. **83**, 47001 (2008)
68. J. Fink, S. Thirupathaiah, R. Ovsyannikov, H.A. Dürr, R. Follath, Y. Huang, S. de Jong, M.S. Golden, Y.-Z. Zhang, H.O. Jeschke, R. Valentí, C. Felser, S.D. Farahani, M. Rotter, D. Johrendt, Phys. Rev. **B79**, 155118 (2009)
69. L.X. Yang, Y. Zhang, H.W. Ou, J.F. Zhao, D.W. Shen, B. Zhou, J. Wei, F. Chen, M. Xu, C. He, Y. Chen, Z.D. Wang, X.F. Wang, T. Wu, G. Wu, X.H. Chen, M. Arita, K. Shimada, M. Taniguchi, Z.Y. Lu, T. Xiang, D.L. Feng, Phys. Rev. Lett. **102**, 107002 (2009)
70. C. Liu, G.D. Samolyuk, Y. Lee, N. Ni, T. Kondo, A.F. Santander-Syro, S.L. Bud'ko, J.L. McChesney, E. Rotenberg, T. Valla, A.V. Fedorov, P.C. Canfield, B.N. Harmon, A. Kaminski, Phys. Rev. Lett. **101**, 177005 (2008)
71. T. Yoshida, I. Nishi, S. Ideta, A. Fujimori, M. Kubota, K. Ono, S. Kasahara, T. Shibauchi, T. Terashima, Y. Matsuda, H. Ikeda, R. Arita, Phys. Rev. Lett. **106**, 117001 (2011)
72. D.J. Singh, Phys. Rev. **B78**, 094511 (2008)
73. H. Shishido, A.F. Bangura, A.I. Coldea, S. Tonegawa, K. Hashimoto, S. Kasahara, P.M.C. Rourke, H. Ikeda, T. Terashima, R. Settai, Y. Ōnuki, D. Vignolles, C. Proust, B. Vignolle, A. McCollam, Y. Matsuda, T. Shibauchi, A. Carrington, Phys. Rev. Lett. **104**, 057008 (2010)
74. R. Houdré, C. Hermann, G. Lampel, P.M. Frijlink, A.C. Gossard, Phys. Rev. Lett. **55**, 734 (1985)
75. T. Miller, A. Samsavar, G.E. Franklin, T.-C. Chiang, Phys. Rev. Lett. **61**, 1404 (1988)
76. K. Garrison, Y. Chang, P.D. Johnson, Phys. Rev. Lett. **71**, 2801 (1993)
77. C. Carbone, E. Vescovo, O. Rader, W. Gudat, W. Eberhardt, Phys. Rev. Lett. **71**, 2805 (1993)
78. J.E. Ortega, F.J. Himpsel, Phys. Rev. Lett. **69**, 844 (1992)
79. F. Patthey, W.-D. Schneider, Phys. Rev. **B50**, 17560 (1994)
80. F.J. Himpsel, J.E. Ortega, G.J. Mankey, R.F. Willis, Adv. Phys. **47**, 511 (1998)
81. T.-C. Chiang, Surf. Sci. Rep. **39**, 181 (2000)
82. J.J. Paggel, T. Miller, T.-C. Chiang, Science **283**, 1709 (1999)

Chapter 7
High Resolution Soft X-ray Angle-Integrated and -Resolved Photoelectron Spectroscopy of Correlated Electron Systems

As discussed in Chap. 4 and Chap. 5, the bulk sensitivity becomes essential in photoelectron spectroscopy when one studies the electronic structure of strongly correlated electron systems where the effect of the short range Coulomb repulsion U becomes decisive. The relative magnitude between the U and the band width W is the key factor to affect the itinerancy or localized character of the electronic structure. Since U/W is usually larger in the surface region than in the bulk, less itinerant electronic structures are often realized in the surface region. The thickness of this region depends on the individual materials. In order to reduce the relative weight from the surface in the photoemission spectra, a longer probing depth is desirable. The bulk sensitivity is also required to study the electronic structures of solids with surface reconstruction and/or surface relaxation. In comparison with conventional PES and ARPES with electron kinetic energies E_K of $\sim$15–200 eV, the soft X-ray PES and -ARPES (SXPES, SXARPES) beyond 500 eV are more plausible to detect the bulk electronic structures of such materials. Typical examples of SXPES are given in Sect. 7.1 on various rare earth (RE) compounds represented by Ce and Yb compounds and transition metal (TM) compounds represented by V, Mn and Fe oxides. Then some SXARPES results are given in Sect. 7.2 on Ce compounds, high T_C cuprates, ruthenates as well as other materials such as V_6O_{13}, $SrCuO_2$, VSe_2, $LaRu_2P_2$ and BiTeI. Finally photoelectron spectroscopy by mean of the standing wave method (SW) is very briefly introduced. In the case when these results are still not enough for understanding their bulk electronic structures, hard X-ray PES and ARPES (HAXPES, HAXARPES) should be performed as explained later in Chap. 8.

7.1 Angle-Integrated Soft X-ray Photoelectron Spectroscopy

In the case of 4f rare earth (RE) compounds, RPES was often employed to extract the RE 4f contributions buried in or mixed with other electronic states with high or comparable PICS. Gradually recognized, however, were clear discrepancies

S. Suga and A. Sekiyama, *Photoelectron Spectroscopy*, 155
Springer Series in Optical Sciences 176, DOI: 10.1007/978-3-642-37530-9_7,

between the 4f electronic structure derived by the 4d-4f RPES and the electronic properties derived from the electron conductivity, specific heat and other bulk sensitive measurements. Such discrepancies were clearly known in the case of Ce compounds. So we start the discussion from the case of Ce systems then move to Yb systems and further move to several transition metal (TM) compounds.

7.1.1 Ce Compounds

From the viewpoint of strongly correlated electron systems, the understanding of the Ce 4f electronic states in solids or compounds may be one of the most important issues, since the decisive effects of the electron–electron interactions, rather than those of electron–phonon interactions, lattice distortions, orbital degrees of freedom, and so on, were clearly seen in most Ce compounds. In addition, such intriguing phenomena were reported as unconventional superconductivity (found in 1979 for $CeCu_2Si_2$ [1] before the discovery of high-temperature superconductors [2]) in the vicinity of magnetic ordering in the phase diagram for several Ce compounds [3]. The nominal 4f electron number is one for the Ce^{3+} ion, which might yield the competition between hybridization and correlation effects in the Ce compounds. In the spectroscopic field, the 4f electronic structure of intermetallic Ce compounds was examined by Ce 3d core-level photoemission (photoelectron kinetic energy $E_K \sim 600$ eV by XPS in home laboratory with using an Al-Kα line hν = 1,486.6 V), inverse photoemission by bremsstrahlung isochromat spectroscopy (BIS, incident electron energy of ~1,500 eV) and 4d-4f RPES ($E_K \sim 120$ eV) in the 1980s [4–7]. Meanwhile theories including such as the so-called GS (Gunnarsson-Schönhammer) model were developed [6–9]. However, various theoretical analyses failed to simultaneously reproduce all the spectra by using a unique set of parameters for each compound, especially the 4d-4f RPES spectra directly representing the Ce 4f states. In the 1990s, mutually different valence-band RPES spectral shapes were reported under the 4d-4f and 3d-4f ($E_K \sim 880$ eV) resonance conditions for Ce compounds as shown in Fig. 7.1 [10, 11]. The difference in the α-Ce spectra at these two hν (or E_k) was thought to be due to the much different surface sensitivities. Considering its larger bulk sensitivity due to the longer photoelectron inelastic mean free path λ_{mp} [12, 13], the Ce 3d-4f resonance photoemission is thought to be a very promising technique to probe bulk Ce 4f states. The high-resolution SXPES (total energy resolution of ~100 meV at hν ~ 1,000 eV) became feasible since 2000 [14, 15] at SPring-8, and then data on high resolution Ce 3d-4f resonance photoemission spectra were accumulated with resolving the fine structures near E_F resulting from the Ce 4f spin–orbit interaction in intermetallic Ce compounds.

Since the 4f electron number is really small in Ce systems, the Ce 4f contributions in the valence-band spectra of alloys and compounds were rather difficult to be estimated because of the strong overlap with those from other valence-band orbital components and hence could not be extracted by photoemission at hν under

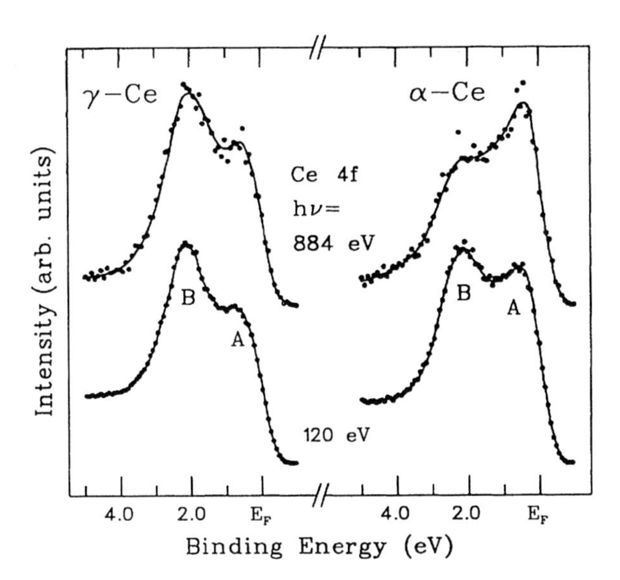

Fig. 7.1 Resonant valence-band photoemission spectra of γ- and α-Ce metals, taken at the $4d$-$4f$ ($h\nu = 120$ eV) and $3d$-$4f$ ($h\nu = 884$ eV) resonances, respectively as reported in 1991 [10]. In order to facilitate direct comparison, the energy resolutions were set equal to both (=700 meV, corresponding to the experimental total energy resolution in the $3d$-$4f$ resonance photoemission)

off-resonance conditions below the Ce core-level absorption threshold. This situation results from the weak Ce 4f PICS relative to those of other orbitals in many cases [16]. In order to reliably evaluate the Ce 4f photoemission spectral weight in the valence bands, RPES was employed. The detailed process of the RPES and 4f-photoexcited final states are now discussed. If the photon energy is on the core absorption threshold (either the Ce 3d or 4d levels), at least two photoexcitation processes can take place as schematically shown in Fig. 7.2. The most decisive process under the resonance excitation condition is the direct recombination type Auger decay process between the photoexcited electron and the core hole (Fig. 7.2(a)) after the core hole excitation to the unoccupied 4f state, where the photoexcited 4f electron and the initially occupied $4f^n$ electrons are no more discriminated. Another process is the direct photoemission process (Fig. 7.2(b)). For the Ce 3d-4f RPES, the former process is expressed as

$$\left|3d^{10}4f^{n}\right\rangle + h\nu \rightarrow \left|3d^{9}4f^{n+1}\right\rangle \rightarrow \left|3d^{10}4f^{n-1}\right\rangle + \varepsilon l, \tag{7.1}$$

where εl denotes the electron to be emitted into the vacuum. Meanwhile the direct photoemission process is expressed as

$$\left|3d^{10}4f^{n}\right\rangle + h\nu \rightarrow \left|3d^{10}4f^{n-1}\right\rangle + \varepsilon l, \tag{7.2}$$

as already explained in Sect. 3.6.2. Since both final states are mutually identical and cannot be discriminated, quantum interference between these two processes occurs, inducing a Fano-type resonance [17]. Then the so-called resonance enhancement of the contributions from the specific orbital (4f orbital in this case) can be seen in the photoemission spectra.

As an example, the hν dependence of the valence-band photoemission spectra near the 3d-4f absorption threshold as well as the absorption spectrum (XAS) itself

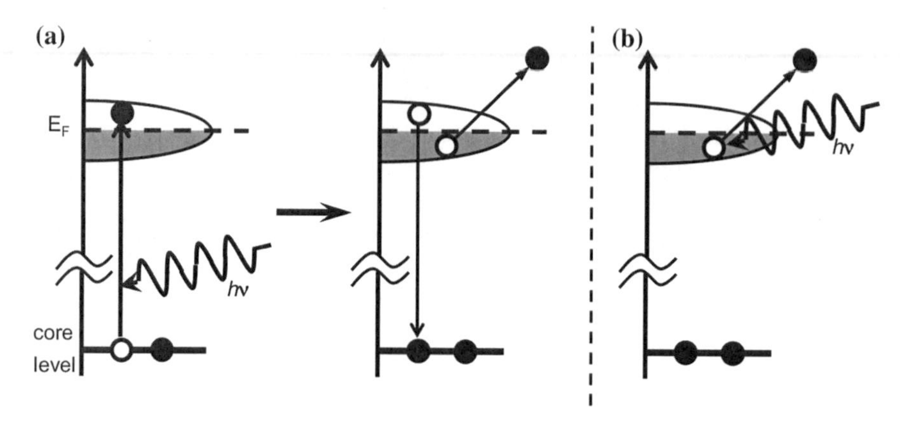

Fig. 7.2 Schematic representation of **a** The direct recombination process of the photoexcited electron–hole pair, and **b** the direct valence-band photoemission process

are shown for CePdSn in Fig. 7.3 [18], where the photoemission intensities were normalized by the photon flux. It is known that CePdSn undergoes an antiferromagnetic (AFM) transition at 7 K [19] due to the RKKY (Ruderman-Kittel-Kasuya-Yosida) interactions between the localized 4f electrons [20–22]. In the absorption spectrum, two peaks and some shoulder structures are observed at $h\nu$ = 881.6 (D) and 882.6 (G) and around 880 eV in Fig. 7.3(a). In the PES shown in Fig. 7.3(b) the off-resonance spectrum A taken at $h\nu = 875$ eV shows a broad peak at $E_B \sim 3.5$ eV, which is ascribed to the Pd 4d states [23]. The intensity of the spectrum B taken at $h\nu = 879.8$ eV, corresponding to the small shoulder in the XAS spectra, was rather similar to that of the spectrum A except for the region near E_F. Going from the spectrum B to D, the photoemission intensities enhance greatly due to the Ce 3d-4f resonance. The spectral weight is once reduced in the entire E_B region in the spectrum E, but it is again enhanced in the spectra F and G (corresponding to the resonance maximum near $h\nu = 882.6$ eV). It is noticed that the degree of the resonance enhancement is related to the XAS intensity at the excitation $h\nu$.

Under the resonance enhancement condition, two comparable peaks at 2.0 and 0.3 eV are prominent in Fig. 7.3(b). The former was ascribed to the "$4f^0$" final states while the latter was due to the "$4f^1$" final states, which were briefly explained by a simplified single impurity Anderson model (SIAM) shown in Fig. 7.4. Here, it is assumed that the valence band, which could be hybridized with the 4f level (with degeneracy $N_f = 14$ with omitting the spin–orbital interactions), is only located at E_F for simplicity, and the Coulomb repulsion energy between the 4f electrons $U_{ff} \rightarrow \infty$ for simplicity. This situation corresponds to the states with forbidden double 4f occupancy. In SIAM, the ground state $|g\rangle$ is represented as the linear combination of the $|4f^0\rangle$ and $|\underline{c}4f^1\rangle$ states, where $\underline{c}$ shows a hole in the valence band. The former represents the state with all electrons in the valence band at E_F with no 4f electron ("Fermi sea" state) and the latter denotes the state with

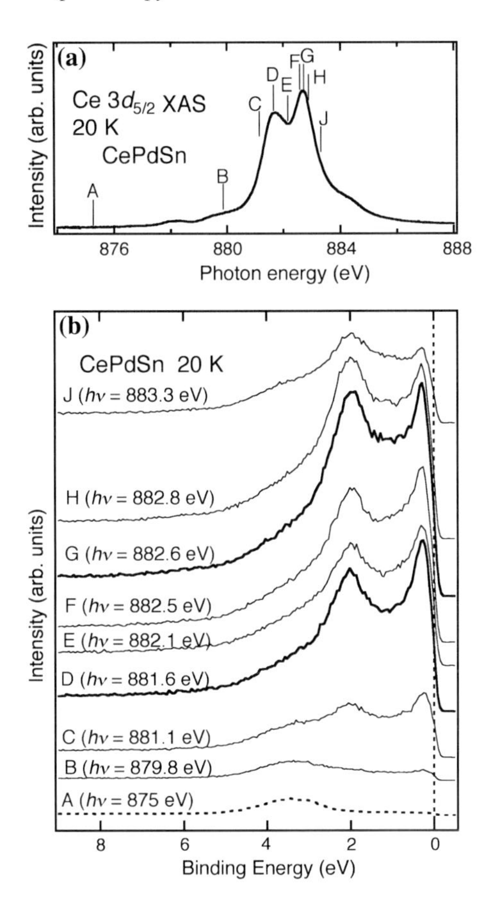

Fig. 7.3 **a** Ce $3d_{5/2}$ core absorption (XAS) spectrum of CePdSn with a resolution <200 meV. **b** Ce 3d-4f resonance photoemission spectra of CePdSn with a resolution of 250 meV. The *dashed line* spectrum A corresponds to the off-resonance spectrum. The *bold solid lines* show the resonance spectra D and G taken at hν where the strong peaks are seen in the XAS spectra [18]

one electron transferred from the valence band to the 4f level at ε_f (<0 for the occupied side) relative to E_F, as

$$|g\rangle = a|4f^0\rangle + b|\underline{c}4f^1\rangle, \tag{7.3}$$

where $a^2 + b^2 = 1$. The 4f occupation number n_f is thus $|b|^2$ (<1). The Hamiltonian for this model is written as

$$\mathcal{H} = \begin{pmatrix} 0 & \sqrt{N_f}V \\ \sqrt{N_f}V & \varepsilon_f \end{pmatrix}, \tag{7.4}$$

where V denotes the hybridization strength between the 4f and valence levels. The eigen energies are obtained as

$$E_{\pm} = \frac{\varepsilon_f \pm \sqrt{\varepsilon_f^2 + 4N_f V^2}}{2}. \tag{7.5}$$

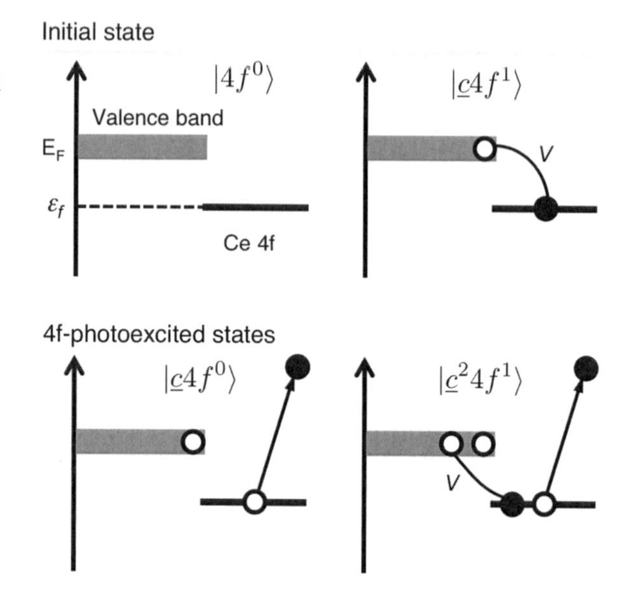

Fig. 7.4 Schematic representation of the electronic states of the initial and 4*f*-photoexcited final states for Ce compounds by using a simplified single impurity Anderson model

The 4f-photoexcited final states are approximately expressed using the same Hamiltonian as (7.4) by

$$|``4f^0"\rangle = b|\underline{c}4f^0\rangle - a|\underline{c}^24f^1\rangle, \quad (7.6)$$

$$|``4f^1"\rangle = a|\underline{c}4f^0\rangle + b|\underline{c}^24f^1\rangle, \quad (7.7)$$

respectively, where the total energy is "E_+" for the "$4f^0$" final state and "E_-" for the "$4f^1$" final state. Here the 4f-orbital degeneracy is much larger than 1 and hence $\sqrt{N_f = 14}$ is used in the off-diagonal components instead of $\sqrt{N_f - 1}$ for simplicity. Then "E_+" and "E_-" become equal to E_+ and E_-. Since the ground state energy is E_- in this model, the "$4f^0$" final state appeared at the binding energy (E_B) of "E_+" $-$ $E_- = \sqrt{\varepsilon_f^2 + 56V^2}$ with the spectral weight of $|b^2|^2 = b^4$ in the spectra (note that the initial $|\underline{c}4f^1\rangle$ component is directly connected with the final $|\underline{c}4f^0\rangle$ state in the 4f-removal process) whereas the "$4f^1$" final state appears at E_B of "E_-" – $E_- = 0$ (E_F) with the spectral weight of $|ab|^2 = a^2b^2$. The total spectral weight is of course equivalent to n_f. When V and ε_f are assumed to be 0.21 and -1.3 eV, respectively, "E_+", "E_-", a, b and n_f are calculated as 0.37 eV, –1.67 eV, 0.426, –0.905 and 0.82. Therefore, the "$4f^0$" final state would appear at the E_B = "E_+" – E_- of ~2.0 eV with the weight of b^4 ~67 %, which qualitatively explains the peak structures in the Ce 3d-4f RPES spectra of CePdSn. Although this explanation was rather crude, it facilitated the understanding of the origin of the peaks in the spectra. For an advanced analysis of the 4f spectra, a spectral simulation, in which the energy dependence of the hybridization strength

$\rho V^2(E_B)$, the 4f spin–orbit splitting Δ_{SO}, and the crystal field splitting Δ_{CEFS} in the $4f_{5/2}$ level are taken into account, should be performed as described later.

The possible effects of an Auger emission (in a narrow sense discriminated here from the direct recombination process) should be commented here for the present RPES spectra. The spectral line shapes D–H in Fig. 7.3 are similar, indicating no recognizable structure caused by an Auger emission. However, the relative height of the "$4f^0$" peak at 2.0 eV to the "$4f^1$" peak at 0.3 eV changes a little with $h\nu$. This might be due to a possible overlap of finite Auger emission [24].

Then bulk and surface contributions will be discussed through a comparison between Ce 3d-4f and 4d-4f RPES. It was generally seen that the observed Ce 4f spectra at the bulk-sensitive 3d-4f resonance were different from the surface-sensitive 4d-4f resonance besides the data for Ce metal given in Fig. 7.1. For the Ce compounds, the strength of hybridization is often represented by the Kondo temperature (T_K). When the hybridization is stronger, T_K is higher. $CeRu_2Si_2$ discussed here is a typical heavy fermion system with $T_K \sim 20$ K [25] and a Sommerfeld (electronic specific heat) coefficient $\gamma \sim 400$ mJ/(mol·K^2) [26] without showing magnetic ordering in ambient pressure and zero magnetic field. In a more strongly hybridized compound, on the other hand, the Ce ions are strongly valence-fluctuating between the Ce^{3+} and Ce^{4+} ($4f^0$) states. In this case, it is considered that the 4f electrons are no longer localized and regarded as fairly itinerant "4f-bands". $CeRu_2$ is a typical material of the strongly valence-fluctuating 4f systems, where T_K is of the order of 1,000 K with much smaller $\gamma \sim 30$ mJ/(mol · K^2) [27] than for $CeRu_2Si_2$.

The Ce 4f spectra of $CeRu_2Si_2$ and $CeRu_2$ in a wide E_B region for the Ce 3d-4f as well as 4d-4f resonance are shown in Fig. 7.5 [14], where the 4d-4f resonance photoemission was measured at BL3B in Photon Factory. For $CeRu_2Si_2$, there is a sharp peak near E_F and a broad tail extending to $E_B \sim$ 4 eV in the 3d-4f RPES spectrum. According to the view from the SIAM, the former corresponds to the

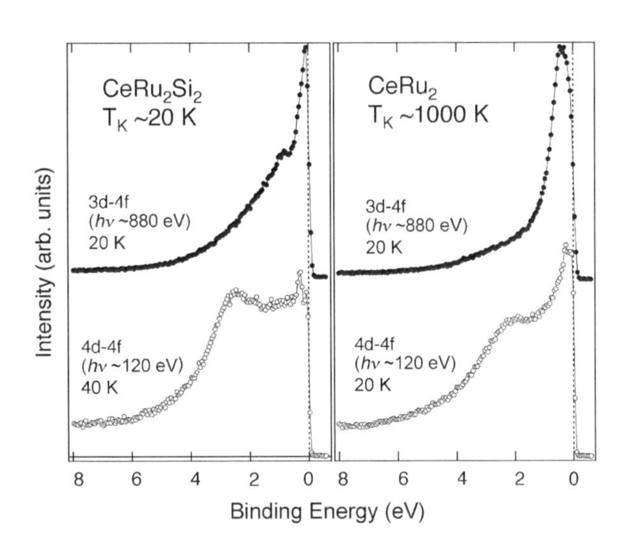

Fig. 7.5 Ce 4*f* spectra of $CeRu_2Si_2$ and $CeRu_2$ obtained from Ce 3*d*-4*f* and 4*d*-4*f* resonance photoemission, with an energy resolution of 200 and 80 meV, respectively [14]. The Ce 4f contributions were estimated by subtracting the off-resonance spectra taken at $h\nu = 875$ (114) eV from the on-resonance ones at $h\nu = 882.6$ (122) eV for the Ce 3*d*-4*f* (4*d*-4*f*) resonance. The relative intensity between the 3*d*-4*f* and 4*d*-4*f* resonance photoemission is arbitrary

contribution of the $|4f^1\rangle$ final states. The latter is ascribed to the $|4f^0\rangle$ final state. In the 4d-4f RPES spectrum on the other hand, the $|4f^0\rangle$ final state located at 2.5 eV is much stronger than in the 3d-4f RPES spectrum, reflecting the localized character of the surface 4f states.

In the case of the 3d-4f RPES spectrum of $CeRu_2$, on the other hand, the $|4f^0\rangle$ broad tail with E_B larger than 2 eV seems to have almost vanished, indicating that the bulk Ce 4f states are much more hybridized than those in $CeRu_2Si_2$. However, the $|4f^0\rangle$ final state is still clearly seen as a shoulder at 2 eV in the 4d-4f RPES spectrum as confirmed by the preceding study [28]. One may consider that the different resonance processes could yield such different spectral line shapes. However, these striking differences between the 3d-4f and 4d-4f RPES spectra could not be explained by a mere different resonance process. It should be noted that the sum of the bulk and surface 4f contributions calculated by the SIAM with considering the difference of surface sensitivities between the Ce 4f spectra derived by the 3d-4f and 4d-4f RPES excellently reproduced both RPES spectra of $CeNi_{1-x}Co_xGe_2$ [29], in which the different resonance process was not taken into account.

In the high-resolution 3d-4f and 4d-4f RPES spectra near E_F shown in Fig. 7.6 [14], one might notice that the 4d-4f RPES spectral line-shapes are qualitatively similar between $CeRu_2Si_2$ and $CeRu_2$, though T_Ks are much different. Namely, there are two comparable structures located just below E_F and near $E_B \sim 0.3$ eV. However, the 3d-4f RPES spectra of these compounds are drastically different

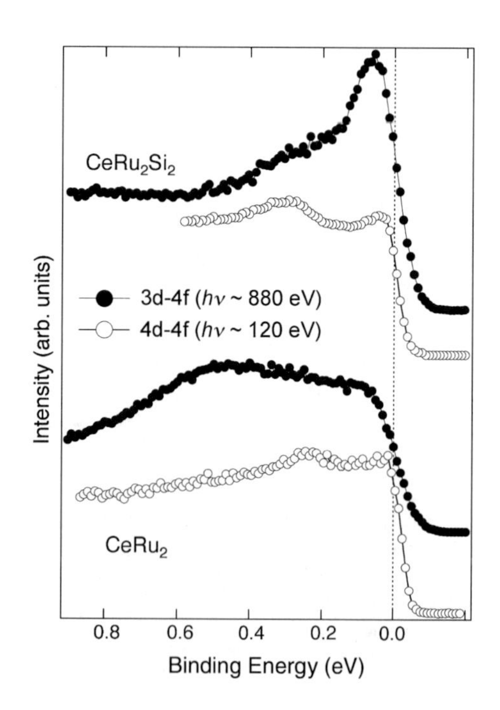

Fig. 7.6 Ce 4*f* spectra near E_F at the 3*d*-4*f* (resolution of 100 meV) and 4*d*-4*f* (resolution of 50 meV) resonances of $CeRu_2Si_2$ and $CeRu_2$. The relative spectral intensity between the 3*d*-4*f* and 4*d*-4*f* resonances and between the compounds is arbitrary. The raw spectra taken at $h\nu = 882.6$ eV are displayed for the Ce 3*d*-4*f* resonance photoemission spectra since the Ce 4f resonance enhancement is predominant. The Ce 4*f* contributions to the 4*d*-4*f* resonance are obtained by subtracting the off-resonance spectra from the on-resonance ones [14]

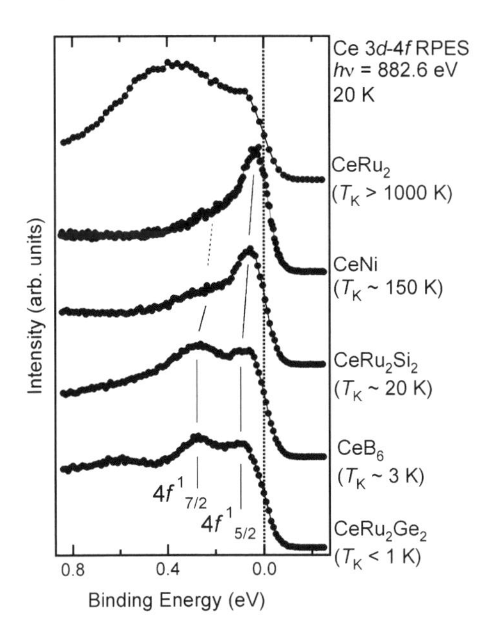

Fig. 7.7 Ce $3d$-$4f$ on-resonance photoemission spectra near E_F of $CeRu_2Ge_2$, CeB_6, $CeRu_2Si_2$, CeNi and $CeRu_2$ [30–34]

from each other, reflecting the different bulk 4f states. There are one strong and narrow peak in the vicinity of E_F and a weak structure as a shoulder at $\sim$ 0.3 eV in $CeRu_2Si_2$. On the other hand, no characteristic feature is seen except for a broad peak in the E_B region from 0.3 to 0.7 eV in $CeRu_2$. As discussed later, the spectral difference between $CeRu_2Si_2$ and $CeRu_2$ definitely reflects the different bulk electronic structures.

In Fig. 7.7 are shown representative Ce 4f spectra obtained by bulk-sensitive 3d-4f RPES with a resolution of $\sim$100 meV of heavy fermion systems from "low-" to "intermediate-" T_K and a 4f-itinerant system ("very high-" T_K) [30–34]. The samples are $CeRu_2Ge_2$ ($T_K < 1$ K) [35, 36], CeB_6 ($T_K \sim 3$ K) [37], $CeRu_2Si_2$ ($T_K \sim 20$ K), CeNi ($T_K \sim 150$ K) [38] and $CeRu_2$ ($T_K > 1{,}000$ K). There are two clear peaks with comparable intensity in the vicinity of E_F and at 0.2–0.3 eV in the spectra of $CeRu_2Ge_2$ and CeB_6. As described in Chap. 2 and in Fig. 7.8(a), the Ce 4f levels are split by $\sim$300 meV into the $4f_{7/2}$ and $4f_{5/2}$ levels due to the spin–orbit interaction, where the $4f_{5/2}$ level is further split into two or three levels by the crystalline electric field (CEF) with energy differences ranging from a few to several hundred K in solids. A similar splitting might also be observable for the $4f_{7/2}$ level, but this is less important than the $4f_{5/2}$ level. For discussion of the 4f ground state and low-energy excitations, the consideration of the $4f_{5/2}$ levels is often enough (Fig. 7.8(b)). However, all 4f levels and the hybridization effects shown in Fig. 7.8(c) should be considered to understand the experimental spectral features as well as to estimate T_K from the spectroscopic data as discussed later. The PES peak in the vicinity of E_F was ascribed to the mixture of the tail of the Kondo resonance located above E_F and its CEF partners (both components resulting from $4f^1_{5/2}$ final

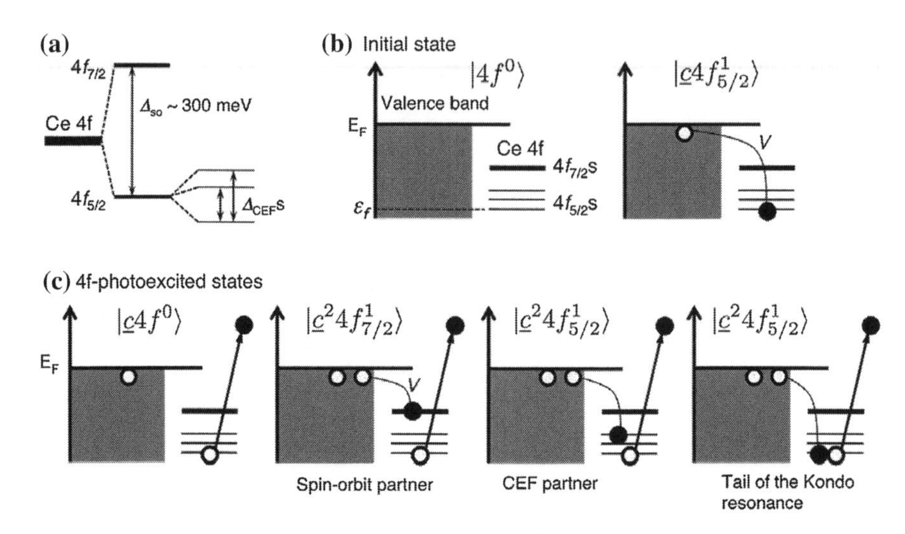

Fig. 7.8 Schematic representations of **a** 4*f* levels of a Ce^{3+} ion split into the $4f_{7/2}$ and $4f_{5/2}$ levels by the spin–orbit interaction (~300 meV), where the 4*f* levels are further split by the crystalline electric field (CEF) depending on materials ($4f_{7/2}$ splitting is omitted). **b** Electronic states of the initial state of Ce compounds by using the single impurity Anderson model. **c** Same as (**b**) but of the 4*f*-photoexcited final states. The total electron energy becomes lower (photoelectron kinetic energy becomes larger) on going from the *left* to the *right*

states) while the peak at E_B ~ 0.2–0.3 eV corresponds to the $4f^1_{7/2}$ final states. In the case of CeNi and $CeRu_2Si_2$, the tail of the Kondo resonance is prominent and its spin–orbit partner ($4f^1_{7/2}$ final states) is markedly suppressed and appears as a shoulder. The relative intensity of the $4f^1_{5/2}$–$4f^1_{7/2}$ final states becomes stronger on going from low-T_K ($CeRu_2Ge_2$) to intermediate-T_K (CeNi) compounds. In the spectrum of $CeRu_2$, on the other hand, the $4f^1_{5/2}$–$4f^1_{7/2}$ final states are no more observed. Instead, a rather conventional Fermi cut-off, namely, a tiny shoulder in the vicinity of E_F and a broad peak centered at E_B ~ 0.4 eV are observed. According to the SIAM, one would have simply expected that the tail of the Kondo resonance in the vicinity of E_F should be much stronger for the very strongly valence-fluctuating system like $CeRu_2$. However, the experimental results that the spectrum of $CeRu_2$ shows a rather conventional Fermi cut-off and the broad peak near 0.4 eV could not be even qualitatively deduced from the SIAM.

In order to further clarify these bulk Ce 4f contributions, we have calculated spectral functions based on the SIAM using the non-crossing approximation (NCA) [39–41]. The calculated spectra could properly reproduce the experimental 3d-4f on-resonance photoemission spectra near E_F as shown in Fig. 7.9, where the spin-orbit splitting of the 4f levels was adjusted to be 310 meV for CeNi and $CeRu_2X_2$ (X = Si, Ge), 315 meV for CeB_6. Other optimized parameters such as the bare (unhybridized) energy of the lowest $4f_{5/2}$ level ($\varepsilon_f < 0$ for the occupied side), the CEF splitting (Δ_{CEF}), the average hybridization strength defined by

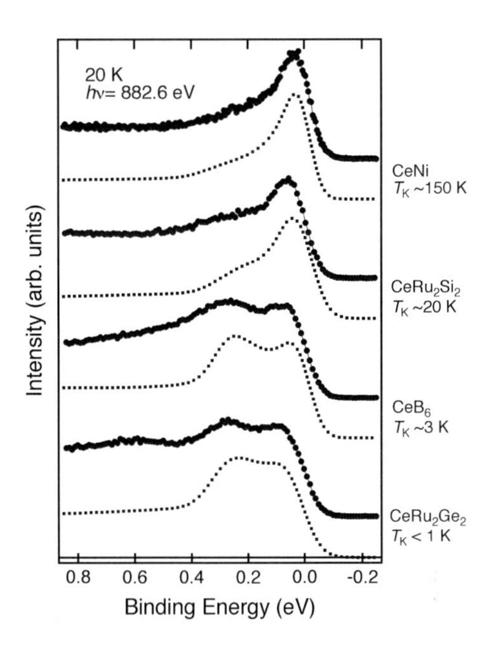

Fig. 7.9 Comparison of the 3d-4f resonance photoemission spectra near E_F with the spectra calculated using the non-crossing approximation based on the SIAM for $CeRu_2Ge_2$, CeB_6, $CeRu_2Si_2$ and CeNi [34]

$\delta \equiv (\pi/B) \int_{-B}^{0} \rho V^2(\varepsilon) d\varepsilon$ where $\rho V^2(E_B)$ is the energy dependence of the hybridization strength and B is the bottom binding energy ($E_B > 0$) of the hybridized valence band, and the hybridization strength at E_F ($\rho V^2(E_F)$) are summarized in Table 7.1 [34]. As discussed later, finite values of the CEF splitting are inevitable for the fitting of the photoemission spectra, though the CEF partners are not clearly seen with a resolution of $\sim$100 meV. For CeNi and $CeRu_2X_2$, the $4f_{5/2}$ level is split into three doublets while it is split into one quartet (Γ_8) and one doublet (Γ_7) for CeB_6. The values of Δ_{CEF} of CeNi, $CeRu_2Ge_2$ and CeB_6 listed in Table 7.1 were set to be consistent with those estimated elsewhere [36, 42, 43]. As for $CeRu_2Si_2$, no CEF peaks were observed within $\sim$450 K by an experimental inelastic-neutron-scattering study [44]. It was suggested that the CEF splitting should be close to 1,000 K to explain the large magnetic anisotropy of $CeRu_2Si_2$ [45]. As shown in Fig. 7.9, the NCA calculation well reproduces the experimental spectra near E_F. Such a successful calculation indicates the applicability of the SIAM to the bulk Ce 4f angle-integrated photoemission spectra of weakly and moderately hybridized systems.

Table 7.1 Optimized physical parameters for CeNi, $CeRu_2Si_2$, $CeRu_2Ge_2$ and CeB_6 by the NCA calculation [34]

	ε_f (eV)	Δ_{CEF} (K)	δ (meV)	ρV^2 (E_r)(meV)
CeNi	−1.58	580, 1200	103	33
$CeRu_2Si_2$	−1.6	600, 800	72	28
CeB_6	−1.8	530	73	26
$CeRu_2Ge_2$	−1.58	500, 800	57	22

In order to evaluate T_K, magnetic excitation spectra were simulated by using NCA with the same parameters as employed for fitting the RPES spectra. The obtained T_K from the magnetic excitation spectra were 150, ~16, ~8 and <1 K for CeNi, $CeRu_2Si_2$, CeB_6 and $CeRu_2Ge_2$, respectively, which were quite comparable to the real T_K of ~150, ~20, ~3 and <1 K. These results suggested that we properly chose the parameters for these compounds. Meanwhile, the NCA calculations were also performed to estimate T_K under the extreme assumption that the CEF splitting was 0 meV, which is equivalent to a sextet $4f_{5/2}$ ground state. We can fit all spectra here even under such a condition with employing different hybridization strength. However, thus estimated T_K with $\Delta_{CEF} = 0$ meV (defined as T_K^h) was >1,000 K for CeNi, >200 K for $CeRu_2Si_2$ and of the order of 40 K for CeB_6 and $CeRu_2Ge_2$. These values were too much deviated from the T_K revealed by other bulk sensitive measurements, indicating that the finite CEF effects should be taken into account for the proper analysis of the 4f photoemission spectra.

On the other hand, such high T_K^h themselves were not fully meaningless. In the T region comparable to these T_K^h, the 4f state could be regarded as six-fold degenerate, where the proper Kondo temperature T_K^h is represented as

$$T_K^h = [T_K \Delta_{CEF}(1) \Delta_{CEF}(2)]^{1/3} \text{ for CeNi and } CeRu_2X_2, \tag{7.8}$$

$$T_K^h = \left[T_K^2 \ \Delta_{CEF}\right]^{1/3} \text{ for } CeB_6 \tag{7.9}$$

from a scaling theory [46]. The validity of T_K^h was really seen in the T dependence of the resistivity in $Ce_xLa_{1-x}Al_2$ [47]. Under the present experimental condition, the measuring T was far below the assumed CEF splitting. However, the finite energy resolution (~100 meV) was beyond these values. Thus experimentally evaluated T_K^0 under the assumption of $\Delta_{CEF} = 0$ meV could have some meaning in our experiment. The formula (7.8) and (7.9) gave T_K^h of $CeRu_2Si_2$ as ~200 K using $T_K \sim 16$ K, $\Delta_{CEF}(1) = 600$ K and $\Delta_{CEF}(2) = 800$ K. The $T_K^0 \geq 200$ K estimated for $CeRu_2Si_2$ for $\Delta_{CEF} = 0$ meV may thus correspond to the T_K^h. As for CeB_6, T_K^h was estimated as 32 K if we employed our estimated $T_K \sim 8$ K. On the other hand, T_K for $CeRu_2Ge_2$ was derived as 0.2 K using $[T_K^h, \Delta_{CEF}(1), \Delta_{CEF}(2)] = (40, 500, 800)$ K. Thus the above relation of the scaling theory is fairly satisfied. $T_K^h \sim 200$ K for $CeRu_2Si_2$ might correspond to a broad peak at 220 K in the observed temperature dependence of the thermopower [48]. Therefore these analyses demonstrated that the NCA fitting of the experimental bulk 4f spectra with $T_K \leq 150$ K could yield a realistic T_K by taking the CEF splitting into account.

From the viewpoint of the itinerant limit, only the lowest "$4f^1_{5/2}$" final state was regarded as the quasi-particle band, in which the heavy quasi-particles are formed by the hybridization between the originally occupied $4f_{5/2}$ orbital and conduction bands in the vicinity of E_F. Taking the CEF partners into account, the real quasi-particle spectral weight in the 4f spectra for these itinerant Ce compounds is very weak relative to the total 4f spectral weight as schematically shown in Fig. 7.10. One could then notice that the electron correlation is much stronger for Ce

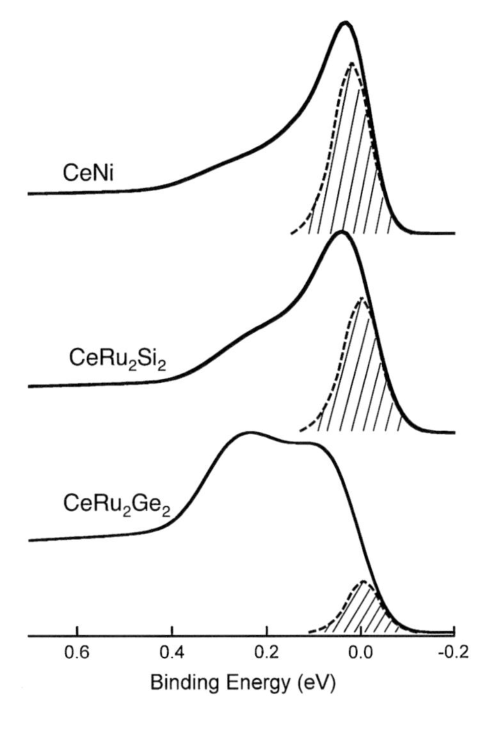

Fig. 7.10 Roughly expected lowest $4f^1_{5/2}$ final-state spectral weight equivalent to the quasi-particle spectral weight represented by the shaded area (below the *handwritten dashed curve*) for the Ce compounds compared to the total 4f spectral functions obtained by the NCA calculations

compounds than for such metallic transition metal oxides as V_2O_3 and $Sr_{1-x}Ca_xVO_3$ as discussed later in this Sect. 7.1.

As already mentioned, the 4f spectral line shape for $CeRu_2$ is qualitatively different from that for other compounds with $T_K \leq 150$ K. Furthermore, the origin of the peak at 0.4 eV was not clear at all within the SIAM. Considering the fact that the 4f electrons behave as itinerant in $CeRu_2$, this peak might originate from a rather itinerant 4f-band structure. Consequently the comparison of the spectra to the PDOS predicted from a band-structure calculation would tell us something about the electronic structure. Figure 7.11 shows the comparison of the experimentally derived Ru 4d and Ce 4f spectra to the individual PDOSs obtained from the band-structure calculation [49], which well explained the Fermi surface topology of $CeRu_2$ [50]. Here the calculated PDOS was convoluted with use of both energy-dependent Lorentzian lifetime broadening with the maximum full width at half maximum (FWHM) of 0.4 eV and Gaussian broadening with a fixed FWHM corresponding to the instrumental resolution.

A detailed comparison of the 3d-4f on-resonance photoemission spectrum near E_F with the f PDOS is shown in Fig. 7.12. Although there are some discrepancies between the spectra, the calculation qualitatively explains the essential features of the experimental 4f spectrum. In particular, the broad peak near 0.4 eV in the Ce 4f spectrum, which was not simply predicted from the SIAM, corresponds well to the band-structure calculation. A peak in the vicinity of E_F in the f PDOS might correspond to the tiny shoulder near E_F in the experimental spectrum.

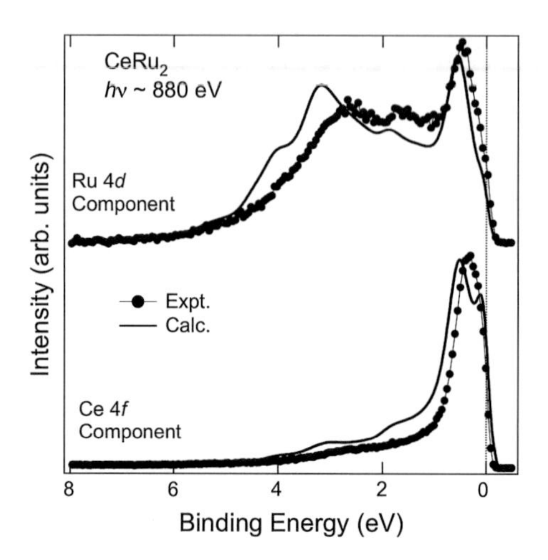

Fig. 7.11 Comparison of the off-resonance spectral shape (dominated by the Ru 4d component) and the Ce 4*f* spectral shape derived from the RPES measured with an resolution of ~200 meV to broadened Ce 4f and Ru 4d partial densities of states, respectively, obtained by band-structure calculations for $CeRu_2$ [31]. The off-resonance spectrum was measured at hν = 875 eV. The Ce 4f spectral weight was obtained by subtracting this off-resonance spectrum from the on-resonance spectrum obtained at hν = 882.6 eV

The presence of the peak at ~0.4 eV in both bulk (Ru) d and (Ce) f partial densities of state in Fig. 7.11 indicates that this peak is due to the very strong hybridization (band effects). It was found that the band-structure calculation from the itinerant picture was a good starting point to analyze bulk 4f photoemission spectra of very strongly hybridized systems. It was reported for another strongly hybridized system $CeRh_3$ that a 4f inverse-photoemission spectrum was fairly reproduced by a band-structure calculation [51].

Although the calculated f PDOS based on the one-electron band theory fairly reproduced the Ce 4f photoemission spectrum in $CeRu_2$, some quantitative disagreements were seen. The calculation did not give a proper Sommerfeld coefficient of $\gamma \sim 30$ mJ/(mol · K^2), suggesting remaining electron correlation effects. Indeed, a reasonable γ could be deduced by a calculation using perturbation expansion with respect to electron–electron interactions, where the mass-enhancement factor of the f electrons was calculated as 5.77. Therefore, we could conclude that the discrepancies between the experimental and calculated spectra seen in Fig. 7.12 originated from the Ce 4f correlation effects.

Then we discuss the possible enhancement of the photoemission from the Ce 5d state in the Ce 3d-4f RPES. So far it was implicitly assumed that only the enhancement of the 4f-contribution is predominant. On the other hand, an Auger type process with the 5d-electron emission could occur also as a result of the direct recombination between the photoexcited 4f electron and 3d core hole, with the identical final state reached by the direct 5d-photoexcitation process, although the enhancement factor is unknown. There was, however, no experimental test for

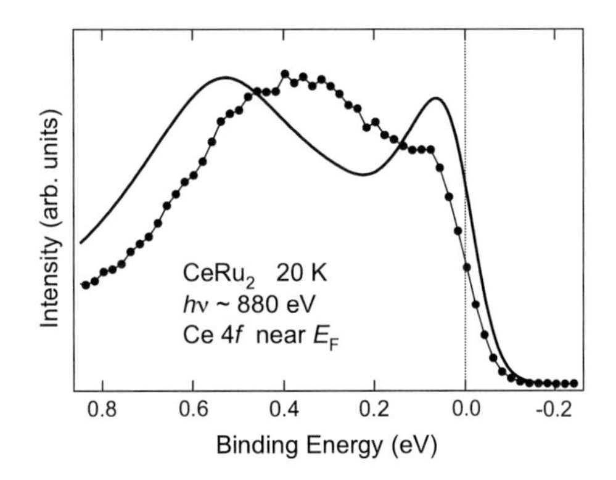

Fig. 7.12 Comparison of the Ce 3*d*-4*f* resonance photoemission spectra near E_F (resolution of ~ 100 meV) representing the 4*f* spectral function to the calculated f partial density of states for $CeRu_2$ [31]

the 5d-emission enhancement at the 3d-4f RPES except for a comparison of data between LaNiSn and CeNiSn [18] as shown in Fig. 7.13. These spectra were normalized by the following way: first the off-resonance spectrum of LaNiSn taken at hν = 866 eV was normalized to that of CeNiSn as shown in the inset of Fig. 7.13. Then both La 3d-4f and Ce 3d-4f on- and off-resonance photoemission spectra were normalized by the photon flux. Thus the on-resonance spectral intensity of LaNiSn could be compared to that of CeNiSn. In the resonance-

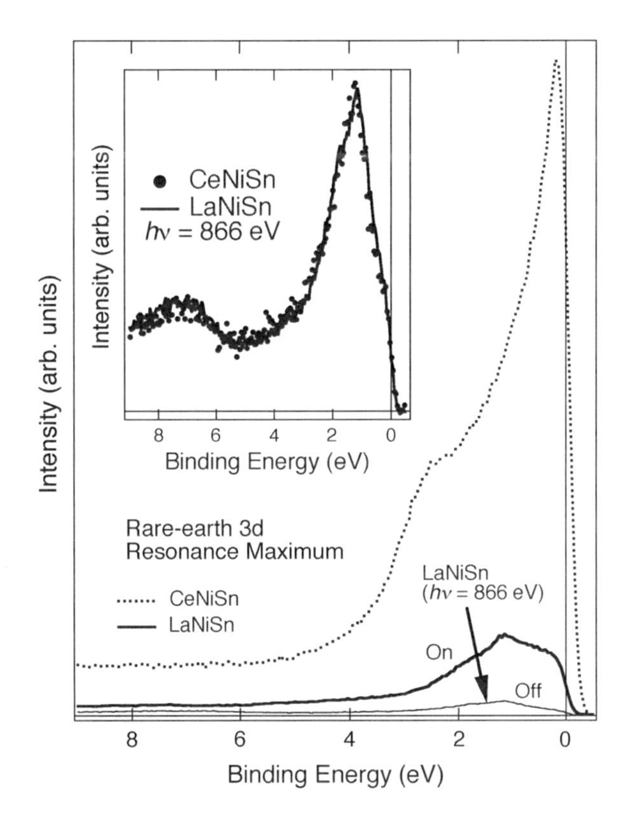

Fig. 7.13 Comparison of the La 3*d*-4*f* resonance-maximum spectrum of LaNiSn with the Ce 3*d*-4*f* resonance-maximum spectrum of CeNiSn together with the off-resonance spectra of LaNiSn. The inset shows the off-resonance photoemission spectra of LaNiSn and CeNiSn taken at $h\nu$ = 866 eV [18]

maximum spectrum of LaNiSn, a broad peak centered at 1.2 eV was observed. One might consider that this resonance enhancement in LaNiSn is mainly due to the valence bands (Ni 3d or Sn 5p states) hybridized with the empty 4f states as previously reported for semi-metallic LaSb [52], where the occupied La 5d electron number is close to zero since the Sb ion is nominally trivalent. It was as well pointed out that the 5d emission is observed in the RPES, though its contribution is much smaller than that of the 4f emission [52, 53]. Since the observed resonance enhancement in LaNiSn was significantly larger than that in LaSb, the enhancement in LaNiSn should be ascribed to the La 5d components hybridized with the Ni 3d or Sn 5p states. The spectral weight of the resonance-maximum spectrum of LaNiSn was surely enhanced in comparison with that of the off-resonance spectrum, but it is still only $\sim$15 % of the resonance-maximum overall spectral weight of CeNiSn.

7.1.2 Yb Compounds

4f electronic states in Yb systems were also attracting much attention in regard to various intriguing phenomena such as valence fluctuation, heavy Fermion behavior, Kondo resonance, phase transitions, quantum critical phenomena. Here we first discuss the 4f electronic states of high purity $YbAl_3$ single crystals. SXPES was measured at BL25SU of SPring-8 [54]. Clean surfaces were obtained by in situ fracturing. The E_B was calibrated to be within $\pm$5 meV before and after the measurement at each T. The spectrum measured at T = 20 K for hν = 700 eV is shown in Fig. 7.14(a) in a wide E_B region. In the Yb^{2+} 4f spectral region, very prominent and sharp doublet components ascribable to the $4f^{13}$ final states with J = 7/2 and 5/2 bulk components were observed. The surface components were strongly suppressed compared to the low hν PES [55–60]. Therefore a much more reliable evaluation of the bulk spectral shape and intensity was feasible. Three spectra measured at 200, 75 and 20 K are displayed in Fig. 7.14(b) in the region of the $4f^{13}$ (from the Yb^{2+} initial state) and the first $4f^{12}$ (Yb^{3+}) component $^3H_6$ of the final state. All $4f^{12}$ multiplet components (not shown) were confirmed to stay within $\pm$5 meV with T change in this T range. On the contrary, the T dependence of the $4f^{13}$ components is prominent. The peak energy shifts of both $4f^{13}$ J = 7/2 and 5/2 states from 20 to 200 K are $\sim$16 meV. The integrated intensities of both J = 7/2 and 5/2 $4f^{13}$ peaks at 20 K are $\sim$40 % larger than those at 200 K.

The surface $4f^{13}$ components were deconvoluted by assuming two sets of weak surface components [54, 61] neglecting any impurity contribution [62]. The presence of the second surface component was confirmed by low hν SXPES. The asymmetry parameter α for the Mahan's lineshape [63] was assumed to be T independent for the bulk component. An example of deconvolution at 200 K is shown in Fig. 7.15(a). The whole $4f^{12}$ multiplet spectrum was well reproduced by an atomic multiplet calculation [64]. According to this deconvolution, both the

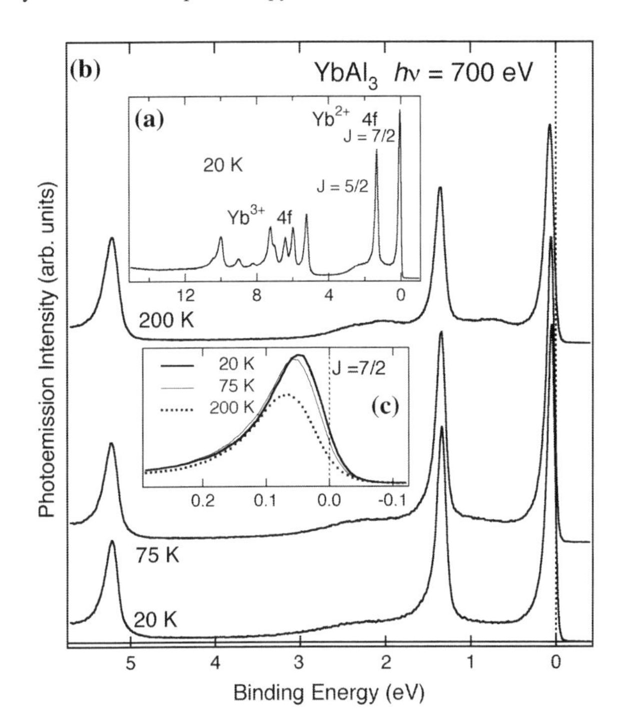

Fig. 7.14 **a** High resolution SXPES of $YbAl_3$ measured at 20 K and $h\nu$ = 700 eV on a fractured single crystal surface with a total energy resolution of 115 meV. **b** Temperature dependence of the Yb 4f spectra measured at 20, 75 and 200 K with a total energy resolution of 75 meV at $h\nu$ = 700 eV. The spectra were normalized by the integrated intensity of the $4f^{12}$ final states between E_B = 4.5 and 12 eV. **c** Detailed comparison of the temperature dependence of the $4f^{13}$ (Yb^{2+}) J = 7/2 peak [54]

J = 7/2 peak energy and the Yb mean valence were accurately evaluated as shown in Fig. 7.15(b).

The Yb mean valences evaluated from our SXPES as 2.71 (200 K)–2.65 (20 K) (Fig. 7.15(b)) were much smaller than the values (2.88 (300 K) $\sim$ 2.77 (10 K) [59] and 2.78 $\pm$ 0.03 (300 K) [58]) evaluated from the low hν (102–100 eV) PES on scraped surfaces but larger than the almost T independent value $\sim$2.63 in the region of 80–300 K for cleaved single crystals [55]. The valence value was confirmed to be 2.71 $\pm$ 0.02 at 180 K by Yb $3d_{5/2}$ core level HAXPES performed at BL ID32 of ESRF at hν = 2,450 and 5,450 eV at 180 K shown in Fig. 7.15(c), supporting the validity of the evaluation made in Fig. 7.15(a), (b).

In order to discuss the Yb 4f spectra, non crossing approximation (NCA) calculations based on the SIAM were performed by considering the f^{14}, f^{13} and f^{12} initial states. A possible CEF splitting below 10 meV of the J = 7/2 state was neglected and its degeneracy N_f was taken as 8. By representing the $4f^{12}$ multiplet structures by a single peak with the whole multiplet intensity for simplicity, the T dependence of the spectra was calculated for various sets of parameters (details are given in Ref. [54]).

The experimental results at 200 K were well reproduced by NCA with respect to the energy positions of the $4f^{13}$ and $4f^{12}$ final states and the Yb valence of 2.71 by the parameters $\varepsilon_f = 0.292$ eV, $\Delta = \pi\rho V^2 = 28.6$ meV and $U_{ff} = 7.12$ eV, where ε_f is the bare Yb 4f energy relative to E_F and its plus sign means its location above E_F. Then the Kondo temperature T_K evaluated from the calculated magnetic

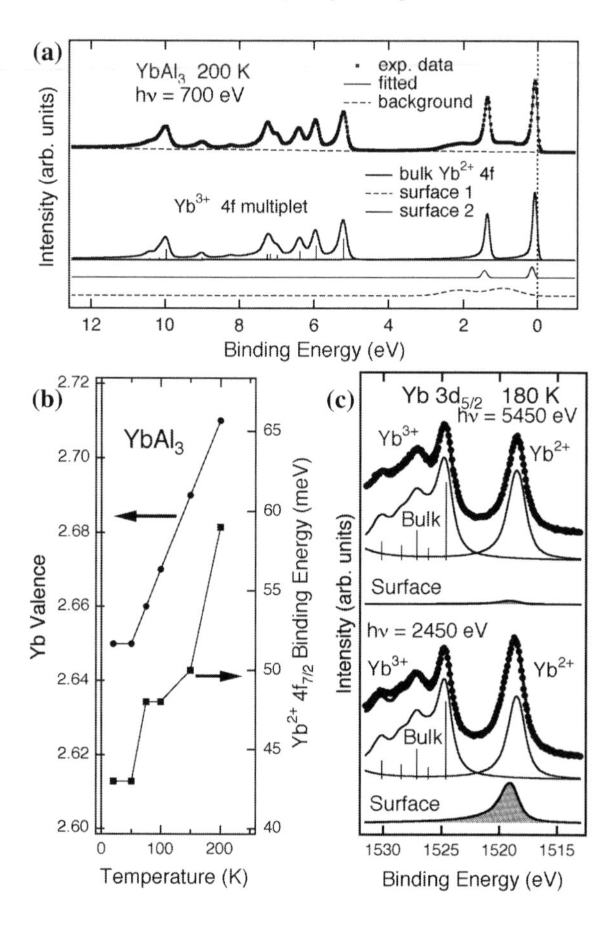

Fig. 7.15 **a** Deconvolution of the SXPES spectrum measured at 200 K and $h\nu = 700$ eV into surface (surface 1), subsurface (surface 2) and bulk components. The $4f^{12}$ final states were fitted by the atomic multiplet calculation. The valence was estimated to be 2.71 at 200 K. **b** T dependence of thus estimated Yb valence and E_B of $4f_{7/2}$ component. **c** Yb $3d_{5/2}$ HAXPES spectra at 180 K measured at 5.45 and 2.45 keV. The spectra were deconvoluted into the bulk Yb^{3+} and surface Yb^{2+} components [54]

excitation was 740 K. If the spectrum at 20 K was calculated with keeping these parameters constant, clear discrepancies from the experimental results were noticed [54]. Although the predicted mean valence 2.64 was close to the experimental value of 2.65, the peak E_B shift of the $4f^{13}$ component was only -5 meV and the peak shift of the $4f^{12}$ component was +50 meV when T was reduced from 200 K to 20 K, whereas the experimental shifts were -16 meV and 0 ± 5 meV. Thus the NCA (SIAM) calculation with constant parameters for different T could not explain the experimental SXPES results and suggested the predominance of the intrinsic Kondo lattice effects beyond SIAM for the bulk 4f states in $YbAl_3$. The deviation of magnetic and thermal properties from the prediction by SIAM was also independently reported [65].

Within NCA (SIAM) calculations, this discrepancy could tentatively be explained by changing the parameters used in the calculation. The results at 20 K are, for example, reproduced by $\varepsilon_f = 0.275$ eV, $\Delta = 24.8$ meV and $U_{ff} = 7.08$ eV [54]. Here T_K and the Yb valence were calculated as 560 K and 2.65. The SXPES results at 50, 75, 100 and 150 K were apparently explicable by ε_f, Δ and U_{ff} between

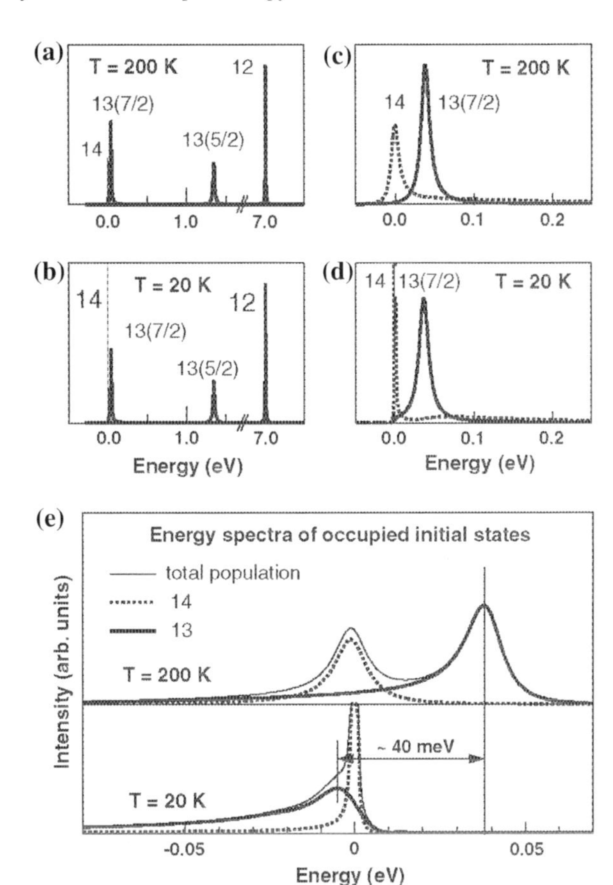

Fig. 7.16 Details of a NCA (SIAM) calculation for a parameter set of $\varepsilon_f = 0.12$ eV, $\Delta = 13$ meV and $U_{ff} = 7.0$ eV. Resolvent spectra in (**a**) and (**b**) represent the contribution of the f^{14}, $f^{13}_{7/2}$ (abbreviated as 13(7/2), for simplicity), $f^{13}_{5/2}$ and f^{12} configurations to all possible states including all excited states. The energy of the f^{14} peak was set to 0 eV. Intensities of the $f^{13}_{7/2}$, $f^{13}_{5/2}$ and f^{12} configurations were 8, 6, and 91 times larger than those in this figure due to the degeneracy. **c** and **d** show the same spectra in a rather narrow energy region. **e** Energy spectra of occupied initial states obtained by multiplying the Boltzmann factor $\exp(-E/k_BT)$ to the resolvent spectra. The degeneracy of $f^{13}_{7/2}$ state was included [54]

those at 200 and 20 K. The valence decrease with T in $YbAl_3$ was now recognized to be accompanied by a shift of ε_f closer to E_F as well as the decrease of Δ and U_{ff}.

In order to conceptually understand the origin of the T dependence of the Yb 4f spectra and the physics involved in the NCA (SIAM) calculation, configuration-resolved energy spectra of all possible states for a parameter set of $\varepsilon_f = 0.12$ eV, $\Delta = 13$ meV and $U_{ff} = 7.0$ eV, which could provide the Kondo resonance peak at $E_B = 45$ meV or $T_K = 520$ K, are shown in Fig. 7.16. By comparing the calculated results at 20 K (b and d) and 200 K (a) and (c), it was noticed that the T dependence was confined in the lowest energy region below 0.1 eV. At 20 K, the f^{14} spectrum (d) had a sharp peak near 0 eV and a dip around 25 meV. The $f^{13}_{7/2}$ spectrum had a small shoulder structure near 0 eV, indicating that the ground state near 0 eV is the Kondo singlet with a strong configuration mixing. The Kondo temperature T_K was defined by the energy difference between the Kondo singlet state and the lowest magnetic state [8], namely, the energy difference between the f^{14} and $f^{13}_{7/2}$ peaks in Fig. 7.16(d). At 200 K (Fig. 7.16(c)), the f^{14} and f^{13} peaks were more independent from each other, indicating much smaller mixing.

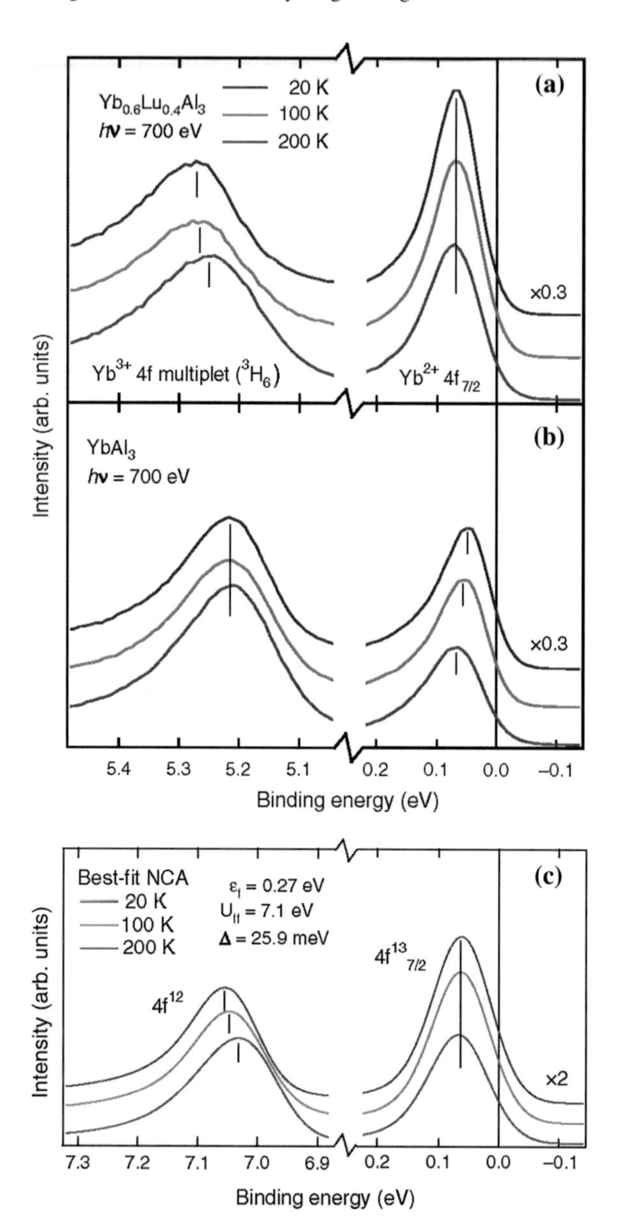

Fig. 7.17 Temperature dependence of the Yb 4f spectra of $Yb_{0.6}Lu_{0.4}Al_3$ (**a**) [66] compared with those of $YbAl_3$ (**b**) [54]. For the $4f^{12}$ final state multiplets from the Yb^{3+} state, only the $^3H_6$ component is shown for simplicity. (**c**) Best-fit NCA (SIAM) calculation for $Yb_{0.6}Lu_{0.4}Al_3$ by $\varepsilon_f = 0.27$ eV, $\Delta = 25.9$ meV and $U_{ff} = 7.1$ eV [66]

Besides, Fig. 7.16(e) shows the configuration resolved energy spectra of the realized initial state (not the PES final state). At 20 K, the proximity between the f^{14} and $f^{13}_{7/2}$ peaks near 0 eV is consistent with the formation of the Kondo singlet. The magnetic $f^{13}_{7/2}$ states near 40 meV are not occupied at all. On the other hand, the magnetic $f^{13}_{7/2}$ states are remarkably populated at 200 K indicating that the

Kondo singlet was destroyed because T was of the order of the T_K. The $f^{13}_{7/2}$ peak shifted by about 40 meV to higher energies with T from 20 to 200 K, corresponding to the qualitative change in character of the realized $f^{13}_{7/2}$ state, namely from the $f^{13}_{7/2}$ Kondo singlet state to the magnetic $f^{13}_{7/2}$ state. The transition from the f^n component in Fig. 7.16(e) to the f^{n-1} component in Fig. 7.16(a–d) could be observed in PES. Therefore, the f^{12} PES peak results from the $f^{13}_{7/2}$ component in Fig. 7.16(e) and the f^{12} component in Fig. 7.16(a) and (b). The predicted (but not observed) $\sim$40 meV shift of the f^{12} peak to lower (smaller) E_B in PES when T was increased to 200 K was interpreted as due to the effect that the $f^{13}_{7/2}$ peak in Fig. 7.16(e) shifts by $\sim$40 meV to higher energies and that the f^{12} peak position in Fig. 7.16(a) and (b) is almost T independent. In contrast to this, the shift of the f^{13} final states was predicted to be much smaller because the shift of the f^{14} peak in Fig. 7.16(e) is much smaller. In this way the SIAM could clearly predict the temperature dependence of both the $4f^{13}$ and $4f^{12}$ final states. Therefore the validity of the SIAM could be checked by the results of both the $4f^{13}$ and $4f^{12}$ peaks of the Yb 4f spectra. In the case of $YbAl_3$, however, it was demonstrated as above that SIAM could not explain the T dependence of the bulk 4f electronic structures. The applicability and limitation of SIAM must always be kept in mind.

The SXPES was performed later for an in situ fractured single crystal $Yb_{0.6}Lu_{0.4}Al_3$ at $h\nu = 700$ eV with a total energy resolution of 65 meV [66]. The spectra at 200, 100 and 20 K are compared with those of $YbAl_3$ [54] in Fig. 7.17(a), (b). In contrast to the case of $YbAl_3$, a $4f^{12}$ peak shift is clear towards larger E_B with decreasing T in contrast to the almost constant E_B of the $4f^{13}$ peak or the Kondo peak. The bulk valence was evaluated by deconvoluting both the Yb 4f surface and subsurface contributions. It was found that the T dependence of the Yb 4f spectra in $Yb_{0.6}Lu_{0.4}Al_3$ was relatively well reproduced by NCA (SIAM) (Fig. 7.17(c)) in contrast to the case of undoped $YbAl_3$. A similar situation was observed also in the case of Lu doped YbB_{12}, where the T dependence of the Yb 4f spectra of the $Yb_{0.875}Lu_{0.125}B_{12}$ could be well reproduced by NCA (SIAM), though the result in undoped YbB_{12} could never be consistently reproduced by NCA (SIAM) as shown later in Fig. 8.10. These results revealed that NCA (SIAM) cannot simply be applied to systems with noticeable Kondo lattice effects, though it can explain the 4f electronic structure including their T dependence in systems with negligible 4f lattice coherence. In this respect, the development of a periodic Anderson model to handle the bulk sensitive PES of systems with non negligible 4f lattice coherence is desired.

7.1.3 Transition Metal Compounds

As the first example of SXPES of transition metal (TM) compounds, V_2O_3 and its family materials are discussed. $(V_{1-x}Cr_x)_2O_3$ displays a complex phase diagram

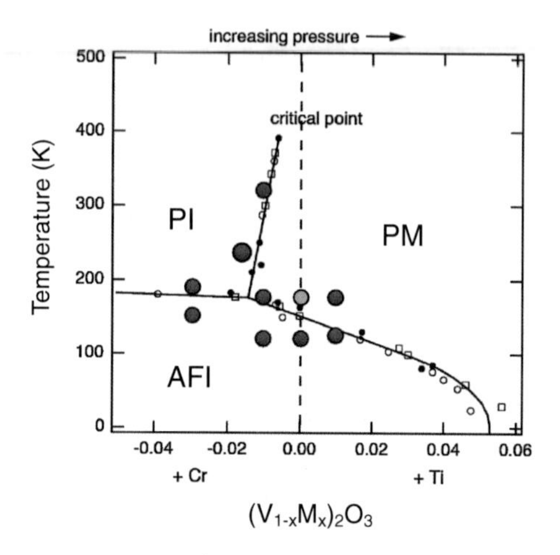

Fig. 7.18 Phase diagram of Cr or Ti doped V_2O_3 in T-P (pressure) or T-doping phase space. The Cr doped phase diagram is from Ref. [67]. The *large circles* show the composition of the samples and the temperatures of SXPES measurements performed [73, 75]

with phases of a paramagnetic metal (PM), a paramagnetic insulator (PI) and an antiferromagnetic insulator (AFI) as shown in Fig. 7.18 [67]. This PM → AFI first order phase transition is accompanied by a structural change from the corundum structure in the PM phase to a monoclinic structure in the AFI phase. However, the same corundum structure is kept on the PM → PI phase transition due to slight Cr substitution on the V sites. So two types of metal-to-insulator transitions (MIT) can be studied in this system. The transition from PM to PI was thought to be a paradigm of the Mott MIT [68–70]. Understanding its bulk electronic structure in the metallic phase is of fundamental importance. SXPES with use of circularly polarized high resolution soft X-rays was first performed on well annealed single crystalline V_2O_3 cleaved in situ with a hexagonal $(10\bar{1}2)$ surface at BL25SU of SPring-8 [71]. The sample temperature was held at 175 K slightly above the PM → AFI transition. Figure 7.19(a) shows the k averaged PES spectrum of V_2O_3 compared to that at $h\nu = 700$ eV [71]. The inset shows a cross section of the k space BZ stacking normal to the cleavage plane together with the spherical surface traversed as the detector angle was varied about the normal for several $h\nu$ between 310 and 700 eV. Namely, the abscissa and ordinate corresponded to $k_{\|}$ and $k_{\perp}$. Two radial lines show the $k_{\|}$-region covered by the long slit with a total acceptance of ±6° corresponding to ± 1.38 Å^{-1} at 700 eV, which covers more than one BZ. The resonance enhancement of the V 3d state took place for $h\nu$ in the range between 510 and 560 eV corresponding to the V 2p core excitation. In order to be free from any Auger effect for the discussion of accurate line shapes of the V 3d state, this energy region was intentionally avoided. As shown in Fig. 7.19(a), the V 3d spectral weight is well separated from the O 2p emission and a sharp peak is observed near E_F. The latter corresponds to the coherent part or a quasiparticle (QP) peak and the bump near 1.2 eV is ascribed to the incoherent part corresponding to the lower Hubbard band (LHB). Figures 7.19(b–d) compare the V 3d spectra in the PM and AFI phases of V_2O_3 and in the PI phase of $(V_{0.985}Cr_{0.015})_2O_3$ at different $h\nu$ after removing a Shirley-type background and normalizing by the intensity of the incoherent peak [71–75]. It was noticed that the intensity of the coherent part in the bulk was much stronger than the result reported at $h\nu = 60$ eV [72] and the relative intensity between the coherent and incoherent parts increased with the increase of $h\nu$ in accordance with the increased bulk sensitivity. When the detection angle θ was changed from the surface normal by rotating the sample up to 60° in steps of 15°, the intensity of the coherent peak decreased monotonically and became comparable to the spectrum at 60 eV, confirming that the surface electronic structure in V_2O_3 systems is much less itinerant than in the bulk [74]. By taking the λ_{mp} of 14 Å at 700 eV [12, 13] and a surface layer thickness of 2.24 Å, for example, the surface spectral weight was estimated to be ~15 % of the whole spectral weight. When the extracted bulk spectral weight from two SXPES experiments were compared with the really bulk sensitive HAXPES at ~8 keV, quite similar spectral shapes were seen in all cases of PM, AFI and PI phases in V_2O_3 and $(V_{1-x}Cr_x)_2O_3$ [75].

LDA + DMFT (QMC) calculations performed at 700 and 300 K are compared with the spectrum in the PM phase in Fig. 3 of Ref. [71], in which typical results are given for U = 5 eV and a Hund's rule coupling J = 0.93 eV. The theory could predict that the QP peak becomes narrowed with decreasing temperature and clearly separates from the lower Hubbard band. If the experimental result at 700 eV and 175 K was compared with the calculated result at 300 K (Fig. 3 of [71]), clear discrepancies were still recognized. Namely the coherent peak in the experiment was broader and had more spectral weight than the theoretical prediction. Even when the temperature in the calculations was further reduced, the spectral weight of the quasiparticle was not expected to change significantly even though the spectrum would become sharper. The reduced U value might partly explain the discrepancy but could not be reconciled with the MIT under Cr-doping. Therefore the discrepancy between the experiment and the DMFT calculation remained unexplained.

When the bulk components, either extracted from the SXPES spectra or directly obtained by HAXPES at ~8 keV, were compared among V_2O_3 in the metal (PM at 175 K) and insulator (AFI at 125 K) phases as well as $(V_{0.985}Cr_{0.015})_2O_3$, the Fermi level crossing in the PM phase and the gap opening in both AFI and PI phases were clearly confirmed. The coherent peak near 0.2 eV in the metal phase was strongly suppressed in the insulator phases. One also noticed a slight but clear difference between the extracted bulk spectra in the AFI and PI phases as shown in Fig. 7.19(e) in agreement with the HAXPES results. Namely the gap was smaller in the PI phase by about ~100 meV while the peak near $E_B = 0.65$ eV was broader with a relatively lower peak height than in the AFI phase [75]. A very similar tendency was observed in the SXPES at $h\nu = 500$ eV of $(V_{0.988}Cr_{0.012})_2O_3$, which showed the phase transitions of PI → PM → AFI with decreasing the T from 300 K [74]. In the latter case, the leading edge near E_F was shifted by ~150 meV between the PI and AFI phases while the bump near $E_B = 0.7$ eV was enhanced in the AFI phase. The increase of the gap through the PI → AFI transition was predicted from the t-J model in the large U limit [76]. DMFT based on the one-band Hubbard model also predicted such an increase of the gap size and the enhanced peak at the upper edge in the AFI phase, which was

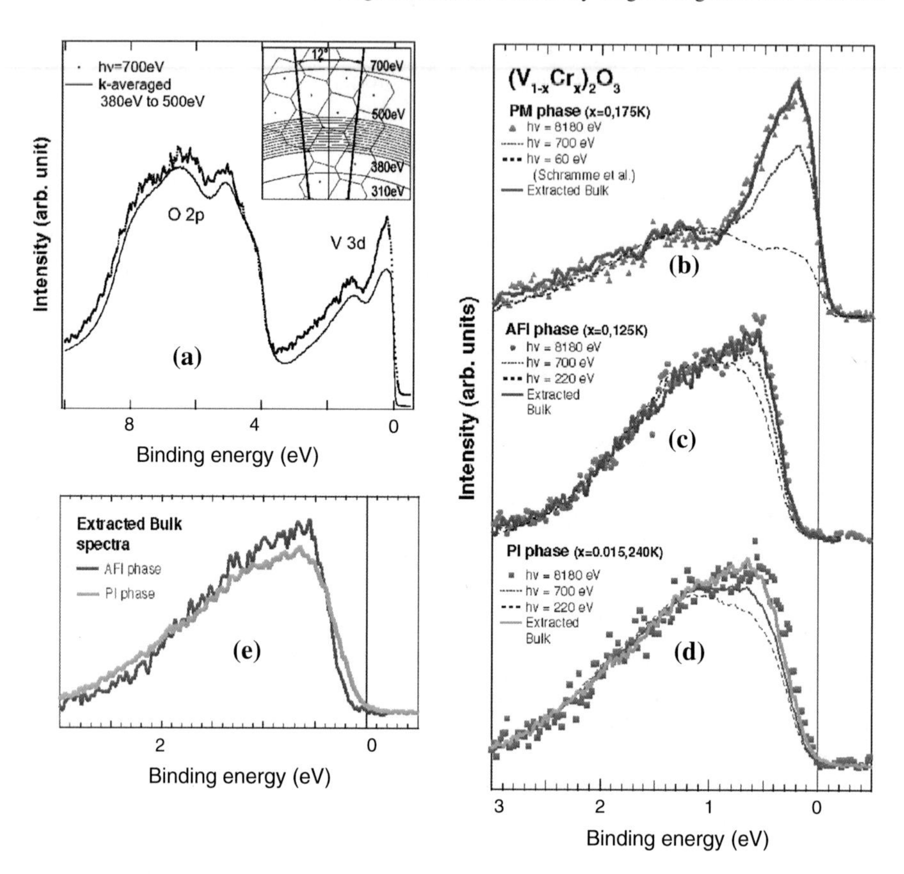

Fig. 7.19 **a** SXPES of V_2O_3 measured on the $(10\bar{1}2)$ surface of single crystals at $h\nu = 700$ eV and at $T = 170$ K [71]. **b** Spectrum in PM phase of V_2O_3 at $h\nu = 700$ eV compared with those at $h\nu = 60$ eV [72] and 8,180 eV [75]. Bulk component extracted from those spectra at 700 and 60 eV is shown by the thick solid line. **c** Spectra in the AFI phase of V_2O_3 at 125 K. The bulk component extracted from the spectra at $h\nu = 700$ and 220 eV (*thick solid line*) is compared with the spectrum at $h\nu = 8{,}180$ eV. **d** Spectra in the PI phase of $(V_{0.985}Cr_{0.015})_2O_3$ at 240 K. Spectra at $h\nu = 700$ and 220 eV as well as the extracted bulk component are compared with that at $h\nu = 8{,}180$ eV. **e** Comparison of the SXPES-extracted bulk spectra between the AFI and PI phases [75]

identified as the formation of the (quasiparticle(QP)-)renormalized Slater peak [77] induced by the long range spin coherence. It is seen in Fig. 7.19(b–d) that the PES intensity is increased with $h\nu$ in the region of $E_B = 0.8$–0.4 eV in both the AFI and PI phases, though the degree of increase is much less than in the case of the QP or the coherent peak in the PM phase. This suggests that the bulk electronic structure even in the AFI and PI phases is less localized compared with that in the surface region. The increase of the spectral weight near E_F in the PI phase might be related to the short range magnetic ordering reported by neutron scattering [78].

Since two structures were certainly observed in the AFI phase in $E_B = 2$–0.5 eV, we tried to subtract the bulk QP spectral weight in the PM phase from the bulk AFI spectrum after shifting the former and adjusting the threshold spectral shape to the AFI spectrum to estimate the incoherent component (or LHB) in the AFI spectrum. Then the LHB peak in the AFI phase was estimated to be located at 1.2 ± 0.15 eV. This value was almost equivalent to the peak energy of the LHB in the PM phase as well as the LHB feature in the PI phase seen in Fig. 7.19(e) [75]. It is experimentally seen in Fig. 7.19(b–d) that the LHB peak position is almost independent of the hν employed. From this experimental analysis, it was concluded that U is essentially constant through the MIT in $(V_{1-x}Cr_x)_2O_3$. Then the scenario of an orbital selective Mott transition, in which U_c/W changes through the MIT due to the enhancement of the effective trigonal splitting [79, 80], is experimentally supported.

The filling of the Mott gap with increasing T was a rather interesting subject in strongly correlated electron systems. The gap filling behavior with increasing T was expected to be very different from that in the band insulator. $(V_{0.972}Cr_{0.028})_2O_3$ shows the PI phase at room temperature. The gap filling behavior up to 800 K was measured on this sample by PES after careful degassing the sample holder for a long time to prevent surface contamination during the high T measurement [73]. The experiment was performed not in the SX region but by use of HeI and II light laboratory sources. Although the surface electronic structure probed by these sources is less itinerant than the bulk electronic structure [71, 74, 75], the study of the high T behavior of correlated insulators was thought to be not spoilt much by this constraint. Scraping was employed to allow multiple T cycles through 300 and 800 K under the scrape-cleaned condition. The gap filling behavior of the surface correlated electronic structure was traced. A gradual change was observed above ~550 K corresponding to the entering into the crossover regime [73]. The thermal behavior was found to be not explicable by phonon broadening or ordinary thermal broadening of the band insulator gap. The HeII spectra at 300 and 750 K were compared with the V 3d spectrum calculated by LDA + DMFT for the PI phase at 1,160 K with $U = 5$ eV. The calculated result was in close resemblance to the 750 K spectrum. The high T behavior of the PI phase was thought to be well described by the crossover regime of the DMFT [73].

In metallic but correlated TM oxides, it was recognized that the lower and upper Hubbard bands remain as the so-called incoherent parts reflecting electron correlations as seen in V_2O_3. A renormalized band crossing E_F (coherent part) is located between the occupied and unoccupied incoherent parts [81]. $Sr_{1-x}Ca_xVO_3$ is one of the metallic Mott–Hubbard systems, not showing a Mott MIT. The occupied 3d electronic structure was expected to be simple because of the nominal $3d^1$ (V^{4+}) configuration. So far reported photoemission studies at low h$\nu \leq 120$ eV [82–84] indicated the presence of a pronounced LHB, which could not be explained by conventional band-structure theory, showing contradictory results to transport and thermodynamic properties. While the thermodynamic and transport properties (Sommerfeld coefficient, resistivity, and paramagnetic susceptibility) did not change much with x, low-hν photoemission showed drastic differences for varying x as shown in Fig. 7.20(a). In fact, these spectroscopic data seemed to imply that $Sr_{1-x}Ca_xVO_3$ is on the verge of a Mott transition for $x \sim 1$.

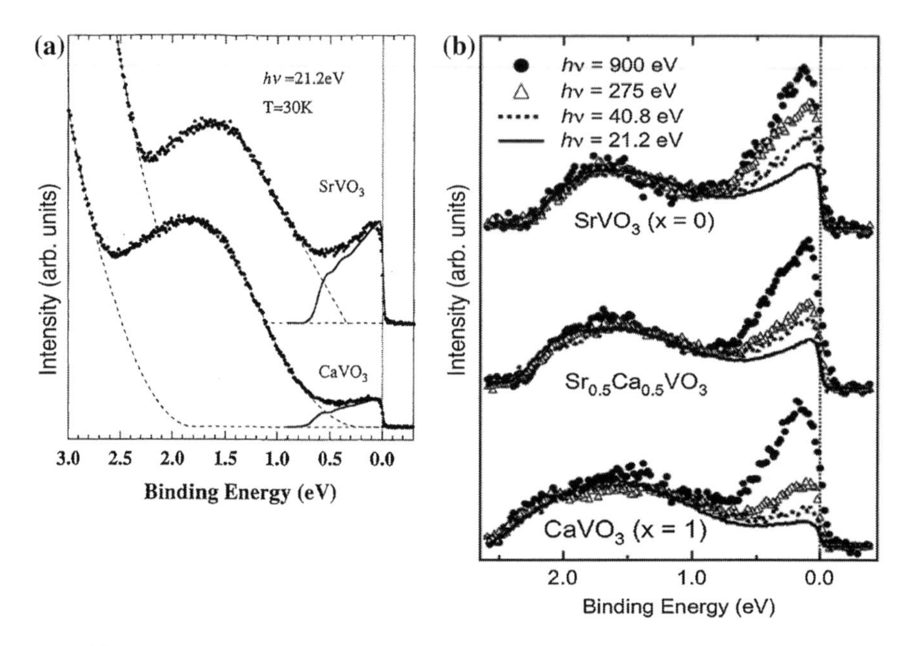

Fig. 7.20 **a** Valence-band photoemission spectra of $SrVO_3$ and $CaVO_3$ at $h\nu = 21.2$ eV with an energy resolution of 35 meV on scraped surfaces [84]. **b** $h\nu$ dependence of the V 3*d* spectral weight for $Sr_{1-x}Ca_xVO_3$, which were obtained by subtracting the fitted tails of the O 2*p* contributions from the raw spectra. In order to clarify the evolution of the quasi-particle peak near E_F as a function of $h\nu$, the V 3*d* spectra shown here were normalized by the integrated intensities of the incoherent part ranging from 0.8 to 2.6 eV [87]

This was in concordance with the widespread theoretical expectation that the bandwidth W should decrease strongly with increasing x since the V–O–V angle is reduced from $\theta = 180°$ for cubic $SrVO_3$ [85] to $\theta = 154.3°$ (rotation angle), 171° (tilting angle) for orthorhombic $CaVO_3$ [86].

The bulk and surface $3d^1$ spectral weights near E_F were then obtained by SXPES [87, 88]. The SXPES was performed at BL25SU in SPring-8. Single crystals of $SrVO_3$ and $Sr_{0.5}Ca_{0.5}VO_3$, and polycrystalline $CaVO_3$ were employed for the measurements. The overall energy resolution was set to about 140 and 80 meV at $h\nu = 900$ and 275 eV, respectively. The results were compared with the low $h\nu$ PES taken at $h\nu = 40.8$ and 21.2 eV by using a He discharge lamp in the laboratory. The energy resolution was set to 50–80 meV. The samples were cooled down to 20 K for all measurements. Clean surfaces were obtained by fracturing the samples in situ at the measuring temperature.

Figure 7.20(b) shows the V 3d dominated valence band photoemission spectra near E_F measured at low- and high-$h\nu$ excitations [87, 88]. In all the spectra, the peak near E_F and the broad peak centered at about 1.6 eV correspond to the coherent (QP peak) and incoherent (LHB) parts, respectively. In contrast to low-$h\nu$ photoemission of $Sr_{1-x}Ca_xVO_3$ reported so far, including the data shown here, the spectral variation with x is much smaller in the bulk sensitive spectra measured at

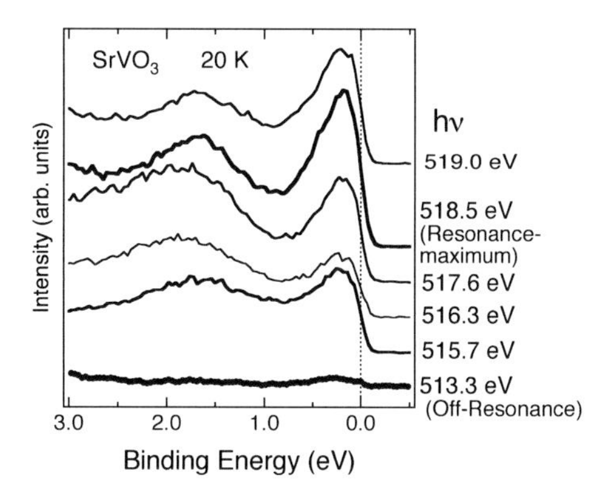

Fig. 7.21 Resonance photoemission spectra of $SrVO_3$ near the V 2p → 3d core excitation threshold with an energy resolution of 200 meV [88]. The spectral intensity was normalized by the photon flux

hν = 900 eV. The coherent spectral weight increases drastically with hν for all compounds. One might consider that the relatively strong incoherent spectral weight of the 1.6 eV peak at low-hν originates possibly from a large O 2p weight (consequently weak V 3d weight) in this peak because the low hν PES of TM oxides is generally more sensitive to the O 2p states than the TM 3d states. In order to verify this possibility, the V 2p-3d RPES was also performed for $SrVO_3$ as shown in Fig. 7.21 [88]. The result shows that both coherent and incoherent parts ranging from E_F to ~2.5 eV are drastically enhanced in the V 2p-3d RPES, and therefore assigned to the predominance of V 3d states. Since the O 2p hybridization with these predominantly 3d states is very minor [89], the hν dependence of the relative V 3d/O 2p PICS, which changes by only 8 % from hν = 275–900 eV [16, 90], could be neglected. Therefore, the monotonous increase of the coherent part with hν must be attributed to the increased bulk sensitivity. Namely, the V 3d spectra at low-hν mainly reflect the surface electronic states which are more localized than the bulk states caused by the broken V–O–V topological connectivity and/or structural relaxations in the surface.

The bulk V 3d photoemission spectra of $Sr_{1-x}Ca_xVO_3$ in Fig. 7.22(a) and (d) were derived from the data at hν = 900 and 275 eV by the following procedure: (1) The inelastic mean free path λ_{mp} was calculated as ~17 and ~7 Å at hν = 900 and 275 eV [12]. (2) The bulk weight R (<1, depending on hν) should be determined as $\exp(-s/\lambda_{mp})$, where s is the "surface thickness". Therefore Rs at 900 eV (R_{900}) and 275 eV (R_{275}) are related as $R_{275} = R_{900}^{2.4}$. (3) The observed V 3d spectral intensity at hν = 900 eV was represented as $I_{900}(E_B) = I_B(E_B) \cdot R_{900} + I_S(E_B)(1-R_{900})$ while $I_{275}(E_B) = I_B(E_B)R_{900}^{2.4} + I_S(E_B)\left(1 - R_{900}^{2.4}\right)$, where $I_B(E_B)$ and $I_S(E_B)$ are the bulk and surface 3d spectra and E_B stands for the binding energy. (4) If s is assumed to be 7.5 Å corresponding to about twice the V–O–V distance, R_{900} (R_{275}) is ~0.64 (~0.34). With this R_{900} we obtain the 3d bulk spectral weight $I_B(E_B)$ as shown in Fig. 7.22(a) and (d). It was noticed that $I_B(E_B)$ hardly changes (<15 % at E_F) even when we assumed s to range from 5.4 to 11 Å. In contrast to the low-hν photoemission spectra, the bulk V 3d spectral functions near E_F are almost equivalent among the three compounds, indicating that the

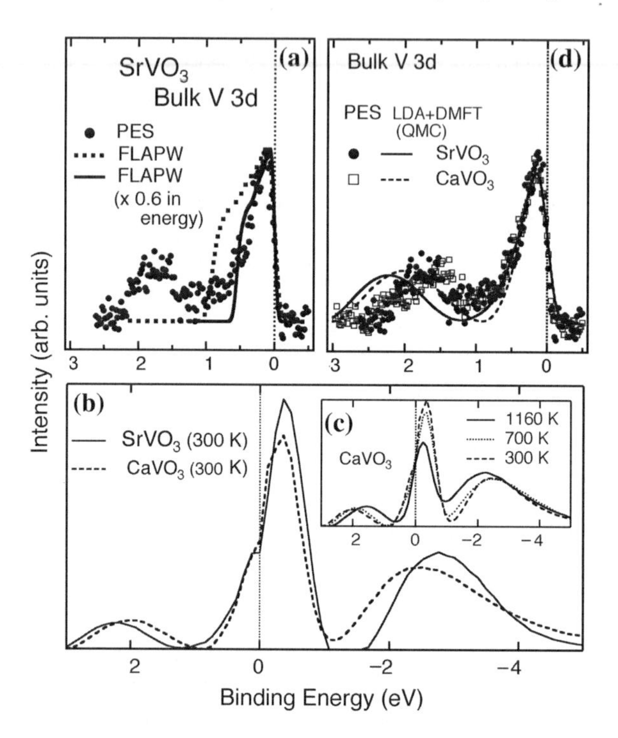

Fig. 7.22 Comparison of the bulk sensitive SXPES of $SrVO_3$ and $CaVO_3$ to theoretical simulations [87]. Bulk V 3*d* photoemission spectral weight near E_F of $Sr_{1-x}Ca_xVO_3$ as obtained from the data in Fig. 7.20(b) by the procedure described in the text [87, 88]. **a** Comparison of the bulk V 3*d* photoemission spectra of $SrVO_3$ (*dots*) to the 3*d* DOS obtained by using the FLAPW method (*dashed curve*), which was multiplied by the Fermi function at 20 K and then broadened by the experimental resolution of 140 meV. The *solid curve* shows the same 3*d* DOS but the energy was scaled down by a factor of 0.6. **b** LDA + DMFT (QMC) spectrum of $SrVO_3$ (*solid line*) and $CaVO_3$ (*dashed line*) calculated at 300 K. **c** Effect of the temperature in the case of $CaVO_3$. **d** Comparison of the bulk photoemission spectra with the calculated LDA + DMFT spectra for $SrVO_3$ and $CaVO_3$

V–O–V distortion does not much influence the occupied bulk V 3d states in $Sr_{1-x}Ca_xVO_3$. This is consistent with the thermodynamic properties. It should be noticed that on scraped sample surfaces nearly x-independent bulk spectral functions could not be obtained [91, 92].

In order to investigate the genuine effect of the electron correlations on the bulk spectra, we first compare in Fig. 7.22(a) the bulk spectrum thus derived with the V 3d partial density of states (DOS) for $SrVO_3$ obtained by the full-potential linearized augmented plane wave (FLAPW) method. Here, the partial DOS broadened by the instrumental resolution (dashed curve) was normalized to the bulk 3d spectral function of $SrVO_3$ by the integrated intensity from E_F to 2.6 eV. This comparison shows that the band-structure calculation cannot reproduce the incoherent spectral weight at all and that the width of the observed coherent part is about 60 % of the predicted value. Clearly, these deviations originate from electron correlations which are not treated correctly in this band-structure calculation. For a more quantitative analysis, theoretical calculations within the local density approximation + dynamical mean field theory scheme using quantum Monte Carlo simulations (LDA + DMFT (QMC)) was applied as a non-perturbative approach [87, 93–97]. The resulting LDA + DMFT (QMC) spectra for $SrVO_3$ and $CaVO_3$ in Fig. 7.22(b) shows genuine correlation effects, i.e., the formation of the LHB at about 2 eV and the upper Hubbard bands (UHB) at about −2.5 eV, with well pronounced QP peaks (coherent part) near E_F. The results indicated that both $SrVO_3$ and $CaVO_3$ are strongly correlated metals, but not on the verge of the Mott MIT. The 4 % difference in the LDA bandwidth between these compounds is reflected only in some additional transfer of spectral weight from the coherent to incoherent parts. LDA + DMFT (QMC) spectra of the two systems were seen to be quite similar, especially in the occupied states. Figure 7.22(c) shows that the effect of temperature T on the spectrum is small for $T \leq 700$ K. The slight differences in the QP peaks between $SrVO_3$ and $CaVO_3$ lead to different effective masses, namely $m^*/m_0 = 2.1$ for $SrVO_3$ and 2.4 for $CaVO_3$, which agrees with $m^*/m_0 = 2$–3 for $SrVO_3$ and $CaVO_3$ as obtained from thermodynamics [85] and de Haas–van Alphen experiments [98].

In Fig. 7.22(d) are compared the experimental bulk V 3d spectral functions with the LDA + DMFT (QMC) spectra (300 K), which were multiplied with the Fermi function at the experimental temperature (20 K) and broadened with the experimental resolution of ~140 meV. The QP peaks in theory and experiment are in very good agreement. In particular, their height and width are almost identical for both $SrVO_3$ and $CaVO_3$. There is some difference in the positions of the incoherent part which might be partly due to uncertainties in the ab initio calculations or in the O 2p subtraction procedure. Nonetheless, the good overall agreement confirms that the high $h\nu$ SXPES indeed facilitates the evaluation of bulk spectra.

Next we discuss Mn compounds. Manganese perovskites attracted much interest due to many intriguing phenomena such as colossal magnetoresistance and the charge/orbital-ordering [99]. In these systems, the electronic states were basically interpreted in terms of the double exchange mechanism arising from the strong Hund coupling between an e.g. conduction electron and t_{2g} local spins on the Mn site [100–102]. Since the discovery of a variety of phases including charge/orbital ordering, however, the importance of the Jahn–Teller effect was pointed out from many experimental and theoretical studies [99, 103]. These reports and references therein suggested that the simple double exchange model could not describe the various electronic states of manganites. In order to facilitate the discussion of their electronic structure, it is demonstrated here that the Mn 2p-3d RPES ($h\nu \sim 640$ eV : $\lambda_{mp} \sim 13$ Å, leading to a predominant bulk contribution of 60 % as discussed below) is very powerful to reveal their bulk electronic structure.

Mn 2p-3d RPES was performed at BL25SU in SPring-8 for single crystalline $Nd_{1-x}Sr_xMnO_3$, which was grown by the floating-zone method. The resolution was set to about 100 meV at $h\nu \sim 643$ eV. A clean sample surface was obtained by in situ fracturing. For estimating the bulk and surface contributions in the spectra, the

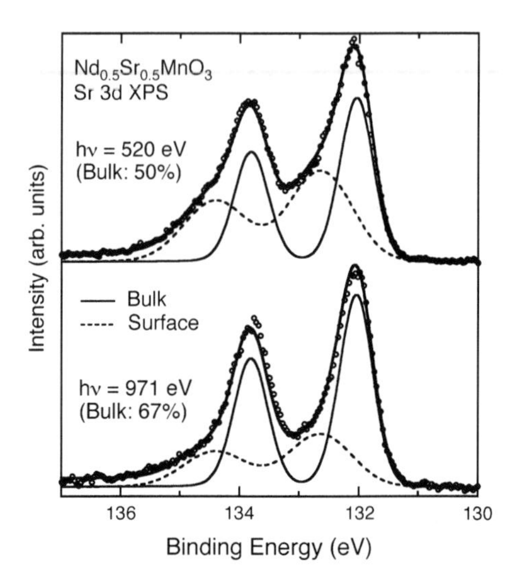

Fig. 7.23 *hν* dependence of the Sr 3*d* core-level spectra of $Nd_{0.5}Sr_{0.5}MnO_3$ at 50 K. The deconvoluted components into bulk and surface contributions in each spectrum are also shown by the *solid* and *dashed lines*

Sr 3d core-level spectra were very useful since the Sr 3d levels showed a clear surface core-level shift [104]. Figure 7.23 shows the hν dependence of the Sr 3d core-level spectra of $Nd_{0.5}Sr_{0.5}MnO_3$. In addition to the peaks at 132 and 134 eV originating from the bulk 3d spin-orbit splitting, a weak shoulder structure appears at $\sim$135 eV in the spectra at $h\nu = 520$ eV (photoelectron kinetic energy E_K of $\sim$390 eV). Since this structure is clearly reduced in the spectra at $h\nu = 971$ eV (E_K of $\sim$840 eV), one could notice that the shoulder structure was from the surface Sr $3d_{3/2}$ state. The spectral weight is also suppressed at $\sim$133 eV at $h\nu = 971$ eV relative to that at $h\nu = 520$ eV, being originating from the surface Sr $3d_{5/2}$ state. The bulk and surface contributions could be quantitatively estimated by the analysis of core-level spectra based on a least-square fitting as shown in this figure. For the analysis, a symmetric line shape (Lorentzian with Gaussian broadening) was used for each component. The Lorentzian width was assumed to be independent of the bulk and surface components (0.06 eV) while the Gaussian width was changed depending on components, though the origin of the difference was not clear. Since there is no Sr-derived conduction electron near E_F, the line shapes of the Sr 3d core level are almost symmetric. The same parameters were used for the spectra at $h\nu = 520$ and 971 eV except for the intensity of bulk and surface components. Then the effective surface thickness was evaluated as $\sim$6 Å corresponding to the length between one and two MnO_6 octahedra (O–Mn–O). The bulk contribution was then estimated as $\sim$60 % at $E_K \sim 640$ eV in the Mn 2p-3d RPES.

Figure 7.24 shows the entire valence-band RPES spectra of $Nd_{0.6}Sr_{0.4}MnO_3$ at 20 K whereas the inset depicts the Mn $2p_{3/2}$-3d X-ray absorption spectrum (XAS) [105]. The excitation hν for the spectra labeled A–E is indicated in the inset. The spectral weight from E_F to 1.5 eV is thought to be derived from the Mn 3d eg electron character (Mn 3d eg–O 2p anti-bonding state). The peak at about 2.0 eV and the broad peak structure at $\sim$7 eV are ascribed to the Mn 3d t_{2g}–O 2p anti-bonding

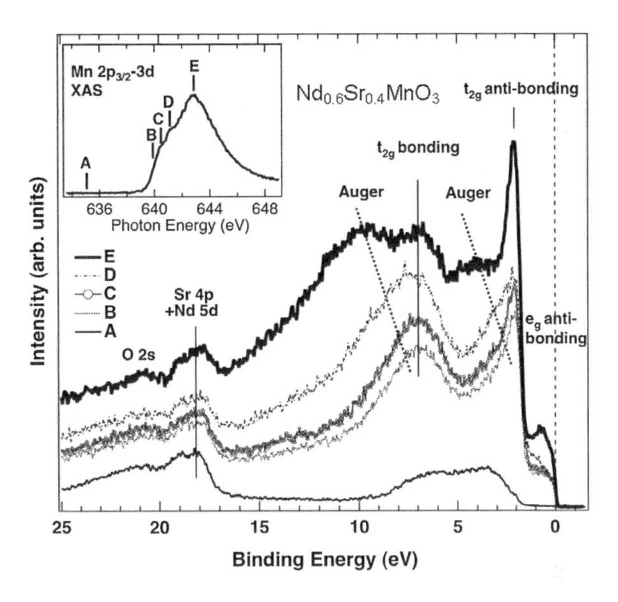

Fig. 7.24 Mn 2p-3d resonance photoemission spectra of $Nd_{0.6}Sr_{0.4}MnO_3$ measured at 20 K [105]. The inset shows the Mn 2p-3d X-ray absorption (XAS). The spectra are normalized by the photon flux

and bonding states. The structures at about 18 and 21 eV are attributed to the states with Sr 4p + Nd 5d and O 2s character [106]. Two structures shifted linearly to higher E_B with increasing $h\nu$ as indicated by dashed lines are due to Auger contributions. However, they became negligible within the energy region from E_F to 3 eV at $h\nu = 643$ eV (E: Mn 2p-3d resonance maximum). All these qualitative spectral features are in agreement with other results in the literature [106, 107].

Figure 7.25(a) demonstrates the Mn 2p-3d RPES spectra [108] of $Nd_{0.5}Sr_{0.5}MnO_3$ with a paramagnetic insulator (PMI)-to-ferromagnetic metal (FMM) transition at $T_C \sim 255$ K, while a charge-ordered insulator (COI) phase was seen below $T_{COI} \sim 160$ K accompanied by a structural phase transition [101–103]. The off-resonance Mn 3d spectral weight at 120 K in the COI phase is drastically enhanced at the Mn $2p_{3/2}$ absorption threshold (XAS maximum corresponding to E in Fig. 7.24). The detailed T dependence of the Mn 3d spectra measured at $h\nu = 643$ eV are reproduced in Fig. 7.25(b). In the COI phase, the peak structure around 1.0 eV represents the localized e_g state due to the charge ordering. Although the spectra showed a marginal change with the PMI–FMM transition, they exhibit a large and abrupt shift towards the higher E_B side across the FMM–COI transition. Especially, the e_g spectral line shape changes drastically with the FMM–COI transition compared with that across the PMI–FMM transition. The size of the energy gap in the COI phase is of the order of 100 meV. In this way, the bulk Mn 3d spectral functions could be experimentally revealed by the soft X-ray RPES.

The last example of SXPES is magnetite Fe_3O_4, which displays a very unique electronic phase transition called Verwey transition at $T_V \sim 123$ K [109–111]. A first order phase transition takes place at this T, where the resistivity dropps by two orders of magnitude from an insulator phase at low T to a "bad metal" phase at high

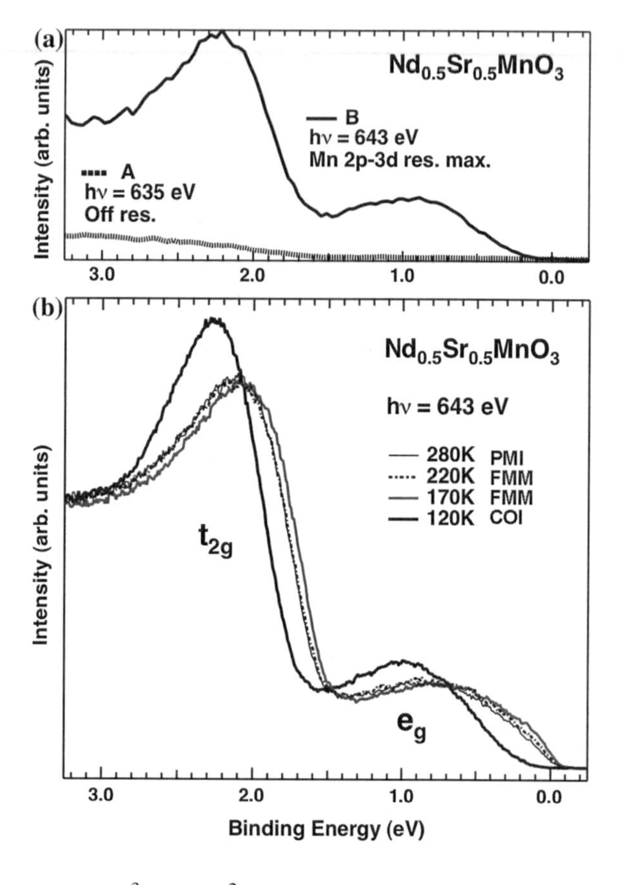

Fig. 7.25 Mn 2p-3d resonance photoemission (RPES) spectra of $Nd_{0.5}Sr_{0.5}MnO_3$ [108]. **a** Mn 2p-3d resonance-maximum ($h\nu = 643$ eV) and off-resonance ($h\nu = 635$ eV) spectra measured at 120 K in the COI phase. **b** Temperature dependence of the Mn 3d spectra measured at $h\nu = 643$ eV

T. In the scenario given by Verwey, Fe^{3+} and Fe^{2+} ions occupying the octahedral B sites of the inverse spinel structure with equal weight become ordered below T_V (charge ordering), whereas they are randomly distributed above T_V. The driving force was suggested to be the inter-atomic Coulomb interaction [69]. However, strong coupling to the lattice was later pointed out. This subject was studied for decades but fundamental questions are not yet answered. For example, charge disproportionation was found by powder diffraction to be very small of the order of $\sim \pm 0.1$e [112] and even the absence of charge ordering was concluded from resonance X-ray diffraction [113] in the insulator phase. In the high temperature phase, finite spectral weight was observed at E_F by high resolution PES excited by a conventional HeI laboratory source and a metallic phase was proposed [114]. In contrast, PES excited at $h\nu = 110$ eV by SR showed a finite gap even in the high temperature phase, although the magnitude of the gap changed notably at T_V [115].

The PES results obtained at low $h\nu$ by HeI were strongly dependent on the surface orientation and preparation, resulting from the surface sensitivity. Therefore bulk sensitive measurements were necessary to discuss the details of the Verwey transition. The SXPES results at $h\nu = 707.6$ eV at 100 K obtained by utilizing the Fe 2p $\rightarrow$ 3d RPES [116] were reproducible and much less dependent

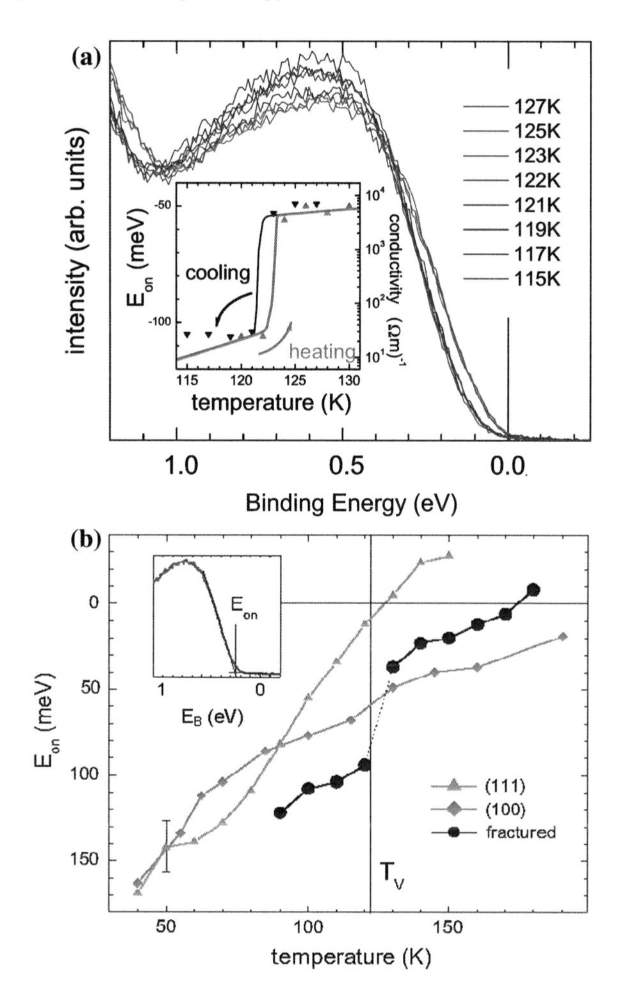

Fig. 7.26 **a** T dependence of the resonance enhanced SXPES spectra on a fractured Fe_3O_4 surface measured at $h\nu = 707.6$ eV in the vicinity of $T_V \sim 123$ K. **b** T dependence of the spectral onset energies (see text for definition), where the results for annealed (111) and (100) surfaces are added [116]

on the surface preparation and orientation. Three peak structures were observed between E_F and $E_B = 5$ eV in Fig.1(b) of [116]. The first peak near $E_B = 0.6$ eV closest to E_F ascribable to the $d^6 \rightarrow d^5$ transition at the B site Fe^{2+} ions had decreasing intensity towards E_F. The detailed T dependence of this peak was measured on the fractured surface at 707.6 eV under the on-resonance condition with steps of 1 or 2 K near T_V as shown in Fig. 7.26(a) [116]. Below T_V the spectrum is consistent with a small energy gap as expected for an insulator. With the increase of temperature through T_V, the spectral onset becomes abruptly shifted towards E_F. However, no indication of a metallic Fermi edge was observed in the high T phase just above T_V.

For a quantitative discussion, the spectral onset (E_{on}) was defined as the intersection of the leading edge of the spectrum with the zero intensity base line as

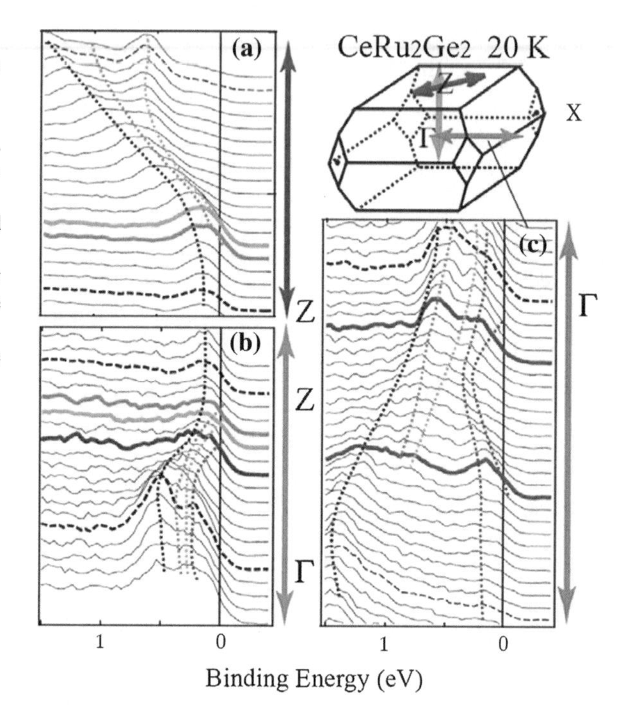

Fig. 7.27 Soft X-ray (*hv:* 735–840 eV) ARPES spectra of $CeRu_2Ge_2$ in the paramagnetic phase together with the Brillouin zone [120]. To obtain the data, an angle-scan was performed for the graph (**a**) at $hv = 755$ eV and (**c**) at $hv = 820$ eV whereas the graph (**b**) was obtained by an *hv*-scan. The arrows in the Brillouin zone correspond to the scanned directions for the graphs (**a**–**c**). The dashed lines representing each band are guides to the eye

shown in the inset of Fig. 7.26(b) and its T dependence is summarized for a fractured surface as well as annealed (111) and (100) surfaces in Fig. 7.26(b). Interestingly, a discrete jump up to 50 meV is seen at T_V only for the fractured surface. In agreement with the previous report [115], the Verwey transition in the bulk is definitely an insulator to insulator transition. It was also noticed that the discontinuous jump of the onset energy was superimposed on top of a continuous energy shift of the onset over a wide T range of tens K. The results for (111) and (100) surfaces show a continuous shift of the onset with T. On these surfaces prepared by thermal annealing, the surface configuration might be relaxed in thermal equilibrium. On the other hand the surface fractured at cryogenic temperatures remains in non thermal equilibrium and the surface relaxation is hindered at low T even though the topmost surface is disturbed to some extent.

The absence of a well defined QP peak near E_F and the absence of a sharp Fermi edge are the intrinsic feature of bulk Fe_3O_4. One of the possibilities to interpret the observed bulk results is to consider the strong electron-lattice coupling leading to the formation of polaronic quasiparticles [117]. A Jahn–Teller effect can be expected for the high spin Fe^{2+} d^6 configuration and might be responsible for this effect. Hard X-ray PES on Fe_3O_4 was consistently understood by the polaron QP model [118].

7.2 Angle Resolved Soft X-ray Photoelectron Spectroscopy

The soft X-ray ARPES (SXARPES) by using SR has an advantage due to the tunability of the excitation photon energy hν. Consequently we can select the momentum (wave vector) $k_\perp$ or k_z perpendicular to the sample surface by tuning hν in addition to momenta parallel to the surface by defining both polar and azimuthal angles (θ, ϕ) of the photoelectron detection. Then one can survey the electronic dispersions in reciprocal space. Namely, the momentum-resolved bulk electronic structure can be investigated in three-dimensional (3D) reciprocal space by SXARPES. This is also owing to the high bulk sensitivity of SXARPES since the momentum resolution normal to the sample surface is represented by $1/\lambda_{mp}(E_k)$, where $\lambda_{mp}(E_k)$ denotes the photoelectron inelastic mean free path as a function of kinetic energy E_k. Namely, SXARPES became a complementary and powerful technique for probing the bulk 3D Fermi surface (FS) topology compared to the so far performed quantum-oscillation measurements using de Haas van Alphen or Shubnikov de Haas effects. Advantages of Fermiology by the SXARPES with respect to the quantum-oscillation measurements are

1. The FS topology can be revealed by the SXARPES data alone (although the information of the band-structure calculations is very helpful for the analysis).
2. All FS sheets can be probed in principle, even an open Fermi surface, which cannot be detected by the quantum oscillations.
3. One can detect the place in the 3D **k** space where the FS sheets are located.
4. Measurements are feasible even at rather high T with noticeable momentum broadening effects [119].
5. The experimental results are much less influenced by the impurities and possible lattice defects.

On the other hand, drawbacks of the SXARPES for Fermiology are

1. The precision of the determination of the cross-sectional Fermi surface is poorer than that by the quantum oscillation measurements.
2. Measurements are not feasible under external magnetic field.
3. Measurements are not feasible under pressure.
4. The lowest measuring T is usually higher than those for the quantum oscillation measurements.

7.2.1 *Ce Compounds*

Since many strongly correlated Ce compounds have 3D electronic structures, the hν-tunable SXARPES would be very powerful to examine their details. However, high symmetry points in k_Z direction must be determined by tuning the hν even though the photon wave vector is properly taken into account because the inner

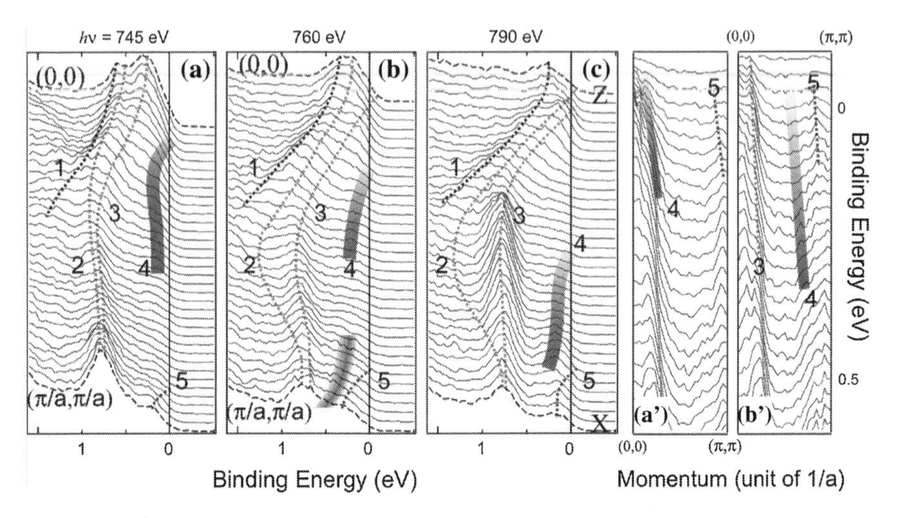

Fig. 7.28 *hν*-dependent SXARPES spectra of $CeRu_2Si_2$ along the (0, 0)–(π/a, π/a) direction at 20 K [121]. The value of k_z are (**a**) (4/13)($2\pi/c$) and (**b**) (7/13)($2\pi/c$) recorded at the photon energies of hν = 745 and 760 eV, respectively. (**c**) Along the high symmetry line *Z-X* recorded at 790 eV ($k_z = 2\pi/c$). (**a**′) and (**b**′) are the expanded MDC (*momentum distribution curves*) figures corresponding to (**a**) and (**b**), respectively. The *dashed lines* representing each band are guides to the eye. The *thick lines* are given for band 4 whose intensity is weaker than others

potential V_o is not known in advance. If one would like to measure the SXARPES under the resonance condition, the portion of the reciprocal space which can be scanned becomes rather limited because the hν is fixed and $k_{||}$ and $k_{\perp}$ are linked to the photon wave vector by (2.2) and (2.3). In this sense, the 3D SXARPES of Ce systems is mainly applicable to the non-4f electronic structures, in which the information of the 4f states is indirectly included. Here a few examples of 3D SXARPES applied to Ce compounds are presented for $CeRu_2Si_2$ and $CeRu_2Ge_2$, of which the angle-integrated RPES study was already presented in Sect. 7.1.1. $CeRu_2Si_2$ is a heavy fermion system with $T_K \sim 20$ K and $CeRu_2Ge_2$ is a ferromagnet below $T_C \sim 8$ K. SXARPES measurements were performed at 20 K for both materials. So the paramagnetic phase was probed for $CeRu_2Ge_2$. The details of the experimental conditions are given in Refs. [120, 121].

Figure 7.27 shows the SXARPES results (energy distribution curves, EDCs) of $CeRu_2Ge_2$ in the paramagnetic phase. One can recognize that the peaks in the spectra disperse as a function of hν (Fig. 7.27(b)) as well as angle (Fig. 7.27(a), (c)), which reflects that these data reveal the 3D electronic structure of this material. Figure 7.28 shows the hν-dependent SXARPES results of $CeRu_2Si_2$ parallel to the (0, 0)–(π/a, π/a) direction. The shape of each band and k_F were comprehensively evaluated by both EDCs (Fig. 7.28(a–c)) and MDCs (momentum distribution curves) (Fig. 7.28(a′), (b′)). As shown in Fig. 7.28(a) and (a′), the band 5 crosses E_F near the (π/a, π/a) point. Although the spectral weight derived from the band 4 was weak, we could trace its dispersion crossing E_F. Figure 7.28(a–c) also indicate that the k_F for the band 4 shifts towards (π/a, π/a) region when hν is away from 745 towards 790 eV, although the intensity of the band 4 at hν = 760 eV is so weak that k_F could not be determined so accurately. Therefore, the position of the band 4 is tentatively given by grey-shaded zones to allow a possible experimental ambiguity. The lines of the guide to the eye are drawn to pass through adjacent (in the sense of wave vector in EDCs and energy in MDCs) noticeable structures in order to compromise with the experimental statistics. When hν was increased from $k_z \sim 0$ (hν = 725 eV) to 2π/c (790 eV) through → (a) → (b) → (c), we could recognize that the bands 3 and 2 cross E_F sequentially. At hν = 745 eV (k_z = (4/13)(2π/c)), the bands 1–3 are on the occupied side at the (0, 0) point. When k_z = (7/13)(2π/c) was chosen at hν = 760 eV (b), only band 3 crosses E_F near the (0, 0) point, while bands 1 and 2 are fully occupied. At hν = 790 eV (k_z = 2π/c), both bands 2 and 3 cross E_F near the Z (0, 0, 2π/c) point, while band 1 is still on the occupied side at the binding energy of 0.27 eV, indicating that the band 1 does not form the FS sheet.

The results for both $CeRu_2Si_2$ and $CeRu_2Ge_2$ in Fig. 7.29(a), (c) and (b), (d) show expanded EDCs near the X and Z points at $k_z \sim 2\pi/c$. In Fig. 7.29(c), (d) the peak position of band 1 at the Z point for $CeRu_2Si_2$ is located at $E_B \sim 0.27$ eV, whereas that of $CeRu_2Ge_2$ is at about 0.14 eV. Figure 7.29(a) and (b) shows a difference of the bottom positions of band 4 and band 3 near the X point between $CeRu_2Si_2$ and $CeRu_2Ge_2$. The position of both bands 3 and 4 of $CeRu_2Si_2$ lies at higher E_B than in the case of $CeRu_2Ge_2$. A prominent difference of the band structure between these $CeRu_2X_2$ appears for the band 5, as shown in Fig. 7.29(e) and (f) at $k_z \sim 0$. This mutual difference of band 5 obtained in SXARPES is in good agreement with the difference of the band structure calculations between $LaRu_2Ge_2$ and $CeRu_2Si_2$ [122, 123]. The band 5 of $LaRu_2Ge_2$ crosses E_F three times in the region along the Γ-X direction, while that of $CeRu_2Si_2$ crosses only once in the same region. When we take a different perspective, it is possible to think that E_F of $CeRu_2Si_2$ is shifted to lower (smaller) E_B or to the unoccupied side compared to $CeRu_2Ge_2$ along the Γ-X direction.

These differences of the band position in $CeRu_2X_2$ (X:Si, Ge) can roughly be understood if E_F of $CeRu_2Si_2$ is energetically higher than that of $CeRu_2Ge_2$ in the paramagnetic phase. The E_F shift of $CeRu_2Si_2$ with respect to $CeRu_2Ge_2$ in the paramagnetic phase could be caused by the increased number of electrons contributing to the bands forming the Fermi surface sheets in $CeRu_2Si_2$ due to the hybridization with the 4f electrons. As confirmed by the 3d-4f RPES in Sect. 7.1.1, the hybridization is stronger for $CeRu_2Si_2$ than for $CeRu_2Ge_2$ indicating the E_F shift mentioned above. For a more precise understanding of the difference in the electronic structure between $CeRu_2Si_2$ and $CeRu_2Ge_2$, this "rigid-band-like" energy shift should not be sufficient because the band structures themselves could be modified due to the different hybridization strengths at different k values. Indeed, such a modification is seen in Fig. 7.29(a–d). For instance, the bands 3–5 are nearly degenerate at the X point in $CeRu_2Si_2$, while they are energetically

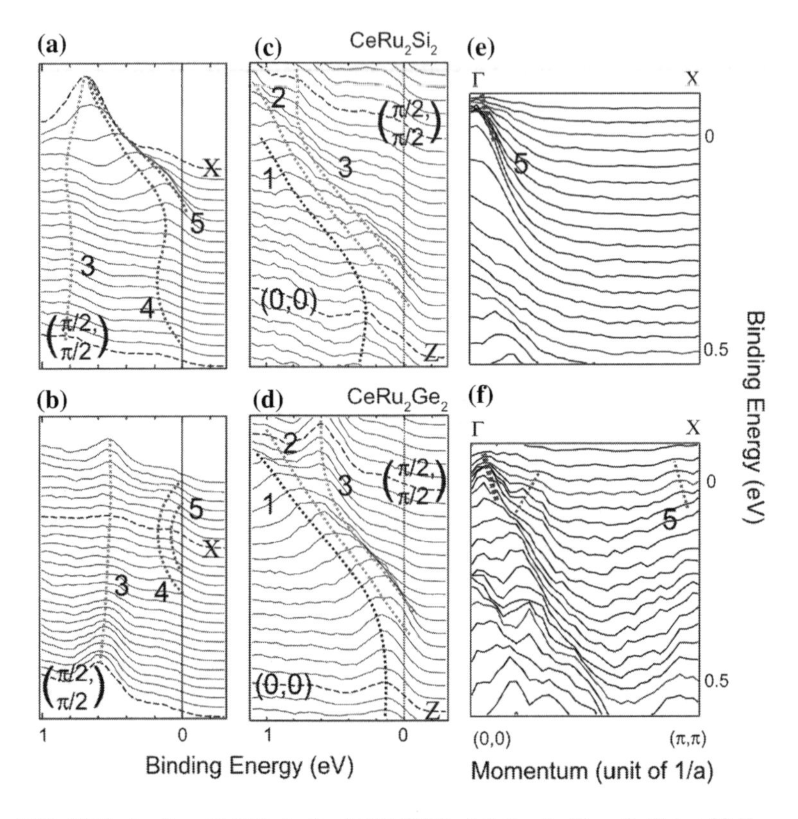

Fig. 7.29 EDCs (**a–d**) and MDCs (**e, f**) of SXARPES of $CeRu_2X_2$ (X = Si, Ge) at 20 K near E_F. Spectra are shown in the limited regions of BZ from the X point to X + (π/(2a), π/(2a)) [Figs. (**a, b**)] as well as from the Z point to Z + (π/(2a), π/(2a)) [Figs. (**c, d**)]. Upper (*lower*) 3 figures are for $CeRu_2Si_2$ ($CeRu_2Ge_2$) in the paramagnetic phase [121]. The numbered *dashed lines* are guides to the eye. (**a–d**) are EDCs along the *Z-X* direction at $k_z \sim 2\pi/c$. The hν are 725 eV for (**a**) and (**c**) and 755 eV for (**b**) and (**d**), respectively. The hν difference results from the difference of the lattice constant for and inner potential V_0. MDCs [(**e, f**)] are along the *Γ-X* direction at $k_z \sim 0$. The hν are (**e**) 725 eV and (**f**) 820 eV, respectively. Details for $CeRu_2Ge_2$ are in Ref. [120]. In accordance with the 3D BZ shown in Fig. 7.27, one can probe the electronic structure at $k_z \sim 0$ in the next BZ at a different polar angle θ even at the same hν = 725 eV corresponding to $k_z \sim 2\pi/c$ in (**a**) and (**c**) for $CeRu_2Si_2$, because the next BZ is shifted by $\Delta k_Z = 2\pi/c$

separated in $CeRu_2Ge_2$. The band 5 dispersion along the Γ-X direction shown in Fig. 7.29(e) and (f), especially near the Γ point, is also essentially modified because the simple E_F shift would lead to a shift of k_F due to band 5 indicated by the bold dotted line in Fig. 7.29(e) and (f) towards the Γ point for $CeRu_2Si_2$, which is inconsistent with the experimental results. The 3D Fermi surface topologies of these two materials revealed by SXARPES are shown in Refs. [120, 121]. In this way, the essential differences in the band structures between $CeRu_2Si_2$ and $CeRu_2Ge_2$ were experimentally clarified by bulk sensitive SXARPES.

7.2.2 $La_{2-x}Sr_xCuO_4$ and $Nd_{2-x}Ce_xCuO_4$

The electronic structure of the high temperature superconductors (HTSCs) are usually approximated by a strongly correlated 2D electron system because of their layered crystal structures. The conducting network is mainly formed by the in-plane Cu $3d_{x^2-y^2}$ orbitals strongly hybridized with the neighbouring O 2p states (hereafter referred as "effective Cu $3d_{x^2-y^2}$ band") within the CuO_2 layer. Therefore, it was thought that the observed electronic states of the top CuO_2 layer (surface CuO_2 layer) underneath the topmost insulating layer probed by surface-sensitive low-energy ARPES were almost equivalent to those in the bulk CuO_2 layers. The electron-like FS was so far reported for highly overdoped ($x \geq 0.2$) $La_{2-x}Sr_xCuO_4$ (LSCO) by low-energy ARPES [124–126]. Here reported high-energy ARPES has advantages in the bulk sensitivity [104, 127], the relatively higher sensitivity to the Cu 3d than to the O 2p states [16], and the small energy dependence of the transition matrix elements. In order to reveal the bulk electronic structure and Fermi surfaces (FSs) of HTSCs, we performed ARPES at $h\nu = 500$ eV [128] for LSCO, which is an ideal system in the sense that the crystal structure is simple with one CuO_2 layer and no super-structure modulation, and the amount of doping can be precisely controlled by the Sr concentration x [129, 130]. Since a wider k region is accessible by SXARPES compared to low $h\nu$ ARPES below 100 eV, no symmetrisation is usually required to discuss the band dispersions in the full BZ.

Figure 7.30(a), (b) shows the high-energy ARPES spectra measured at $h\nu = 500$ eV and 20 K along the (0, 0)–(0, π/a) direction for $x = 0.16$ [128]. The energy distribution curves (EDCs), momentum distribution curves (MDCs) and the QP dispersion obtained by the second derivative of the EDCs are shown in Fig. 7.30(a–c). All these results show a dispersive feature of the QP with the peak reaching E_F near (0, $\pm 0.85\pi$/a) indicating a FS crossing between (0, 0) and (0, $\pm\pi$/a). The spectral weight within $E_F \pm 100$ meV, which approximately represents the FS, and Fermi wave vectors (k_F) estimated from the MDCs for $x = 0.16$ are shown in Fig. 7.30(e). The bulk FS which is depicted in Fig. 7.30(e) is electron-like centred at (0, 0) in strong contrast to the hole-like FSs reported for the optimally doped HTSCs including LSCO by low-energy ARPES [124, 125, 131–135]. The area enclosed by the FS for $x = 0.16$ was estimated to be 83 ± 8 % of the half of the BZ area, which satisfies Luttinger's sum rule ($1-x$).

We also investigated FSs of LSCO for overdoped $x = 0.23$ and slightly underdoped $x = 0.14$. The observed FS for $x = 0.23$ was also electron-like (Fig. 7.30(d)) and the estimated FS area was 77 ± 5 % of the half BZ area, also satisfying Luttinger's sum rule. In the case of $x = 0.14$ (Fig. 7.30(f)), however, the character of the bulk FS could be not determined because the FS crossing of the QP could not be identified within the limited energy resolution of the high-energy ARPES measurement. Still it was clear that a FS was located in the vicinity of (0, π/a) for $x = 0.14$. Therefore, the doping dependence of the electronic structure was qualitatively understood by considering a rigid-band-like shift of E_F due to the

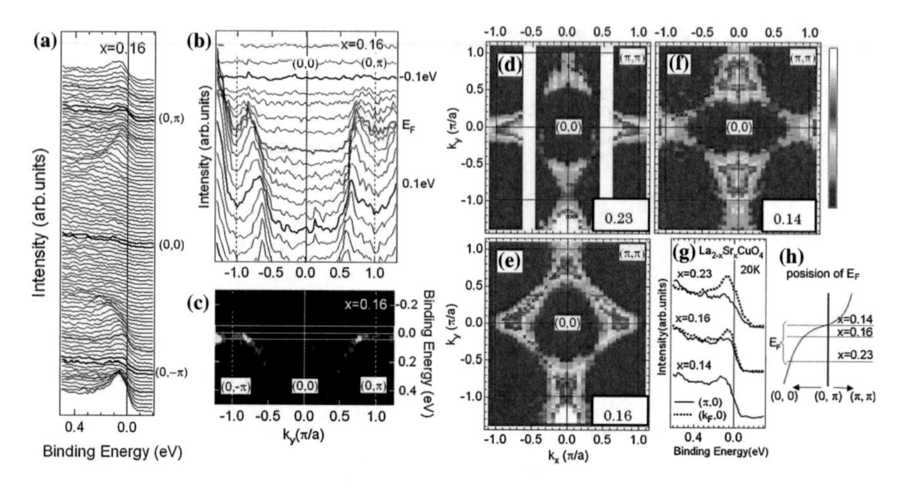

Fig. 7.30 SXARPES data for $La_{2-x}Sr_xCuO_4$ obtained at hν = 500 eV and at T = 20 K on an in situ cleaved surface at this T with angular resolution of ±0.1° (± 0.024π/a). **a** EDCs along (0, 0)–(0, π/a) for nearly optimally doped x = 0.16. **b** Corresponding MDCs. **c** QP dispersion derived from (**a**) by means of the second energy derivative. **d** Spectral weight integrated over the window of E_F ± 100 meV and k_F (*red marks*) estimated from the MDCs for x = 0.23. **e** Same as (**d**) but for x = 0.16. **f** Same as (**d**) but for x = 0.14. **g** EDCs at (π/a, 0) compared to those at k_F between (0, 0) and (π/a, 0) for x = 0.16 and 0.23. The energy resolution of EDCs was 100 meV for x = 0.14 and 0.16 but 170 meV for x = 0.23. **h** Schematically drawn QP dispersion along the (0, 0)–(0, π/a)–(π/a, π/a) directions and doping dependence of E_F for LSCO

hole doping as suggested in Fig. 7.30(h). The presence of a near-E_F peak for x = 0.14 suggests that the boundary between the electron-like and hole-like FSs in the bulk is close to the optimally doped x of ∼0.15, while the exact boundary in LSCO is still in controversy [126].

The so-called stripe effect was reported for x = 0.15 and x = 0.22 by low-energy ARPES as a presence of "straight segments" in the spectral weight in the vicinity of E_F near (0, π/a) [126, 135]. In the present study, such a straight segment was not observed in the spectral weight near E_F for x = 0.14, 0.16 and 0.23 whereas a straight segment was seen for x = 0.07 and 0.1 (not shown here), which ruled out strong contributions from the proposed site-centred stripes [136] for x ≥ 0.14 including the optimally doped region.

Figure 7.31(a) shows the hν and polar angle (θ) dependence of the angle-integrated valence-band spectra for x = 0.1 in a wider E_B region. The bulk sensitivity of the measurement is suppressed at lower hν and larger θ [87, 104]. In the deeper E_B region in the spectrum measured at hν = 500 eV and θ = 0°, a shoulder appears at E_B = 1 eV in addition to a peak at about E_B = 1.7 eV. This shoulder is noticeably reduced for θ = 60° (hν = 500 eV) and hν = 220 eV (θ = 0°) and then understood to be reflecting bulk electronic states. In addition, these spectral features were found in SXARPES to disperse clearly as a function of momentum as shown in Fig. 7.31(b) for x = 0.07. Considering the larger PICS of the Cu 3d states with respect to that of the O 2p states [16] at high hν, we concluded that the

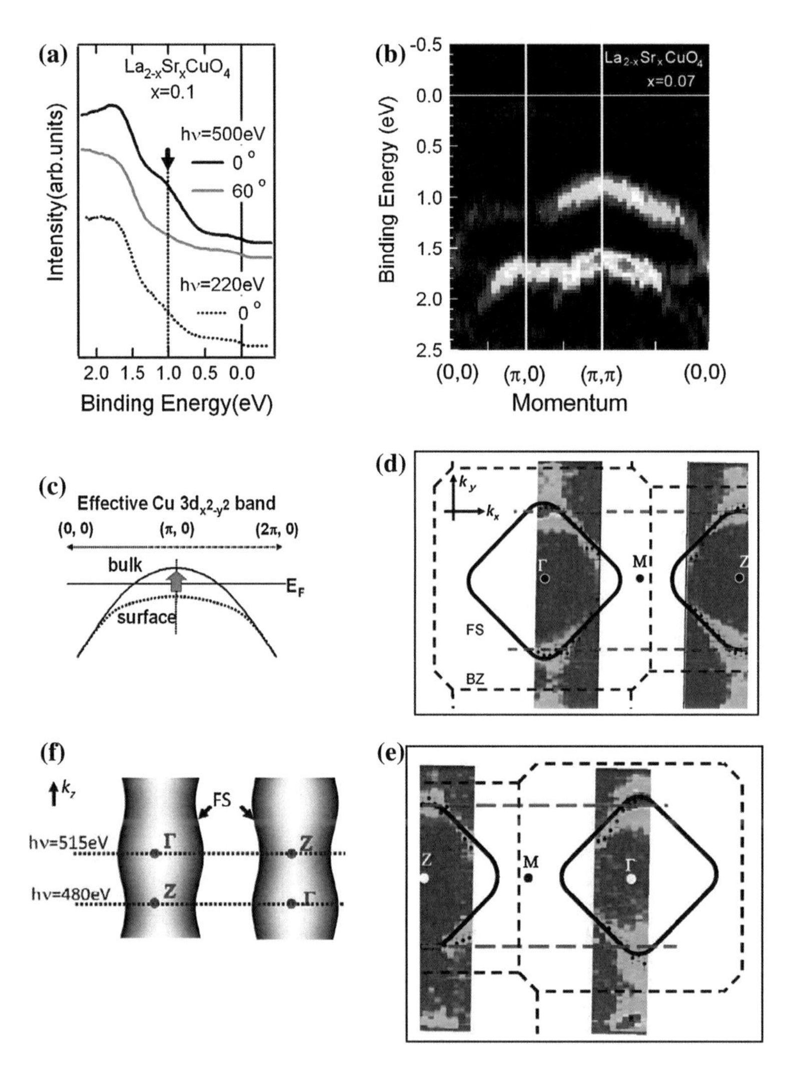

Fig. 7.31 (**a**) hν and θ dependence of the angle-integrated PES spectra from E_F to E_B ~ 2 eV for $La_{1.9}Sr_{0.1}CuO_4$ (x = 0.1). (**b**) Second energy derivative of the EDCs at hν = 500 eV along the (0, 0)–(0, π/a)–(π/a, π/a) directions for $La_{1.93}Sr_{0.07}CuO_4$ (x = 0.07). (**c**) Schematic picture of the different QP dispersion near (π/a,0) between the bulk and surface CuO_2 layers due to different hopping transfer $t_\perp$. (**d**) FSs of LSCO for x = 0.16 at hν = 515 eV, where FS(Γ) > FS(Z). (**e**) FSs at hν = 480 eV, where FS(Z) < FS(Γ). (**f**) Schematic FS topology along k_Z for optimally doped LSCO in accordance with the results in (**d**) and (**e**) [142]

dispersion near 1.7 eV and that near 1 eV were ascribable to the "effective Cu 3d xy, yz and zx bands" and the "effective Cu $3d_{3z^2-r^2}$ band", respectively, which are hybridized with the O 2p states. The features of the dispersion for the effective Cu $3d_{3z^2-r^2}$ band are consistent with the results of band-structure calculations [137, 138]. The suppression of the bulk $3d_{3z^2-r^2}$ spectral weight in the surface-sensitive spectra suggested that the effective $3d_{3z^2-r^2}$ state is remarkably different between the bulk and surface. These dispersive structures were likewise observed in the SXARPES spectra for x = 0.14, 0.16 and 0.23.

A possible origin of the qualitative difference in the effective $3d_{x^2-y^2}$ (QP) states between the bulk and surface CuO_2 planes was then discussed. If the doping level deviates from the nominal value x to an underdoped level in the surface, the hole-like FS should be favoured for x = 0.16. Judging from the single CuO_2 layered crystal structure, however, we consider that such a doping level shift in the surface is not likely to occur. On the other hand, the difference could be explained without assuming a possible change of the doping level as follows. A finite hopping transfer perpendicular to the CuO_2 layers, $t_\perp$, which is often omitted in discussing the main physics of HTSCs, is supposed to decrease in the surface layer, making the two-dimensionality of the QP stronger in the surface. Namely, the bulk QP actually has a finite three-dimensionality with a strong anisotropy in LSCO. According to theoretical studies [139, 140], the QP bandwidth becomes larger in the presence of the apical oxygen, from which a finite $t_\perp$ is derived. Band-structure calculations [138] show that the effective $3d_{x^2-y^2}$ band is widened along the (0, 0)–(0, π/a) direction due to a mixture of the $3d_{3z^2-r^2}$ state especially near (0, π/a), while this mixture was not seen along the (0, 0)–(π/a, π/a) direction. From these results it can be inferred that the QP band dispersion is modified especially near (0, π/a) in the surface compared to the bulk due to the difference in $t_\perp$. It was thus confirmed that the QP band in the bulk crosses E_F between (0, 0) and (0, π/a), leading to the electron-like FS of LSCO for x = 0.16. In the surface, on the other hand, the QP width becomes narrower along the (0, 0)–(0, π/a) cut and its top energy is lowered near (0, π/a) giving rise to the QP in the occupied states at (0, π/a) in the surface of the optimally doped LSCO as illustrated in Fig. 7.31(c). E_F may be slightly shifted in order to keep the same carrier number as the bulk. A 2D electronic state was assumed in the stripe model [136] for HTSCs. However, the stripe effect should be weakened by a finite $t_\perp$ in the bulk. The finite three-dimensionality of the QP in LSCO has been observed as metallic behaviour of the temperature-dependent resistivity perpendicular to the CuO_2 plane [141] for x > 0.13. According to the three-dimensionality, the stripe effect near (0, π/a) would also be suppressed in SXARPES. Figure 7.31(d) and (e) shows the in-plane FS shapes measured at two different k_z containing the Γ and Z points in the BZ. It is clearly demonstrated that the electron FS is larger around the Γ point than around the Z point, confirming the k_z dependence of the FS size or the three dimensionality of the FS topology in the bulk of LSCO with x = 0.24 as illustrated in Fig. 7.31(f) [142]. Since the hν used for the results in Fig. 7.30 (500 eV) was close to the centre between 515 and 480 eV, the FS shape and size must be close to the average of those obtained at the Γ and Z points. In this way SXARPES is very powerful to study the Fermi surface topology of some high T_C cuprates, too.

In an electron-doped HTSC $Nd_{2-x}Ce_xCuO_4$(NCCO), SXARPES was first performed at hν = 400 eV by circularly polarized light. The obtained bulk sensitive results showed a hole-like FS centred at (π/a, π/a) [143] as seen in low-energy

ARPES [134, 144, 145]. The symmetrized FS was compared with the FS obtained by low hν ARPES and the mutual difference was discussed [143]. The hole-like FS was generally favoured by the electron doping for the HTSCs in both bulk and surface, inducing the FS crossing between (0, π/a) and (π/a, π/a). Later we performed SXARPES with fully circularly polarized light (more than 98 %) at hν = 460 and 500 eV for NCCO with x = 0.075 and 0.15 with better statistics [146] in addition to an angle integrated measurement of the Cu 2p core level at hν = 1,504 eV and T = 20 K in order to confirm the real bulk sensitivity from its θ dependence. The FS shape was obtained without symmetrization as shown in Fig. 7.32(a), (c). The observed bulk FS shape is rather circular in contrast to the square like FS shape obtained after symmetrisation at 400 eV [143] and clearly deviates from those of low hν ARPES [145] in the sense that k_F are much closer to the ($-\pi$/a, 0)–(π/a, 0) and (0, $-\pi$/a)–(0, π/a) axes in SXARPES. In contrast to low hν ARPES [145], a suppression of intensity in the intersection regions between the FS and the antiferromagnetic BZ boundary (AF BZ boundary) was not confirmed in our bulk sensitive results. In addition, the dispersion of the QP measured with a total energy resolution of 100 meV clearly shows a nodal kink structure in Fig. 7.32(e) around E_B ~ 50–80 meV in contrast to the most surface sensitive low hν ARPES at hν = 55 eV [147]. In addition, no clear antinodal kink was observed as seen in Fig. 7.32(f), again in strong contrast to low hν ARPES. Neither the AF pseudo gap behaviour nor the near-E_F QP mass enhancement were observed along the antinodal direction by the present hν = 500 eV bulk sensitive SXARPES. The electron–phonon coupling constant λ' estimated from the present SXARPES is 1.2–1.6 for NCCO with x = 0.15 comparable to those in hole doped high T_c cuprates. All these results are close to those of the nodal "kink" reported for several families of hole-doped high T_c cuprates and suggest that the electron–phonon coupling in NCCO is as strong as in the hole doped high T_c cuprates, implying that the existence of the nodal "kink" behaviour is inherent to high T_c cuprates irrespective of the types of doped carriers. The possibility that the surface electronic structure deviates much from the bulk could stabilize antiferromagnetism in the surface region even for optimally doped NCCO. Then the electron-phonon interactions might be masked at the surface.

7.2.3 Layered Ruthenates $Sr_{2-x}Ca_xRuO_4$

Then we discuss the FSs of the superconductor Sr_2RuO_4 and the non-superconductor $Sr_{1.8}Ca_{0.2}RuO_4$ (paramagnetic metal: PMM) probed for the first time by bulk-sensitive SXARPES. It was known that Sr_2RuO_4 shows "triplet" superconductivity with T_c ~0.93 K [148, 149], which disappears with a very small amount of Ca-substitution [150]. For $Sr_{2-x}Ca_xRuO_4$ with x > 1.8, Mott insulator behavior occurs. The combination of quantum oscillation measurements and band-structure calculations suggested one hole-like FS sheet centered at (π/a, π/a) (α sheet) and two electron-like FS sheets centered at (0, 0) (β and γ sheets) in Sr_2RuO_4

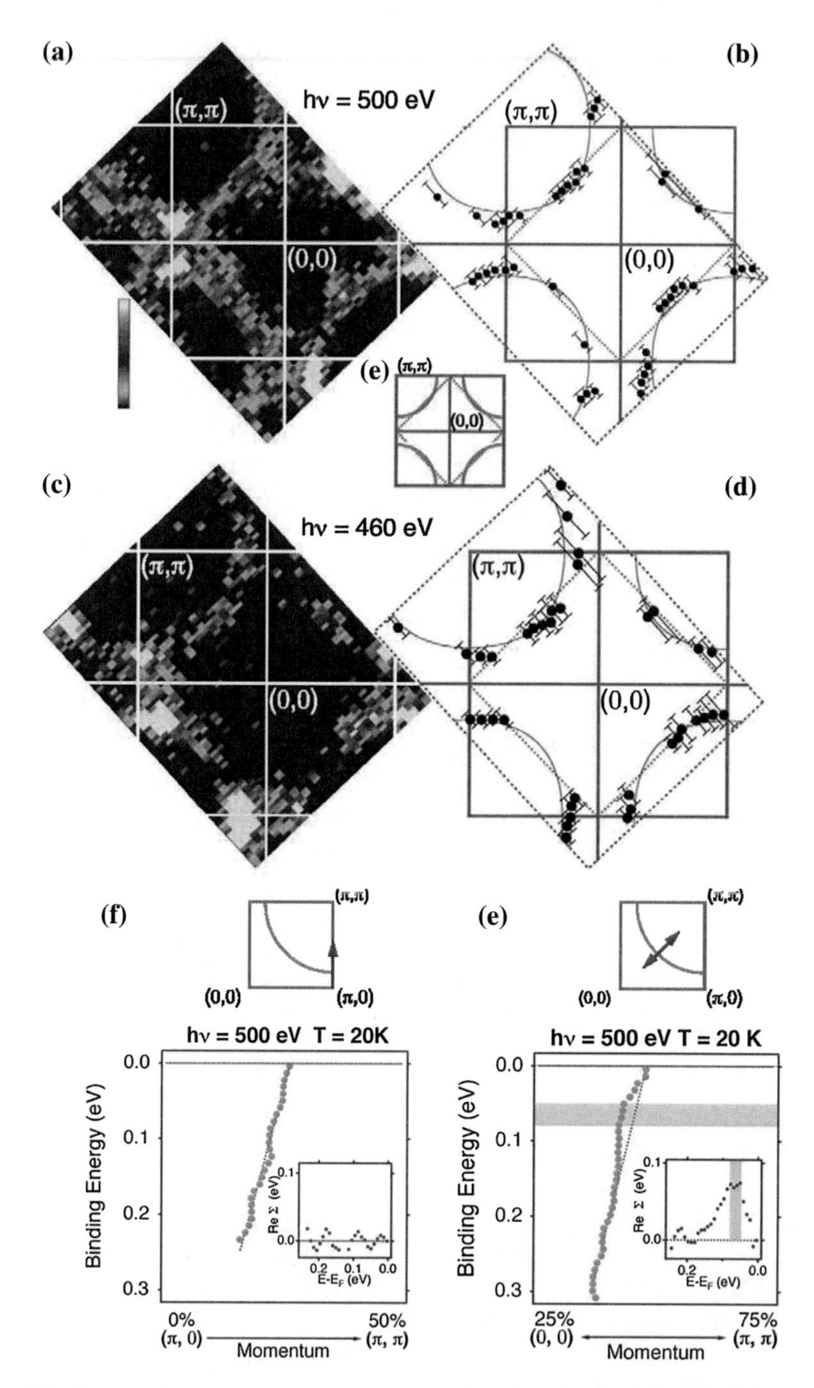

Fig. 7.32 Raw Fermi surface of $Nd_{1.85}Ce_{0.15}CuO_4$ obtained at hν = 500 and 460 eV at T = 20 K with a total energy resolution of 200 meV (**a**–**d**). **e**, **f** QP dispersion along the nodal and antinodal directions determined from SXARPES performed with a total energy resolution of 100 meV [146]

[151–153] (Fig. 7.33(h)). On the other hand, the so far reported results of low hν ARPES for Sr_2RuO_4 were controversial. Two hole-like FSs and one electron-like FS were concluded by Yokoya et al. [154]. However, the following ARPES studies [134, 155, 156] suggested that the earlier finding originated from surface states, and that the bulk FSs were qualitatively similar to the result of the band-structure calculation. It was also reported that a lattice distortion took place at the surface [157, 158], giving FSs different from the bulk. Thus the character and shape of the two-dimensional bulk FSs of Sr_2RuO_4 were experimentally still unclear because the reported shapes of the FSs from low hν ARPES depended on the surface preparation and the excitation photon energies [155, 156]. Low hν ARPES on Sr_2RuO_4 showed that the FSs shapes measured on the "degraded" surface obtained by cleavage at 180 K and fast cooled down seemed to be similar to the prediction of band-structure calculations in contrast to those obtained on clean surfaces prepared by cleavage at 10 K [134, 156]. In general, photoemission data on cleaner surfaces prepared by cleavage at lower temperatures, at which surface desorption and diffusion of atoms from inside are suppressed, are more reliable. Thus the mere similarity of the FSs as obtained by low hν ARPES and theory cannot guarantee that the genuine bulk FSs of Sr_2RuO_4 were really established. Besides, the FSs of lightly Ca-substituted $Sr_{1.8}Ca_{0.2}RuO_4$ were not yet clarified.

Single crystals of $Sr_{2-x}Ca_xRuO_4$ (x = 0, 0.2) were used for the SXARPES measurements at hν = 700 eV performed at BL25SU in SPring-8. The (001) clean surface was obtained by cleaving the samples in situ at the measuring temperature of 20 K. Photoelectrons within a polar acceptance angle of about ±6° with respect to the normal of the sample surface were simultaneously collected. A larger $k_{||}$ region than a whole BZ was simultaneously detected along the direction of the analyzer slit. $k_{||}$ perpendicular to the slit was measured by rotating the sample. The overall energy resolution was set to ~120 and ~200 meV for high-resolution measurements and FS mapping, respectively. The angular resolution was ±0.1°(±0.15°) for the perpendicular (parallel) direction to the analyzer slit. These values correspond to a momentum resolution of ±0.024 $Å^{-1}$ (±0.035 $Å^{-1}$) (6 and 9 % of π/a, where a is the lattice constant of Sr_2RuO_4, 3.87 Å [148]) at hν = 700 eV. We also measured core-level spectra with an energy resolution of 200 meV.

From the polar-angle (θ) dependence of the Sr 3d core-level spectra of Sr_2RuO_4 at hν = 700 eV, the surface contributions in the spectra were clearly confirmed [104]. Namely, in addition to peaks at E_B = 132 and 134 eV originating from the 3d spin–orbit splitting, two shoulder structures appeared on the higher E_B side of the main peaks in the spectrum at $\theta = 0°$. The intensity of these shoulders was remarkably enhanced in the more surface-sensitive spectrum at $\theta = 60°$. Therefore the main peaks and the additional structures were ascribed to the bulk and surface components, respectively. A more detailed analysis showed that the bulk contribution in the valence-band spectra at $\theta \sim 0°$ at E_K of ~700 eV was 63 %. So the SXARPES at hν = 700 eV were expected to mainly provide information on the bulk electronic structure.

The SXARPES spectra of Sr_2RuO_4 and $Sr_{1.8}Ca_{0.2}RuO_4$ along the (π/a, 0)–(π/a, π/a) and (0, 0)–(π/a, 0) directions are given in Fig. 7.33(a–d) in the form of EDC.

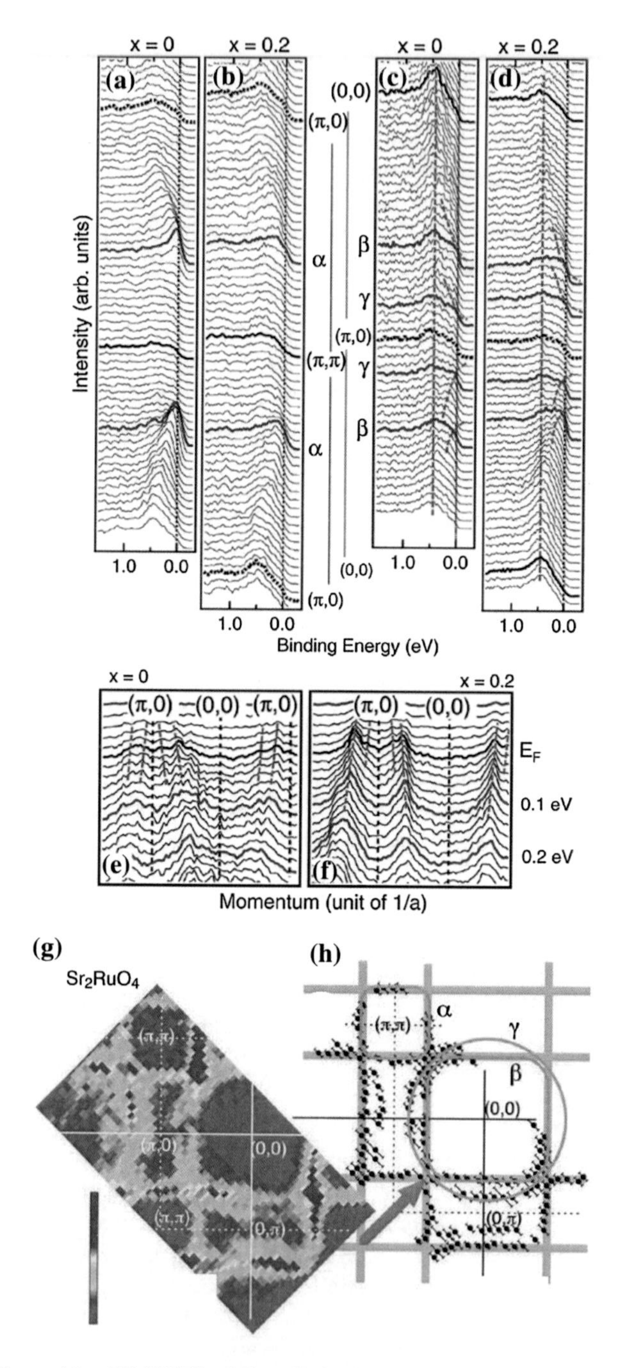

Fig. 7.33 Bulk sensitive SXARPES of $Sr_{2-x}Ca_xRuO_4$ with x = 0 and 0.2, in situ cleaved at T = 20 K and measured at hν = 700 eV and T = 20 K. **a–d** EDC, (**e**, **f**) MDC, (**g**) FS for x = 0 integrated between E_F and −100 meV (unoccupied side). **h** *Small full circles*: k_F results for x = 0 determined from EDC and MDC [104]

The MDC for the (0, 0)–(π/a, 0) direction are given in Fig. 7.33(e) and (f). The bulk FS obtained by integrating the intensity from E_F to −0.1 eV (unoccupied side) is shown in Fig. 7.33(g). Two electron like FSs and one hole-like FS were confirmed around the (0, 0) and (π/a, π/a) point in the raw data. The experimentally estimated k_F values from EDC and MDC are shown in Fig. 7.33(h) as full circles with error bars. The square-like FSs around the (0, 0) and (π/a, π/a) points are known as the β and α FS, whereas the circular FS around the (0, 0) point is known as the γ branch. A similar behavior was observed in $Sr_{1.8}Ca_{0.2}RuO_4$ except that the γ FS was less prominent than in Sr_2RuO_4.

SXARPES spectra of $Sr_{2-x}Ca_xRuO_4$ along the (0, 0)–(π/a, 0) cut shown in Fig. 7.33(c), (d) are rather complicated because there are three QP bands below E_F. For both compounds, the band forming the α sheet is located at E_B ~ 0.5 eV, shifting hardly between (0, 0) and (π/a, 0), while the other two bands forming the β and γ sheets show dispersion and cross E_F. The behavior of the E_F crossing of the β and γ branches (abbreviated as β and γ crossing) was also confirmed by the momentum distribution curves (MDCs) in Fig.7.33(e) and (f). These shapes reflect that the γ sheet is mainly composed of a rather ideally two-dimensional d_{xy} band while the other square-shaped β and α sheets are due to the d_{yz} and d_{zx} bands, which are to some extent one-dimensional. The estimated area of each sheet is comparable to the results obtained from the quantum oscillations experiments for Sr_2RuO_4. The obtained FSs of $Sr_{1.8}Ca_{0.2}RuO_4$ are similar to those of Sr_2RuO_4. The two-dimensional topology of the FSs is also consistent with the prediction from band-structure calculations.

Additional SXARPES was performed for the paramagnetic metal $Sr_{0.5}Ca_{1.5}RuO_4$ between hν = 650 and 730 eV with changing the hν by 5 eV steps. Clear differences of the QP dispersions were observed for both β and γ branches between Fig. 7.34(b) and (c), which correspond to two different k_z, namely parallel to the Z(0, 0, 2π/c)–(0, π/a, 2π/c) or Γ(0, 0, 0)–(0, π/a, 0) direction (shown by two horizontal red arrows in Fig. 7.34(a)). In a plane including both directions, FS shapes were found to be noticeably dependent on k_z as recognized in Fig. 7.34(e) [159]. Such results were double checked by comparing the results in the left and right halves of Fig. 7.34(e). If the electronic structure was highly 2D, k_f should not depend on k_z. However, noticeable k_z dispersion of the γ branch was surprising, in contrast to the much less k_z dispersion of the β branch. Thus the electronic structure of $Sr_{0.5}Ca_{1.5}RuO_4$ is found to be much more 3D like than those of Sr_2RuO_4. This result is in consistence with the fact that the anisotropy of the resistivity is smaller for $Sr_{2-x}Ca_xRuO_4$ with $x > 1.1$ than for Sr_2RuO_4 and that coherent metallic conduction is expected along k_z at T = 20 K for x = 1.5 from transport experiments [160]. We also measured the FS of $Sr_{0.5}Ca_{1.5}RuO_4$ by SXARPES at hν = 708 eV corresponding to k_z ~ 0. The result was very similar to the bulk sensitive results of Sr_2RuO_4 and $Sr_{1.8}Ca_{0.2}RuO_4$ but qualitatively different from the surface sensitive low hν ARPES of $Sr_{0.5}Ca_{1.5}RuO_4$ at hν = 32 eV [161, 162].

A plausible scenario for the 3D character of the observed FSs in $Sr_{0.5}Ca_{1.5}RuO_4$ is to take into account the rotation of the RuO_6 octahedron within the conduction

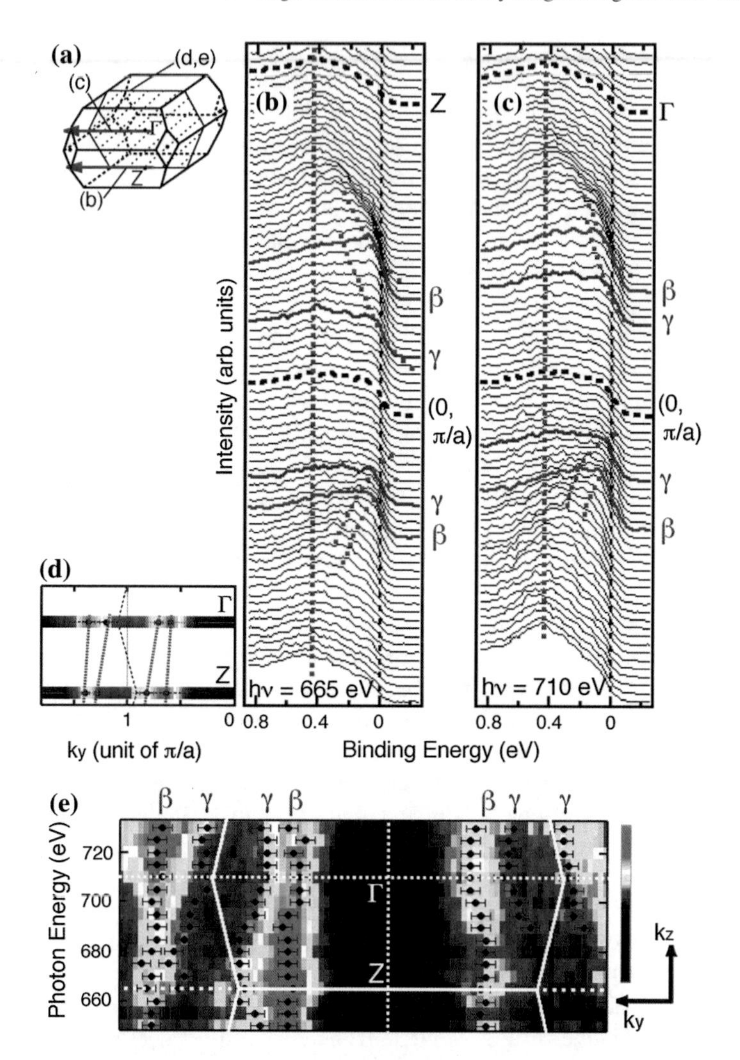

Fig. 7.34 SXARPES of $Sr_{0.5}Ca_{1.5}RuO_4$ measured from hν = 650 to 730 eV with 5 eV steps. **a** Schematic BZ with the *arrows* indicating the region covered by SXARPES in (**b**, **c**). **b** SXARPES at hν = 665 eV along Z(0, 0, 2π/c)–(0, π/a, 2π/c) ∥ (Ru–O) direction in the tetragonal BZ. **c** SXARPES at hν = 710 eV along the Γ(0, 0, 0)–(0, π/a, 0) direction. The total energy resolution was set to 110 meV for both (**b**) and (**c**). A difference depending on k_z (or hν) is clearly recognized. **e** Intensity map and k_F determined from the EDC and MDC spectra in the $k_x = 0$ plane (hexagonal plane shown as (**d**, **e**) in (**a**)) shown by the *full dots*. Clear k_z dependence of FS of both β and γ branches are seen [159]

plane. Then small but finite σ-bonding is induced with the O $2p_{x/y}$ orbitals, which are further hybridized with the Ru $4d_{x^2-y^2}$ and $4d_{3z^2-r^2}$ orbitals of the next Ru sites in similarity to LSCO, where such σ-bonding leads to 3D behavior. Although it might be weaker than in LSCO, the Ru $4d_{xy}$ derived band in the distorted crystal structure of $Sr_{0.5}Ca_{1.5}RuO_4$ can have 3D character. It should also be noted that such σ-bonding is absent between the Ru $4d_{yz/zx}$ and neighboring O 2p orbitals after the RuO_6 in-plane rotation. The mixing of σ-bonding is not expected for systems without in-plane rotation as Sr_2RuO_4 and lightly-substituted $Sr_{1.8}Ca_{0.2}RuO_4$ [160]. Thus the 3D FSs of $Sr_{0.5}Ca_{1.5}RuO_4$ were clarified owing to the bulk sensitivity of SXARPES.

7.2.4 V_6O_{13} and $SrCuO_2$

V_6O_{13} is a layered material with quasi-1D behavior due to the V chains along the b axis [163–165]. This material is metallic (M) above $T_c \sim 145$ K and insulating (I) below it (antiferromagnetic below $T_N = 55$ K). Valence mixing is taking place in the M phase with 3 inequivalent V sites of nominally $V^{4+}(3d^1)$ and $V^{5+}(3d^0)$ (with a ratio of 2 to 1). Three kinds of distorted VO_6 octahedra are running along the b axis by edge-sharing. The angle integrated SXPES was first performed with a resolution of 80 meV under the resonance excitation condition at $h\nu = 515.7$ eV, where no Auger feature was observed. Under this condition the V 3d spectral weight is one order of magnitude larger than that of the O 2p state [16]. The spectra in the metallic (M) and insulating (I) phases showed clear differences. In the M phase at 180 K, E_F crossed the rising part of the EDC. In addition, a strong peak was observed at $E_B = 0.70 \sim 0.75$ eV. In the I phase at 100 K, however, a band gap of $\sim$120 meV was observed [166]. This gap was noticeably smaller than the gap of $\sim$0.2 eV revealed by the low $h\nu$ ARPES [167]. A strong peak was observed in the I phase at 0.5–0.7 eV. In addition, a clear hump was reproducibly observed near 1.5 eV, whereas such a feature was absent in the low $h\nu$ PES [163] and ARPES [167].

We then performed SXARPES at $h\nu = 515.7$ eV along the b axis of V_6O_{13} with an energy resolution of 160 meV. The polar angle θ for the ARPES shown in Fig. 7.35 covers the region from -6 to $+5°$. The raw energy distribution curves (EDC) of SXARPES in both the M and I phase are given in Fig. 7.35(a), (b), while no noticeable dispersion was seen perpendicular to the b axis. The intensity plot of the ARPES data and the corresponding momentum distribution curves (MDC) clearly show that a branch crosses E_F with noticeable intensity on both sides of $\theta = 0°$ near $k_{//} = 0.30$ π/b. This structure corresponds to a QP or a coherent peak. Although the decrease of the PES intensity towards E_F is often observed for quasi-1D systems, the result in the M phase in Fig. 7.35(d) together with the finite EDC intensity at E_F in Fig. 7.35(a) are different from what is predicted for a Tomonaga-Luttinger-liquid (TLL) [168] for a 1D system and therefore suggests a non-negligible 3D-character of the relevant electronic states in the M phase of V_6O_{13}.

In the MDC in Fig. 7.35(d) was found that the dispersion below E_F is larger than 0.5 eV with its bottom located at the Γ point ($k_{//} = 0$). The second energy derivatives are displayed in Fig. 7.35(e) and (f). The dispersive QP peak in the M phase coexists in the same k region with another broad peak centered at

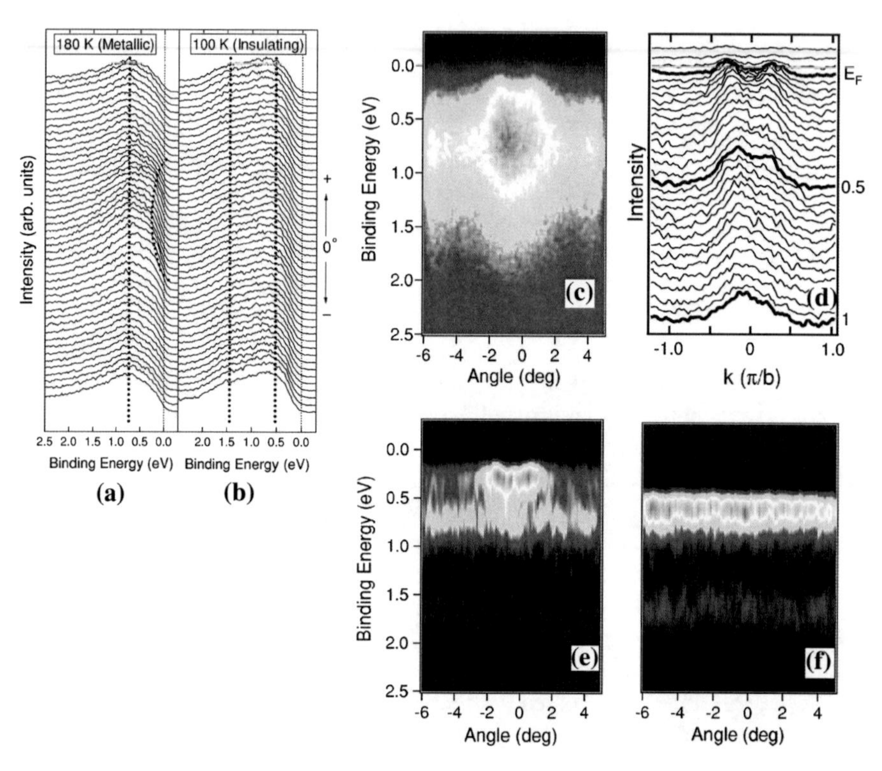

Fig. 7.35 SXARPES of V_6O_{13} at hν = 517.5 eV measured in the M phase at T = 180 K (**a**) and I phase at T = 100 K (**b**). (**c**) Intensity plot for the M phase. **d** MDC for the M phase. **e**, **f** Dispersions derived from the second energy derivative for the M and I phases [166]

0.7–0.75 eV as shown in Fig. 7.35(e). In the I phase, however, the QP peak near E_F is collapsed and a broad structure is observed at 0.5–0.7 eV. In addition, an incoherent peak is observed near 1.5 eV in Fig. 7.35(f). Although the broad peak at $E_B \sim 0.7$ eV in both M and I phases was once interpreted as the incoherent peak [166], further bulk sensitive results at h$\nu \sim 8$ keV revealed that both could be resulting from the coherent peak [169] as explained later in detail in Chap. 8 (Fig. 8.11) of this book and the incoherent peak is located at $E_B \sim 1.5$ eV through VO_2, V_nO_{2n-1} (n = 2, 3, 4, 5) and V_6O_{13}. Then the complex structures in the coherent part in the M phase in V_6O_{13} might be resulting from the crystallographically non equivalent V sites with essentially different valences.

In the case of the 1D antiferromagnetic insulator (AFMI) $SrCuO_2$, there are longstanding discussions whether the spin-charge separation is a proper scenario to interpret the ARPES results. In this material, the double Cu–O chains are lined parallel to the c axis as already illustrated in Fig. 6.8 [170, 171]. The inter-chain coupling is much weaker than the intra-chain interaction. Cleavage can be done between two Sr–O planes. Then the Cu–O chains correspond to the 2nd and 3rd planes as well as to the 6th and 7th planes and so on from the surface. The

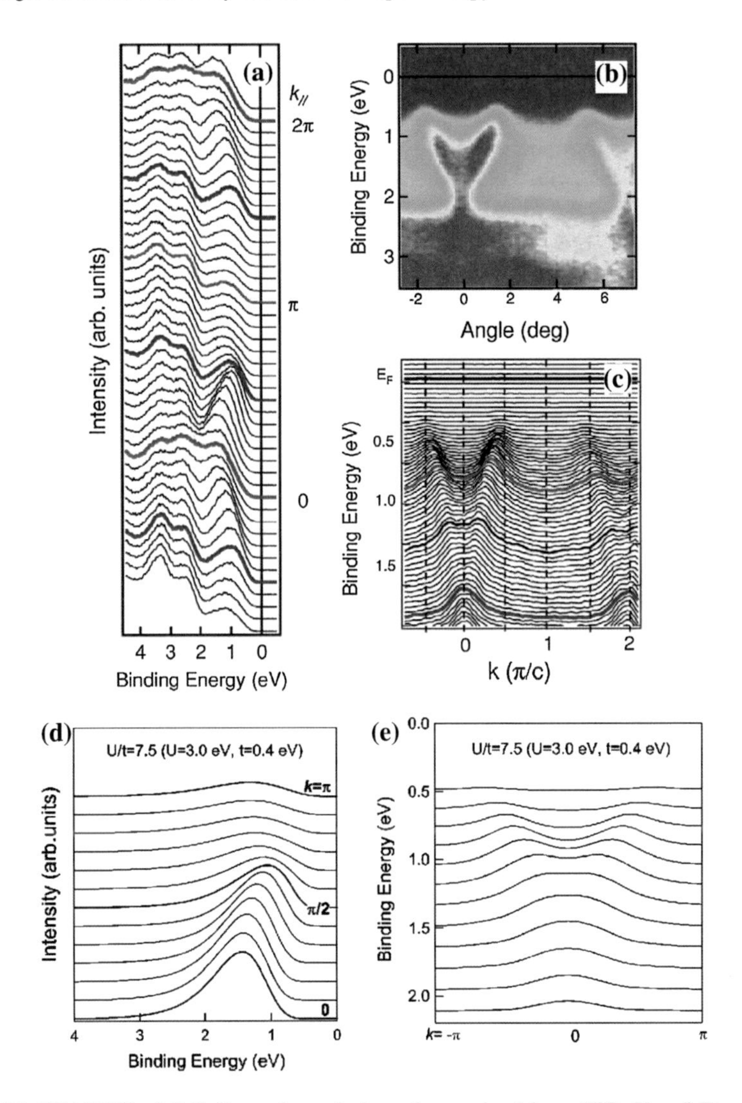

Fig. 7.36 SXARPES of $SrCuO_2$ performed along the c-axis at hν = 700 eV and T = 300 K. **a** EDC. **b** Intensity plot. **c** MDC [166]. **d** and **e** Calculated results for $SrCuO_2$ within the one-band 1D half-filled Hubbard model based on quantum Monte Carlo calculations

experiments were performed at hν = 700 eV [166], where the results are much more bulk sensitive than the results at hν = 85–200 eV [172]. The relative PICS between the Cu 3d and O 2p states is expected to be larger at hν = 700 eV than at hν = 85–200 eV [16]. The SXARPES measured parallel to the chain axis at hν = 700 eV is summarized in Fig. 7.36(a). These EDC show that there is a clear band gap between the top of the observed bands and E_F. The double peak structures in the region of 2 ~ 4 eV were ascribed to the so called "O 2p bands" in which the Cu 3d components are strongly hybridized. The structure within 2 eV from E_F shows a prominent dispersion as seen in the intensity density plot in Fig. 7.36(b). The shallowest peak is seen in Fig. 7.36(a) around $E_B = 0.95$ eV near the angle corresponding to ($\pi/(2c)$, 0). The peak shift from (0, 0) to ($\pi/(2c)$, 0) is 0.5–0.6 eV. The momentum distribution curve (MDC) is shown in Fig. 7.36(c). The dispersion in the MDC is V shaped with the bottom at the $\Gamma(0, 0)$ point around $E_B = 1.9$ eV and the gap was estimated to be less than 0.5 eV from E_F. No dispersion was observed in the corresponding E_B region to the direction perpendicular to the c axis.

The present SXARPES results at $h\nu = 700$ eV were obtained by circularly polarized light excitation. In the MDC, no dispersive feature was noticed in the ($\pi/(2c)$, 0)–(π/c, 0)–($3\pi/(2c)$, 0) region. Besides, no trace of any additional component was detected in the EDC and MDC in the ($-\pi/(2c)$, 0)–(0, 0)–($\pi/(2c)$, 0) region of SXARPES despite sufficient resolution in energy and $k_{//}$ as well as the high Cu 3d sensitivity in the present experiment, in contrast to the results at lower $h\nu$ [172].

In order to interpret the present ARPES results of $SrCuO_2$, we employed the one band, 1D half-filled Hubbard model [173] based on QMC calculations, where the quantum fluctuation effects on the response function are completely included. The parameters to be employed are the transfer energy t (to the nearest neighbor site) and the on site Coulomb repulsive energy U. Parameters were selected to reproduce the behavior of the EDC (width, energy positions and dispersion) as a function of $k_{//}$ as well as of the MDC. The results at 400 K for $U/t = 7.5$ ($U = 3.0$ eV and $t = 0.4$ eV) are shown in Fig. 7.36(d), (e) in good agreement with the bulk sensitive experimental results, where the peak shift of 0.5 eV and the shallowest peak at 0.95 eV are reproduced. The MDC also shows the V shaped dispersion in the ($-\pi/(2c)$, 0)–(0, 0)–($\pi/(2c)$, 0) region. In this model, the broad width of the main EDC peak corresponds to multi-magnon excitations coupled with the photo-created hole. In the infinite U case, as is well-known, the spin and charge excitations are exactly decoupled. However, the present results show that we still have significant spin-charge coupling in such a highly correlated electron system with $U/t = 7.5$ in the one band 1D half-filled Hubbard model. Since SXPES at $h\nu = 700$ eV is nearly 30 times more sensitive to the Cu 3d states than to the O 2p states [16], the results were mostly ascribed to the Cu 3d components, which are strongly renormalized by U, t and the magnon excitations. According to this model, the weaker structures observed in ($\pi/(2c)$, 0)–(π, 0) correspond to the incoherent component induced by the effect of electron correlation U.

The lower $h\nu$ ARPES at 85–200 eV was interpreted by the t-J model as introduced in Sect.6.4, in which a double occupancy of the same site was excluded by the large U. As a result, the dispersions from (0, 0) to ($\pi/(2c)$, 0) were ascribed to both spinon and holon as well as their mixed excitations, whereas only the holon dispersion was predicted in the ($\pi/(2c)$, 0)–(π, 0) region. The values of $t \sim 0.65$ and $J \sim 0.27$ eV were employed there for $SrCuO_2$ (U, approximated by $4t^2/J$, becomes as large as 6.26 eV, resulting in $U/t \sim 9.6$) [172]. The controversy whether the spin and charge are still coupled or fully decoupled in the real system with finite U/t is not settled yet.

7.2.5 Other Materials (VSe_2, $LaRu_2P_2$, BiTeI)

The 3D SXARPES with proper energy and momentum resolution are very useful for studying the 3D electronic structure of materials. VSe_2 grown by chemical vapor deposition and cleaved in situ was studied by this method at the ADRESS beam line of the Swiss Light Source (SLS) [174], which can provide linearly or circularly polarized light in the range from 300 to 1,600 eV with high photon flux up to 10^{13} photons/s per 0.01 % band width at 1 keV. A PHOIBOS-150 analyzer was used for the ARPES measurements at $h\nu = 880$–960 eV. VSe_2 is one of the layered TM dichalcogenide materials with wide van der Waals gaps between the chalcogen-TM-chalcogen trilayers and has quasi-2D properties. Nevertheless, the out-of-plane Se $4p_z$ orbital can induce a 3D character with finite k_z dispersion perpendicular to the cleaved surface. In conventional ARPES at $h\nu < 100$ eV, Δk_z is broadened to such a degree comparable to the BZ height, and a k_z dispersion cannot be probed. VSe_2 shows a CDW transition at $T_{CDW} \sim 110$ K and the corresponding wave vector $\mathbf{q}_{CDW}$ is known to have a large out-of-plane component. Clear dispersion was observed around $h\nu = 900$ eV, despite the reduction of the PICS by a factor of ~1,800 and ~34 for the V 3d and Se 4p orbitals compared with those at $h\nu = 50$ eV, respectively. Images of SXARPES covering an angle of ±6° with a total energy resolution of ~120 and ~70 meV could be detected in ~7 and ~40 min, with rather good statistics [174]. With use of the empirical V_0 of 7.5 eV, it is found that $h\nu = 885$ eV corresponds to $k_z = 0$ (Γ point) in the case of normal emission. Slices of the FS perpendicular to k_z were obtained with good statistics and the k_z dispersion was quantitatively evaluated along the full k_z BZ. As a result the 3D $\mathbf{q}$ vector matching the 3D CDW was experimentally confirmed for the first time in VSe_2.

A stoichiometric pnictide superconductor with T_c of ~4 K was also studied by SXARPES [175] at $h\nu = 615$ and 300–550 eV on a surface cleaved at 10 K and the results were compared with those measured at $h\nu = 28$–88 eV as well as those reported on Fe pnictide superconductors. SXARPES of $LaRu_2P_2$ revealed that the electronic structure in the normal state is in good agreement with the calculated results by density functional theory (DFT). Renormalization effects on the band width were found to be negligible in $LaRu_2P_2$ in strong contrast to high T_c Fe pnictide superconductors. Consequently a different origin of the superconductivity in $LaRu_2P_2$ from Fe pnictides superconductors is concluded. Namely the superconductivity seems to be more conventional in $LaRu_2P_2$. Increased bulk sensitivity in SXARPES was essential to derive such a conclusion.

SXARPES measurements on BiTeI at $h\nu = 310$–850 eV were compared with conventional ARPES at $h\nu = 20$–63 eV and SP-ARPES at 24 eV [176]. This material has a layered structure and strong spin–orbit coupling leading to a giant Rashba spin splitting. 3D bulk FSs were observed by SXARPES. The bulk electronic structure was disentangled from the 2D surface electronic structure probed by high-resolution ARPES performed in $h\nu = 24$–30 eV and SP-ARPES. The 2D Rashba-split state reported previously [177] was found to be not due to a bulk quantum well state (QWS) but rather a surface state. The combination of SXARPES and conventional ARPES is very useful to discuss the bulk 3D and surface 2D character of the electronic structure.

7.3 Standing Wave

Soft X-ray and X-ray standing wave (SXSW and XSW) techniques were useful in materials sciences as well as spectroscopy of solids. SX- or X-ray interference between the incoming and reflected waves takes place at a Bragg reflection creating a standing wave. The phase between the incoming and reflected wave changes by π over the angular range of maximum reflectivity (the Darwin curve). Then the maxima and minima of the interference field move through the unit cell of the diffracting crystal by tuning the Bragg reflection condition as a function of angle or wavelength [178]. When the wavelength of the standing wave matches with the inter-atomic distance, one can obtain information not only on atomic structure but also on site-selective electronic and magnetic properties. If combined with photoelectron spectroscopy, element- and site-specific as well as chemical sensitive information can be obtained. In the case of large spacing multilayers, SXSW facilitates various spectroscopic techniques to probe, e.g., photoelectrons, Auger electrons, fluorescence, or photo-desorbed ions. Among the wide variety of applications of SXSW and XSW techniques [179], we will here focus on its application to photoelectron spectroscopy.

Site specific valence band structure of GaAs was measured by utilizing the spatial dependence of the electric field (**E**) of the standing wave (SW) at 1,900 eV [180]. The probability of photoelectron emission from an atom in an external **E** field is proportional to its intensity at the location of the atom. If the photoemission intensity is recorded by tuning the **E** intensity within the unit cell, spatially resolved information of the valence band DOS might be experimentally obtained. At this hν, the maximum **E** intensity was close to the Ga atomic plane for the GaAs(111) reflection and it was close to the As atomic planes for the GaAs($\bar{1}\bar{1}\bar{1}$) reflection as illustrated in Fig. 7.37(a). Under these conditions, significant contrast was expected between the Ga and As contributions to both the valence band and core level photoelectron spectra as shown in Fig. 7.37(b). The noticeable intensity of the As 3d (Ga 3d) core level PES for the **E** intensity maximized near the Ga (As) atomic plane, was due to the finite (non-zero) **E** intensity near the As (Ga) atomic plane as illustrated in Fig. 7.37(a). Therefore a rather careful analysis was required for discussing the PDOSs in the valence band region, though useful information on the valance band PDOSs of Ga and As states and the charge transfer effects would be obtained from the SW photoemission. By use of SW PES technique, the Ti and O PDOSs in the valence band of rutile TiO_2 were likewise investigated at hν $\sim$ 2,700 eV [181]. The spectra were recorded at two different hν near the rutile TiO_2 (200) Bragg back-reflection condition to

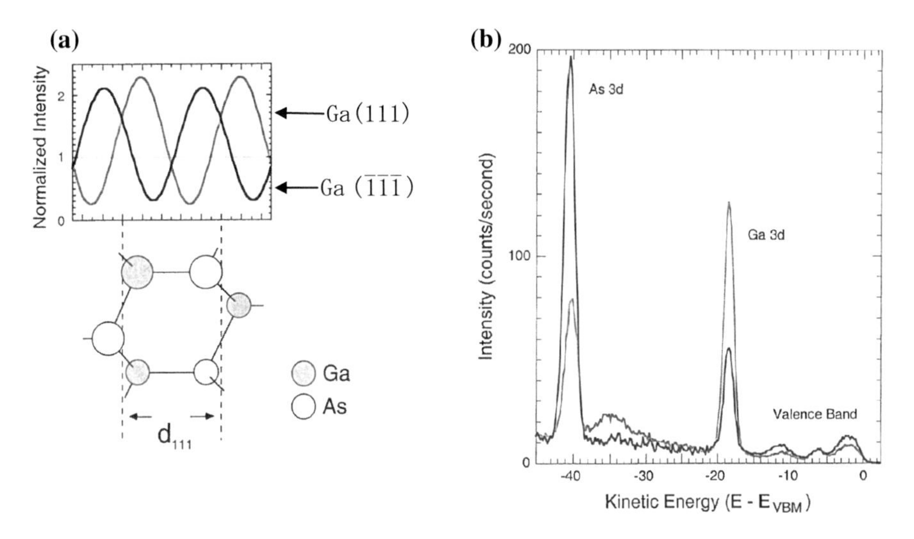

Fig. 7.37 Normalized electric field intensity for GaAs(111) and GaAs ($\bar{1}\bar{1}\bar{1}$) under Bragg back reflection conditions calculated at $h\nu = 1900$ eV (**a**). The spatial positions of the **E** intensity within the GaAs unit cell are shown relative to the Ga and As atomic planes. Core level and valence band spectra (**b**), where the Ga 3d and As 3d peaks are stronger for GaAs(111) and GaAs($\bar{1}\bar{1}\bar{1}$) reflection [180], respectively

maximize the **E** intensity on either the Ti or O atomic plane. The hybridization of the Ti 3d, 4s, 4p and O 2s, 2p states on each site was experimentally revealed.

Understanding of the electronic and magnetic properties of buried solid-state interfaces is quite important nowadays in the field of nano-science and technology including applications such as magnetic storage devices. One of the approaches is illustrated in Fig. 7.38, where a B_4C/W multilayers mirror (40 periods of a B_4C/W unit layer each with $d_{ML} = 40.5$ Å thickness in this case) worked as a SW generator. For the 1st order Bragg reflection, $\lambda = 2\ d_{ML} \cdot \sin\theta_{inc}$ is satisfied where θ_{inc} is the angle of the incident X-rays measured from the layer plane, and the period of E^2 of the SW would be given by $\lambda_{SW}(|\mathbf{E}|^2) = (\lambda/2) \cdot \sin\theta_{inc}$. A high reflectivity of the multilayer mirror was achieved by a grazing-incidence total reflection geometry. On top of the multilayer mirror were grown a Cr wedge and an Fe layer with constant thickness, which were sometimes further covered by a protective Al or Al_2O_3 cap. With the movement of the incident SX beam or the sample towards the x direction, the Fe/Cr interface passed across the SW by virtue of the wedge structure, though the SW **E** intensity modulation was pinned in phase to the top of the multilayer [182–185]. The interface of Fe/Cr, which was a prototype giant magnetoresistance (GMR) system, was studied by this setup. Well focused soft X-ray synchrotron radiation (SX-SR) was incident on the multilayer mirror and strong SW was achieved above the mirror. The slope of the wedge was arranged in such a way as its height changed by a few times the standing wave period d_{ML} over the full sample length along x (typically a few mm). The beam size can be <0.1 mm.

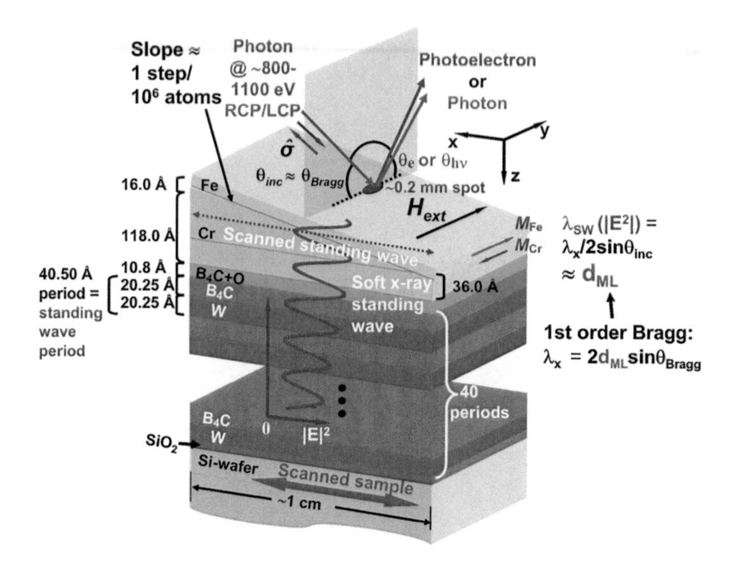

Fig. 7.38 Illustration of the standing wave/wedge method for probing buried interfaces via photoelectron or photon emission. A wedge-shaped bilayer sample (Cr wedge and Fe thin film) is grown on a multilayer-mirror standing wave generator. The buried interface between Fe and Cr is selectively studied by scanning the sample position along the x direction. Then the interface passes through the standing wave [182–186]

With measuring the intensity of both Cr 3p and Fe 3p core level PES, the sample position was scanned along the x direction. The intensity ratio between both is plotted in Fig. 7.39(b), which shows a sinusoidal oscillation. Figure 7.39(c) shows the results of a rocking curve scan at a fixed x or fixed Cr thickness by rotating the sample along the x axis. The experimental results shown by the triangle marks in Fig. 7.39(b) and full circles in Fig. 7.39(c) are well reproduced by a simple model calculation (smooth line). Such information on interface as wedge-position dependent layer thicknesses, interface mixing and roughness could be obtained from such experiments. In addition, magnetic circular dichroism of PES from 2p and 3p core levels of Fe and Cr were also measured by use of circularly polarized SR, providing the information on magnetic properties of the Fe/Cr interface and its Cr thickness dependence [186]. Thus these SW wedge techniques were very powerful for the characterization of buried interfaces. SX-PEEM is nowadays widely used to study lateral magnetic structures. If combined with the SW method, depth sensitivity will be added to the lateral resolving potential of PEEM. SW technique is also useful in the hard X-ray region for imaging and investigation of geometrical structures, chemical composition as well as electronic structures of surfaces and interfaces [187–189].

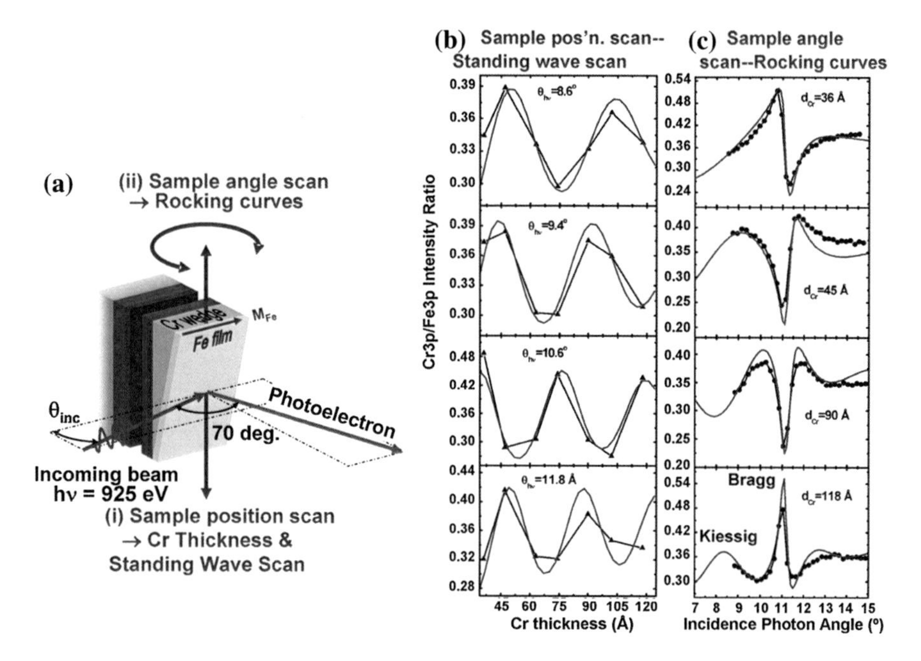

Fig. 7.39 **a** Examples of SXSW studies of an Fe/Cr wedge interface. Sample position scan as explained in detail in the text is shown in (**b**) and measurement for the sample angle rotation along x axis is shown in (**c**). In both cases, the PES intensity of the Cr 3p and Fe 3p core levels is measured and their ratio is plotted in (**b**) and (**c**) which are well reproduced by a simple model calculation [182–186]

References

1. F. Steglich, J. Aarts, C.D. Bredl, W. Lieke, D. Meschede, W. Franz, H. Schäfer, Phys. Rev. Lett. **43**, 1892 (1979)
2. J.G. Bednorz, K.A. Müller, Z. Phys. B **64**, 189 (1986)
3. Y. Onuki, Y. Kitaoka (eds.), Specitial topics: Frontiers of novel superconductivity in heavy fermion compounds. J. Phys. Soc. Jpn. **76**, 051001–051013 (2007)
4. J.C. Fuggle, F.U. Hillebrecht, Z. Zołnierek, R. Lässer, Ch. Freiburg, O. Gunnarsson, K. Schönhammer, Phys. Rev. B **27**, 7330 (1983)
5. F.U. Hillebrecht, J.C. Fuggle, G.A. Sawatzky, M. Cmpagna, O. Gunnarsson, K. Schönhammer, Phys. Rev. B **30**, 1777 (1984)
6. J.W. Allen, S.J. Oh, O. Gunnarsson, K. Schönhammer, M.B. Maple, M.S. Torikachvili, I. Lindau, Adv. Phys. **35**, 275 (1986)
7. K.A. Gschneidner Jr., L. Eyring, S. Hüfner (eds.), *Handbook on the Physics and Chemistry of Rare Earth*, vol. 10 (1987)
8. O. Gunnarsson, K. Schönhammer, Phys. Rev. B **31**, 4815 (1985)
9. T. Jo, A. Kotani, J. Phys. Soc. Jpn. **55**, 2457 (1986)
10. E. Weschke, C. Laubschat, T. Simmons, M. Domke, O. Strebel, G. Kaindl, Phys. Rev. B **44**, 8304 (1991)
11. L. Duò, S. De Rossi, P. Vavassori, F. Ciccacci, G.L. Olcese, G. Chiaia, I. Lindau, Phys. Rev. B **54**, R17363 (1996)
12. S. Tanuma, C.J. Powell, D.R. Penn, Surf. Sci. **192**, L849 (1987)

13. S. Tanuma, C.J. Powell, D.R. Penn, Surf. Inter. Anal. **35**, 268 (2003) and Refs. 4.8–13 in Ch. 4
14. A. Sekiyama, T. Iwasaki, K. Matsuda, Y. Saitoh, Y. Onuki, S. Suga, Nature **403**, 396 (2000)
15. Y. Saitoh, H. Kimura, Y. Suzuki, T. Nakatani, T. Matsushita, T. Muro, T. Miyahara, M. Fujisawa, K. Soda, S. Ueda, H. Harada, M. Kotsugi, A. Sekiyama, S. Suga, Rev. Sci. Instrum. **71**, 3254 (2000)
16. J.J. Yeh, I. Lindau, At. Data Nucl. Data Tables **32**, 1 (1985)
17. U. Fano, Phys. Rev. **124**, 1866 (1961)
18. A. Sekiyama, S. Suga, T. Iwasaki, S. Ueda, S. Imada, Y. Saitoh, T. Yoshino, D.T. Adroja, T. Takabatake, J. Electron Spectr. Relat. Phenom. **114**, 699 (2001)
19. D.T. Adroja, S.K. Malik, B.D. Padalia, R. Vijayaraghavan, Solid State Commun. **66**, 1201 (1988)
20. M. Ruderman, C. Kittel, Phys. Rev. **96**, 99 (1954)
21. T. Kasuya, Prof. Theor. Phys. **16**, 45 (1956)
22. K. Yosida, Phys. Rev. **106**, 893 (1957)
23. A. Sekiyama, S. Suga, Y. Saitoh, S. Ueda, H. Harada, T. Matsushita, T. Nakatani, T. Iwasaki, K. Matsuda, M. Kotsugi, S. Imada, T. Takabatake, T. Yoshino, D.T. Adroja, R. Takayama, O. Sakai, H. Harima, T. Nanba, Solid State Commun. **111**, 373 (1999)
24. E.-J. Cho, R.-J. Jung, B.-H. Choi, S.-J. Oh, T. Iwasaki, A. Sekiyama, S. Imada, S. Suga, T. Muro, J.-G. Park, Y.S. Kwon, Phys. Rev. B **67**, 155107 (2003)
25. A. Amato, D. Jaccard, J. Sierro, F. Lapierre, P. Haen, P. Lejay, J. Flouquet, J. Magn. Magn. Mater. **76–77**, 263 (1988)
26. J.D. Thompson, J.O. Willis, C. Godard, D.E. Maclaughlin, J.C. Gupta, Solid State Commun. **56**, 169 (1985)
27. A.D. Huxley, C. Paulsen, O. Laborde, J.L. Tholence, D. Sanchez, A. Junod, R. Calemczuk, J. Phys.: Condens. Matter **5**, 7709 (1993)
28. J.-S. Kang, C.G. Olson, M. Hedo, Y. Inada, E. Yamamoto, Y. Haga, Y. Onuki, S.K. Kwon, B.I. Min, Phys. Rev. B **60**, 5348 (1999)
29. H. J. Im, T. Ito, J. B. Hong, S. Kimura and Y. S. Kwon, Phys. Rev. B **72**, 220405(R) (2005)
30. A. Sekiyama, K. Kadono, K. Matsuda, T. Iwasaki, S. Ueda, S. Imada, S. Suga, R. Settai, H. Azuma, Y. Onuki, Y. Saitoh, J. Phys. Soc. Jpn. **69**, 2771 (2000)
31. A. Sekiyama, S. Suga, S. Ueda, S. Imada, K. Matsuda, T. Iwasaki, M. Hedo, E. Yamamoto, Y. Haga, Y. Onuki, Y. Saitoh, T. Matsushita, T. Nakatani, M. Kotsugi, S. Tanaka, H. Harima, Solid State Commun. **121**, 561 (2002)
32. A. Sekiyama, S. Suga, Phys. B **312**, 634 (2002)
33. A. Sekiyama, K. Kadono, T. Iwasaki, S. Imada, S. Kasai, S. Suga, S. Araki, Y. Onuki, Acta Phys. Pol., B **34**, 1105 (2003)
34. A. Sekiyama, S. Imada, A. Yamasaki, S. Suga, in *Very High-resolution Photoelectron Spectroscopy*, ed. by S. Hüfner, Lecture Notes in Physics (Springer, Heidelberg 2007), vol. 715, p. 351
35. A. Loidl, G. Knopp, H. Spille, F. Steglich, A.P. Murani, Physica B **156–157**, 794 (1989)
36. A. Loidl, K. Knorr, G. Knopp, A. Krimmel, R. Caspary, A. Böhm, G. Sparn, C. Geibel, F. Steglich, A.P. Murani, Phys. Rev. B **46**, 9341 (1992)
37. N. Sato, A. Sumiyama, S. Kunii, H. Nagano, T. Kasuya, J. Phys. Soc. Jpn. **54**, 1923 (1985)
38. S. Araki, R. Settai, Y. Inada, Y. Onuki, H. Yamagami, J. Phys. Soc. Jpn. **68**, 3334 (1999)
39. Y. Kuramoto, Z. Phys. B: Condens. Matter **53**, 37 (1983)
40. N.E. Bickers, D.L. Cox, J.W. Wilkins, Phys. Rev. B **36**, 2036 (1987)
41. N.E. Bickers, Rev. Mod. Phys. **59**, 845 (1987)
42. E. Zirngiebl, B. Hillebrands, S. Blumenröder, G. Güntherodt, M. Loewenhaupt, J.M. Carpenter, K. Winzer, Z. Fisk, Phys. Rev. B **30**, 4052 (1984)
43. R. Felten, G. Weber, H. Rietschel, J. Magn. Magn. Mater. **63–64**, 383 (1987)
44. A. Severing, E. Holland-Moritz, B.D. Rainford, S.R. Culverhouse, B. Frick, Phys. Rev. B **39**, 2557 (1989)
45. P. Haen, J. Flouquet, F. Lapierre, P. Lejay, G. Remenyi, J. Low Temp. Phys. **67**, 391 (1987)

46. K. Hanzawa, K. Yamada, K. Yosida, J. Magn. Magn. Mater. **47–48**, 357 (1985)
47. Y. Onuki, Y. Furukawa, T. Komatsubara, J. Phys. Soc. Jpn. **53**, 2734 (1984)
48. C. Godart, A.M. Umarji, L.C. Gupta, R. Vijayaraghavan, Phys. Rev. B **34**, 7733 (1986)
49. S. Tanaka, H. Harima, A. Yanase, J. Phys. Soc. Jpn. **67**, 1342 (1998)
50. M. Hedo, Y. Inada, T. Ishida, E. Yamamoto, Y. Haga, Y. Onuki, M. Higuchi, A. Hasegawa, J. Phys. Soc. Jpn. **64**, 4535 (1995)
51. E. Weschke, C. Laubschat, R. Ecker, A. Höhr, M. Domke, G. Kaindl, L. Severin, B. Johansson, Phys. Rev. Lett. **69**, 1792 (1992)
52. C.G. Olson, P.J. Benning, M. Schmidt, D.W. Lynch, P. Canfield, D.M. Wieliczka, Phys. Rev. Lett. **76**, 4265 (1996)
53. J.M. Lawrence, A.J. Arko, J.J. Joyce, R.I.R. Blyth, R.J. Bartlett, P.C. Canfield, Z. Fisk, P.S. Riseborough, Phys. Rev. B **47**, 15460 (1993)
54. S. Suga, A. Sekiyama, S. Imada, A. Shigemoto, A. Yamasaki, M. Tsunekawa, C. Dallera, L. Braicovich, T.-L. Lee, O. Sakai, T. Ebihara, Y. Onuki, J. Phys. Soc. Jpn. **74**, 2880 (2005)
55. J.J. Joyce, A.B. Andrews, A.J. Arko, R.J. Bartlett, R.I.R. Blyth, C.G. Olson, P.J. Benning, P.C. Canfield, D.M. Poirier, Phys. Rev. B**54**, 17515 (1996) and references therein
56. F. Patthey, J.-M. Imer, W.-D. Schneider, H. Beck, Y. Baer, B. Delly, Phys. Rev. B **42**, 8864 (1990) and references therein
57. F. Reinert, R. Claessen, G. Nicolay, D. Ehm, S. Hüfner, W.P. Ellis, G.-H. Gweon, J.W. Allen, B. Kindler, W. Assmus, Phys. Rev. B **58**, 12808 (1998)
58. S.-J. Oh, S. Suga, A. Kakizaki, M. Taniguchi, T. Ishii, J.-S. Kang, J.W. Allen, O. Gunnarsson, N.E. Christensen, A. Fujimori, T. Suzuki, T. Kasuya, T. Miyahara, H. Kato, K. Schonhammer, M.S. Torikachvili, M.B. Maple, Phys. Rev. **B37**, 2861 (1988)
59. L.H. Tjeng, S.-J. Oh, E.-J. Cho, H.-J. Lin, C.T. Chen, G.-H. Gweon, J.-H. Park, J.W. Allen, T. Suzuki, M.S. Makivic, D.L. Cox, Phys. Rev. Lett. **71**, 1419 (1993)
60. L.H. Tjeng, S.-J. Oh, C.T. Chen, J.W. Allen, D.L. Cox, Phys. Rev. Lett. **72**, 1775 (1994)
61. Y. Baer, W.-D. Schneider, Ch. 62 in K.A. Gschneidner Jr., L. Eyring, S. Hüfner (eds.), *Handbook on the Physics and Chemistry of Rare Earth*, vol. 10 (1987)
62. T. Ebihara, Y. Inada, M. Murakawa, S. Uji, C. Terakura, T. Terashima, E. Yamamoto, Y. Haga, Y. Onuki, H. Harima, J. Phys. Soc. Jpn. **69**, 895 (2000)
63. G.D. Mahan, Phys. Rev. **163**, 612 (1967)
64. F. Gerken, J. Phys. **F13**, 703 (1983)
65. A.L. Cornelius, J.M. Lawrence, T. Ebihara, P.S. Riseborough, C.H. Booth, M.F. Hundley, P.G. Pagliuso, J.L. Sarrao, J.D. Thompson, M.H. Jung, A.H. Lacerda, G.H. Kwei, Phys. Rev. Lett. **88**, 117201 (2002)
66. J. Yamaguchi, A. Sekiyama, S. Imada, A. Yamasaki, M. Tsunekawa, T. Muro, T. Ebihara, Y. Onuki, S. Suga, New J. Phys. **9**, 317 (2007)
67. D.B. McWhan, J.P. Remeika, Phys. Rev. **B2**, 3734 (1970)
68. N.F. Mott, Rev. Mod. Phys. **40**, 677 (1968)
69. N.F. Mott, *Metal-Insulator Transitions* (Taylor & Francis, London, 1990)
70. F. Gebhard, *The Mott Metal-Insulator Transition* (Springer, Berlin, 1997)
71. S.-K. Mo, J.D. Denlinger, H.-D. Kim, J.-H. Park, J.W. Allen, A. Sekiyama, A. Yamasaki, K. Kadono, S. Suga, Y. Saitoh, T. Muro, P. Metcalf, G. Keller, K. Held, V. Eyert, V.I. Anisimov, D. Vollhardt, Phys. Rev. Lett. **90**, 186403 (2003)
72. M. Schramme, Ph. D Thesis, Universität, Augsburg (2000)
73. S.-K. Mo, H.-D. Kim, J.W. Allen, G.-H. Gweon, J.D. Denlinger, J.-H. Park, A. Sekiyama, A. Yamasaki, S. Suga, P. Metcalf, K. Held, Phys. Rev. Lett. **93**, 076404 (2004)
74. S.-K. Mo, H.-D. Kim, J.D. Denlinger, J.W. Allen, J.-H. Park, A. Sekiyama, A. Yamasaki, S. Suga, Y. Saitoh, T. Muro, P. Metcalf, Phys. Rev. **B74**, 165101 (2006)
75. H. Fujiwara, A. Sekiyama, S.-K. Mo, J.W. Allen, J. Yamaguchi, G. Funabashi, S. Imada, P. Metcalf, A. Higashiya, M. Yabashi, K. Tamasaku, T. Ishikawa, S. Suga, Phys. Rev. **B84**, 075117 (2011)
76. W. Metzner, P. Schmit, D. Vollhardt, Phys. Rev. **B45**, 2237 (1992)

77. G. Sangiovanni, A. Toschi, E. Koch, K. Held, M. Capone, C. Castellani, O. Gunnarsson, S.-K. Mo, J.W. Allen, H.-D. Kim, A. Sekiyama, A. Yamasaki, S. Suga, P. Metcalf, Phys. Rev. **B73**, 205121 (2006)
78. W. Bao, C. Broholm, G. Aeppli, P. Dai, J.M. Honig, P. Metcalf, Phys. Rev. Lett. **78**, 507 (1997)
79. E. Pavarini, S. Biermann, A. Poteryaev, A.I. Lichtenstein, A. Georges, O.K. Andersen, Phys. Rev. Lett. **92**, 176403 (2004)
80. A.I. Poteryaev, M. Ferrero, A. Georges, O. Parcollet, Phys. Rev. B **78**, 045115 (2008)
81. A. Fujimori, I. Hase, H. Namatame, Y. Fujishima, Y. Tokura, H. Eisaki, S. Uchida, K. Takegahara, F.M.F. de Groot, Phys. Rev. Lett. **69**, 1796 (1992)
82. Y. Aiura, F. Iga, Y. Nishihara, H. Ohnuki, H. Kato Phys, Rev. B **47**, 6732 (1993)
83. I.H. Inoue, I. Hase, Y. Aiura, A. Fujimori, Y. Haruyama, T. Maruyama, Y. Nishihara, Phys. Rev. Lett. **74**, 2539 (1995)
84. K. Morikawa, T. Mizokawa, K. Kobayashi, A. Fujimori, H. Eisaki, S. Uchida, F. Iga, Y. Nishihara, Phys. Rev. B **52**, 13711 (1995)
85. I.H. Inoue, O. Goto, H. Makino, N.E. Hussey, M. Ishikawa, Phys. Rev. B **58**, 4372 (1998)
86. M.J. Rey, Ph Dehaudt, J.C. Joubert, B. Lambert-Andron, M. Cyrot, F. Cyrot-Lackmann, J. Solid State Chem. **86**, 101 (1990)
87. A. Sekiyama, H. Fujiwara, S. Imada, S. Suga, H. Eisaki, S.I. Uchida, K. Takegahara, H. Harima, Y. Saitoh, I.A. Nekrasov, G. Keller, D.E. Kondakov, A.V. Kozhevnikov, T.H. Pruschke, K. Held, D. Vollhardt, V.I. Anisimov, Phys. Rev. Lett. **93**, 156402 (2004)
88. A. Yamasaki, A. Sekiyama, S. Imada, M. Tsunekawa, A. Higashiya, A. Shigemoto, S. Suga, Nucl. Instrum. Meth. A **547**, 136 (2005)
89. V.I. Anisimov, D.E. Kondakov, A.V. Kozhevnikov, I.A. Nekrasov, Z.V. Pchelkina, J.W. Allen, S.-K. Mo, H.-D. Kim, P. Metcalf, S. Suga, A. Sekiyama, G. Keller, I. Leonov, X. Ren, D. Vollhardt, Phys. Rev. B **71**, 125119 (2005)
90. Detailed hν dependence of the relative V 3d/O 2p PICS is shown in S.-K. Mo, H.-D. Kim, J.D. Denlinger, J.W. Allen, J.-H. Park, A. Sekiyama, A. Yamasaki, S.Suga, Y. Saitoh, T. Muro, P. Metcalf, Phys. Rev. B**74**, 165101 (2006)
91. K. Maiti, P. Mahadevan, S. Sarma, Phys. Rev. Lett. **80**, 2885 (1998)
92. K. Maiti, D.D. Sarma, M.J. Rozenberg, I.H. Inoue, H. Makino, O. Goto, M. Pedio, R. Cimino, Europhys. Lett. **55**, 246 (2001)
93. V.I. Anisimov, A.I. Poteryaev, M.A. Korotin, A.O. Anokhin, G. Kotliar, J. Phys.: Condens. Matter **9**, 7359 (1997)
94. A.I. Lichtenstein, M.I. Katsnelson, Phys. Rev. B **57**, 6884 (1998)
95. K. Held, I.A. Nekrasov, G. Keller, V. Eyert, N. Blümer, A.K. McMahan, R.T. Scaletter, T.H. Pruschke, V.I. Anisimov, D. Vollhardt, Psi-k Newsl. **56**, 65 (2003), www.psi-k.org/newsletters/News_56/Highlight_56.pdf
96. E. Pavarini, A. Yamasaki, J. Nuss, O.K. Andersen, New J. Phys. **7**, 188 (2005)
97. O. Gunnarsson, O.K. Andersen, O. Jepsen, J. Zaanen, Phys. Rev. B **39**, 1708 (1989)
98. I.H. Inoue, C. Bergemann, I. Hase, S.R. Julian, Phys. Rev. Lett. **88**, 236403 (2002)
99. Y. Tokura, N. Nagaosa, Science **288**, 462 (2000)
100. C. Zener, Phys. Rev. **82**, 403 (1951)
101. H. Kuwahara, Y. Tomioka, A. Asamitsu, Y. Moritomo, Y. Tokura, Science **270**, 961 (1995)
102. H. Kawano, R. Kajimoto, H. Yoshizawa, Y. Tomioka, H. Kuwahara, Y. Tokura, Phys. Rev. Lett. **78**, 4253 (1997)
103. R. Kajimoto, H. Yoshizawa, K. Kawano, H. Kuwahara, Y. Tokura, K. Ohoyama, M. Ohashi, Phys. Rev. B **60**, 9506 (1999)
104. A. Sekiyama, S. Kasai, M. Tsunekawa, Y. Ishida, M. Sing, A. Irizawa, A. Yamasaki, S. Imada, T. Muro, Y. Saitoh, Y. Onuki, T. Kimura, Y. Tokura, S. Suga, Phys. Rev. B **70**, 060506(R) (2004)
105. H. Fujiwara, A. Sekiyama, A. Higashiya, K. Konoike, A. Tsunekawa, A. Yamasaki, A. Irizawa, S. Imada, T. Muro, K. Noda, H. Kuwahara, Y. Tokura, S. Suga, J. Electron Spectr. Relat. Phenom. **144–147**, 807 (2005)

106. A. Sekiyama, S. Suga, M. Fujikawa, S. Imada, T. Iwasaki, K. Matsuda, T. Matsushita, K.V. Kaznacheyev, A. Fujimori, H. Kuwahara, Y. Tokura, Phys. Rev. B **59**, 15528 (1999)
107. J.-S. Kang, J.H. Kim, A. Sekiyama, S. Kasai, S. Suga, S.W. Han, K.H. Kim, E.J. Choi, T. Kimura, T. Muro, Y. Saitoh, C.G. Olson, J.H. Shim, B.I. Min, Phys. Rev. B **68**, 012410 (2003)
108. H. Fujiwara, A. Sekiyama, A. Higashiya, K. Konoike, A. Yamasaki, S. Imada, T. Muro, K. Noda, H. Kuwahara, Y. Tokura, S. Suga, Physica B **378–380**, 515 (2006)
109. E.J.W. Verwey, Nature **144**, 327 (1939)
110. M. Imada, A. Fujimori, Y. Tokura, Rev. Mod. Phys. **70**, 1039 (1998)
111. F. Walz, J. Phys.: Condens. Matter **14**, R285 (2002)
112. J.P. Wright, J.P. Attfield, P.G. Radaelli, Phys. Rev. Lett. **87**, 266401 (2001)
113. G. Subias, J. Garcia, J. Blasco, M.G. Proietti, H. Renevier, M.C. Sanchez, Phys. Rev. Lett. **93**, 156408 (2004)
114. A. Chainani, T. Yokoya, T. Morimoto, T. Takahashi, S. Todo, Phys. Rev. **B51**, 17976 (1995)
115. J.-H. Park, L.H. Tjeng, J.W. Allen, P. Metcalf, C.T. Chen, Phys. Rev. **B55**, 12813 (1997)
116. D. Schrupp, M. Sing, M. Tsunekawa, H. Fujiwara, S. Kasai, A. Sekiyama, S. Suga, T. Muro, V.A.M. Brabers, R. Claessen, Europhys. Lett. **70**, 789 (2005)
117. L. Perfetti, S. Mitrovic, G. Margaritondo, M. Grioni, L. Forro, L. Degiorgi, H. Höchst, Phys. Rev. **B66**, 075107 (2002)
118. M. Kimura, H. Fujiwara, A. Sekiyama, J. Yamaguchi, K. Kishimoto, H. Sugiyama, G. Funabashi, S. Imada, S. Iguchi, Y. Tokura, A. Higashiya, M. Yabashi, K. Tamasaku, T. Ishikawa, T. Ito, S. Kimura, S. Suga, J. Phys. Soc. Jpn. **79**, 064710 (2010)
119. N. Kamakura, Y. Takata, T. Tokushima, Y. Harada, A. Chainani, K. Kobayashi, S. Shin, Phys. Rev. B **74**, 045127 (2006)
120. M. Yano, A. Sekiyama, H. Fujiwara, T. Saita, S. Imada, T. Muro, Y. Onuki, S. Suga, Phys. Rev. Lett. **98**, 036405 (2007)
121. M. Yano, A. Sekiyama, H. Fujiwara, Y. Amano, S. Imada, T. Muro, M. Yabashi, K. Tamasaku, A. Higashiya, T. Ishikawa, Y. Onuki, S. Suga, Phys. Rev. B **77**, 035118 (2008)
122. H. Yamagami, A. Hasegawa, J. Phys. Soc. Jpn. **63**, 2290 (1994)
123. H. Yamagami, A. Hasegawa, J. Phys. Soc. Jpn. **62**, 592 (1993)
124. A. Ino, C. Kim, T. Mizokawa, Z.-X. Shen, A. Fujimori, M. Takaba, K. Tamasaku, H. Eisaki, S. Uchida, J. Phys. Soc. Jpn. **68**, 1496 (1997)
125. A. Ino, C. Kim, M. Nakamura, T. Yoshida, T. Mizokawa, A. Fujimori, Z.-X. Shen, T. Kakeshita, H. Eisaki, U. Uchida, Phys. Rev. B **65**, 094504 (2002)
126. T. Yoshida, X.J. Zhou, M. Nakamura, S.A. Kellar, P.V. Bogdanov, E.D. Lu, A. Lanzara, Z. Hussain, A. Ino, T. Mizokawa, A. Fujimori, H. Eisaki, C. Kim, Z.-X. Shen, T. Kakeshita, S. Uchida, Phys. Rev. B **63**, 220501(R) (2001)
127. A. Sekiyama, S. Suga, J. Electron Spectrosc. Relat. Phenom. **137–140**, 681 (2004)
128. S.Kasai, A. Sekiyama, M. Tsunekawa, P. T. Ernst, A. Shigemoto, A. Yamasaki, A. Irizawa, S. Imada, M. Sing, T. Muro, T. Sasagawa, H.Takagi, S. Suga, J. Electron Spectorsc. Rel. Phenom. **144**, 507 (2005) and unpublished results
129. H. Takagi, T. Ido, S. Ishibashi, M. Uota, S. Uchida, Y. Tokura, Phys. Rev. **B40**, 2254 (1989)
130. H. Takagi, B. Batlogg, H.L. Kao, J. Kwo, R.J. Cava, J.J. Krajewski, W.F. Peck Jr, Phys. Rev. Lett. **69**, 2975 (1992)
131. D.S. Dessau, Z.-X. Shen, D.M. King, D.S. Marshall, L.W. Lombardo, P.H. Dickinson, A.G. Loeser, J. DiCarlo, C.-H. Park, A. Kapitulnik, W.E. Spicer, Phys. Rev. Lett. **71**, 2781 (1993)
132. H.M. Fretwell, A. Kaminski, J. Mesot, J.C. Campuzano, M.R. Norman, M. Randeria, T. Sato, R. Gatt, T. Takahashi, K. Kadowaki, Phys. Rev. Lett. **84**, 4449 (2000)
133. S.V. Borisenko, M.S. Golden, S. Legner, T. Pichler, C. Dürr, M. Knupfer, J. Fink, G. Yang, S. Abell, H. Berger, Phys. Rev. Lett. **84**, 4453 (2000)
134. A. Damascelli, Z. Hussain, Z.-X. Shen, Rev. Mod. Phys. **75**, 473 (2003)

135. X.J. Zhou, T. Yoshida, S.A. Kellar, P.V. Bogdanov, E.D. Lu, A. Lanzara, M. Nakamura, T. Noda, T. Kakeshita, H. Eisaki, S. Uchida, A. Fujimori, Z. Hussain, Z.-X. Shen, Phys. Rev. Lett. **86**, 5578 (2001)
136. M.G. Zacher, R. Eder, E. Arrigoni, W. Hanke, Phys. Rev. Lett. **85**, 2585 (2000)
137. K. Takegahara, H. Harima, A. Yanase, Jpn. J. Appl. Phys. **26**, L352 (1987)
138. A.J. Freeman, J. Yu, Physica B+C **150**, 50 (1988)
139. L.F. Feiner, J.H. Jefferson, R. Raimondi, Phys. Rev. Lett. **76**, 4939 (1996)
140. R. Raimondi, J.H. Jefferson, L.F. Feiner, Phys. Rev. **B53**, 8774 (1996)
141. T. Kimura, S. Miyasaka, H. Takagi, K. Tamasaku, H. Eisaki, S. Uchida, K. Kitazawa, M. Hiroi, M. Sera, N. Kobayashi, Phys. Rev. B**53**, 8733 (1996)
142. A. Sekiyama, S.Suga, et al., to be published
143. T. Claesson, M. Månsson, C. Dallera, F. Venturini, C. De Nadaï, N.B. Brookes, O. Tjernberg, Phys. Rev. Lett. **93**, 136402 (2004)
144. T. Sato, T. Kamiyama, T. Takahashi, K. Kuruhashi, K. Yamada, Science **291**, 1517 (2001)
145. N.P. Armitage, D.H. Lu, C. Kim, A. Damascelli, K.M. Shen, F. Ronning, D.L. Feng, P. Bogdanov, Z.-X. Shen, Y. Onose, Y. Taguchi, Y. Tokura, P.K. Mang, N. Kaneko, M. Greven, Phys. Rev. Lett. **87**, 147003 (2001)
146. M. Tsunekawa, A. Sekiyama, S. Kasai, S. Imada, H. Fijiwara, T. Muro, Y. Onose, Y. Tokura, S. Suga, New J. Phys. **10**, 073005 (2008)
147. N.P. Armitage, D.H. Lu, C. Kim, A. Damascelli, K.M. Shen, F. Ronning, D.L. Feng, P. Bogdanov, X.J. Zhou, W.L. Yang, Z. Hussain, P.K. Mang, N. Kaneko, M. Greven, Y. Onose, Y. Taguchi, Y. Tokura, Z.-X. Shen, Phys. Rev. **B68**, 064517 (2003)
148. Y. Maeno, H. Hashimoto, K. Yoshida, S. Nishizaki, T. Fujita, J.G. Bednorz, F. Lichtenberg, Nature **372**, 532 (1994)
149. K. Ishida, H. Mukuda, Y. Kitaoka, K. Asayama, Z.Q. Mao, Y. Mori, Y. Maeno, Nature **396**, 658 (1998)
150. S. Nakatsuji, Y. Maeno, Phys. Rev. Lett. **84**, 2666 (2000)
151. T. Oguchi, Phys. Rev. B **51**, 1385 (1995)
152. A.P. Mackenzie, S.R. Julian, A.J. Diver, G.J. McMullan, M.P. Ray, G.G. Lonzarich, Y. Maeno, S. Nishizaki, T. Fujita, Phys. Rev. Lett. **76**, 3786 (1996)
153. Y. Yoshida, R. Settai, Y. Onuki, H. Takei, K. Betsuyaku, H. Harima, J. Phys. Soc. Jpn. **67**, 1677 (1998)
154. T. Yokoya, A. Chainani, T. Takahashi, H. Katayama-Yoshida, M. Kasai, Y. Tokura, Phys. Rev. Lett. **76**, 3009 (1996)
155. A.V. Puchkov, Z.-X. Shen, T. Kimura, Y. Tokura, Phys. Rev. B **58**, R13322 (1998)
156. A. Damascelli, D.H. Lu, K.M. Shen, N.P. Armitage, F. Ronning, D.L. Feng, C. Kim, Z.-X. Shen, T. Kimura, Y. Tokura, Z.Q. Mao, Y. Maeno, Phys. Rev. Lett. **85**, 5194 (2000)
157. R. Matzdorf, Z. Fang, X. Ismail, J. Zhang, T. Kimura, Y. Tokura, K. Terakura, E.W. Plummer, Science **289**, 746 (2000)
158. A.P. Mackenzie, Y. Maeno, Rev. Mod. Phys. **75**, 657 (2003)
159. A. Sekiyama, S. Suga, axXiv:0711.2160 to be published
160. S. Nakatsuji, Y. Maeno, Phys. Rev. **B62**, 6458 (2000)
161. J. Zhang, Ismail, R.G. Moore, S.-C. Wang, H. Ding, R. Jin, D. Mandrus, E.W. Plummer, Phys. Rev. Lett. **96**, 066401 (2006)
162. S. Shin, S. Suga, M. Taniguchi, M. Fujisawa, H. Kanzaki, A. Fujimori, H. Daimon, Y. Ueda, K. Kosuge, S. Kachi, Phys. Rev. B **41**, 4993 (1990)
163. S.-C. Wang, H.-B. Yang, A.K.P. Sekharan, S. Souma, H. Matsui, T. Sato, T. Takahashi, C. Lu, J. Zhang, R. Jin, D. Mandrus, E.W. Plummer, Z. Wang, H. Ding, Phys. Rev. Lett. **93**, 177007 (2004)
164. P.D. Dernier, Mat. Res. Bull. **9**, 955 (1974)
165. M. Itoh, H. Yasuoka, Y. Ueda, K. Kosuge, J. Phys. Soc. Jpn. **53**, 1847 (1984)
166. S. Suga, A. Shigemoto, A. Sekiyama, S. Imada, A. Yamasaki, A. Irizawa, S. Kasai, Y. Saitoh, T. Muro, N. Tomita, K. Nasu, H. Eisaki, Y. Ueda, Phys. Rev. B **70**, 155106 (2004)
167. R. Eguchi, T. Yokoya, T. Kiss, Y. Ueda, S. Shin, Phys. Rev. B **65**, 205124 (2002)

168. H. Ishii, H. Kataura, H. Shiozawa, H. Yashioka, H. Otsubo, Y. Takayama, T. Miyahara, S. Suzuki, Y. Achiba, M. Nakatake, T. Narimura, M. Higashiguchi, K. Shimada, H. Namatame, M. Taniguchi, Nature **426**, 540 (2003)
169. S. Suga, A. Sekiyama, M. Obara, J. Yamaguchi, M. Kimura, H. Fujiwara, A. Irizawa, K. Yoshimura, M. Yabashi, K. Tamasaku, A. Higashiya, T. Ishikawa, J. Phys. Soc. Jpn. **79**, 044713 (2010)
170. N. Motoyama, H. Eisaki, S. Uchida, Phys. Rev. Lett. **76**, 3212 (1996)
171. C. Kim, Z.-X. Shen, N. Motoyama, H. Eisaki, S. Uchida, T. Tohyama, S. Maekawa, Phys. Rev. B **56**, 15589 (1997)
172. B.J. Kim, H. Koh, E. Rotenberg, S.-J. Oh, H. Eisaki, N. Motoyama, S. Uchida, T. Tohyama, S. Maekawa, Z.-X. Shen, C. Kim, Nat. Phys. **2**, 397 (2006)
173. N. Tomita, K. Nasu, The peak position and the magnitude of dispersion are sensitive to U and t, respectively. Phys. Rev. B**60**, 8602 (1999)
174. V.N. Strocov, M. Shi, M. Kobayashi, C. Monney, X. Wang, J. Krempasky, T. Schmitt, L. Patthey, H. Berger, P. Blaha, Phys. Rev. Lett. **109**, 086401 (2012)
175. E. Razzoli, M. Kobayashi, V.N. Strocov, B. Delly, Z. Bukowski, J. Karpinski, N.C. Plumb, M. Radovic, J. Chang, T. Schmitt, L. Patthey, J. Mesot, M. Shi, Phys. Rev. Lett. **108**, 257005 (2012)
176. G. Landolt, S.V. Eremeev, Y.M. Koroteev, B. Slomski, S. Muff, T. Neupert, M. Kobayashi, V.N. Strocov, T. Schmitt, Z.S. Aliev, M.B. Babanly, I.R. Amiraslanov, E.V. Chulkov, J. Osterwalder, J.H. Dil, Phys. Rev. Lett. **109**, 116403 (2012)
177. K. Ishizaka, M.S. Bahramy, H. Murakawa, M. Sakano, T. Shimojima, T. Sonobe, K. Koizumi, S. Shin, H. Miyahara, A. Kimura, K. Miyamoto, T. Okuda, H. Namatame, M. Taniguchi, R. Arita, N. Nagaosa, K. Kobayashi, Y. Murakami, R. Kumai, Y. Kaneko, Y. Onose, Y. Tokura, Nature Mater. **10**, 521 (2011)
178. B.W. Batterman, H. Cole, Rev. Mod. Phys. **36**, 681 (1964)
179. S.-H. Yang, B.S. Mun, C.S. Fadley, Synchrotron Radiat. News **17**(3), 24 (2004)
180. J.C. Woicik, E.J. Nelson, T. Kendelewicz, P. Pianetta, M. Jain, L. Kronik, J.R. Chelikowsky, Phys. Rev. **B63**, 041403(R) (2001)
181. J.C. Woicik, E.J. Nelson, L. Kronik, M. Jain, J.R. Chelikowsky, D. Heskett, L.E. Berman, G.S. Herman, Phys. Rev. Lett. **89**, 077401 (2002)
182. S.-H. Yang, B.C. Sell, C.S. Fadley, J. Appl. Phys. **103**, 07C519 (2008)
183. B.C. Sell, S.B. Ritchey, S.-H. Yang, S.S.P. Parkin, M. Watanabe, B.S. Mun, L. Plucinski, N. Mannella, A. Nambu, J. Guo, M.W. West, F. Salmassi, J.B. Kortright, C.S. Fadley, J. Appl. Phys. **103**, 083515 (2008)
184. S.-H. Yang, B.S. Mun, A.W. Kay, S.-K. Kim, J.B. Kortright, J.H. Underwood, Z. Hussain, C.S. Fadley, Surf. Sci. **461**, L557 (2000)
185. C.S. Fadley, J. Elect Spectrosc, Rel. Phenom. **178–179**, 2 (2010)
186. S.-H. Yang, B.S. Mun, N. Mannella, S.-K. Kim, J.B. Kortright, J. Underwood, F. Salmassi, E. Arenholz, A. Young, Z. Hussain, M.A.V. Hove, C.S. Fadley, J. Phys. Condens. Matter. **14**, L407 (2002)
187. J. Zegenhagen, B. Detlefs, T.-L. Lee, S. Thiess, H. Isern, L. Petit, L. Andre, J. Roy, Y. Mi, I. Joumard, J. Electron Spectrosc. Rel. Phenom. **178**, 258 (2010)
188. J. Zegenhagen, Surf. Sci. **554**, 77 (2004)
189. M. Drakopoulos, J. Zegenhagen, A. Snigirev, I. Snigireva, M. Hauser, K. Eberl, V. Aristov, L. Shabelnikov, V. Yunkin, Appl. Phys. Lett. **81**, 2279 (2002)

Chapter 8
Hard X-ray Photoelectron Spectroscopy

In order to realize as high bulk sensitivity as possible in PES studies of strongly correlated electron systems (SCES materials), hard X-ray photoelectron spectroscopy (HAXPES) at hν and E_K above $\sim$a few keV is now recognized to be inevitable in spite of rather low count rates compared with PES and ARPES even in the soft X-ray region. Although sample surface cleanliness is almost always required for SXPES as well as lower energy conventional PES, untreated surfaces being exposed to air can be studied for some chemically stable materials by HAXPES. The count rate is, however, often much reduced compared to that for clean surfaces. In such a case, a clean surface is desirable for high accuracy measurements. Grazing incidence photon excitation can be utilized for HAXPES studies, where the photon penetration depth from the surface match the electron escape depth of the order of 50–100 Å. Then the small PICS in HAXPES can be compensated to a large extent (sometimes two orders of magnitude). This grazing incidence technique is available if the photon beam size on the sample is small enough judged from the width and the length of the slit of the electron analyzer and the sample surface is sufficiently flat. Really grazing incidence photon excitation is not so easy in the case of fractured or cleaved surfaces with surface steps and it is not useful for scraped surfaces even though the damage by scraping might not be very serious for HAXPES. In this sense, grazing incidence excitation will be most suitable to films grown in situ. However, thick enough films will be required to be not much influenced by the strain induced by the lattice mismatch with the substrate.

Although there are more than five HAXPES stations in the world, our system at BL19LXU of SPring-8 is introduced as an example. In this system, an A1-HE analyzer of MB Scientific is mounted to a UHV analyzer chamber, which was double shielded by μ-metals. The retardation voltage is up to 14 kV and the stability of the high voltage power supply is $\pm$8 mV over tens of hours at 8 kV. The high voltage is confined within the individual power supplies to realize high security to users. The analyzer input lens axis lies in the horizontal plane and is set to 60° from the photon incidence direction. Several straight and curved slits are mounted in the analyzer. The analyzer slit is oriented horizontally and the energy

S. Suga and A. Sekiyama, *Photoelectron Spectroscopy*,
Springer Series in Optical Sciences 176, DOI: 10.1007/978-3-642-37530-9_8,

analysis is achieved along the vertical direction on a 2D MCP. The focus of the analyzer is very sharp for the magnification of 12 and the fine remote control of the sample and analyzer positions is inevitable to tune the instrument to the best condition. A total resolution of better than 55 meV was so far achieved at 8 keV and a pass energy of 50 eV with a slit width of 0.2 mm.

So far the best realized resolution was worse than the predicted value from the expected hν resolution and the analyzer performance. This might be mostly due to the recoil effects of ions on photoelectron emission as later discussed in detail in Sect. 8.6. Sometimes, full utilization of the incident photon polarization dependence of photoelectron emission spectra provides useful information on complex electronic structures in PES. Such an experiment is possible by (1) rotating the electron analyzer around the sample, (2) using two analyzers set with their energy dispersive planes parallel and perpendicular to the linear polarization of the incident light or (3) utilizing a diamond phase plate. In the case of (3) one can relatively easily change the polarization from horizontal polarization to vertical, left- or right-handed circularly polarized X-rays. From the incident light polarization dependence one can discuss the symmetry of the orbitals in the valence band and dichroism of PES spectra as explained later in Sect. 8.8. Before these subjects, we introduce the HAXPES results on LSMO, LSCO, NCCO, Sm-, Pr-, Yb-compounds, as well as V oxides.

8.1 $La_{1-x}Sr_xMnO_3$, $La_{2-x}Sr_xCuO_4$ and $Nd_{2-x}Ce_xCuO_4$

As is known for $Nd_{1-x}Sr_xMnO_3$ (hereafter abbreviated as NSMO) already, the electronic structure of hole doped manganese oxides with perovskite structure, $R_{1-x}A_xMnO_3$ (R: trivalent rare earth atom, A: divalent alkaline earth atom), exhibits abundant intriguing physical properties. Therefore, understanding of the valence band together with the inner core level electronic structure was strongly required. The interplay between spin, charge, orbital and lattice degrees of freedom is the key issue in such systems. In order to probe the bulk electronic structure with high reliability, however, hard X-ray PES (HAXPES) was required because the photoelectron kinetic energies E_K for deep core levels are inevitably smaller than those for the valence electrons by the amount corresponding to the core level binding energies E_B. The advantage of HAXPES in the region above ~6 keV is its long λ_{mp} beyond ~50 Å even for various core electrons with $E_B \approx 1$ keV.

The first example is a single crystalline thin film of $La_{1-x}Sr_xMnO_3$ (LSMO) [1] grown epitaxially on $SrTiO_3$ (STO) substrates. The parent compound $LaMnO_3$ (LMO) is an antiferromagnetic insulator (AFMI). With substituting Sr for La, hole doping is achieved and the system becomes a ferromagnetic metal (FMM) with colossal magnetoresistance behavior. The compound with $x = 0.4$ shows the Curie temperature T_C of ~360 K. For further doping with $x > 0.5$ the system becomes an antiferromagnetic metal (AFMM). Due to the strain induced by the substrate, the critical temperature and resistivity are slightly different from those in the bulk crystal. For more than 10 unit cell thick films (>30 Å), physical properties qualitatively similar to the bulk were experimentally confirmed, though thinner films showed the suppression of metallicity, Tc, and magnetization. The results of HAXPES of Mn $2p_{3/2}$ core level states performed at BL29XU and BL47XU of SPring-8 on untreated 100 ML ($\sim$400 Å) thick samples grown by laser MBE and transported in air are shown in Fig. 8.1(a–d). Measurements were performed at $h\nu = 5.95$ keV and at both 300 and 40 K. The most characteristic feature is seen on the lower E_B side of the main peak. Namely, a small but sharp spike structure is observed at $E_B = 639$ eV at 40 K for x = 0.2 and 0.4, whereas it is almost absent for x = 0 and 0.55. This spike structure is remarkably suppressed at 300 K [1]. A corresponding structure is almost absent in SXPES at $h\nu = 800$ eV even for a clean surface of a sample grown in situ. Besides, the intensity of this spike structure is suppressed with increasing electron emission angle θ from the sample surface normal. Therefore, this structure in LSMO was understood to be a bulk intrinsic feature. An insulator–metal transition takes place between 300 and 40 K in x = 0.2 thin films. x = 0.4 thin films are metallic below 300 K and the metallicity increases at low temperatures with a FM state. An AFMI phase is known for x = 0 bulk LMO. A tiny hump was observed for x = 0 thin film LMO in this E_B region at 300 K for nearly normal incidence angle $\theta \sim 0°$. Although x = 0 stoichiometric samples are insulating, hole doping was induced by the excess oxygen during the growth of thin films. An AFMM phase was expected for x = 0.55, in which a weak spike was recognized at 40 K. Under these circumstances, it was conjectured that the observed bulk intrinsic spike feature was resulting from a well screened state induced by the doping-induced high density of states near E_F corresponding to the ferromagnetism and metallicity.

Meanwhile, a hump structure corresponding to such a spike structure was observed on a fractured single crystal NSMO with x = 0.5 and 0.47 at $h\nu = 1{,}750$ eV as shown in Fig. 8.1(e) and (f) [2]. In NSMO with x = 0.5, the high T ferromagnetic metal (FMM) phase changes into an antiferromagnetic insulator (AFMI) phase due to the onset of the charge ordering below 150 K. On the other hand NSMO with x = 0.47 is in the ferromagnetic metal (FMM) phase below $\sim$270 K in which a hump feature was observed at $h\nu = 1{,}340$ and 1,750 eV though it was no more observed at $h\nu = 970$ eV. This might be due to the increased surface sensitivity with the reduction of E_K below $\sim$330 eV.

In addition to the Mn 3d and ligand O 2p states taken into account in a MnO_6 ($3d^4$) cluster model with D_{4h} symmetry, new states C at E_F were considered in the calculation, representing the doping-induced states. Expressing the ligand hole as $\underline{L}$, four initial states, $|3d^4\rangle$, $|3d^5\underline{L}\rangle$, $|3d^3C\rangle$ and $|3d^5\underline{C}\rangle$ (due to the charge transfer from the C states at E_F), were considered for the high spin configuration. By adjusting two new parameters, the charge transfer energy between the Mn 3d and C states (Δ^*) and the hybridization between the Mn 3d and C states (V^*), with leaving the other parameters fixed as U = 5.1 eV, $\Delta = 4.5$ eV, V = 2.94 eV, $U_{dc} = 5.4$ eV and 10Dq = 1.5 eV, the HAXPES results were rather well

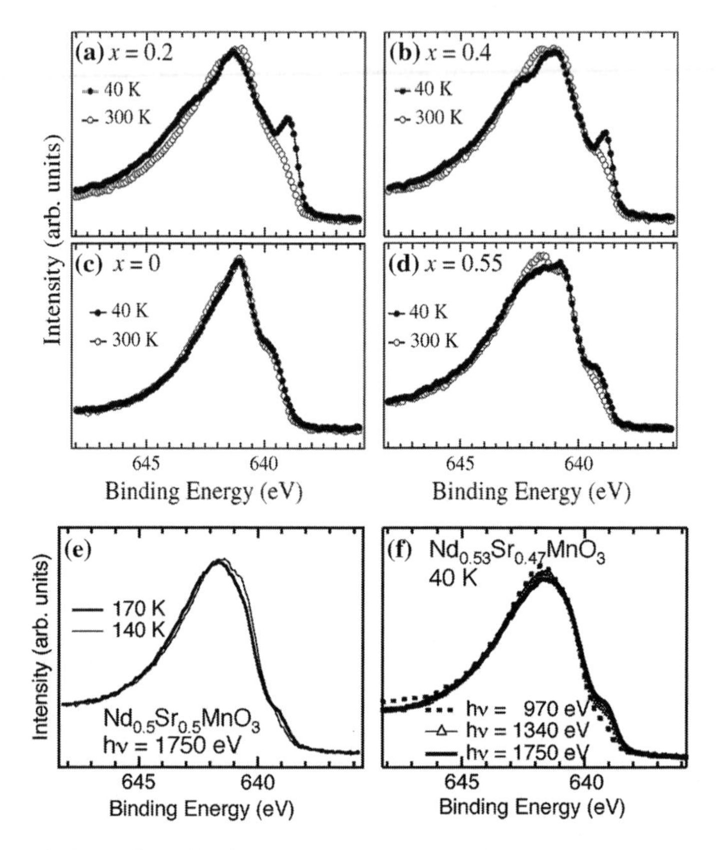

Fig. 8.1 **a–d** HAXPES on thin films (~100 ML) of $La_{1-x}Sr_xMnO_3$ grown by laser MBE on $SrTiO_3$ measured at hν ~ 5.95 keV with a total resolution of 300 meV. Samples were transferred through the air and the measurement was done on untreated surfaces [1]. SXPES spectra of single crystal $Nd_{1-x}Sr_xMnO_3$ [2] fractured in situ in UHV and measured at hν = 1,750 eV and at both 170 and 140 K for x = 0.5 (**e**), and those for x = 0.47 measured at 970, 1,340, and 1,750 eV at 40 K (metal phase) (**f**)

reproduced. The well screened peak was understood to originate from the $|2p^5 3d^5 \underline{C}\rangle$ final state. Larger V* was found for the FM phase (x = 0.2 and 0.4) compared to the AFM phase (x = 0 and 0.55).

As the next example the high T_c superconductors $Nd_{2-x}Ce_xCuO_4$ (NCCO) and $La_{2-x}Sr_xCuO_4$ (LSCO) are discussed. x = 0.15 NCCO and LSCO are typical electron- and hole-doped cuprate high T_c superconductors with a $d_{x^2-y^2}$ type superconducting gap. x = 0 Nd_2CuO_4 (NCO) and La_2CuO_4 (LCO) were recognized as charge-transfer insulators, where the on-site Coulomb repulsive energy (U ~ 8 eV) is much larger than the O2p-Cu3d charge transfer energy (Δ ~ 2 eV). Cu $2p_{3/2}$ HAXPES of single crystal NCCO (x = 0.15, T_c = 22 K), LSCO (x = 0.15, T_c = 36 K) and undoped NCO and LCO were performed at hν ~ 5.95 keV on BL29XU of SPring-8 [3]. All samples were fractured in situ and the surface-normal emission geometry was employed. NCCO and LSCO were measured at 35 K, whereas NCO and LCO were measured at room temperature to avoid the charging up effects.

Cu $2p_{3/2}$ HAXPES spectra of these materials are compared with the SXPES spectra measured at $h\nu = 1.5$ keV with a total energy resolution of ~ 0.3 eV on BL17SU of SPring-8 in Fig. 8.2(a). It was found that NCCO shows a sharp spike structure α on the low E_B side of the main peak, which might be ascribed to the $|2p^5 3d^{10}\underline{L}\rangle$ state. Such a feature was neither observed in the SXPES nor HAXPES spectra of NCO (Fig. 8.2(b)). Its energy position was also different from that of the Zhang-Rice singlet (ZRS) feature in LCO induced by the non-local screening. In contrast, ZRS was absent in NCO, in which the sharp peak α observed in NCCO HAXPES was also missing. By means of the SIAM, the spectrum for NCO was first calculated by considering the $|3d^9\rangle + |3d^{10}\underline{L}\rangle$ states (dotted line in Fig. 8.2(c)) and the spectrum for NCCO by using a linear combination composed of 85 % $|3d^9\rangle + |3d^{10}\underline{L}\rangle$ and 15 % $|3d^{10}\rangle$ states as shown in Fig. 8.2(c) by the solid line, where the relative energy shift between $|3d^{10}\underline{L}\rangle$ and $|3d^{10}\rangle$ states was assumed to be 1.5 eV. However, the NCCO HAXPES spectrum was not well reproduced by

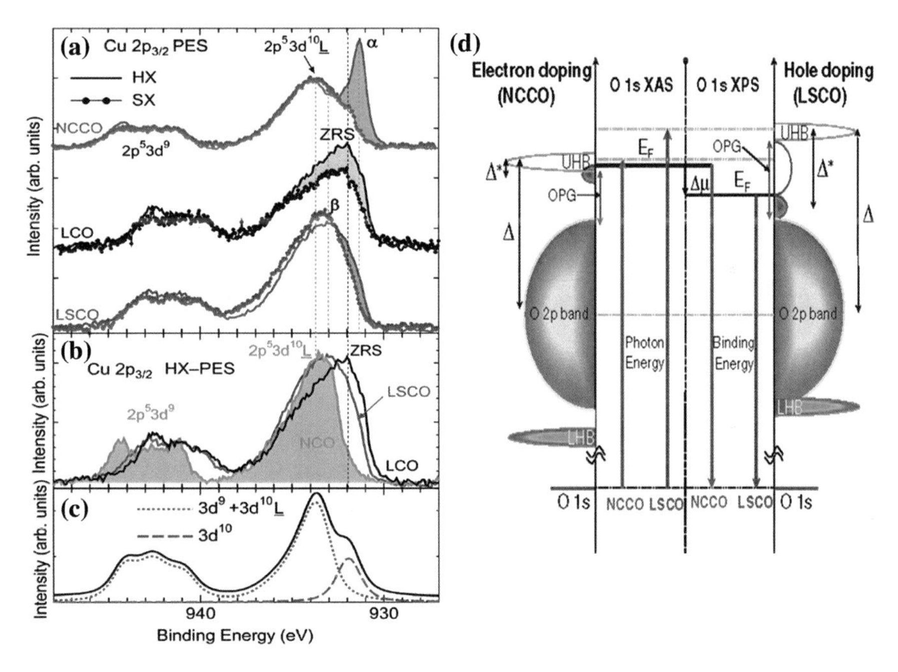

Fig. 8.2 **a** Comparison of Cu 2p HAXPES at $h\nu = 5.95$ keV (*solid line*) and SXPES at 1.5 keV (*line with symbols*) for electron-doped NCCO, undoped LCO, and hole-doped LSCO. Clear differences are seen in the shaded area. **b** Cu 2p HAXPES of NCO, LSCO, and LCO. **c** Calculated spectrum with $3d^9 + 3d^{10}\underline{L}$ (*dotted line*, representing the NCO spectrum) and that with $3d^{10}$ (*dashed line*). The sum is shown by the *solid line*. **d** Illustration of the energy levels for NCCO and LSCO used for the single impurity Anderson model calculation. OPG stands for the optical gap in undoped NCO and LCO [3]

this model with respect to the feature α. Then charge transfer from the doping-induced states C at E_F to the upper Hubbard band (UHB) was further considered, where Δ^* was defined as $E(|3d^{10}\underline{C}\rangle) - E(|3d^9\rangle)$ in comparison with the usual charge transfer energy Δ from the O 2p band to UHB defined as $E(|3d^{10}\underline{L}\rangle) - E(|3d^9\rangle)$ (Fig. 8.2(d)). Although this SIAM calculation cannot reproduce the ZRS feature, the main peak and the satellite structure on the low E_B side of NCCO are relatively well reproduced by employing $\Delta^* = 0.25$ eV for NCCO, where the structure α was interpreted as due to the $|2p^5 3d^{10}\underline{C}\rangle$ state, screened by the doping-induced states at E_F just below UHB. Δ^* was evaluated as 1.35 eV for LSCO, revealing that the doping-induced states lie far below the UHB and near the top of the valence band in LSCO. Based on this interpretation, the small chemical shift of the O 1s HAXPES as well as O 1s XAS were now consistently understood [3].

8.2 Sm Compounds

Besides Ce and Yb compounds, many PES studies were reported so far on Sm compounds, showing interesting behavior with respect to its valence instability in the bulk and surface [4]. Among various Sm compounds, Sm pnictides Sm_4X_3 are known to have a variety of physical properties such as mixed valence and charge ordering in the case of X = Bi and ferromagnetic order in X = As and Sb. So far PES showed Sm divalent peaks in both Sm 3d XPS and Sm 4f low hν PES in these compounds. For a reliable discussion of the bulk electronic states, it is necessary to know the contribution from the surface spectral weight in PES. First the hν dependence of PES of Sm_4As_3 and its bulk electronic structure are discussed. Sm_4As_3 is known to show a ferromagnetic transition at $T_C \sim 160$ K, around which the Kondo-like anomaly is observed in the electric resistivity.

The PES measurement was performed on in situ fractured surfaces of single crystal Sm_4As_3 between hν = 220 and 2,445 eV. The measurement below 1,750 eV was made at BL25SU of SPring-8 at 20 K and that at hν = 2,445 eV and 180 K at ID32 of ESRF. The total energy resolution was set to $h\nu/\Delta E \sim 4{,}000 - 5{,}000$ in both cases [4]. The valence band PES is shown in Fig. 8.3(a). The PICS of the Sm 4f states is more than 60 times larger than those of As 4s, 4p and Sm 5d states at 220 eV [5], where the structures in E_B within 4.4 eV from E_F and in $E_B = 5.5-10.5$ eV were assigned to the $|4f^5\rangle$ and $|4f^4\rangle$ final states resulting from the Sm^{2+} and Sm^{3+} states, respectively. Three structures located at $E_B = 11.4$, 3.2 and 0.5 eV become enhanced with the increase of hν and are interpreted as derived from the As 4s, As 4p and Sm 5d states from the behavior of PICS. The As 4p and Sm 5d states are hybridized resulting in bonding and antibonding states. Although the 4f multiplet components of Sm^{3+} are observable at high hν, the Sm^{2+} 4f features are strongly suppressed with increasing hν, suggesting their surface origin.

The Sm 3d core level spectra are summarized in Fig. 8.3(b) at three hν, where the spectra were normalized by the Sm^{3+} $|3d^9 4f^5\rangle$ peak. The intensity of the Sm^{2+}

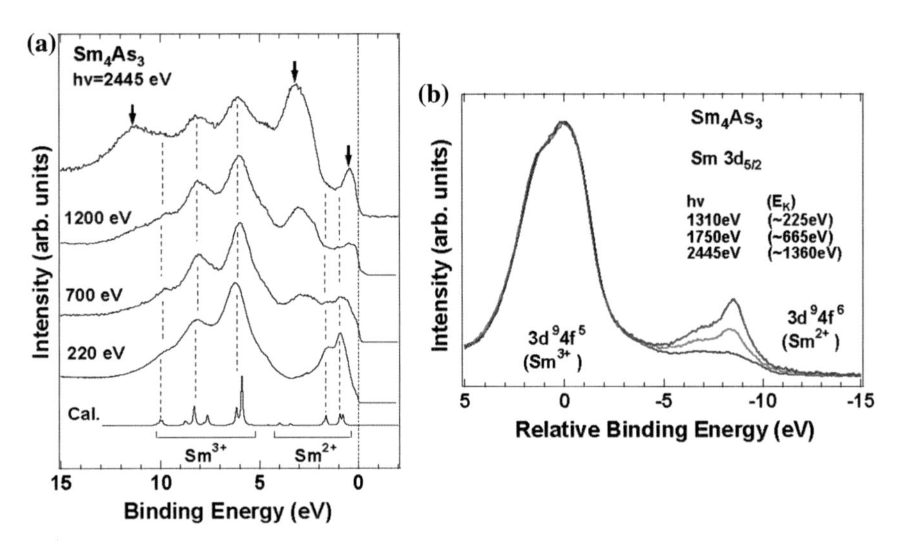

Fig. 8.3 hν dependence of the valence band (**a**) and Sm $3d_{5/2}$ core level PES (**b**) of a fractured single crystal Sm_4As_3 measured at 20 K (only the 2,445 eV spectrum was measured at 180 K). Spectra are normalized in the main peak region in the Sm^{3+} multiplet. At the bottom of (**a**) are shown the calculated $4f^4$ and $4f^5$ spectra. The arrows in (**a**) indicate the contribution from the As 4s, 4p and Sm 5d states [4]

$|3d^94f^6\rangle$ peak is found to decrease with increasing hν in consistence with its surface origin. The Sm 3d XAS spectrum (Fig. 3 of [4]) was also consistent with the theoretical XAS calculation assuming a pure Sm^{3+} ground state in the bulk.

The surface layer thickness s was assumed to be s $\sim$ 2.2 Å, a quarter of the lattice constant a (= 8.833 Å) in the anti-Th_3P_4 structure. Only the topmost Sm atoms are contained in this region having divalent character. The second and deeper Sm atoms from the surface were thought to have bulk electronic properties with trivalent character. Then one can calculate the Sm 4f and 3d core level spectra by the sum of the surface and bulk components considering the theoretically calculated inelastic mean free path λ_{mp}. The Sm 4f spectrum thus obtained by the full atomic multiplet calculation is in fair agreement with the experimental spectrum at hν = 220 eV, where the bulk spectral weight was 77 % (surface spectral weight is 23 %) for λ_{mp} = 8.5 Å. Although $\sim$5 % surface divalent contribution was still expected at hν = 2,445 eV in the Sm 4f spectrum, it was smeared out by the prominent Sm 5d contribution in the present case. Under this situation, an equivalent analysis must be made for the Sm 3d core level spectrum. Then the bulk spectral weight was estimated as 77 − 80 % at hν = 1,310 eV (E_K $\sim$ 225 eV) and $\sim$93 % at 2,445 eV (E_K $\sim$ 1,360 eV) in qualitative agreement with the experimental results. The bulk trivalent and surface divalent features were thus confirmed in Sm_4As_3 [4]. In this way the measurements of PES in a wide hν region and of both valence and core levels combined with core level XAS could fully reveal the bulk and surface electronic structure of strongly correlated 4f electron systems.

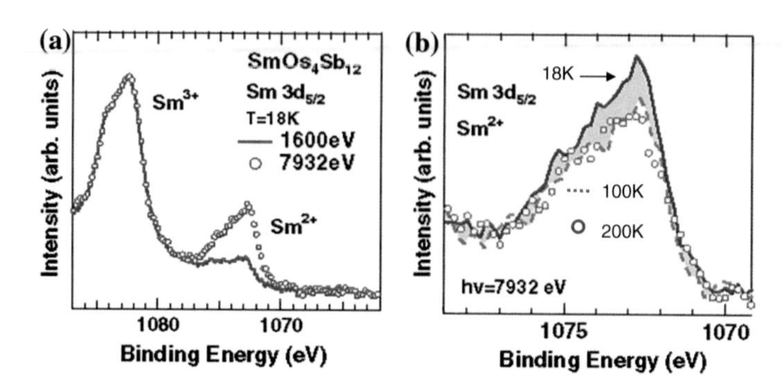

Fig. 8.4 Sm $3d_{5/2}$ core level spectra of $SmOs_4Sb_{12}$ [6] measured at hν = 1,600 and 7,932 eV (**a**), where the spectra are normalized at the Sm^{3+} peak intensity. Temperature dependence at hν = 7932 eV is shown in (**b**)

The next example is $SmOs_4Sb_{12}$, which is a heavy fermion (HF) filled skutterudite. Among many heavy fermion compounds recently discovered in lanthanide systems $SmOs_4Sb_{12}$ filled skutterudite attracts much attention due to such properties as Kondo anomaly and weak ferromagnetism. If ROs_4Sb_{12} is compared with RFe_4P_{12} (where R is a rare earth (RE) metal atom), ROs_4Sb_{12} generally has a lower superconducting temperature for the La system, smaller hybridization gap for the Ce system and lower Curie temperature (T_C) for Nd and Eu systems, suggesting a relatively weaker hybridization than RFe_4P_{12} between the conduction and 4f electrons and/or a weaker direct exchange interaction between R atoms due to the larger lattice constant in ROs_4Sb_{12}. T_C in $SmOs_4Sb_{12}$ is, however, 2–3 K being larger than that of $SmFe_4P_{12}$ (1.6 K).

In order to probe the bulk electronic states of this material, photoelectron spectroscopy was performed at hν = 1,070 – 1,600 eV (BL25SU of SPring-8) and at 7,932 eV (BL19LXU of SPring-8) in addition to Sm 3d-4f XAS (BL25SU) on in situ fractured surfaces of single crystals in UHV at pressures $<7 \times 10^{-8}$ Pa [6]. When the Sm 3d-4f XAS spectrum was compared with that of Sm_4As_3 a noticeable shoulder was recognized in the region of h$\nu \sim$ 1,076.5 eV, where no corresponding structure was observed in Sm_4As_3, strongly suggesting the contribution from divalent Sm and the mixed valence character of $SmOs_4Sb_{12}$. Sm $3d_{5/2}$ core level PES spectra measured at hν = 1,600 and 7,932 eV at 18 K are shown in Fig. 8.4(a), for which λ_{mp} was around 12 and 87 Å. The two peaks were ascribed to Sm^{3+} and Sm^{2+} with $|3d^9 4f^5\rangle$ and $|3d^9 4f^6\rangle$ final states. It is very clear that the intensity of the Sm^{2+} component is enhanced at higher hν in contrast to the cases of Sm_4As_3 and many other Sm compounds. This result revealed that much more divalent atoms exist in the bulk than on the surface. Namely, a mixed valence state is realized in the bulk of $SmOs_4Sb_{12}$. Sm 3d-4f XAS and Sm 3d core level HAXPES were calculated by the SIAM, considering atomic multiplets with configuration interaction. Experimental spectra were well reproduced and the Sm valence at $\sim$20 K was evaluated as $\sim$2.73 from the Sm 3d HAXPES [6]. The effective charge transfer energy Δ_{eff} defined as $E(4f^6\underline{L})$–$E(4f^5)$ in the ground state and the effective hybridization energy V_{eff} were evaluated as ~65 and ~310 meV at ~20 K, whereas V_{eff} ~ 237 meV at 100 K. Both Δ_{eff} and V_{eff} were much smaller than those for $PrFe_4P_{12}$, where Δ_{eff} and V_{eff} were evaluated as 2.3 and 1.2 eV. The much smaller Δ_{eff} satisfies the condition for strongly mixed valence in $SmOs_4Sb_{12}$ in spite of the weak hybridization.

Figure 8.4(b) shows the T dependence of the Sm 3d core level HAXPES spectra at $h\nu = 7{,}932$ eV. Although the relative spectral weight of the Sm^{2+} to Sm^{3+} does not change above 100 K, it is noticeably increased (by ~15 %) at 18 K. Considering the Kondo temperature of T_K ~ 20 K, this behavior was understood as the developing Kondo coherence with lowering T. In fact the V_{eff} at ~20 K was found to be appreciably larger than at 100 K. In addition, 3d-4f RPES had shown a significant Sm 4f contribution at E_F. Thus, the mixed valence states are the characteristic feature of $SmOs_4Sb_{12}$, and the Sm 4f states are responsible for the HF behavior in this material. The coexistence of these behaviors is due to the small energy difference between the $|4f^6\underline{L}\rangle$ and $|4f^5\rangle$ states [6].

8.3 Pr Compounds

Pr filled skutterdites have also attracted wide interest due to such anomalous properties as heavy fermion (HF) superconductivity in $PrOs_4Sb_{12}$ in contrast to conventional superconductivity in $PrRu_4Sb_{12}$, antiferro-quadrupole order in $PrFe_4P_{12}$, metal-insulator and structural phase transitions in $PrRu_4P_{12}$ and so on. The Pr 4f states are more localized and less hybridized with the conduction electron states than in Ce systems. No HF Pr compound was known until the discovery of $PrInAg_2$ with γ ~ 6.5 J/mol·K^2. A heavy electron mass was also found in $PrFe_4P_{12}$ under high magnetic field (~1.2 J/mol·K^2 under B = 6T). In both systems, a non-magnetic non-Kramers doublet CEF ground state was suggested with electric quadrupole degree of freedom. The HF behavior was therefore thought to be resulting from the quadrupolar Kondo effect [7]. $PrFe_4P_{12}$ shows a phase transition around 6.5 K and Kondo anomalies in transport properties in the high T phase. The Kondo temperature T_K was suggested to be of the order of 10 K.

In order to probe the electronic structure related to the Pr 4f states, the RPES is a very useful technique as described in Sect. 3.6.2 [8, 9]. Although the 3d-4f RPES belongs to the SXPES, this study is summarized here because it is very helpful to discuss the 4f states handled in HAXPES. RPES for $PrFe_4P_{12}$ was first performed for the Pr 4d-4f resonance excitation around $h\nu = 124$ eV [10]. When the off resonance- or resonance minimum- spectrum at $h\nu = 115$ eV was subtracted from the resonance maximum spectrum at $h\nu = 124$ eV, the contribution from the Pr 4f component could be quantitatively evaluated. However, there were two fatal drawbacks in this 4d-4f RPES compared to the 3d-4f RPES as already discussed in the case of Ce compounds. First, the degree of the resonance enhancement was much less than in the case of 3d-4f RPES. Second and the most essential was the surface sensitivity of 4d-4f RPES, which meant that the bulk electronic structure

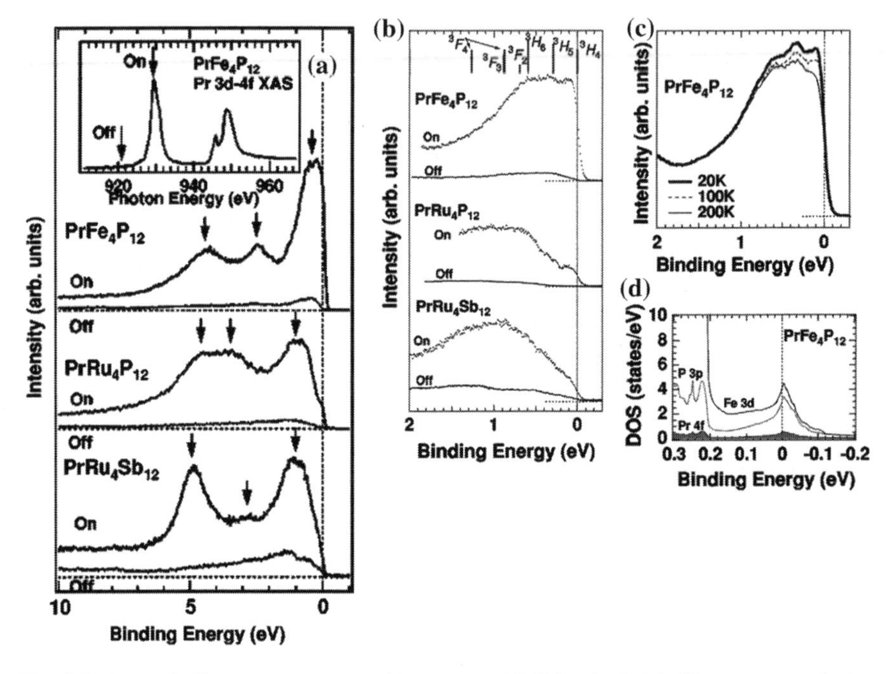

Fig. 8.5 On- and off-resonance spectra of $PrFe_4P_{12}$ at 20 K for the Pr 3d-4f inner core excitation [11]. **a** 3d-4f XAS spectrum of $PrFe_4P_{12}$ is shown at the *top*. Both the on- and off-RPES spectra are normalized by the photon flux. **b** Pr 3d → 4f on- and off-RPES near E_F at 20 K measured with a resolution of 80 meV. Vertical lines show the atomic $4f^2$ multiplets with the ground state at E_F. **c** Temperature dependence of the on-RPES. **d** Calculated partial density of states (*PDOS*)

was very difficult to be directly probed by this technique in most 4f RE compounds as already described for some Ce systems. Therefore, the 3d 4f RPES was mandatory for the study of light rare earth 4f related bulk electronic structures and HAXPES was required to complement them via bulk sensitive core level spectra.

The Pr 3d-4f RPES was performed on in situ fractured surfaces of single crystalline $PrFe_4P_{12}$ and $PrRu_4Sb_{12}$ and polycrystalline $PrRu_4P_{12}$ [11]. XAS and RPES measurements were performed at BL25SU of SPring-8. Except for the measurement of the T dependence, the samples were kept at T = 20 K. High resolution 3d-4f RPES measurements were performed with a total energy resolution of 80 meV. The results are shown in Fig. 8.5, where the Pr 3d-4f XAS of $PrFe_4P_{12}$ is given in the top inset in Fig. 8.5(a), suggesting the Pr^{3+} ($4f^2$) character in the initial state. XAS of the other two samples were very similar to this spectral shape. The on-resonance spectra were measured at $h\nu = 929.4$ eV, whereas the off-resonance spectra were measured at $h\nu = 921$ and 825 eV. The latter two spectra were quite similar to each other. The difference between the on- and off-resonance spectra in Fig. 8.5(b) reflects the Pr 4f contribution and the off-resonance spectra are ascribed to the Pr 5d and 4f, transition metal (T) 3d or 4d and chalcogen p orbitals in the valence band ranging from 0 to 7 eV, among which the transition metal d orbitals are dominating [5]. The off-resonance spectral features of all 3 materials were well reproduced by the calculation based on full potential linearized augmented-plane wave (FLAPW) and local density approximation (LDA + U) band-structure calculations [12], where $U = 5.4$ eV was employed.

The on-resonance spectra in Fig. 8.5(a) are characterized by three major 4f peak structures whose relative intensity depends sharply upon the individual materials. The 4f peak near E_F is the most prominent in $PrFe_4P_{12}$ and located closer to E_F compared with the other two materials. The bulk spectra were calculated by the cluster model. The part of the valence band to hybridize strongly with the Pr 4f states was expected to be similar between PrT_4X_{12} and LaT_4X_{12}. Since the occupied La f states below E_F were allowed only through the hybridization with the valence band (v), the La f PDOS of LaT_4X_{12} at a certain energy was thought to roughly represent the v-f hybridization strength at that energy. For simplicity we simulated the La f PDOS by two levels v_1 and v_2 at the energies $E(v_1)$ and $E(v_2)$. The calculation was performed by assuming the $|f^2>$ initial state. The PES final states were linear combinations of $|f^1\rangle$, $|(f^2)^*\underline{v_1}\rangle$, $|(f^2)^*\underline{v_2}\rangle$ final states, where $\underline{v_1}$ and $\underline{v_2}$ represent a hole in the valence band. Here $(f^2)^*$ contains all the excited states and is different from the initial ground state f^2. The average excitation energy $E((f^2)^*)$–$E(f^2)$ was ~1.4 eV dominated by the exchange interaction. The hybridization between the $|f^1\rangle$ and $|(f^2)^*\underline{v_1}\rangle$ or $|(f^2)^*\underline{v_2}\rangle$ is represented by V_1 or V_2. The weight of $|f^1\rangle$ in each final state corresponds to the Pr 4f PES intensity. It was found that the calculated results qualitatively well reproduce systematic changes of the experimental Pr 4f spectral weight with respect to the mutual intensity among the three components indicated by down arrows in Fig. 8.5(a) and their energy positions. The state in the vicinity of E_F was then interpreted as due to the bonding state between $|f^1\rangle$ and $|(f^2)^*\underline{v_1}\rangle$ [11].

In the spectra in a narrow E_B region measured with a resolution of 80 meV (Fig. 8.5(b)), a clear peak was seen at $E_B \sim 100$ meV in $PrFe_4P_{12}$, whereas the other two materials show only a weak hump in this region. This result strongly suggests pure $|4f^2\rangle$ dominance in the $PrRu_4P_{12}$ and $PrRu_4Sb_{12}$ and a finite contribution of the $|4f^1\rangle$ or $|4f^3\rangle$ state to the $|4f^2\rangle$ dominant Kondo state in $PrFe_4P_{12}$. The observed structures in $PrFe_4P_{12}$ were assigned to the Kondo resonance and satellites, corresponding to excitations from the $^3H_4$ ground state to several excited states ($^3H_5$, $^3H_6$ and so on) of the $4f^2$ states. The Pr 4f spectrum in $PrFe_4P_{12}$ shows a clear T dependence, where the intensity increases with lowering T down to 20 K. The increase in both the Kondo resonance peak and its spin–orbit partner at low T was known for Yb Kondo resonance systems. In the case of the Kondo tail in Ce compounds, the weights of the Kondo tail and its spin–orbit partner decrease with lowering temperature in contrast to the behavior observed for $PrFe_4P_{12}$. Therefore, the initial state in $PrFe_4P_{12}$ could be understood as a Kondo resonance peak dominated by $c_2\ |f^2\rangle + c_3\ |f^3\rangle$. The Kondo peak expected around ~1 meV ($T_K \sim 10$ K) seems to be broadened by the energy resolution of 80 meV in the present case.

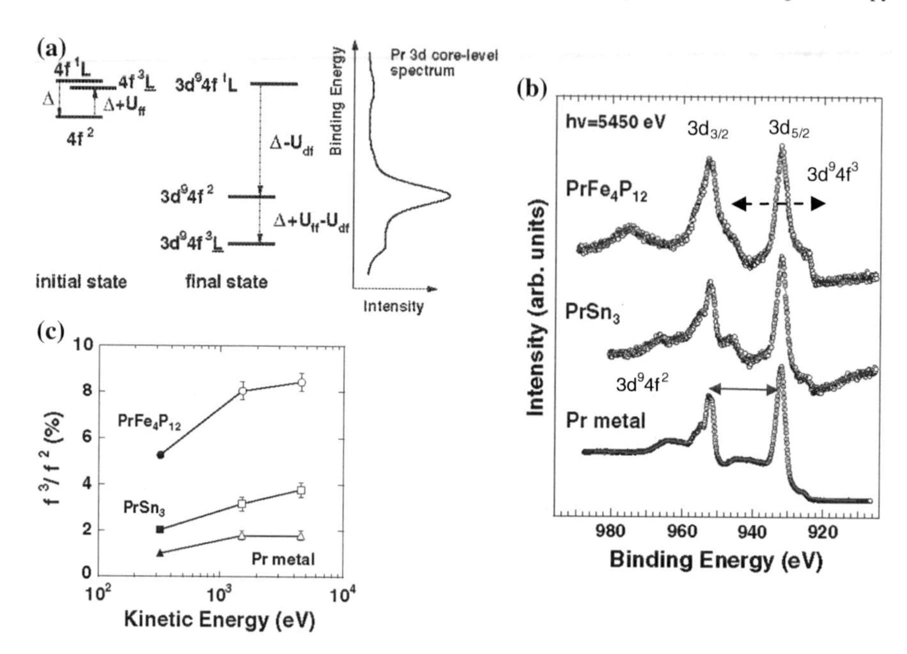

Fig. 8.6 Energy level diagram for calculating the Pr 3d core level spectrum (**a**). HAXPES of fractured single crystals of $PrFe_4P_{12}$ as well as an in situ scraped single crystal of $PrSn_3$ and polycrystalline Pr metal measured at ~180 K at $h\nu = 5{,}450$ eV (**b**), where *empty circles* represent the experimental results and thin solid lines show the calculated spectra [13]. E_K dependence of the f^3/f^2 ratio is shown in (**c**), where the solid marks at $E_K \sim 300$ eV are from [10] by XPS

The mean valence of Pr could be derived from core level HAXPES. The results for in situ fractured single crystal $PrFe_4P_{12}$ at $h\nu = 5{,}450$ eV and ~180 K at BL ID32 in ESRF were shown in Fig. 8.6 [13]. The two main peaks at $E_B \sim 952$ and 931 eV were ascribed to the $3d_{3/2}$ and $3d_{5/2}$ spin–orbit split Pr $|3d^9 4f^2\rangle$ peaks. On the lower E_B side around $E_B = 924$ eV were observed weaker shoulder structures resulting from the $|3d^9 4f^3\rangle$ states, suggesting strong hybridization. One should pay attention to the possible energy loss features expected at 22.8 eV above the E_B of the $|3d^9 4f^2\rangle$ main peaks in $PrFe_4P_{12}$.

Then the Pr 3d spectra were compared with configuration-interaction cluster model calculations based on the SIAM [13]. The parameters employed for the calculation were $U_{ff} = 7.7$ eV, $U_{df} = 9.2$ eV. By fitting the experimental spectrum, the effective hybridization $V_{eff}(\sqrt{12}\,V) = 1.2$ eV and the charge transfer energy $\Delta = E(4f^2) - E(4f^1L) = -4.0$ eV were evaluated for $PrFe_4P_{12}$, suggesting stronger hybridization between the Pr 4f and P 3p states in this material than the Pr 4f hybridization in $PrSn_3$ and Pr metal. From this fitting at $h\nu = 5{,}450$ eV, it was found that the ratio of the $|4f^1L\rangle$, $|4f^2\rangle$ and $|4f^3\underline{L}\rangle$ in the initial state in $PrFe_4P_{12}$ was 0.008:0.915:0.077. The number of the 4f electron came out as 2.07, noticeably deviated from 2.0 for the trivalent Pr due to the strong hybridization in the bulk. The mixture between the $|4f^2\rangle$ and $|4f^3\underline{L}\rangle$ components is thus responsible for the Kondo resonance in $PrFe_4P_{12}$ [13].

8.4 Yb Compounds

In the case of Yb compounds, RPES is not useful in contrast to the cases of Sm, Pr, Ce compounds because the 4f level is almost occupied. Therefore, careful measurements of the T dependence of the spectral shapes with high energy resolution, the dependence of the spectra on atom substitution as well as bulk sensitive core level PES at different hν are required. hν dependent measurements facilitate the semi-quantitative analysis of the surface contribution besides revealing the genuine bulk electronic structure.

The first example given here is $YbInCu_4$, which attracted wide attention because of the first-order bulk valence-transition around $T_v \sim 42$ K [14]. Accompanied with this transition, sudden changes in the lattice constant, electric resistivity, magnetic susceptibility have been observed without change of the crystal structure. Although rather many photoemission studies were accumulated up to now, the behavior of the genuine bulk Kondo resonance and the valence change were not yet established because of the presence of the so-called subsurface region in addition to the surface [15]. For example, inelastic neutron scattering showed a broad magnetic signal at around 40 meV (2.3 meV) below (above) T_v [16]. The signal at $\sim$40 meV was widely recognized as the Kondo resonance, whereas there is a controversy on the signal at $\sim$2 meV. As for the valence value, however, the experimental evaluations were scattered among thermodynamic data, Yb 2p core absorption, and PES experiments [17, 18]. Since the Yb valence estimated from PES measurements was much lower than the values estimated by other bulk sensitive measurements and the valence change with temperature near T_v was gradual than predicted for the first order transition, the identification of the bulk electronic behavior by confirming the surface and subsurface contributions was desired.

It is conceivable that the surface scraping cleaning induced a modification of the electronic structure in the near-surface region by the applied stress and increased surface area compared with the fractured surface. It became also known that the repeated heat cycles through T_v suppress the sharp valence transition. So SXPES was performed at hν = 800 eV on the single crystal fractured at 100 K and the measurement was done only on the first cooling cycle [19]. Still the valence evaluated from the Yb $4f^{13}$ and $4f^{12}$ spectral components by de-convoluting the surface and subsurface components was lower than those evaluated by other bulk sensitive measurements even though a jump of the E_B by $\sim$50 meV was observed for the $|4f^{12}\rangle$ final states near T_v. The mean valence change with T was also not sharp enough in contrast to the first order transition. Since the subsurface layer was thought

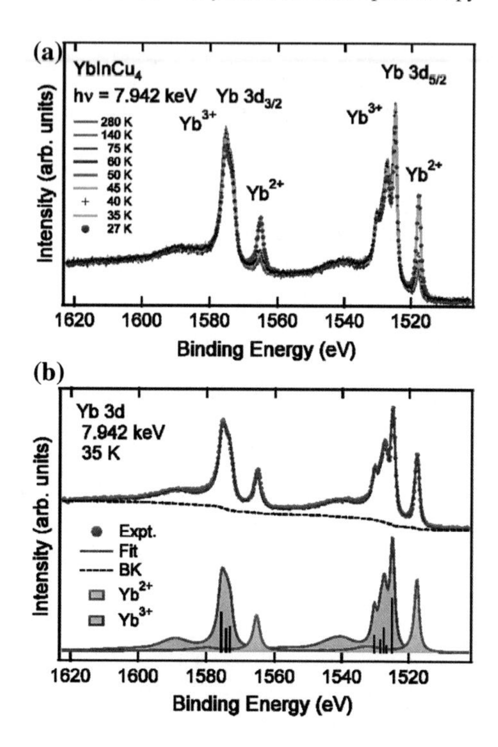

Fig. 8.7 Temperature dependence of the Yb 3d core level spectra measured at $h\nu = 7.942$ keV with a resolution of 120 meV. **a** A gradual intensity increase in the Yb^{2+} component is seen with decreasing temperature from 140 K down to 40 K, below which a sudden increase of the intensity is observed. **b** Deconvolution of the 35 K spectrum into the Yb^{2+} and Yb^{3+} components, where energy loss features at $\sim$14 eV above the main peaks were taken into account [21]

to be rather thick in this system in comparison with other Yb systems such as $YbAl_3$, PES with higher bulk sensitivity was really desired. HAXPES was performed at 5.95 keV at BL29XU of SPring-8 with a total resolution of 270 meV [20], where a single crystal was fractured at 300 K in a vacuum of 5×10^{-7} Pa and then transferred to an analyzer chamber with a vacuum of 3×10^{-8} Pa. Since the valence values were still lower than those reported by other bulk sensitive experiments, the contributions of the surface and subsurface were thought to be not negligible. Then we performed HAXPES at $h\nu = 7.932$ and 7.942 keV at BL19LXU of SPring-8 with a total energy resolution of 120 meV [21]. The spectra were measured in the $Yb3d_{5/2}$ and $3d_{3/2}$ core regions with decreasing the sample T down to 27 K. The sharp peaks at $E_B \sim 1{,}520$ and 1,567 eV in Fig. 8.7(a) are ascribed to the J = 5/2 and 3/2 components of the Yb^{2+} $|3d^9 4f^{14}\rangle$ final states. The broad bumps near 1,543 and 1,591 eV are attributed to energy loss structures. The multiplet structures between the $|3d^9 4f^{14}\rangle$ final states and the energy loss structures are ascribed to the Yb^{3+} $|3d^9 4f^{13}\rangle$ final state structures. The $|3d^9 4f^{14}\rangle$ peak was found to increase suddenly between the nominal temperature of 40 and 35 K (real sample T might be slightly higher). On careful inspection, however, this peak intensity decreases still rather gradually between 75 and 40 K as shown in Fig.8.8(a). With the help of atomic multiplet calculations, the Yb 3d spectrum was deconvoluted into the $|3d^9 4f^{14}\rangle$ and $|3d^9 4f^{13}\rangle$ final states by considering their Lorentzian broadening and Gaussian broadened energy loss peaks in addition to a Shirley type background.

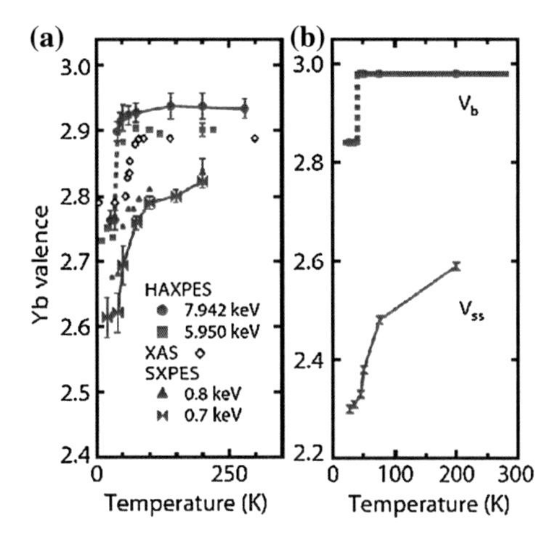

Fig. 8.8 Temperature dependence of the Yb valence in $YbInCu_4$. The results in **a** at $h\nu = 7.942$ keV and 5.950 keV were obtained from the Yb 3d core level spectra and those at 0.8 and 0.7 keV from the Yb 4f spectra. **b** The bulk valence and the valence in the subsurface were derived as explained in the text [21]

From Yb 3d core level HAXPES, the T dependence of the Yb valence was evaluated as shown in Fig. 8.8. The negligible overlap between the Yb^{2+} and Yb^{3+} components compared with the XAS was a great advantage to precisely evaluate the bulk valence of Yb systems. It was found that the valence evaluated at 7.942 keV was larger than those obtained at 5.950 keV [20], confirming the higher bulk sensitivity at ~8 keV. However, the valence value obtained at $h\nu = 7.942$ keV was still smaller than the bulk values. Moreover, the valence decreases gradually from 75 K down to 40 K. Such a behavior could not be consistently understood without considering the subsurface contribution. As explained in Sect. 4.2, we considered both the surface and subsurface layers with different electronic properties compared with the bulk. For simplicity, the subsurface layer was assumed to be homogeneous within the thickness of ss. The valence in the surface region (V_s) with the thickness of s could be safely assumed to be $V_s = 2.0$. It was possible to experimentally derive the valences in the bulk (V_b) and subsurface (V_{ss}) in addition to the thickness of s and ss from 4 independent experimental results performed at 4 different $h\nu$ by taking into account the $h\nu$ dependent inelastic mean free path λ_{mp} of photoelectrons. Then these quantities for $T > T_v$ were evaluated as s ~ 1.5 Å, ss ~ 6.5 Å, $V_b = 2.98$ as shown in Fig. 8.8(b). This V_b value, which becomes constant above T_V is consistent with the results of othcr bulk sensitive experiments. Under this assumption V_{ss} is T dependent and ~2.33 (2.38) at 45 (50) K. By assuming constant s and ss against T, V_b was evaluated as 2.84 and V_{ss} as ~2.31 (2.30) at 35 (27) K. It was thus confirmed that the electronic properties beneath s + ss ~ 8.0 Å from the surface were really reflecting the bulk electronic properties with the first order valence transition [21].

The valence band HAXPES at $h\nu = 7.932$ keV in the p-polarization configuration is shown in Fig. 8.9(a). $4f^{13}$ and $4f^{12}$ final state peaks are observed together with the Cu 3d and underlying In 5sp valence bands. Between 45 and 35 K is observed a drastic increase in the peak intensities of the $4f^{13}$ peaks. As shown in Fig. 8.9(b),

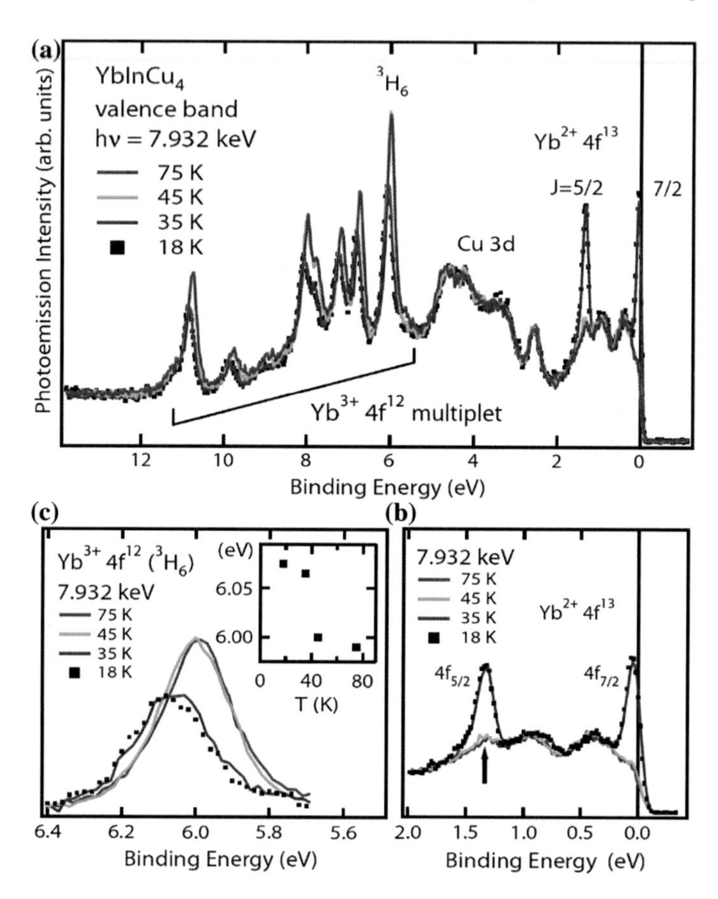

Fig. 8.9 Temperature dependence of the valence band HAXPES of $YbInCu_4$ measured at $h\nu = 7.932$ keV [21]. **a** Wide spectra normalized by the Cu 3d peak. **b** Expanded spectra from E_F to 2.0 eV. **c** Expanded spectra of the $^3H_6$ component of the $4f^{12}$ final states

two prominent peaks are observed at ~42 meV and 1.32 eV at both 35 and 18 K. These peaks are dramatically suppressed at 45 K leaving a small hump at 1.32 eV. A unique subtraction of the Cu 3d, 4s and In 5sp contributions was, however, almost impossible in this E_B region in the p-polarization configuration (see Sect. 8.8). Simultaneously seen was a sudden shift of the $4f^{12}$ multiplet structures as represented by the $^3H_6$ component of the $4f^{12}$ multiplets up to -65 ± 5 meV as shown in Fig. 8.9(c). According to the non crossing approximation (NCA) based on SIAM, the present results for $T < T_v$ could be explained by considering $\varepsilon_f = 0.71$ eV, $\Delta = 117$ meV and $U_{ff} = 8.1$ eV, providing a Kondo temperature of $T_K = 400$ K and the Kondo resonance peak at $E_B \sim 42$ meV. On the other hand, the results for $T > T_v$ including the T dependence of the $4f^{12}$ peak positions were well reproduced by $\varepsilon_f = 0.73$ eV, $\Delta = 67$ meV and $U_{ff} = 8.3$ eV, providing an extremely low $T_K \sim 0$ K. It was thus revealed that the hybridization energy Δ is suddenly decreased when T becomes higher than T_v. The results were consistent with the absence of any spectral feature near $E_B \sim 2$ meV in a very high resolution low energy PES measurements at $h\nu = 7$ eV [22]. The tiny residual peak at $E_B \sim 1.3$ eV at 45 K was consistently interpreted as due to the spin-orbit partner of the Kondo resonance peak in the sub-surface region and the bump between E_F and $\sim E_F + 150$ meV at 45 K was ascribable to the remnant of the broadened Kondo resonance in the sub-surface region according to the SIAM [21]. The spectral features with peaks at $\sim$0.4 and 0.9 eV were clarified as the In sp dominated conduction band structures.

HAXPES studies were also performed for the Kondo semiconductor YbB_{12} and the Lu-substituted $Yb_{1-x}Lu_xB_{12}$ ($x = 1/8$) alloy from 200 down to 20 K to study the Kondo lattice effect [23]. Similar analyses as performed for $YbInCu_4$ on both the Yb 3d core level and 4f states were made to evaluate the physical quantities describing their electronic structure. Experimental results were analyzed by NCA calculations based on SIAM with considering the crystalline electric field effects. T_K of YbB_{12} was suggested to be $\sim$240 K and a narrow gap was reported to be induced below 80 K. Its magnitude was reported to be in the range of $10 \sim 18$ meV. The reported Yb valence values were rather scattered as $\sim$2.86 at 30 K according to PES at $h\nu = 125$ eV, 2.9 at room temperature by 1.25 keV PES and beyond 2.95 and very close to 3 by Yb $2p_{3/2}$ core level XAS at 20 K [24]. The valence evaluation from the Yb $2p_{3/2}$ core level XAS edge was not so reliable because of the strong overlap of Yb^{2+} and Yb^{3+} components. Therefore, more reliable studies and a higher precision evaluation of the valence was desired for discussing the electronic structure of YbB_{12}. In the case of $Yb_{1-x}Lu_xB_{12}$, $T_K \sim 250$ K was predicted for $Yb_{7/8}Lu_{1/8}B_{12}$ and the gap seen in optical spectra in YbB_{12} at low T was reported to disappear rapidly with substitution of a small amount of Lu ($x = 1/8$), whereas a gap was traced up to $x = 0.9$ in INS studies.

HAXPES measurements were performed by using the same setup as employed for $YbInCu_4$, but the spectra in the region of Yb^{2+} $4f_{7/2}$ state were measured with a total energy resolution of 65 meV. T dependent energy shifts of the $4f^{13}$ and $4f^{12}$ components are shown in Fig. 8.10(b) and (c) and the Yb valence in Fig. 8.10(a). The sign of the energy shift is seen to be opposite between the $4f^{13}$ and $4f^{12}$ components. NCA analyses based on SIAM were applied to both materials. It is seen that the T dependence of the experimental results for $Yb_{7/8}Lu_{1/8}B_{12}$ are rather well reproduced by this calculation (calc.1 in [23]) with $\varepsilon_f = 0.91$ eV, $U_{ff} = 8.5$ eV and $\Delta = 64$ meV, where the consideration of the CEF effects was essential. However, we could not find any parameter set to well reproduce the T dependent results for YbB_{12}. Although SIAM was applicable for explaining the experimentally observed HAXPES in Lu doped samples, where the Yb 4f lattice coherence is destroyed, SIAM could not satisfactorily explain the temperature-dependent results of undoped YbB_{12}, where the Kondo lattice effects must be essential. A similar situation was already found in the case of $Yb_{1-x}Lu_xAl_3$ [25, 26]. Since genuine bulk electronic structures were gradually revealed by high $h\nu$ PES in various rare earth compounds, advanced analyses based on the periodic Anderson model are highly desired.

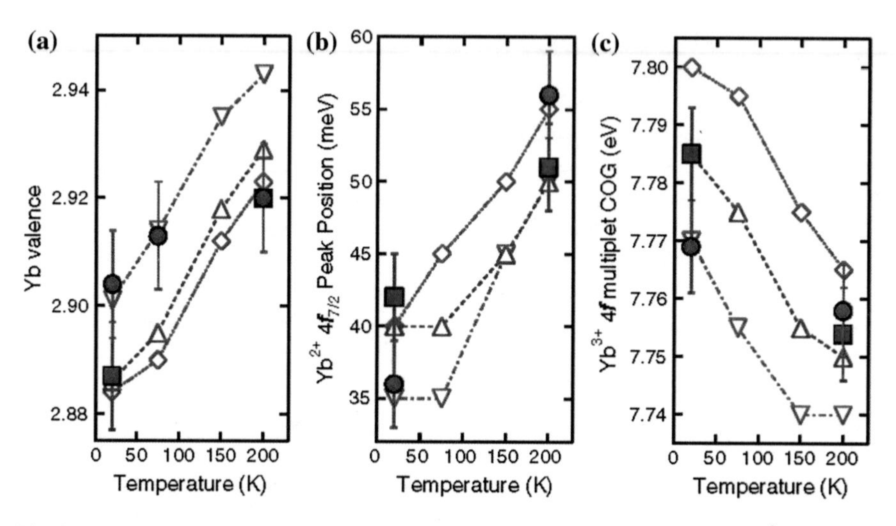

Fig. 8.10 Temperature dependence of the Yb valence and the peak energy of the Yb^{2+} $4f_{7/2}$ state and the center of gravity energy of the Yb^{3+} 4f multiplet of both YbB_{12} (*red full circles*) and $Yb_{7/8}Lu_{1/8}B_{12}$ (*blue squares*) [23]. Empty marks connected by lines (*dashed*, *dot–dashed* and *double dot–dashed*) represent the prediction by NCA calculations based on SIAM for 3 different sets of parameters. The *vertical bar* shows the experimental ambiguity. The best fit results for $Yb_{7/8}Lu_{1/8}B_{12}$ are given by the *empty triangles* calculated with $\varepsilon_f = 0.91$ eV, $U_{ff} = 8.52$ eV and $\Delta = 63.8$ meV, providing $T_K \sim 420$ K.

8.5 V Oxides

Many transition metal oxides show the metal–insulator transition (MIT), which is one of the intriguing properties in SCES. The origins of MIT have intensively been discussed so far [27, 28]. Extensive studies and discussions were given to MIT systems in vanadium oxides. Among them VO_2 has so far attracted wide interest as a $3d^1$ system with a MIT near $T_c \sim 340$ K from a high T metallic rutile (R) structure to an insulating monoclinic (M1) structure [29, 30] in which dimerization of V atoms and their tilting take place. PES studies were very useful to directly observe the change of the valence band electronic structure across the MIT [31–33]. High resolution SXPES spectra had shown a prominent coherent peak and a much weaker incoherent peak in the metal (M) phase compared with the surface sensitive PES below $h\nu = 120$ eV. The SXPES results required an explanation beyond the Peierls transition model and the standard single-band Hubbard model. Based on LDA + CDMFT (cluster extension of dynamical mean field theory) calculations, the importance of the interplay between the strong electronic Coulomb interaction and structural distortions, in particular, the V–V dimerization effect was demonstrated [34–36], as named correlation-assisted Peierls transition [35]. In the analysis of the anisotropy of the V 2p core level XAS spectra, experimental evidence for an orbital switching in the V 3d states across the MIT was obtained [36]. On going from the M to the I phase, the orbital occupation changes in such a way that charge fluctuations and effective band widths are reduced and the system becomes more one-dimensional and more susceptible to a Peierls-like transition. The massive orbital switching could only be made if the system was close to the Mott insulating regime. The MIT in VO_2 was thus understood as an orbital assisted "collaborative" Mott-Peierls transition (or simply orbital assisted MIT) [36]. Contrarily, the results of HAXPES later performed on a thin film (10 nm) of VO_2 grown on TiO_2 were interpreted in the Mott transition scenario [37]. In order to solve such a reappeared controversy, we performed HAXPES on a fractured surface of a single crystal VO_2 [38] and revealed noticeable differences from the results on a thin film, where the electronic structure might be much influenced by the strain induced by the substrate.

A series of HAXPES measurements were made on fractured surfaces of single crystal VO_2 and several other V_xO_y. Since the PICS of the V 3d state is really low at $h\nu = 7.932$ keV, the total resolution was set to 250 meV. The single crystal VO_2 was first fractured in situ in UHV at 260 K. Then various core level spectra including the V 1s core level as well as V 3d dominated valence band spectra were measured in the insulator (I) phase [38]. After E_B calibration by the Au E_F, the sample T was increased and the same series of measurements was repeated in the M phase at 350 K. Nearly surface-normal emission geometry was employed to realize the best bulk sensitivity for a light incidence angle $\theta \sim 60°$ from the surface normal. The change of the whole valence band in VO_2 on the MIT is shown in Fig. 8.11(a) [38]. The V 3d spectra in the metallic and insulating phases in several V_xO_y are compared in Fig. 8.11(b) [39–41].

Before discussing the valence band in detail, we first examine the changes of the core level spectra on the MIT. Clear changes of spectral shapes were observed for the V 1s, O 1s and V 2p core level spectra in VO_2 across the MIT temperature T_V [38]. In the M phase, the low E_B thresholds of the V 1s, O 1s and V 2p core level spectra were apparently shifted by $\sim$0.9, $\sim$0.4 and $\sim$1 eV towards smaller E_B as partly shown in Fig. 8.12. Spectral weight transfer towards the low E_B side was also revealed on the transition from the I phase to the M phase in these core levels. Less prominent but still noticeable changes were also observed for the V 3s, 3p and O 2s core level spectra as well as the O 2p valance band in VO_2 [38]. The spectra in the wide valence band region reproduced in Fig. 8.11(a) shows very clear changes in the whole E_B range of the valence band. First of all, the strong peak near $E_B = 7$ eV could be ascribed to the V 4s component hybridized with the O 2p state. The V 3d component is hybridized in the region of $E_B \sim 5$ eV. The prominent peak around 7.3 eV in the I phase is shifted by 0.3–0.4 eV towards E_F in the M phase. Another peak near 5.2 eV in the I phase is broadened in the M phase and its low E_B tail was shifted by $\sim$0.3 eV toward E_F in the M phase. These shifts are much larger than in the case of the thin film [37]. In addition, it was found that the intensity near $E_B = 9$ eV is noticeably increased in the M phase.

Detailed V 3d dominated valence band spectra are shown in Fig. 8.11(b) together with other V_xO_y systems below and above their MIT temperature. In VO_2, the main peak located at $E_B = 0.25$ eV in the M phase is definitely ascribed to the coherent peak or the QP peak. The rather flat plateau expanding in the region with $E_B = 0.9$–

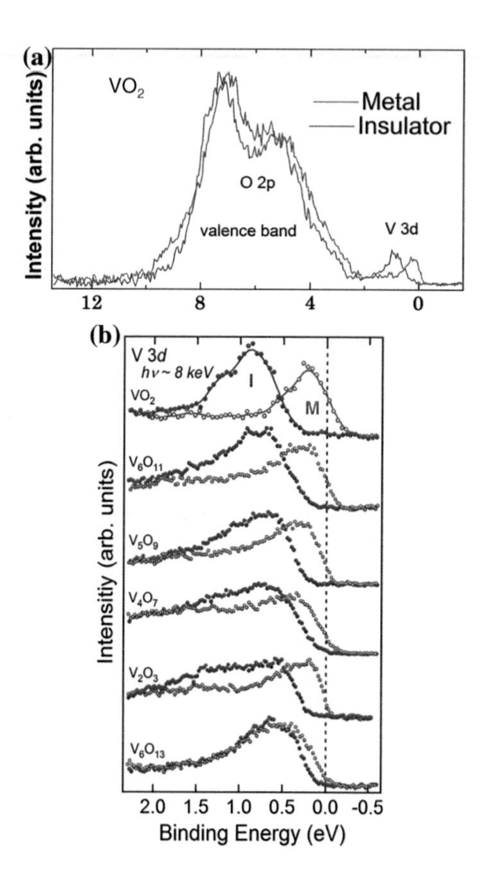

Fig. 8.11 **a** HAXPES of "O 2p" and V 3d valence bands in VO_2 measured at ~8 keV at 260 and 350 K for the I and M phases [38]. Figure **b** shows the comparison of the spectra in the I and M phases of various V_xO_y compounds in the V 3d valence band region [39–41]. The spectrum of V_2O_3 shown here is in the AFI phase

2.2 eV in the M phase could be interpreted as an incoherent component or the LHB. In the I phase, a prominent peak is observed at 0.90 eV. If we extended the lower E_B tail of the "O 2p" band (Fig. 8.11(a)) smoothly to the zero intensity, we could certainly recognize the V 3d contribution in the $E_B = 1.6–2.4$ eV region in the I phase. HAXPES of the V 3d dominated spectra in Magneli phase V_nO_{2n-1} MIT systems [39], V_2O_3 and Wadley phase V_6O_{13} [40, 41] are also shown in Fig. 8.11(b). For Magneli phase V_nO_{2n-1} MIT systems, both coherent and incoherent parts are clearly seen in their M phase. In the I phase as well, the incoherent part is clearly seen in V_2O_3, and recognizable in V_4O_7, V_5O_9 and in V_6O_{11} as well as VO_2. The smaller E_B part of the main peak in the I phase is thus interpreted as due to the coherent part. It was noticed that the MIT shift of the coherent peak is much larger in VO_2 compared with other V_xO_y. The V–V dimerization effects, which are strong in the I phase of VO_2, are still noticeable in the I phase of V_6O_{11} but negligible in V_5O_9 as confirmed later from the results of the V 1s and V 2p core level spectra in Fig. 8.12. No anomalously short V–V distance was reported in the I phase in V_6O_{13}, though the MIT in this material is accompanied with a structural distortion. In V_6O_{13}, spectral changes are much less prominent than in VO_2 in all cases of O 1s, V $2p_{3/2}$, V 1s, V 3s,

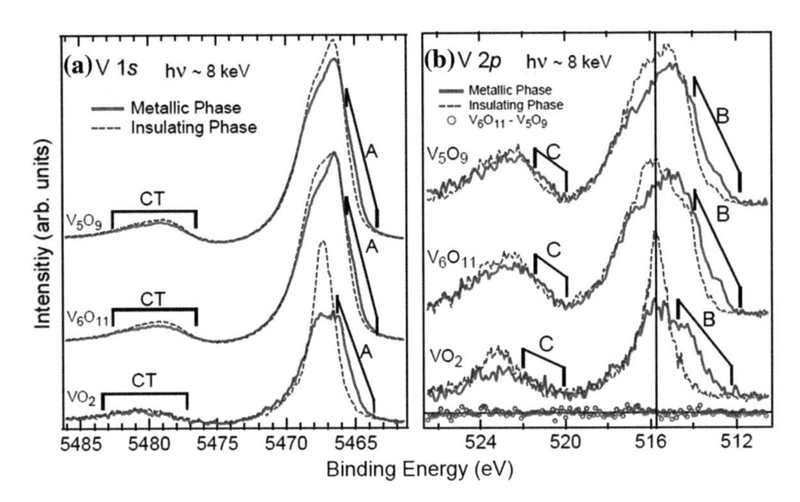

Fig. 8.12 HAXPES spectra of the V 1s (**a**) and V 2p (**b**) core levels measured at ~8 keV in both the I (*dashed lines*) and M phases (*solid lines*) of single crystals of VO_2, V_6O_{11} and V_5O_9 on their fractured surfaces [39]. V 2p spectra in the M phase are very similar between V_6O_{11} and V_5O_9 as shown by the *small empty circles* at the *bottom* of figure (**b**). Spectra are normalized by the integrated intensity of the main peak

V 3p, O 2s as well as the "O 2p" band [41], in which the shifts of the lower E_B threshold of the O 1s and V 1s peaks is only about 0.1 eV and the shift of the coherent peak is <0.2 eV. These results strongly suggest that the V–V dimerization effects are playing an important role in the MIT of VO_2, supporting the scenario of the correlation-assisted Peierls transition model [35]. If we extend the lower E_B tail of the 0.9 eV coherent peak in the I phase of VO_2, it crosses the zero line at $E_B = 0.28$ eV. Corresponding gaps in the I phase of other V_xO_y are much smaller than this value, proving that the gap opening in VO_2 is not solely due to the electron correlation effects. It was additionally argued that the V 3d peak, which has a high DOS at $E_B = 0.9$ eV in the I phase of VO_2, could not be due to the LHB but might have a dispersion up to $E_B = 0.28$ eV in order to explain the gap threshold energy evaluated by infrared absorption (0.68 eV) and the lowest energy broad absorption peak at 0.85 eV in the I phase [38], further supporting the above mentioned scenario [35, 36].

Figure 8.12 summarizes the V 1s and V 2p core level HAXPES spectra measured at $h\nu = 7945$ eV in both I and M phases of VO_2, V_6O_{11} and V_5O_9 [39], which were recognized as fully dimerized, partially dimerized and non dimerized systems in the I phase. The V 1s spectrum in the I phase is composed of a single prominent peak in VO_2 whereas those in V_6O_{11} and V_5O_9 are composed of a main peak and a shoulder on its higher E_B side. In the M phase, however, the intensity on the smaller E_B side of the main peak is increased and an intensity reduction is seen in the main peak region. The intensity reduction of the main peak and the transfer of the spectral weight to smaller E_B in the region A in the M phase are most prominent in VO_2.

In the case of V $2p_{3/2}$ core level spectra in the I phase, VO_2 shows a sharp single prominent peak, whereas V_6O_{11} shows a rather broad peak with a shoulder on the lower (smaller) E_B side of the main peak. V_5O_9 shows also a broad peak but with a shoulder on the higher (larger) E_B side of the main peak. In the M phase, however, the intensity on the lower E_B side (region B) of the $2p_{3/2}$ main peak is slightly enhanced with simultaneous reduction of the intensity in the main peak region in all these three materials. Then the V $2p_{3/2}$ spectral shapes become very similar between V_6O_{11} and V_5O_9 in the M phase as seen from their difference spectrum shown at the bottom. The MIT spectral shape changes are again most prominent in VO_2.

Judging from the similarity of the V 1s, V 2p and VB spectral shapes between V_6O_{11} and V_5O_9 in the M phase, we concluded that the electronic properties of these two materials are rather similar in the M phase. It was noticed that the relative intensity of the two components in each main peak region of both V 1s and V 2p core levels changed much across T_c in V_6O_{11}, while it did not change much in V_5O_9. It was then found that both V 1s and V 2p spectra of V_6O_{11} in the I phase were well reproduced by the weighted sum of the spectra of V_5O_9 and VO_2 with the ratio of 0.9:0.1 for V 1s and 0.85:0.15 for V 2p spectrum, in reasonable agreement with the ratio of the very short chains (~ 17 % [42]) in the I phase of V_6O_{11}. The effects of dimerization are thus confirmed in the V core level HAXPES spectral shapes in the I phase of V_6O_{11}. In these materials the lower E_B shoulders of the main peak of the V 1s and V$2p_{3/2}$ core level HAXPES spectra in the M phase may be due to the long-range or non-local metallic screening. Bulk sensitive HAXPES of both valance band and core levels with high enough resolution for single crystals has thus a high potential to clarify the details of electronic structures and the mechanism of the MIT in various family compounds. Brief discussions on V_2O_3 were already given in Sect.7.1.3.

So far we mainly discussed the potential of HAXPES to study bulk electronic structures discriminating the contributions of surface and subsurface. In this respect, HAXPES is also useful to study interface electronic structure. Artificially grown interface structures are attracting wide attention nowadays on both electronic and magnetic properties. Although the subject is slightly outside V oxides, the light transition metal Ti compounds such as an epitaxial heterostructure made of $LaAlO_3$ (LAO) and $SrTiO_3$ (STO) is attracting wide attention because of the presence of a metallic interface state at room temperature. By pulsed laser deposition (PLD) onto TiO_2 terminated STO (001) surfaces LAO/STO heterostructures were grown with 2, 4, 5 and 6 unit cell thick LAO overlayers [43]. HAXPES was performed at room temperature at $h\nu = 3$ keV with a resolution of 0.5 eV. Ti 2p core level spectra showed weak additional structures on the lower (smaller) E_B side of the main peak (by ~ 2.2 eV), whose intensity increased for larger θ (emission angle measured from the surface normal). Then this structure was ascribed to the Ti^{3+} state in comparison with the Ti^{4+} state in STO. For a quantitative analysis, a simple model was employed where the 2D electron gas extended from the interface to a depth d into the STO substrate. The interface region was assumed to be stoichiometric and characterized by a constant fraction p of Ti^{3+} ions per unit cell. In addition, the escape probability of the photoelectrons to the vacuum was assumed as $e^{-z/\lambda_{mp}}$ (z is the depth). Then the ratio between the Ti^{3+} and Ti^{4+} could be evaluated as a function of θ. λ_{mp} was assumed to be 40 Å. From the fitting to the θ dependence of the experimental results, p and d were experimentally evaluated for different cases of LAO overlayers [43]. It was revealed that the carrier concentration was much less than the expected value of 0.5 e^- and the thickness d of the 2D electron gas was only a few STO unit cells. The band bending was excluded as an origin for the formation of the 2D electron gas in the samples used for HAXPES. Instead a scenario that electronic reconstruction neutralizes the polar catastrophe matches the HAXPES results.

8.6 Recoil Effects

The advantage of HAXPES is undoubtedly its bulk sensitivity and the drawback is the very low PICS for most electronic states. In order to realize a count rate sufficiently high for reliable analyses, the energy resolutions of the photon monochromator and the electron analyzer must be properly set to ~50–400 meV, though the ultimate instrumental resolution can be < a few meV. When we measured the Fermi edge of some metals by HAXPES with the same instrumental resolution as for low hν SXPES, an anomalous broadening of the Fermi edge was sometimes observed. In the case of PES on atoms and molecules, the nucleus (or strictly speaking ion) recoil effects were known [44–46]. The investigation of similar phenomena in solids started when high resolution and high stability HAXPES became available. The recoil effects on the C 1s core level photoemission was soon found in highly oriented pyrolytic graphite (HOPG) in comparison with SXPES [47]. In the case of $Yb_{7/8}Lu_{1/8}B_{12}$, recoil effects were clearly seen for the B 1s core level shift of the order of ~300 meV, though the Yb 4d core level spectra did not show such clear recoil effects as seen in Fig. 8.13(a) [48, 49]. If single-nucleus recoil effect is considered in the photoelectron emission, the recoil energy E_R is predicted to be $\sim E_K \times m/M$, where E_K, m and M are the electron kinetic energy, and the masses of the electron and nucleus. The predicted values for various atoms at different hν are summarized in Fig. 8.13(b) [49]. According to this figure, the possible recoil shift will be <25 meV even at hν ~ 8 keV for Yb as well as Au. The observed recoil shift of the B 1s state is in good agreement with the predicted value.

Then the recoil effects on the valence electron photoemission was first found in solid Al in addition to the recoil effects on the Al 2p core level photoemission [50, 51] as reproduced in Fig. 8.14(a) and (b). For the Al 2p core level, the recoil shift of ~140 meV was confirmed in Fig. 8.14(a). The HAXPES spectrum is in fact slightly broader than the SXPES spectrum. Although the valence band spectra of Al and Au overlap rather well at hν = 275 eV (Fig. 8.14(b)), a slight shift of the Al spectrum towards larger E_B is seen at 700 eV. The shift of Al spectrum is further increased at 8,180 eV. The magnitude of the recoil shift energy is comparable to the predicted

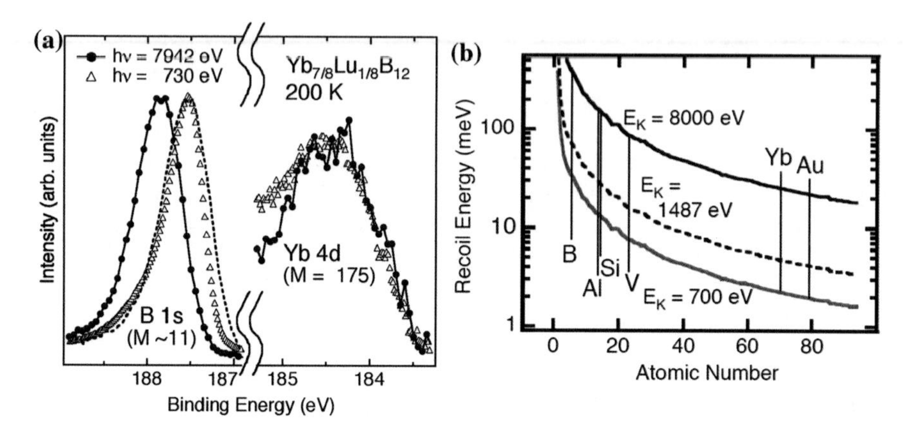

Fig. 8.13 **a** Comparison of HAXPES and SXPES for B 1s and Yb 4d core level spectra of $Yb_{7/8}Lu_{1/8}B_{12}$ at 200 K with the energy resolution of 140 meV. **b** $(m/M)E_K$ as a function of the atomic number [49]

E_R, though the shift became slightly increased with the increase of the PES intensity. Here an interesting question was raised why the momentum of the photoelectron from the valence band, whose wave function extends over the whole crystal, is not shared by the whole crystal, providing negligible E_R. The qualitative answer is illustrated in Fig. 8.14(c). The wave function of the final photoelectron state with very high E_K is rapidly oscillating in space. Although the wave function of the initial valence band electron is rather smoothly oscillating in the middle space between adjacent atoms, it becomes rapidly oscillating in the vicinity of the atom because of the core electron potential [52]. Then the excitation matrix element becomes enhanced in the vicinity of the atom (nucleus). Under this condition, the recoil effects are expected even for the valence band in solids. The recoil effect of Al in the near E_F region was well reproduced later by a simple model of Bloch electrons based on the isotropic Debye model [53].

Systematic photoelectron studies were performed for LiV_2O_4 and in the M phase of V_2O_3, V_5O_9 and VO_2 to check the possible photoelectron recoil effects for the valence band (Fig. 8.15) and core levels [54]. LiV_2O_4 was known to have geometrical spin frustration and shows heavy Fermion behavior below ~20 K. Li 1s core level spectra measured at 20 K showed a recoil shift up to ~530 meV between $h\nu = 1{,}249.5$ and 8,180 eV. O 2s core level spectra likewise showed a recoil shift of about ~230 meV. Although the O 1s spectrum at $h\nu = 1{,}249.5$ eV contained a noticeable contribution of the surface component on the higher E_B side of the main peak, the low E_B threshold was shifted by ~230 meV at 8,180 eV [55]. A recoil shift of ~80 meV was also observed for the V 2p core level spectra. In addition, a clear recoil shift of ~120 meV was seen for the E_F cut-off in LiV_2O_4, where the incoherent peak showed a comparable recoil shift. This recoil shift energy was smaller than those for the Li and O core levels but definitely larger than that for the V core level. Since the valence electron wave function can

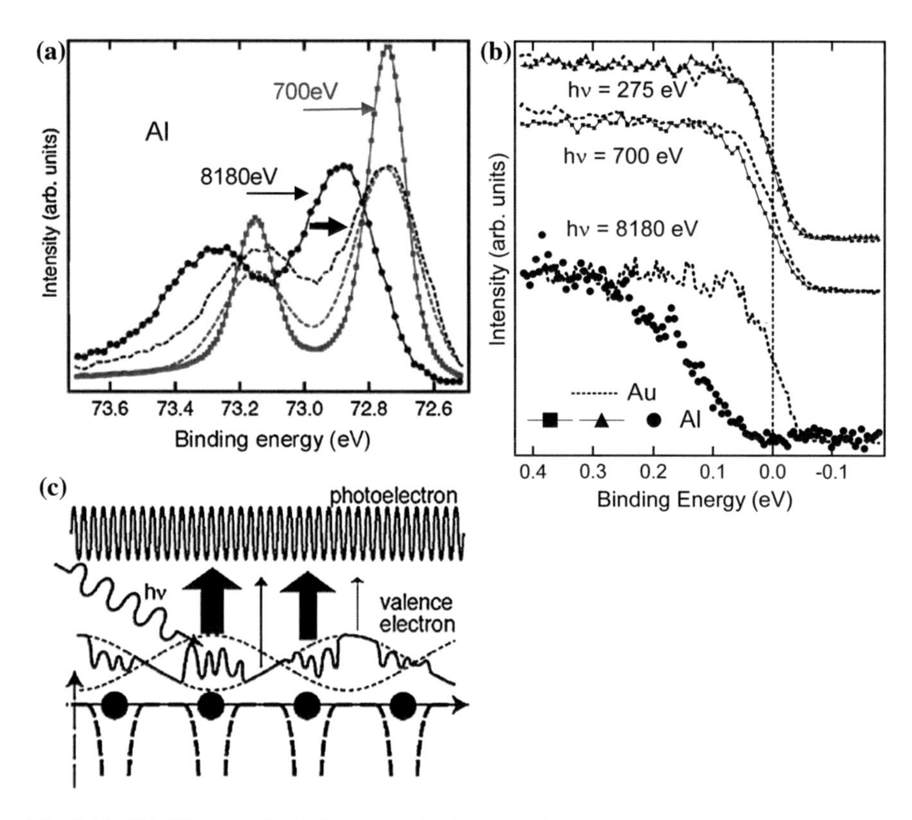

Fig. 8.14 HAXPES and SXPES spectra of Al measured at 20 K [50, 51]. **a** Al 2p core level spectra for which the resolution is ~200 meV (~90 meV) for HAXPES (SXPES). Dots are the raw data and the *dashed lines* are for comparison between the SXPES broadened to the resolution of 200 meV and E_B shifted HAXPES. **b** HAXPES compared with SXPES in the valence band region near E_F. The resolution was set to ~90, ~90 and ~75 meV at $h\nu = 8{,}180$, 700 and 275 eV. Au spectra are added for comparison. **c** Schematic description of the spatially localized photoemission process for the itinerant valence-band electrons in solids at very high kinetic energy. Waves represent the spatially oscillatory behavior of the wave function of the initial and final electron states

be expressed by a linear combination of Wannier functions and the quasi particle is composed of not only the V 3d but also the O 2p and Li 2s states, the magnitude of the observed recoil shift would reflect the different recoil shifts for these states. For example, the E_F edge recoil shift in LiV_2O_4 was well reproduced by considering 66 % of V 3d weight and 34 % of O 2p weight near E_F and the magnitude of the observed core level recoil shift energies, which were in agreement with E_R, under the present experimental instrumental resolution of ~200 meV.

On the contrary, no recognizable recoil shifts were observed in the M phase of VO_2 at 350 K for both the O 1s core level and valence band spectra, though Debye temperatures in both VO_2 and LiV_2O_4 are ~400 K. In order to understand this unexpected result, the atomic configuration must be examined. One recognized

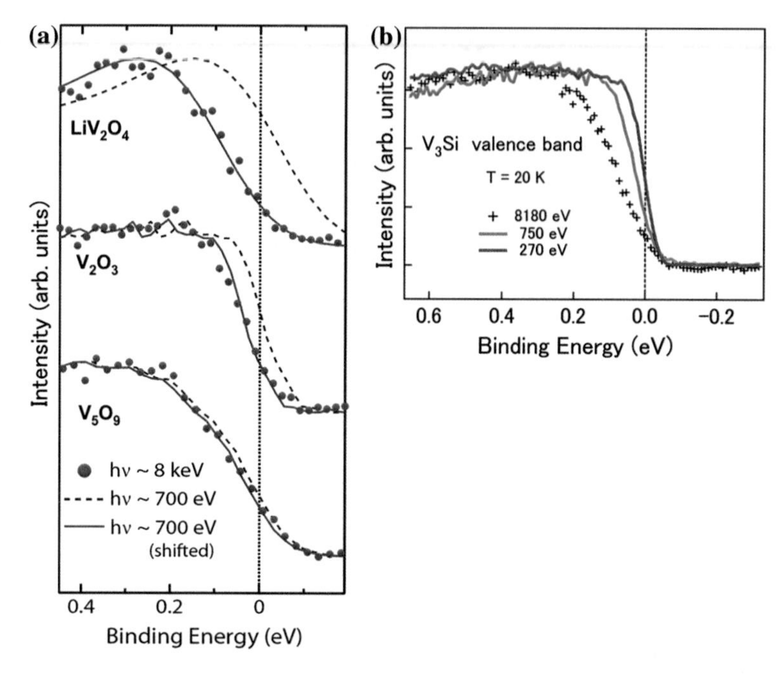

Fig. 8.15 Recoil effects for the valence band in LiV_2O_4, and M phases of V_2O_3 and V_5O_9 (**a**) and V_3Si (**b**) [54–56]. Sample temperatures were 20, 175, 150 and 20 K for LiV_2O_4, V_2O_3, V_5O_9 and V_3Si. The resolution in (**b**) was set to ~100 meV for 8,180 and 750 eV and ~50 meV for 270 eV

that both materials contain V-O_6 octahedra. Li-O_4 tetrahedra are additionally contained in LiV_2O_4. Judging from the magnitude of the recoil shift energies of all core levels in agreement with E_R in LiV_2O_4, all constituent atoms were thought to be subjected to the recoil effects in LiV_2O_4. Namely, all O, Li and V atoms could freely recoil at the moment of photoemission. In contrast to this, neither V nor O could freely recoil on photoemission in VO_2. One noticed here that the minimum V-O distance in LiV_2O_4 is 1.97 Å at 20 K, which is noticeably longer than that of the V-O distance in VO_2 (1.92 Å at 350 K). The inter-atomic distance seems to rather strongly influence the recoil effects in solids. Figure 8.15(a) summarizes the recoil shifts of the valence bands of LiV_2O_4 as well as V_2O_3 and V_5O_9 in the M phase. The recoil shift energies decreased in this order. Since recoil effects predicted by E_R were observed for the O core levels but not observed for the V core levels in both V_2O_3 and V_5O_9, the recoil shifts observed in the valence band in V_2O_3 and V_5O_9 were understood to be reflecting the relative weight of the O 2p components. The observed results suggests a O 2p spectral weight of ~0.2 and ~0.1 for V_2O_3 and V_5O_9 near E_F of the valence bands. Although roughly rigid-shift like recoil shifts were observed for E_F in these V_xO_y and LiV_2O_4, the results of V_3Si show a very contrasting behavior in Fig. 8.15(b) [56]. Namely, the amount of the recoil shift energy increases with the increase in the PES intensity in the near-E_F threshold region, though the total instrumental resolution was set to the same value at hν = 8,180 and 750 eV. In this material, recoil shifts were found to be negligible in both cases of V 2p and 3p spectra, though E_R was predicted to be ~80 meV. Then a negligible recoil shift was suggested for the V 3d spectra. On the other hand, Si 2p spectra showed a clear recoil shift up to ~120 meV, which is a little bit smaller than E_R. So the observed behavior of the recoil shift in the valence band in V_3Si was ascribable to the contribution of the Si component even though the V 4p and 3d components are dominant in the DOS near E_F. The observed recoil behavior in the near-E_F region in V_3Si could only be explained by considering the E_B dependent Si 3p PDOS in the vicinity of E_F, namely the Si 3p PDOS increases with increasing E_B across E_F relative to the PDOSs of V 4p and 3d states in the vicinity of E_F [56].

So far we encountered 3 typical cases of the photoemission recoil effects in solids. Recoil shifts were observed for (1) all core levels of all constituent atoms and the valence band, (2) core levels of some atoms and the valence band with negligible effects for core levels of other atoms and (3) neither any core levels of any atoms nor the valence band. The valence band recoil shifts were found to provide useful information on the relative spectral weight of the electronic states in the valence bands. Even the energy dependence of the PDOS near E_F can be revealed from the E_B dependence of the recoil shifts.

8.7 Angle Resolved Hard X-ray Photoelectron Spectroscopy

The importance of the bulk sensitivity of SXARPES with hν above several hundred eV for probing bulk band dispersions in SCES has intensively been discussed already in Chap. 7. Since HAXPES can realize even higher bulk sensitivity than SXPES, it is important to check its applicability to ARPES with modest energy resolution. First of all, the wave vector **q** of the photon must be properly taken into account for hard X-ray ARPES (HAXARPES) as in the case of SXARPES. Even if the instrumental energy- and wave vector (**k**)-resolutions as well as the counting rate satisfy the condition for ARPES, **k** broadening due to the creation and annihilation of phonons in the photocxcitation process may induce **k** smearing. It is known that the Debye–Waller factor semi-quantitatively describes this **k** broadening process [57]. When this factor is large enough (>0.9, for example), **k** is not much smeared out, whereas **k** smearing is serious if this factor is very small (<0.1). It was already discussed in Sect. 8.6 that the recoil effects might also broaden and shift the valence band spectral features in some materials. So HAXARPES can only be applicable to such systems which do not show noticeable recoil effects on photoelectron emission or just show recoil effects within the energy resolution of the measurements. In this respect, one must be very careful in the case of solids with light element(s). So far reported are the examples of W at

~6 keV and GaAs at ~3.2 keV [57] performed at BL15XU of SPring-8, in which the recoil shifts were predicted as 18 and 23 meV. The Debye–Waller factor for W is 0.45 at 30 K and 0.09 at 300 K. The λ_{mp} for the photoelectrons from the valence band of W at 6 keV is ~60 Å [58] corresponding to ≤20 unit cells of W. The results at 300 K showed non-dispersive valance band features with a modulation in angle. This modulation could be attributed to X-ray photoelectron diffraction (XPD) [59]. The raw spectrum at 30 K showed some signs of dispersive bands. Then this spectrum was corrected by considering the intensity modulation by both the DOS effect and the XPD effect to clarify the band dispersion. After these treatments, dispersing bands could be recognized, which were qualitatively explicable within the free-electron final-state model. The results could better be explained by the results of one-step theory based on a local density approximation (LDA) layer-KKR approach. Highly accurate sample alignment was required in HAXARPES to reduce the artificial deviation of **k** since |**k**| increases in proportion to $\sqrt{E_K}$.

In the case of GaAs (001), HAXARPES was performed at ~3.2 keV because of the lighter atomic mass and lower Debye temperature. The Debye–Waller factor for GaAs is 0.31 at 20 K and 0.01 at 300 K. The λ_{mp} was predicted to be 57 Å, whereas experimentally a value of 32 Å was found [60] corresponding to 10–6 unit cells. In the results at 300 K, the DOS and XPD effects were dominating. The raw spectra at 20 K clearly showed dispersing bands in $E_B = 0-8$ eV. DOS and XPD corrected spectra were compared with the calculated results within the free-electron and one-step models. Most parts of the dispersions were in consistence with the calculated results. Some little deviation recognized between experimental and theoretical results should be further examined with respect to a possible small misalignment of the sample, finite angular resolution and **k** smearing due to inelastic scattering and many-electron effects. It was thus demonstrated that HAXARPES of valence bands at low enough temperatures will facilitate studies of bulk band dispersions in some materials with negligible recoil effects on photoelectron emission.

8.8 Polarization Dependence of Hard X-ray Photoelectron Spectroscopy

HAXPES at $h\nu = 5-10$ keV has unique characteristics beyond the high bulk sensitivity in contrast to SXPES at $h\nu = 0.5-1.5$ keV [61–63] as: (1) The PICS for the s and p states are comparable with those for the d and f states in HAXPES. (2) For linearly polarized light excitation in HAXPES, the photoelectron angular distribution has a characteristic orbital dependence with respect to the angle θ between the two directions of photoelectron detection and the excitation light polarization (electric field **E**) (Fig. 8.16). As an overall tendency, the calculation [61–63] predicts that the photoelectron intensity for the s and ip (i > 4) states is strongly suppressed along the emission direction perpendicular to **E** ($\theta = 90°$, s-polarization for the vertical sample surface) compared to that along or near the

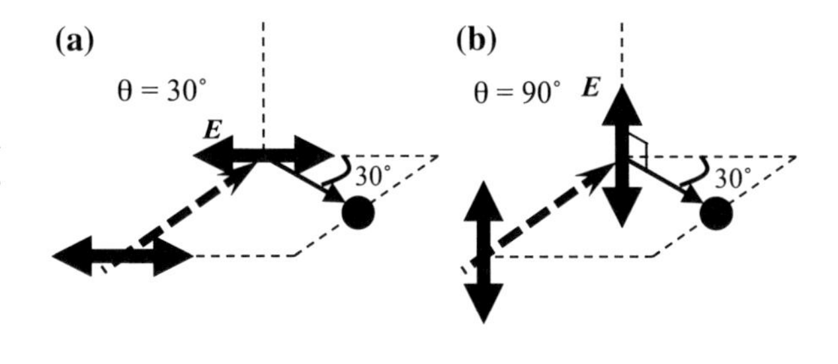

Fig. 8.16 Experimental geometries and their notations with respect to the direction of light propagation (indicated by *dashed arrows*), polarization vector ***E*** (indicated by *bold solid arrows*) and detected photoelectrons (indicated by *filled circles* with thin *solid arrows*)

vector **E** ($\theta \sim$ 0–30°, p-polarization), while the angular dependence of the photoelectron intensity is relatively weak for the d and f states. Namely, HAXPES with linearly polarized excitation at $\theta \sim$ 0–30° (p-polarization in the present case) is rather sensitive to the s and ip states whereas that at $\theta = 90°$ (s-polarization) is highly d- and f-sensitive. Therefore, the extraction of the s and ip contributions as well as that for the d and f contributions in the bulk valence band of solids becomes feasible by linear polarization-dependent HAXPES.

Among the three methods already mentioned in the beginning of Chap. 8 to measure the polarization dependent HAXPES, we employed the method (3) using a thin diamond single crystal. In the hard X-ray region with $h\nu = 5-16$ keV, we can easily switch the light polarization by using a diamond phase retarder (or phase plate). The technique developed at BL19LXU in SPring-8 for switching the polarization from horizontal to vertical direction enabled the experiment schematically shown in Fig. 8.16. A Si (111) double-crystal monochromator selects ~8 keV radiation with linear polarization along the horizontal direction (the degree of linear polarization is $P_L > +0.98$), which is further monochromatized by a channel-cut crystal (Si (444) (440) or (551) reflection). In order to switch the linear polarization of the hard X-rays from the horizontal to vertical direction, a (100) single-crystalline diamond with a thickness of 0.7 mm is used as a phase retarder [64] in the Laue geometry with the (220) reflection, which is placed downstream of the channel-cut crystal. The transmittance of the X-rays at ~8 keV for this diamond was confirmed to be ~35 %. P_L of the X-ray downstream of the phase retarder was estimated to be better than -0.8 (closer to -1), which corresponds to the linear polarization components along the horizontal and vertical directions of <10 and >90 %, respectively. As shown in Fig. 8.16, the emission direction of the photoelectron to be detected is set within the horizontal plane, where θ, defined in this section as in Fig. 8.16 between the light polarization and the electron emission angle, was 30° (90°) for the excitation with horizontal (vertical) linear polarization.

The polarization dependence of the core-level photoemission spectra for polycrystalline gold (Au) is shown in Fig. 8.17 [65]. The photoemission spectral weight at $\theta = 90°$ is strongly suppressed compared with that at $\theta = 30°$ for the 4s, 4p, 5s and 5p core levels with respect to the 4f levels. The intensity ratios $I_{\theta=90°}/I_{\theta=30°}$ for the

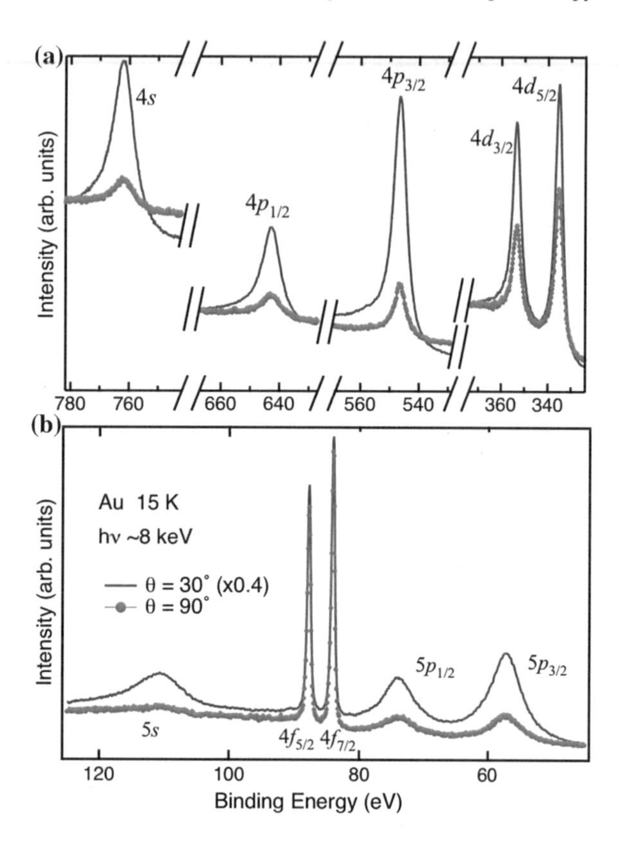

Fig. 8.17 **a** Linear polarization dependence of the hard X-ray excited 4*s*, 4*p* and 4*d* core-level photoemission spectra of polycrystalline Au. **b** The same as (**a**) but for the shallow 5*s*, 4*f* and 5*p* core-level spectra. The experimental geometries and their notations with respect to the directions of light propagation, polarization (electric field) vector ***E*** and detected photoelectrons are shown in Fig. 8.16. The spectral intensity at $\theta = 30°$ is scaled by multiplying a factor of 0.4 for comparison with the spectral intensity at $\theta = 90°$ [65]

core levels estimated from our experimental data are consistent with the calculation as shown in Table 8.1 except for the s states. The experimentally estimated ratios for the 4s and 5s core levels are much larger than the predicted values from the calculation, but still much smaller than those for the 4d and 4f states. On the other hand, the calculation has also predicted that the ratio for the p state depends strongly on the principal quantum number and atomic number, where the ratio should be as large as >0.8 for the 2p state of such light elements as oxygen and aluminum. The polarization dependence of the 2p core-level spectral weight for Al is shown in Fig. 8.18 as an example for light elements. The ratio $I_{\theta=90°}/I_{\theta=30°}$ for the Al 2p spectra is estimated as $\sim$0.8, which is comparable to the predicted value 0.94 from the calculation and much larger than the experimentally estimated ratios for the Au 4p and 5p excitations. These results for Al and Au suggest that the calculation of the photoelectron angular distribution for neutral single atoms well explains the polarization dependence of the HAXPES intensity for solids.

Figure 8.19 shows the polarization dependence of the valence-band spectra for polycrystalline Ag. There is strong spectral weight between $E_B = 4$ and 7 eV in both spectra at $\theta = 30°$ and 90°, which is predominantly ascribed to the Ag 4d contributions. It is found that the experimentally estimated $I_{\theta=90°}/I_{\theta=30°}$ in this energy region is consistent with the calculated ratio for the 4d excitations as shown

Table 8.1 Intensity ratio $I_{\theta=90°}/I_{\theta=30°}$ for the gold core-level excitations at the kinetic energy of ~8 keV. The parameters for the calculations are listed in [62, 63]

	$4s$	$4p_{1/2}$	$4p_{3/2}$	$4d_{3/2}$	$4d_{5/2}$	$4f$	$5s$	$5p_{1/2}$	$5p_{3/2}$
Experiment	0.07	0.11	0.09	0.2	0.24	0.4	0.1	0.16	0.13
Calculation	0.02	0.1	0.07	0.24	0.3	0.54	0.02	0.1	0.07

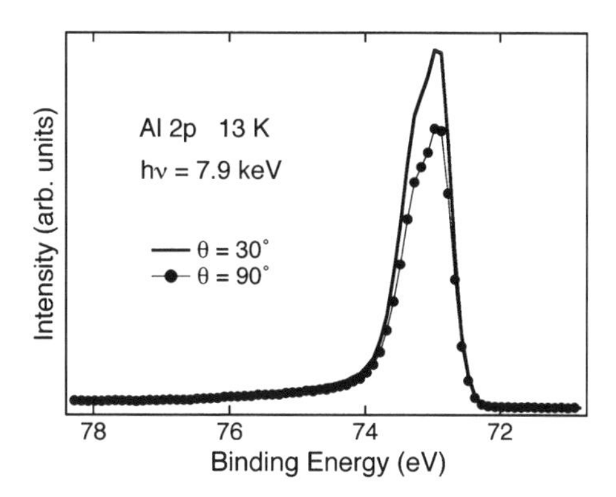

Fig. 8.18 Linear polarization dependence of the hard X-ray excited $2p$ core-level photoemission spectra of polycrystalline Al

in the upper panel of Fig. 8.19(a). The experimental ratio decreases rapidly from $E_B \sim 4$ to ~ 3 eV and then stays almost flat towards E_F, indicating that the spectral weight near E_F is strongly suppressed at $\theta = 90°$ compared with that of the 4d states located in the region of $E_B = 4-7$ eV. The intensity ratio near E_F is larger than the calculated ratio for the 5s state and slightly less than that for the 5p state. The ratio 0.17 employed for the direct comparison near E_F is anyway much smaller than 0.45 employed for the comparison of the Ag 4d state. The slope of the intensity from 2.5 eV to E_F at both $\theta = 30°$ and 90° is qualitatively consistent with that of the partial density of states (PDOSs) with s and p symmetries, which was obtained by our band-structure (local density approximation, LDA) calculation by using the WIEN2k package [66], but incompatible with that of the PDOS with d symmetry. These results reveal that the 4d bands are located far below E_F and well separated from the 5sp conduction band in an energy region from E_F to ~3 eV. It is thus experimentally confirmed that the 4d orbitals are nearly fully occupied in the solid Ag as expected for a long time. The reduction of $I_{\theta=90°}/I_{\theta=30°}$ beyond $E_B = 6.5$ eV compared with that at 4.5 eV is mainly due to the mixture of the 5s state.

The polarization dependence of the valence-band spectra for bulk Au is not only quantitatively but also qualitatively different from that for Ag, as demonstrated in Fig. 8.20. The intensity ratio $I_{\theta=90°}/I_{\theta=30°}$, which changes hardly from 5 to 2 eV in the 5d band region, decreases gradually from 2 eV to E_F without showing a rapid suppression. The ratio $I_{\theta=90°}/I_{\theta=30°}$ in the vicinity of E_F estimated to be ~0.22 is much larger than the calculated values for the 6s and 6p states. Even if $I_{\theta=90°}/I_{\theta=30°}$ for the 6s state is in fact larger than the calculated value and close to the experimentally obtained ratio ~0.1 for the 4s or 5s core-level state (Table 8.1), $I_{\theta=90°}/I_{\theta=30°}$ in the vicinity of E_F is still much larger than these values.

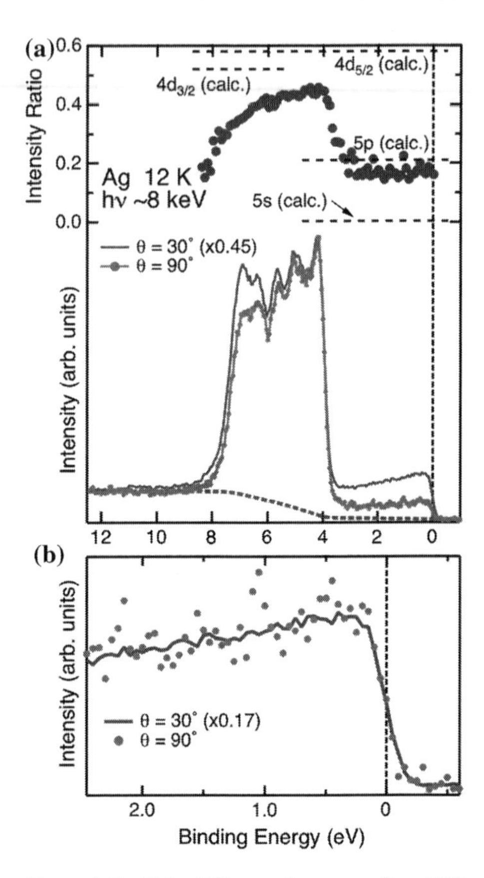

Fig. 8.19 Linear polarization dependence of the valence-band spectra of Ag [65]. **a** Spectra in the whole valence-band region. The spectral intensity at $\theta = 30°$ is scaled by multiplying 0.45 for direct comparison with the spectral intensity at $\theta = 90°$. The top graph represents the ratio $I_{\theta=90°}/I_{\theta=30°}$. The expected ratios for the $4d_{3/2}$, $4d_{5/2}$, $5s$ and $5p$ states from the calculation are shown by *dashed horizontal lines.* **b** Spectra near E_F, where the spectral intensity at $\theta = 30°$ is scaled by multiplying 0.17 for direct comparison with the spectral intensity at $\theta = 90°$

In addition, the spectral line shape from E_F to 1.5 eV is different between $\theta = 30°$ and 90° as shown in Fig. 8.20(b), which is in contrast to that for Ag in Fig. 8.19(b). While the intensity is nearly flat or slightly enhanced from 1.5 eV to E_F at $\theta = 30°$, it gradually decreases to E_F at $\theta = 90°$. The variation of the slope of intensity with θ suggests that the 5d-orbital contribution extends to the region of E_F and the additive 6sp-orbital components contribute in the spectrum at $\theta = 30°$. Namely the polarization-dependent spectra of Au near E_F are thus fully consistent with the result of the band-structure calculation. The spectrum at $\theta = 90°$ is fully reproduced by the d PDOS, while a simple sum of the s, p and d PDOS well simulates the spectrum at $\theta = 30°$. From a detailed analysis by comparison of the spectrum at $\theta = 30°$ with the results of the band-structure calculation, we have estimated the 5d weight to the total density of states at E_F as 40–60 %. On the other hand, we can estimate the 5d contribution from an analysis of $I_{\theta=90°}/I_{\theta=30°}$ in which the relative photoelectron cross sections and angular distributions are taken into account. We have successfully estimated the 5d contribution in Au as 50 ± 30 % from the analysis of $I_{\theta=90°}/I_{\theta=30°}$ in the vicinity of E_F.

The Fermi surface topology of the noble metals, which reflects the nature of the conduction electrons, deviates partially from that expected for free electrons in

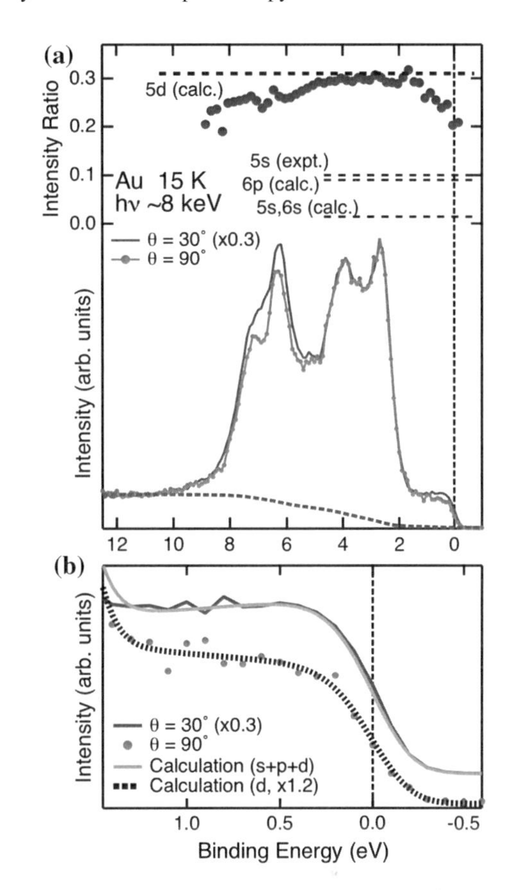

Fig. 8.20 Linear polarization dependence of the valence-band spectra of Au [65]. **a** The same as Fig. 8.19a but for polycrystalline Au. The spectral intensity at $\theta = 30°$ is scaled by multiplying 0.3 for comparison with the spectral intensity at $\theta = 90°$. The expected intensity ratios for the $5d$, $6s$ and $6p$ states from the calculation are shown by horizontal dashed lines. **b** Spectra near E_F at $\theta = 30°$ and 90° in comparison with the sum of s, p and d PDOS, and the only d PDOS obtained by the band-structure calculation, respectively

crystalline solids [67, 68]. This has been shown theoretically to be resulting from the hybridization of the (n + 1)sp band with the nd bands [69–71]. It has also been predicted that the d–sp hybridization near E_F is quantitatively stronger for Au than for Ag due to different energies and different degrees of itinerancy between the 5d and 4d bands. Indeed, the experimentally observed low E_B threshold of the 'nd bands' is closer to E_F for Au (~2 eV) than for Ag (~4 eV) as shown in Figs. 8.19 and 8.20. The strong itinerancy and d–sp hybridization for the 5d orbitals in Au are due to relativistic effects. The prominent 5d contribution to the conduction electrons in Au and the essentially negligible 4d mixture for Ag are understood as the results of the markedly different strength of d–sp mixing near E_F. On the other hand, it should be noted that such a qualitative difference of the d mixture in the conduction band crossing E_F has not been predicted from the band-structure calculations, in which additional electron correlation effects were not taken into account.

The band-structure calculations, in which all the valence-band electrons are treated as itinerant, basically give the results of conduction electrons with

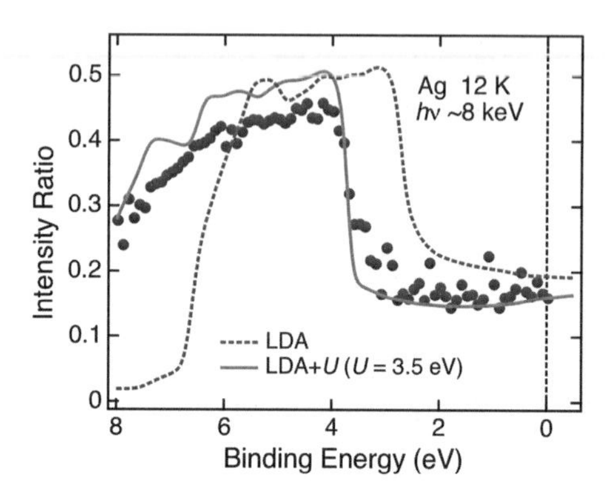

Fig. 8.21 Comparison of the experimental ratio $I_{\theta=90^\circ}/I_{\theta=30^\circ}$ (*dots*) for Ag with both the predicted ratio from the band-structure (LDA) calculation and that from the LDA + U calculation in which an on-site Coulomb interaction value $U = 3.5$ eV was chosen [65]

noticeable d–sp mixing for Ag and Au. Such a calculated result well explains the experimental spectra of Au. For Ag, however, the experimental 4d contribution in the spectra near E_F is much less than that from the calculation. This deviation can be understood by considering the finite 4d electron correlation effects in Ag. It is naturally expected that the correlated orbital contribution to the conduction electrons is suppressed due to the localization when the on-site Coulomb interactions are switched on, as seen for many rare-earth compounds. In order to verify whether the above scenario is correct for Ag or not, LDA + U-like calculations [72] have also been performed by using the linear muffin-tin orbital method [73], in which the on-site Coulomb repulsive interaction value U = 3.5 eV was applied to the 4d orbitals. Figure 8.21 shows the comparison of the ratio $I_{\theta=90^\circ}/I_{\theta=30^\circ}$ for Ag with the predicted ratios from the calculations. One can notice that the LDA + U results better explain the experimental ratio near E_F as well as in a wide valence-band region than the LDA calculation. Thus, it is concluded that the 4d electron correlation effects are responsible for its negligible contribution to the conduction electrons in Ag.

The polarization-dependent HAXPES (in other words linear dichroism in HAXPES) is powerful to resolve the orbital contributions in the valence band. An example is shown in Fig. 8.22 for polycrystalline $LiRh_2O_4$ in the metallic phase, in which the polarization dependence of the valence-band as well as O 2s HAXPES spectra are described [74, 75]. As shown in the figure, the O 2s contribution is dramatically suppressed at $\theta = 90°$. Furthermore, it is recognized that the spectral weight from $E_B = 10$ to 3 eV is also relatively suppressed at $\theta = 90°$ compared with that near the Fermi level. Since the O 2p photoionization cross-section is much weaker than those for the Rh 4d and 5s states, we conclude that the Rh 5s as well as 4d components hybridized with the O 2p states appear in the range of 10 to 3 eV.

The combination of polarization and hν dependent HAXARPES measurements may be used in the near future to unravel complicated valence band structures in various compounds.

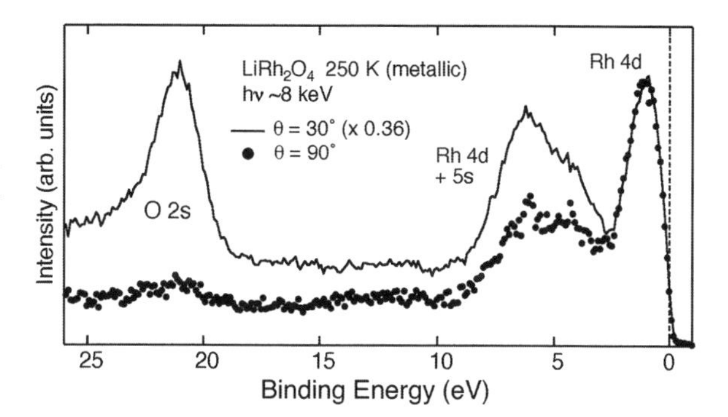

Fig. 8.22 Linear polarization dependence of the whole valence-band and O 2*s* photoemission spectra of $LiRh_2O_4$ [75]. The spectral intensity at $\theta = 30°$ is scaled by multiplying 0.36 for a direct comparison with the spectral intensity at $\theta = 90°$

References

1. K. Horiba, M. Taguchi, A. Chainani, Y. Takata, E. Ikenaga, D. Miwa, Y. Nishino, K. Tamasaku, M. Awaji, A. Takeuchi, M. Yabashi, H. Namatame, M. Taniguchi, H. Kumigashira, M. Oshima, M. Lippmaa, M. Kawasaki, H. Koinuma, K. Kobayashi, T. Ishikawa, S. Shin, Phys. Rev. Lett. **93**, 236401 (2004)
2. A. Sekiyama et al, unpublished
3. M. Taguchi, A. Chainani, K. Horiba, Y. Takata, M. Yabashi, K. Tamasaku, Y. Nishino, D. Miwa, T. Ishikawa, T. Takeuchi, K. Yamamoto, M. Matsunami, S. Shin, T. Yokoya, E. Ikenaga, K. Kobayashi, T. Mochiku, K. Hirata, J. Hori, K. Ishii, F. Nakamura, T. Suzuki, Phys. Rev. Lett. **95**, 177002 (2005)
4. A. Yamasaki, A. Sekiyama, S. Imada, M. Tsunekawa, C. Dallera, L. Braicovich, T.-L. Lee, A. Ochiai, S. Suga, J. Phys. Soc. Jpn. **74**, 2538 (2005)
5. J.J. Yeh, I. Lindau, At. Data Nucl. Data Tables **32**, 1 (1985)
6. A. Yamasaki, S. Imada, H. Higashimichi, H. Fujiwara, T. Saita, T. Miyamachi, A. Sekiyama, H. Sugawara, D. Kikuchi, H. Sato, A. Higashiya, M. Yabashi, K. Tamasaku, D. Miwa, T. Ishikawa, S. Suga, Phys. Rev. Lett. **98**, 156402 (2007)
7. D.L. Cox, Phys. Rev. Lett. **59**, 1240 (1987)
8. J.W. Allen, in *Synchrotron Radiation Research: Advance in Surface and Interface Science*, vol. 1: Techniques, ed. by R.Z. Bachrach (Plenum Press, New York, 1992)
9. S. Suga, S. Imada, H. Yamada, Y. Saitoh, T. Nanba, S. Kunii, Phys. Rev. **52**, 1584 (1995)
10. H. Ishii, K. Obu, M. Shinoda, C. Lee, Y. Takayama, T. Miyahara, T. D. Matsuda, H. Sugawara, H. Sato, J. Phys. Soc. Jpn. **71**, 156 (2002)
11. A. Yamasaki, S. Imada, T. Nanba, A. Sekiyama, H. Sugawara, H. Sato, C. Sekine, I. Shirotani, H. Harima, S. Suga, Phys. Rev. **B70**, 113103 (2004)
12. H. Harima, K. Takegahara, Physica. B **312–313**, 843 (2002)
13. A. Yamasaki, S. Imada, A. Sekiyama, M. Tsunekawa, C. Dallera, L. Braicovich, T.L. Lee, H. Sugawara, H. Sato, R. Settai, Y. Onuki, S. Suga, J. Phys. Soc. Jpn. **74**, 2045 (2005)
14. I. Felner, I. Nowik, D. Vaknin, U. Potzel, J. Moser, G.M. Kalvius, G. Wortmann, G. Schmiester, G. Hilscher, E. Gratz, C. Schmitzer, N. Pillmayr, K.G. Prasad, H. de Waard, H. Pinto, Phys. Rev. B **35**, 6956 (1987)
15. F. Reinert, R. Claessen, G. Nicolay, D. Ehm, S. Hüfner, W.P. Ellis, G.-H. Gweon, J.W. Allen, B. Kindler, W. Assmus, Phys. Rev. B **58**, 12808 (1998)
16. J.M. Lawrence, S.M. Shapiro, J.L. Sarrao, Z. Fisk, Phys. Rev. **B55**, 14467 (1997)
17. J.J. Joyce, A.J. Arko, L.A. Morales, J.L. Sarrao, H. Höchst, Phys. Rev. **B63**, 197101 (2001)
18. S. Schmidt, S. Hüfner, R. Reinert, W. Assmus, Phys. Rev. B **71**, 195110 (2005)
19. H. Sato, K. Yoshikawa, K. Hiraoka, M. Arita, K. Fujimoto, K. Kojima, T. Muro, Y. Saitoh, A. Sekiyama, S. Suga, M. Taniguchi, Phys. Rev. B **69**, 165101 (2004)

20. H. Sato, K. Shimada, M. Arita, K. Hiraoka, K. Kojima, Y. Takeda, K. Yoshikawa, M. Sawada, M. Nakatake, H. Namatame, M. Taniguchi, Y. Takata, E. Ikenaga, S. Shin, K. Kobayashi, K. Tamasaku, Y. Nishino, D. Miwa, M. Yabashi, T. Ishikawa, Phys. Rev. Lett. **93**, 246404 (2004)
21. S. Suga, A. Sekiyama, S. Imada, J. Yamaguchi, A. Shigemoto, A. Irizawa, K. Yoshimura, M. Yabashi, K. Tamasaku, A. Higashiya, T. Ishikawa, J. Phys. Soc. Jpn. **78**, 074704 (2009) (HAXPES spectra are compared with the SXPES spectra measure at 0.7 and 0.8 keV with the resolution of 70–120 meV at BL25SU)
22. K. Yoshikawa, H. Sato, M. Arita, Y. Takeda, K. Hiraoka, K. Kojima, K. Tsuji, H. Namatame, M. Taniguchi, Phys. Rev. **B72**, 165106 (2005)
23. J. Yamaguchi, A. Sekiyama, S. Imada, H. Fujiwara, M. Yano, T. Miyamachi, G. Funabashi, M. Obara, A. Higashiya, K. Tamasaku, M. Yabashi, T. Ishikawa, F. Iga, T. Takabatake, S. Suga, Phys. Rev. B **79**, 125121 (2009)
24. P.A. Alekseev, E.V. Nefeodova, U. Staub, J.-M. Mignot, V.N. Lazukov, I.P. Sadikov, L. Soderholm, S.R. Wassermann, Y.B. Paderno, N.Y. Shitsevalova, A. Murani, Phys. Rev. B **63**, 064411 (2001)
25. S. Suga, A. Sekiyama, S. Imada, A. Shigemoto, A. Yamasaki, M. Tsunekawa, C. Dallera, L. Braicovich, T.-L. Lee, O. Sakai, T. Ebihara, Y. Onuki, J. Phys. Soc. Jpn. **74**, 2880 (2005)
26. J. Yamaguchi, A. Sekiyama, S. Imada, A. Yamasaki, M. Tsunekawa, T. Muro, T. Ebihara, Y. Ōnuki, S. Suga, New J. Phys. **9**, 317 (2007)
27. N.F. Mott, *Metal-Insulator Transition* (Taylor and Francis, London, 1974)
28. M. Imada, A. Fujimori, Y. Tokura, Rev. Mod. Phys. **70**, 1039 (1998)
29. J.B. Goodenough, J. Solid State Chem. **3**, 490 (1971)
30. A. Zylbersztejn, N.F. Mott, Phys. Rev. **B11**, 4383 (1975)
31. C. Blaauw, F. Leenhouts, F. van der Woude, G.A. Sawatzky, J. Phys. C: Solid State Phys. **8**, 459 (1975)
32. S. Shin, S. Suga, M. Taniguchi, M. Fujisawa, H. Kanzaki, A. Fujimori, H. Daimon, Y. Ueda, K. Kosuge, S. Kachi, Phys. Rev. **B41**, 4993 (1990)
33. K. Okazaki, A. Fujimori, M. Onoda, J. Phys. Soc. Jpn. **71**, 822 (2002)
34. T.C. Koethe, Z. Hu, M.W. Haverkort, C. Schüßler-Langeheine, F. Venturini, N.B. Brookes, O. Tjernberg, W. Reichelt, H.H. Hsieh, H.-J. Lin, C.T. Chen, L.H. Tjeng, Phys. Rev. Lett. **97**, 116402 (2006)
35. S. Biermann, A. Poteryaev, A.I. Lichtenstein, A. Georges, Phys. Rev. Lett. **94**, 026404 (2005)
36. M.W. Haverkort, Z. Hu, A. Tanaka, W. Reichelt, S.V. Streltsov, M.A. Korotin, V.I. Anisimov, H.H. Hsieh, H.-J. Lin, C.T. Chen, D.I. Khomskii, L.H. Tjeng, Phys. Rev. Lett. **95**, 196404 (2005)
37. R. Eguchi, M. Taguchi, M. Matsunami, K. Horiba, K. Yamamoto, Y. Ishida, A. Chainani, Y. Takata, M. Yabashi, D. Miwa, Y. Nishino, K. Tamasaku, T. Ishikawa, Y. Senba, H. Ohashi, Y. Muraoka, Z. Hiroi, S. Shin, Phys. Rev. **B78**, 075115 (2008)
38. S. Suga, A. Sekiyama, S. Imada, T. Miyamachi, H. Fujiwara, A. Yamasaki, K. Yoshimura, K. Okada, M. Yabashi, K. Tamasaku, A. Higashiya, T. Ishikawa, New J. Phys. **11**, 103015 (2009)
39. M. Obara, A. Sekiyama, S. Imada, J. Yamaguchi, T. Miyamachi, T. Balashov, W. Wulfhekel, M. Yabashi, K. Tamasaku, A. Higashiya, T. Ishikawa, K. Fujiwara, H. Takagi, S. Suga, Phys. Rev. B **81**, 113107 (2010)
40. H. Fujiwara, A. Sekiyama, S.-K. Mo, J.W. Allen, J. Yamaguchi, G. Funabashi, S. Imada, P. Metcalf, A. Higashiya, M. Yabashi, K. Tamasaku, T. Ishikawa, S. Suga, Phys. Rev. **B84**, 075117 (2011)
41. S. Suga, A. Sekiyama, M. Obara, J. Yamaguchi, M. Kimura, H. Fujiwara, A. Irizawa, K. Yoshimura, M. Yabashi, K. Tamasaku, A. Higashiya, T. Ishikawa, J. Phys. Soc. Jpn. **79**, 044713 (2010)
42. U. Schwingenschlögl, V. Eyert, U. Eckern, Europhys. Lett. **61**, 361 (2003)

43. M. Sing, G. Berner, K. Goß, A. Müller, A. Ruff, A. Wetscherek, S. Thiel, J. Mannhart, S.A. Pauli, C.W. Schneider, P.R. Willmott, M. Gorgoi, F. Schäfers, R. Claessen, Phys. Rev. Lett. **102**, 176805 (2009)
44. F. Gel'mukhanov, P. Salek, H. Agren, Phys. Rev. A **64**, 012504 (2001)
45. E. Kukk, K. Ueda, U. Hergenhahn, X.-J. Liu, G. Prümper, H. Yoshida, Y. Tamenori, C. Makochekanwa, T. Tanaka, M. Kitajima, H. Tanaka, Phys. Rev. Lett. **95**, 133001 (2005)
46. W. Domke, L.S. Cederbaum, J. Electr. Spectrosc. Rel. Phenom. **13**, 161 (1978)
47. Y. Takata, Y. Kayanuma, M. Yabashi, K. Tamasaku, Y. Nishino, D. Miwa, Y. Harada1, K. Horiba, S. Shin, S. Tanaka, E. Ikenaga, K. Kobayashi, Y. Senba, H. Ohashi, T. Ishikawa, Phys. Rev. B**75**, 233404 (2007)
48. S. Suga, Appl. Phys. **A92**, 479 (2008)
49. S. Suga, A. Sekiyama, Euro. Phys. J. **169**, 227 (2009)
50. A. Sekiyama, S. Suga, Japan Unexamined Patent Publication No. 2008-170356 (2008). Jan 12, 2007 submitted, Application No. 2007-005230
51. S. Suga, A. Sekiyama, J. Electron Specrtrosc. Rel. Phenom. **181**, 48 (2010)
52. N.W. Ashcroft, N.D. Mermin, *Solid State Physics* (Saunders College, Philadelphia, 1976)
53. Y. Takata, Y. Kayanuma, S. Oshima, S. Tanaka, M. Yabashi, K. Tamasaku, Y. Nishino, M. Matsunami, R. Eguchi, A. Chainani, M. Oura, T. Takeuchi, Y. Senba, H. Ohashi, S. Shin, T. Ishikawa, Phys. Rev. Lett. **101**, 137601 (2008)
54. S. Suga, A. Sekiyama, H. Fujiwara, Y. Nakatsu, T. Miyamachi, S. Imada, P. Baltzer, S. Niitaka, H. Takagi, K. Yoshimura, M. Yabashi, K. Tamasaku, A. Higashiya, T. Ishikawa, New J. Phys. **11**, 073025 (2009)
55. S. Suga, A. Sekiyama, H. Fujiwara, Y. Nakatsu, J. Yamaguchi, M. Kimura, K. Murakami, S. Niitaka, H. Takagi, M. Yabashi, K. Tamasaku, A. Higashiya, T. Ishikawa, I. Nekrasov, J. Phys. Soc. Jpn. **79**, 044711 (2010)
56. S. Suga, S. Itoda, A. Sekiyama, H. Fujiwara, S. Komori, S. Imada, M. Yabashi, K. Tamasaku, A. Higashiya, T. Ishikawa, M. Shang, T. Fujikawa, Phys. Rev. **B86**, 035146 (2012)
57. A.X. Gray, C. Papp, S. Ueda, B. Balke, Y. Yamashita, L. Plucinski, J. Minár, J. Braun, E.R. Ylvisaker, C.M. Schneider, W.E. Pickett, H. Ebert, K. Kobayashi, C.S. Fadley, Nat. Mater. **10**, 759 (2011)
58. S. Tanuma, C. J. Powell, D. R. Penn, Surf. Interface Anal. **43**, 689 (2011)
59. M.A.V. Alvarez, H. Ascolani, G. Zampieri, Phys. Rev. **B54**, 14703 (1996)
60. C Dallera, L. Duò, L. Braicovich, G. Panaccione, G. Paolicelli, B. Cowie, J. Zegenhagen, Appl. Phys. Lett. **85**, 4532 (2004)
61. M.B. Trzhaskovskaya, V.I. Nefedov, V.G. Yarzhemsky, At. Data Nucl. Data Table **77**, 97 (2001)
62. M.B. Trzhaskovskaya, V.I. Nefedov, V.G. Yarzhemsky, At. Data Nucl. Data Table **82**, 257 (2002)
63. M.B. Trzhaskovskaya, V.K. Nukulin, V.I. Nefedov, V.G. Yarzhemsky, At. Data Nucl. Data Table **92**, 245 (2006)
64. M. Suzuki, N. Kawamura, M. Mizumaki, A. Urata, H. Maruyama, S. Goto, T. Ishikawa, Jpn. J. Appl. Phys. **37**, L1488 (1998)
65. A. Sekiyama, J. Yamaguchi, A. Higashiya, M. Obara, H. Sugiyama, M.Y. Kimura, S. Suga, S. Imada, I.A. Nekrasov, M. Yabashi, K. Tamasaku, T. Ishikawa, New J. Phys. **12**, 043045 (2010)
66. P. Blaha, K. Schwarz, G.K.H. Madsen, D. Kvasnicka, J. Luitz, *WIEN2k, An Augmented Plane Wave+ Local Orbitals Program for Calculating Crystal Properties* (Techn. Universität Wien, Vienna, 2001)
67. A.S. Joseph, A.C. Thorsen, Phys. Rev. **138**, A1159 (1965)
68. A.S. Joseph, A.C. Thorsen, F.A. Blum, Phys. Rev. **140**, A2046 (1965)
69. N.E. Christensen, B.O. Seraphin, Phys. Rev. B **4**, 3321 (1971)
70. N.E. Christensen, Phys. Status Solidi B **54**, 551 (1972)

71. O. Jepsen, D. Glotzel, A.R. Mackintosh, Phys. Rev. B **23**, 2684 (1981)
72. V.I. Anisimov, F. Aryasetiawan, A.I. Lichtenstein, J. Phys. Condens. Matter **9**, 767 (1997)
73. O.K. Andersen, O. Jepsen, Phys. Rev. Lett. **53**, 2571 (1984)
74. Y. Okamoto, S. Niitaka, M. Uchida, T. Waki, M. Takigawa, Y. Nakatsu, A. Sekiyama, S. Suga, R. Arita, H. Takagi, Phys. Rev. Lett. **101**, 086404 (2008)
75. Y. Nakatsu, A. Sekiyama, S. Imada, Y. Okamoto, S. Niitaka, H. Takagi, A. Higashiya, M. Yabashi, K. Tamasaku, T. Ishikawa, S. Suga, Phys. Rev. B **83**, 115120 (2011)

Chapter 9
Very Low Photon Energy Photoelectron Spectroscopy

In contrast to the hard X-ray PES (HAXPES) handled in Chap. 8, we call the low energy PES below h$\nu \lesssim 10$ eV as extremely low energy PES (ELEPES) in this book. According to the roughly calculated results or empirically acquired knowledge, the λ_{mp} or escape depth of photoelectrons in this very low energy region was thought to increase with the decrease in the kinetic energy (E_K). So far a variety of experiments were performed showing really large probing depth in some cases, though short λ_{mp} were still reported from coverage experiments for some materials showing the surface sensitivity even in this energy region. If a large λ_{mp} is realized in a certain material, the low energy photoelectron spectroscopy has the merit to realize very high energy resolution in addition to its bulk sensitivity.

ELEPES and angle-resolved ELEPES (ELEARPES) under steady progress in some laboratories are first introduced, demonstrating its high potential. Then the studies by use of low hν SR are shown since the performance of SR in the low hν region is also under noticeable progress. Then introduced is the small size laboratory sources such as electron cyclotron resonance rare gas lamps, which can also provide high energy resolution and high flux photons at certain hνs and are nowadays promoting the laboratory experiments. Finally the status of multiphoton spectroscopy, in particular, the two photon photoelectron spectroscopy is briefly introduced.

9.1 Angle Integrated and Resolved ELEPES by Laser Excitation

9.1.1 Angle-Integrated Measurements

Here is first described a case of extremely high resolution photoelectron spectroscopy performed slightly below h$\nu = 7$ eV by use of a high repetition laser [1]. Details of this light source were described in Sect. 3.3.e. It was widely known that many lasers could realize high energy resolutions in the ultraviolet region.

S. Suga and A. Sekiyama, *Photoelectron Spectroscopy*,
Springer Series in Optical Sciences 176, DOI: 10.1007/978-3-642-37530-9_9,

Except for the cases of two photon excitation photoelectron spectroscopies described later in Sect. 9.4 in this chapter, the minimum hν required for PES is the energy beyond the work function (ϕ) of individual materials. In order to cover proper E_B region below E_F, the h$\nu \geq \sim$6 eV is usually desirable. Although lasers in this energy region were often supplied in the form of pulsed lasers, very short pulses are not suitable to the extremely high resolution photoelectron spectroscopy, because of the uncertainty principle determining the energy resolution. For example, ~0.66 meV resolution is given by 1 ps pulsed lasers. Therefore, fs lasers were out of question for the use in high energy-resolution PES. The second problem was the space charge effect explained in Sect. 3.3.e. In order to reduce the space charge effect, a low peak intensity of the excitation laser pulse is required. Therefore, quasi-cw laser or even really cw light with high average intensity is ideal for such an experiment.

If one studies the electronic structure near E_F, the broadening of the Fermi edge by the Fermi-Dirac function (FD) up to ~4 k_BT is influential, which is ~600 μeV at 1.8 K. Therefore, the realization of low sample temperatures is one of the key issues of this kind of experiment. Some examples of angle-integrated ELEPES are shown below, first for magnesium diboride (MgB_2) in Fig. 9.1(a), which had attracted wide interest because of its quite high superconducting transition temperature T_c close to the upper limit predicted by the BCS theory [1]. It had gradually been recognized that this material had multiple superconducting gaps originating from two Fermi surfaces with different characters. With fully utilizing the sub-meV resolution of the before-mentioned laser PES, the angle integrated ELEPES was performed on the $Mg(B_{1-x}C_x)_2$ system. With changing the carbon concentration x from 0.0 to 0.075, T_c changed from 38 to 24 K. Figure 9.1(b) reproduced the results of high resolution laser ELEPES on $Mg(B_{1-x}C_x)_2$ [2]. By virtue of the very high energy resolution of the laser at hν = 6.994 eV and the electron analyzer (their sum contributes to about 400 μeV except for thermal broadening by the FD function of about 1.2 meV at 3.5 K), the two different superconducting gaps were clearly confirmed, though the structures were not so clear in the measurement by a He discharge lamp. The energy accuracy of this measurement was better than ±0.2 meV at ~3.5 K. In addition, x dependences were revealed. MgB_2 has two layers containing either Mg or B. Among the four bands crossing E_F, two bands of B 2p σ orbitals have two-dimensional (2D) character and the other two bands of B 2p π orbitals have three-dimensional (3D) character. These σ and π bands were so far observed by ARPES with hν > 21 eV [3–5]. The multiple superconducting gap was explained by the two-band model with considering the electron–phonon coupling, whose strength was thought to depend upon the symmetry of the Fermi surface. In the conventional BCS theory, the decrease in the DOS at E_F results in the reduction of T_c and the accompanying superconducting gap. The measurement was performed for $Mg(B_{1-x}C_x)_2$ polycrystalline samples with electron doping, which shows decreasing T_c with x [2]. Owing to the high resolution, two peaks were clearly observed at E_B = 7.1 and 2.6 meV for x = 0 corresponding to the superconducting gaps in the σ and π bands. The higher E_B component shifts towards smaller E_B with increasing x,

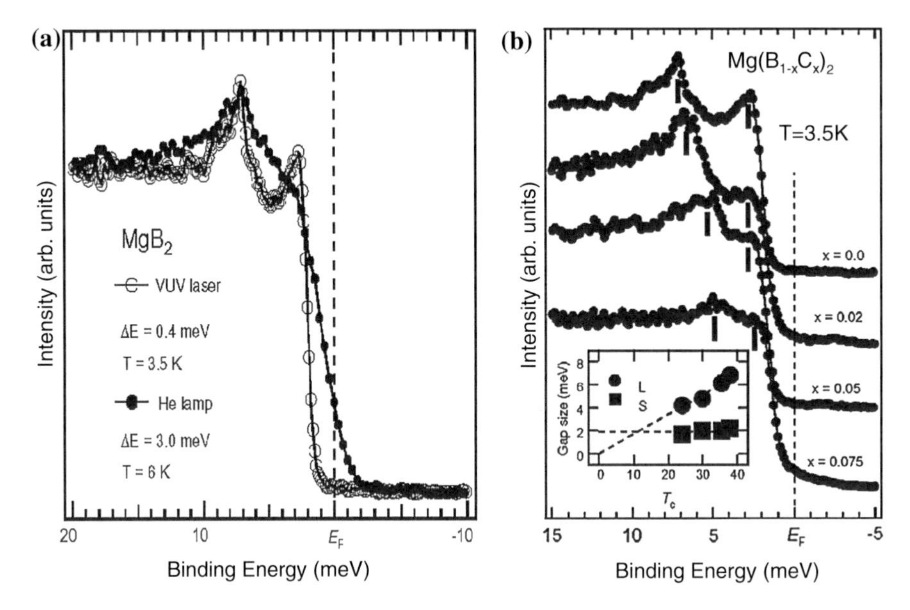

Fig. 9.1 Angle integrated PES raw spectra of polycrystalline $Mg(B_{1-x}C_x)_2$ [2]. **a** MgB_2 measured at $h\nu = 6.994$ eV and 3.5 K (*open circle*) and $h\nu = 21.2$ eV and 6 K (*filled circle*). **b** PES for $Mg(B_{1-x}C_x)_2$ with x = 0.0, 0.02, 0.05 and 0.075 at 3.5 K. Inset shows the estimated two superconducting gap sizes as a function of T_c

while the shift of the lower E_B component is much less. When the spectra are fit by two Dynes functions, the x dependence of the gap size can be evaluated as shown in the inset of Fig. 9.1(b), where the relative ratio of two functions was optimized to reproduce the experimental spectra. The larger gap (Δ_L) decreases with x, while the smaller gap (Δ_s) is almost independent of x up to 0.07. Δ_L was further noticed to be almost linear to T_c. The reduced gap of the σ band, $2\Delta_L/k_BT_c$, which is a measure of the coupling strength of Cooper pairing was found to be independent of T_c (or x). This value was 4.1 and slightly larger than the value expected in the weak coupling limit, suggesting that the electron–phonon coupling is strong in the σ band and T_c is governed by the σ band [2]. The temperature and x (doping) dependences of the two gaps were explained within a simple two-band model based on a mean field theory. Inter- and intra-band coupling constants were deduced and T_c was found to be dominated by the intraband coupling of the σ band, though the interband coupling contributes to further increase of T_c.

The next example is $CeRu_2$, which has a relatively high T_c of 6.2 K among various f-electron superconductors. The Ce 4f character is very strong at E_F [6], and the Ce 4f states hybridize with the Ru 4d states. Since the surface has much different electronic states compared with the bulk in 4f compounds [7], bulk sensitive ELEPES is as powerful as SXPES and HAXPES. Owing to the ultrahigh resolution of ELEPES and effective cooling of the sample, many materials could be tackled. The subject of superconductivity in f electron systems with an extremely small energy scale is one of the typical examples to be effectively studied by laser ELEPES. As explained in detail in [8] there were controversies on the origin of the superconductivity of this material. For example, specific heat measurements early suggested an axial symmetry with a line node but later studies reported an isotropic s-wave gap. The values $2\Delta/k_BT_c$ scattered from 3.3 to 6.6.

The measurement was performed on a high quality $CeRu_2$ single crystal surface fractured in situ at 3.8 K with a system energy resolution (analyzer and laser light) of 520 μeV. The spectra obtained on an uneven surface did not show any angular and polarization (horizontal and vertical) dependences, suggesting substantial angle integration. The work function was estimated to be 3.6 eV. The maximum of the photoelectron wave vector $k_{||}$ parallel to the surface was estimated as 0.94 $Å^{-1}$ for $\theta = 90°$. The lattice constant is a = 7.54 Å and the Brillouin wave vector is $\pi/a = 0.41$ $Å^{-1}$. Therefore, integration over the whole $k_{||}$ region in a Brillouin zone was in principle possible if the measurement with limited angular resolution was repeated by properly rotating the sample (Even in this case, the transmission probability from inside the crystal to outside the crystal depends upon the electron incidence angle to the sample surface. Such an effect should be taken into account when a really angle integrated spectrum is required from a specularly cleaved sample).

The laser ELEPES above and below $T_c = 6.2$ K is reproduced in Fig. 9.2(a), where a clear Fermi edge is observed at 8.0 K in the normal phase. A superconducting (SC) gap and a sharp quasiparticle peak at 1.35 meV are observed at 3.8 K in the SC phase [8]. In the SC phase is seen a weak peak above E_F corresponding to thermally excited electrons across the SC gap, which corresponds well to the peak in the symmetrized spectrum shown in the inset. The spectral shape was analyzed by taking the gap anisotropy effect into account in Fig. 9.2(b). Among the following three models: (1) isotropic gap, (2) anisotropic gap, (3) maximum anisotropy (nodal gap) case corresponding to a point or line node gap, the normalized spectrum in the SC phase was found to be best reproduced by the anisotropic gap model with $\Delta_{min}/\Delta_{max} \sim 0.45$. The T dependent angle averaged gap was found to be $2\Delta_{av}(0)/k_BT_c = 3.72$, suggesting that $CeRu_2$ is between the weak and strong coupling superconductors. The magnitude of Δ_{av} and the anisotropy behavior are consistent with the results of specific heat [9] and nuclear quadrupole resonance [10, 11] but inconsistent with the results of tunneling measurements [12, 13]. This is because tunneling measurements are extremely surface sensitive and easily affected by the surface condition. The consistence with other bulk sensitive measurements strongly supported the bulk sensitivity of laser ELEPES in $CeRu_2$.

Laser ELEPES is also useful to study heavy Fermion systems. It was applied to a 3d transition metal heavy Fermion (HF) material, single crystal LiV_2O_4, which was fractured in situ at 11 K [14]. The electron specific heat γ of this material is ~420 mJ/mol·K^2, which is comparable to that of typical f-electron HF compounds. The characteristic temperature T* with respect to the anomalies of magnetic susceptibility and electron specific heat is T* ~ 20 K in this material. T dependence of the spectra was measured between 70 and 11 K in the E_B range

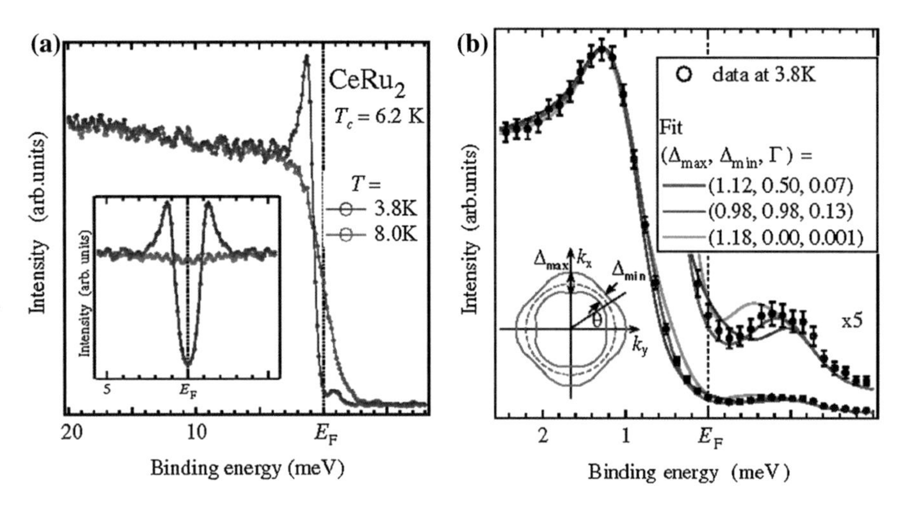

Fig. 9.2 Laser ELEPES of $CeRu_2$ fractured in situ at 3.8 K. **a** Raw spectra at 3.8 and 8.0 K. Inset shows symmetrized spectra. **b** Normalized SC DOS at 3.9 K (*dots*) compared with calculated spectra with the parameters given in the inset [8]

from −30 to +50 meV, where the intensity at E_F was always >60 % of the Fermi edge height in contrast to 50 % in the case of Au, suggesting a higher DOS above E_F. The spectra divided by the FD function convoluted by a Gauss function with a FWHM of 4 meV shows a peak above E_F at around $E_B = -4$ meV for T > 26 K. With further decreasing T, the peak tail closer to E_F increases very rapidly. These behaviors suggest that the peak corresponds to the Kondo resonance peak [14]. The bulk sensitivity was also validated by a comparison with the results of SXPES at 1.25 keV and HAXPES at 8 keV [15].

9.1.2 Angle-Resolved Measurements

One of the advantages of laser ELEPES was the small focus (≤0.1 mm) on the sample owing to its small emittance besides the high energy resolution. The sizes of newly synthesized samples are often rather small and cleavage does not always provide a large specular area with high enough homogeneity. In the case of angle resolved measurements in ELEPES (ELEARPES), sample rotation is usually necessary to cover a wide region in **k** space. The small beam spot (≤0.1 mm) on the sample thus facilitates the measurement of a small cleaved region on the sample. With high enough **k** and E resolutions, the **k**-dependent gap formation upon lowering T across the superconducting transition from the incommensurate CDW phase was better revealed than those so far reported by He lamp excitation for such samples as 2H–$NbSe_2$ ($T_c = 7.2$ K and $T_{CDW} \sim 33$ K) [16, 17]. In fact $E_k(k_{||})$ can experimentally be evaluated in laser ELEARPES with the possible best $|\mathbf{k}|$ resolution of ±0.0008 $Å^{-1}$ [1] for normal emission.

Sharp quasiparticle peaks in $Bi_{2.1}Sr_{1.9}CaCu_2O_{8+\delta}$ (Bi2212) were already reported by using 6 eV laser [18, 19] and 7.57 eV SR [20]. However, the laser ELEARPES system operated at $h\nu = 6.994$ eV could realize much higher resolution [1]. Figure 9.3(a), (b), and (c) show the results measured at T = 4 K for an over-doped Bi2212 with $T_c = 73$ K along the (0, 0)–(π/a, π/a) nodal direction (here a denotes the lattice constant with the body-centered tetragonal symmetry), where the d-wave SC gap becomes closed. MDC at $E = E_F$ and EDC at $k = k_F$ are shown in (b) and (c), respectively [1]. The widths of both MDC (0.0039 $Å^{-1}$) and EDC (4.2 meV) are the narrowest among those so far reported for Bi2212, owing to the good performance of the experimental setup. In optimally doped Bi2212 with $T_c = 90$ K, the FWHM at k_F was evaluated as 8 meV in consistence with the results of THz conductivity [21]. A long standing controversy between ARPES and transport results was solved by this experiment with extremely high **k** and E resolution.

Another advantage of the use of laser comes from its linear polarization. By inserting a λ/2 phase plate, one can easily switch the polarization to orthogonal polarization. (One can also deliver circularly polarized light by inserting a λ/4 phase plate). The s- and p-polarization dependence of ELEARPES on this optimally doped Bi2212 is shown in Fig. 9.3(d), (e), (f) and (g), where the polarization angle $\phi = 0°$ corresponds to the p-polarization perpendicular to the analyzer slit and $\phi = 90°$ corresponds to the s-polarization parallel to the sample plane. The measurement was performed parallel to the (0, 0)–(π/a, π/a) direction but shifted by ~20 % of the (π/a, 0)–(0, π/a) length towards the (0, π/a) direction. For p-polarization ($\phi = 0°$), a kink is observed near 70 meV and the top of the dispersion is seen near 20 meV at ~13° of the detector angle due to the opening of the SC gap as shown in Fig. 9.3(d). For the s-polarization, however, the dispersion in the region of $E_B = 50$ meV to E_F is shifted to ~17°. These dispersions seemed

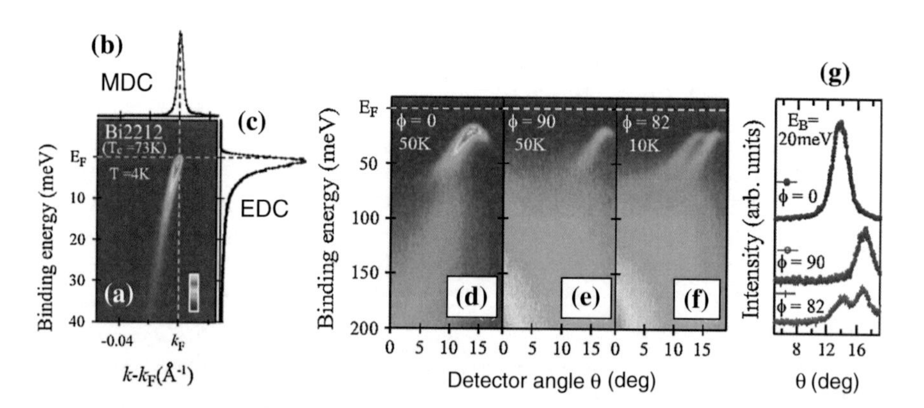

Fig. 9.3 Quasiparticle dispersion in **a**, MDC at $E = E_F$ in **b** and EDC at k_F in **c** of overdoped Bi2212 ($T_c = 73$ K) along the (0, 0)–(π/a, π/a) nodal direction at 4 K measured by laser ELEARPES at $h\nu = 6.994$ eV [1]. Results for optimally doped Bi2212 ($T_c = 90$ K) are shown in **d**–**g**. Incident light polarization was p- or s-polarization at $\phi = 0°$ or 90°. Measurements were made parallel to (0, 0)–(π/a, π/a) but slightly shifted to the (0, π/a) direction. Bilayer splitting is clearly observed for $\phi = 82°$

to correspond to the bilayer splitting or antibonding-bonding state splitting reported already by careful measurements by SR at hν = 7.57 eV [20]. The strong polarization dependence of the matrix element effects was thus demonstrated. At $\phi = 82°$, these two bands are simultaneously observed at 10 K (Fig. 9.3(f)). MDC plots at E_B = 20 meV in Fig. 9.3(g) clearly show this situation.

Later a kink-like feature in the electron self-energy was found in Bi2212 in the E_B region <10 meV separated from the well-known kink at 50–80 meV along the (0, 0)–(π/a, π/a) nodal direction [22] and its systematic doping dependence and T dependence of the nodal Fermi velocity (v_F) were studied in detail by similar laser ELEARPES [23, 24]. So far v_F measured by ARPES along the nodal direction appeared to be independent of the cuprate family or the number of CuO_2 planes in the individual compounds and nearly constant across the phase diagram, though other electronic properties varied with doping. This low energy kink feature was found, however, strongest below T_c and weakened at higher T and the v_F as measured in this range was noticed to vary rapidly with T. v_F(T) appeared to change with doping and in fact a deviation from the "universal" nodal v_F in cuprates was recognized. Measurements performed for Bi2212 with hole doping p in the range between p = 0.076 and 0.14 [24] clearly showed this low energy kink from the MCD-derived nodal dispersions in addition to the prominent kink near 70 meV. The dispersion within 10 meV from E_F deviated noticeably from that extrapolated from the region with E_B = 30–40 meV and the deviation was found to be more prominent for more underdoped sample. By considering both velocities of v_F and v_2, which are perpendicular and tangential to the Fermi surface at the node, respectively, discrepancies between doping-dependent ARPES and thermal conductivity could be resolved [24]. The origin of the low energy kink and its doping and T dependence are, however, left unsolved.

Even for studying the high energy kink feature near 400 meV and the high energy band in the 400–1,000 meV region, laser ELEARPES was applied to optimally doped Bi2212 (T_c = 91 K) [25], where MDC and high resolution second derivative-MDC and -EDC were reported along the nodal as well as some off-nodal directions. Two obvious kink features were observed near 50–70 meV and near 400 meV in the MDC-derived band structure. The latter kink shifted to smaller E_B ($\sim$230 meV) when the momentum cuts were shifted to the antinodal region. In the case of EDC-derived bands, one band was seen around E_B = 0.9 eV and the other dispersing band was observed in the smaller E_B region with a band width of 0.5 eV for the nodal direction. This band width decreased when the momentum cuts were shifted to the antinodal region. A clear difference was seen between the MDC- and EDC-derived band structure in the high E_B region beyond the high energy kink. Then the necessity of a very careful analysis of ARPES in such high E_B region was pointed out [25].

In the case of iron (Fe) pnictides superconductors, laser ARPES was applied to $BaFe_2(As_{0.65}P_{0.35})_2$ and $Ba_{0.6}K_{0.4}Fe_2As_2$ [26]. In comparison with the SR-ARPES of $BaFe_2(As_{1-x}P_x)_2$ performed in the hν region of 46–85 eV [27], laser ELEARPES has enabled one to measure the superconductivity (SC) gap with high precision and shown an orbital-independent SC gap magnitude for the hole Fermi surfaces in contrast to previous studies. The results were not expected from superconductivity associated with spin fluctuations and nesting. It was suggested that the results would be better explained by considering magnetism-induced inter-orbital pairing, orbital fluctuations, or coupling between orbital and spin fluctuations [26].

The T-dependent laser ELEARPES study of URu_2Si_2 [28] tells us the power and limitation of ELEARPES. A new quasi-particle dispersion structure appeared between E_F and E_B of 10 meV in the "hidden order" phase below 17.5 K, which could be a key to solve its mechanism, owing to the ultrahigh energy resolution of ELEARPES. On the other hand, there was relatively huge dispersive spectral weight centered at the normal angle on E_B of several tens meV, which was ascribed to be a surface contribution, requiring a careful consideration of the surface (bulk) sensitivity even in ELEPES.

9.2 ELEPES by Synchrotron Radiation

Although the $h\nu$ resolution of synchrotron radiation (SR) is worse than that of the laser source, the $h\nu$ tunability is a definite advantage of SR. Nowadays SR is often used near and below $h\nu = 10$ eV for PES. Few examples of ELEPES are treated in this section. The first example is the ELEARPES on nearly optimally doped $Bi_2Sr_2CaCu_2O_{8+\delta}$ (Bi2212) with $T_c = 86$ K. Nodal quasiparticles (QP) were studied by SR-ELEARPES at BL-9 helical undulator beam line of HiSOR with use of circularly polarized light and with a total energy resolution of 4 meV in the T range between 9 and 200 K as introduced in Sect. 3.3. ELEARPES spectra were measured at various $h\nu$ along the (0, 0)–(π/a, π/a) direction clearly revealing a kink structure at $E_B \sim 70$ meV.

One example at $h\nu = 7.57$ eV is shown in Fig. 9.4, where a quite small splitting of 0.0075 $Å^{-1}$ is revealed in the MDC. In this case the QP momentum width Δk was estimated as 0.0065 $Å^{-1}$ [20]. This doublet was identified as due to the bonding and antibonding bands of two proximate CuO_2 layers and called bilayer splitting as discussed also in Sect. 9.1. From the momentum splitting of 0.0075 ± 0.001 $Å^{-1}$ between the bonding and antibonding bands and the Fermi velocity v_F of 1.9 eV·Å, the nodal splitting energy was evaluated as 14 ± 1 meV. The spectral function is essentially unchanged below T_c but the QP peak broadens much in the normal state. The T and E_B dependence of the scattering rate was evaluated in their wide ranges, providing the T dependence of the QP renormalization energy. The deviation of dispersion from a straight line could give the real part of the self-energy, $\mathrm{Re}\Sigma'(E_B)$, and the imaginary part of the self-energy, $\mathrm{Im}\Sigma(E_B)$, was derived from the momentum width Δk. The $\mathrm{Im}\Sigma(E_B)$ show a step-like increase feature beyond $E_B \sim 70$ meV (Fig. 4 of [20]). Then $\mathrm{Im}\Sigma'_{KK}(E_B)$ derived from the Kramers–Kronig transformation of the $\mathrm{Re}\Sigma'(E_B)$ was compared with the width-derived $\mathrm{Im}\Sigma(E_B)$. Clear step features were seen in both $\mathrm{Im}\Sigma(E_B)$ and $\mathrm{Im}\Sigma'_{KK}(E_B)$ around 80 meV at 9 K and their mutual difference was ascribed to the electron–electron scattering [20].

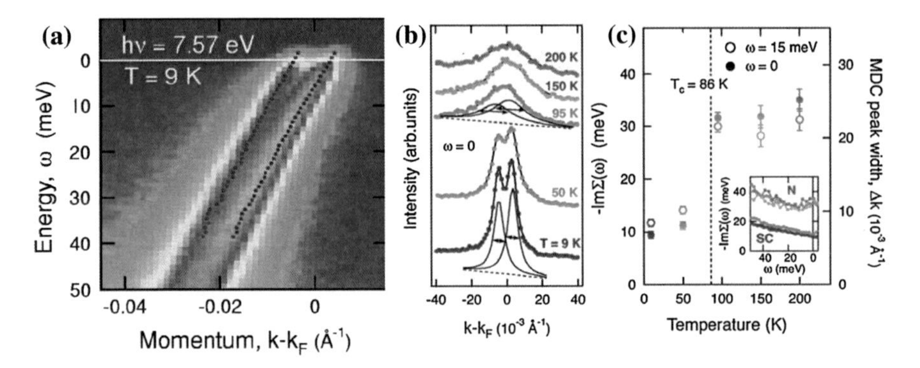

Fig. 9.4 ELEARPES of nearly optimally doped Bi2212 with $T_c = 86$ K. **a** Spectral intensity map in the $E_B : (k - k_F)$ space in the vicinity of E_F. **b** Temperature dependence of the MDC at $E_B = 0$ meV. **c** Temperature dependence of the scattering rate represented by $-\mathrm{Im}\Sigma(E_B)$ determined from the momentum width Δk at $E_B = 0$ and 15 meV. The values in the SC phase and normal phase (N) are clearly different as recognized in the inset [20]

T dependences of the MDC at E_B=0 meV and scattering rate $-\mathrm{Im}\Sigma$ are shown in Fig. 9.4(b) and (c). It was noticed in (c) that the momentum width decreases suddenly across $T_c = 86$ K from the high T side, revealing that the nodal scattering rate is suppressed in the SC phase in consistence with transport results but in contrast to early ARPES results [29]. It was also argued that the reduction of the scattering rate below T_c occurs not only for the inelastic part but also for the elastic part. Figure 9.4(c) thus suggests that the elastic scattering by impurities may be reduced in the SC phase even for the nodal QPs. As for the bilayer resolved scattering rates, the antibonding peak was broader than the bonding peak for $E_B \geq 20$ meV, though they were equivalent at $E_B = 0$ meV. The larger E_B-linear term of the scattering rate for the antibonding peak suggests that the scatterers would be outside the CuO_2 bilayer and possibly ascribable to the antisite defects or excess oxygen in the BiO and SrO layers. For such impurities far away from the conduction plane, the nodal QP might be scattered by small angles into the neighborhood of the node. Then the E_B-linear dependence could be explained within this model. Predominance of the forward scattering at low energies in the SC phase of Bi2212 was thus revealed by SR-ELEARPES [20].

The second example is $(V_{1-x}Cr_x)_2O_3$ (x = 0 and 0.011) measured with a typical instrumental resolution of 50 meV in the hν range between 5 and 26 eV [30]. The angular acceptance was set to 7°, which corresponds to 0.14 Å^{-1} at $E_K = 5$ eV. On a fresh (102) cleaved surface, no pronounced QP peak was observed in the paramagnetic metal (PM) phase of $(V_{0.989}Cr_{0.011})_2O_3$. Although the cleavage along the (001) plane was less easy, the high symmetry ΓZ direction was accessible by tuning hν even in the normal emission configuration, in which intense signal of the QP was seen between hν = 8 and 9 eV at around $E_B = 0.25$ eV in Fig. 9.5(a). The incoherent component or the LHB was also recognized at $E_B \sim 1.2$ eV in the PM phase. The QP intensity was strongest at hν = 9 eV, where the probed region was thought to be on the ΓZ axis. Apart from this region, QP intensity was found to

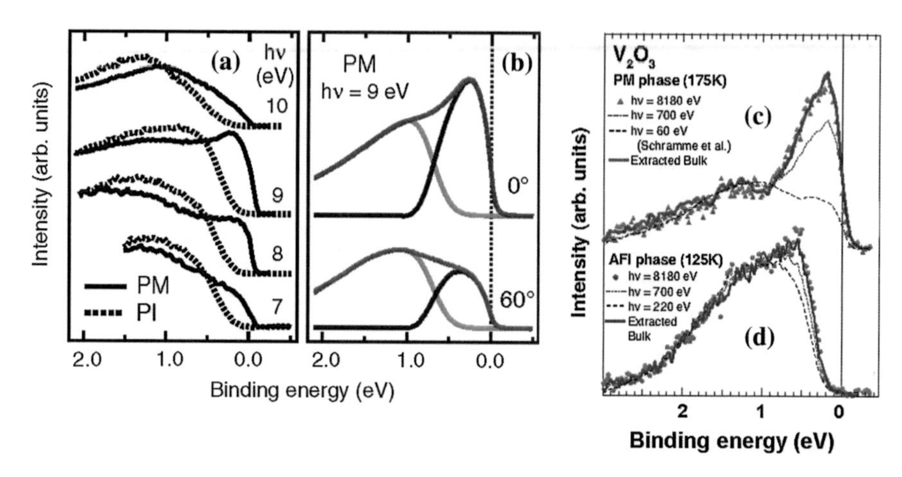

Fig. 9.5 ELEARPES **a** and **b** of $(V_{1-x}Cr_x)_2O_3$ [30] compared with bulk sensitive HAXPES **c** and **d**. Normal emission spectra of $(V_{0.989}Cr_{0.011})_2O_3$ at 300 K (PI) and 200 K (PM) from the (001) surface are shown in **a** at $h\nu = 7$, 8, 9 and 10 eV. PM spectra at 300 K and $h\nu = 9$ eV of V_2O_3 from the (001) surface at normal emission (0°) and grazing emission (60° from the normal) are shown in **b**. Figures **c** and **d** show the $h\nu$ dependence of the angle integrated spectra at higher $h\nu$ of V_2O_3 in the PM phase (175 K) and AFI phase (125 K) as a Ref. [31]

decrease drastically. Angle dependence of the spectrum at $h\nu = 9$ eV on the (001) surface of PM V_2O_3 in Fig. 9.5(b) showed that the QP intensity decreased with increasing polar angle θ from the surface normal in contrast to the much less affected intensity in the LHB region. These results suggested that the intensity of the QP in the angle integrated ELEPES spectrum (if experimentally evaluated) would be relatively much weaker than that of HAXPES [31] and even weaker than that of SXPES [32] as judged from a comparison with Fig. 9.5(c) and (d), revealing that the probing depth of ELE(AR)PES at $h\nu = 9$ eV is not longer than that at $h\nu = 300$–500 eV for the PM phase of V_2O_3. The difference between the surface and bulk electronic structure, the presence of a less itinerant surface layer on PM V_2O_3, and a much shorter λ_{mp} or probing depth of the ELEPES in the region of $h\nu = 7$–10 eV for the PM phase of V_2O_3 compared to SXPES and HAXPES were certainly confirmed in these results.

The third example of SR-ELEPES is on Fe_3O_4 which shows the Verwey transition at $T_V \sim 123$ K. The ELEPES experiments performed at $h\nu = 7.5$ eV in the T range between 140 and 330 K are shown in Fig. 9.6(a) and compared with HAXPES performed at 100–330 K shown in Fig. 9.6(b), (c) [33]. The ELEPES and HAXPES intensity at E_F is still negligible at 140 K even in the high T region just above T_V in consistence with the SXPES results on fractured surfaces [34]. Its intensity at E_F becomes, however, no more negligible at 250 and 330 K. The spectral behavior within $E_B = 0.5$ eV from E_F in 7.5 eV ELEPES is fully consistent with that of HAXPES at $h\nu \sim 8$ keV, demonstrating the bulk sensitivity of ELEPES in the case of Fe_3O_4 with low DOS near E_F. The spectral behavior near E_F in the high T phase above T_V could be well explained by a polaronic model

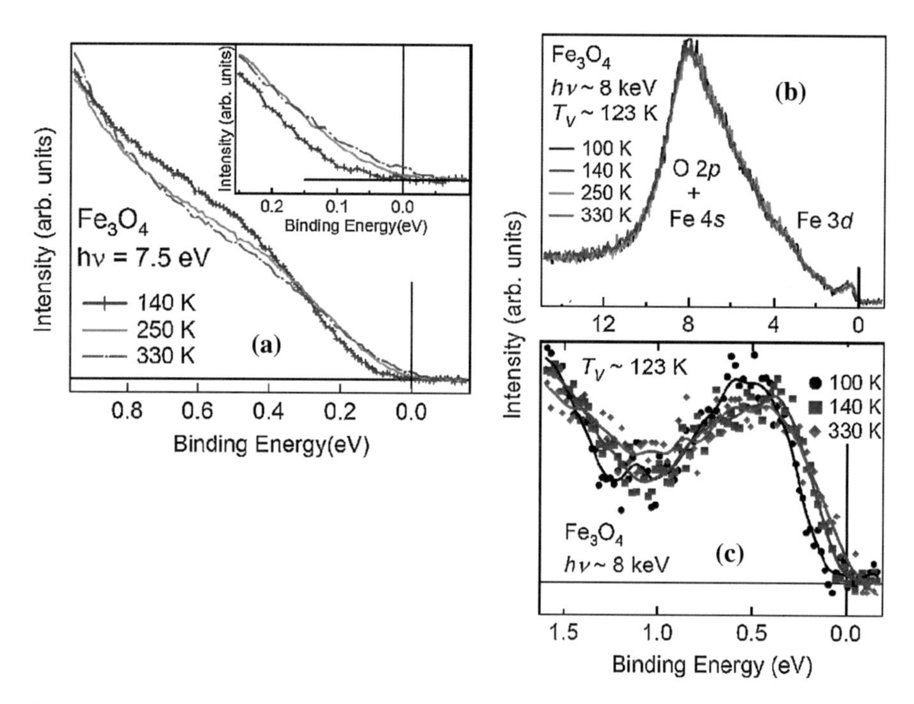

Fig. 9.6 **a** ELEPES at hν = 7.5 eV of Fe_3O_4 fractured in situ at 140 K and measured at 140, 250 and 330 K. **b** HAXPES of Fe_3O_4 measured at ~ 8 keV in a wide E_B range at 100, 140, 250 and 330 K. **c** Expanded spectra near E_F [33]

showing a change from small polarons to large polarons with increasing T through 250–300 K [33].

9.3 ELEPES by Microwave Excited Rare Gas Lamp

As already mentioned in Sect. 3.3.d, the microwave excited electron cyclotron resonance rare gas (Xe, Kr and Ar) lamp can supply very strong and highly energy resolved (<600 μeV) cw photons at hν =~8.4, ~10.0 and ~11.6 eV [35–37]. Since the sample-analyzer chamber can be separated from the light source by a LiF (or MgF_2 or CaF_2) window, measurements can be performed for tens of hours on a clean sample surface without noticeable contamination even at low T. If a MgF_2 or a CaF_2 window is used, Ar line cannot transmit through the window. Some examples of experiments are introduced below.

The first report is for Yb metal excited by the Xe 8.4 eV line [35]. The surface core level shift of the Yb $4f_{7/2}$ component was widely known on clean Yb surfaces. Although the intensity of the surface component was as strong as the bulk component at 21.2 eV, it became much weaker but still clearly observable at hν = 8.4 eV. According to more systematic analyses of the surface component relative to the bulk

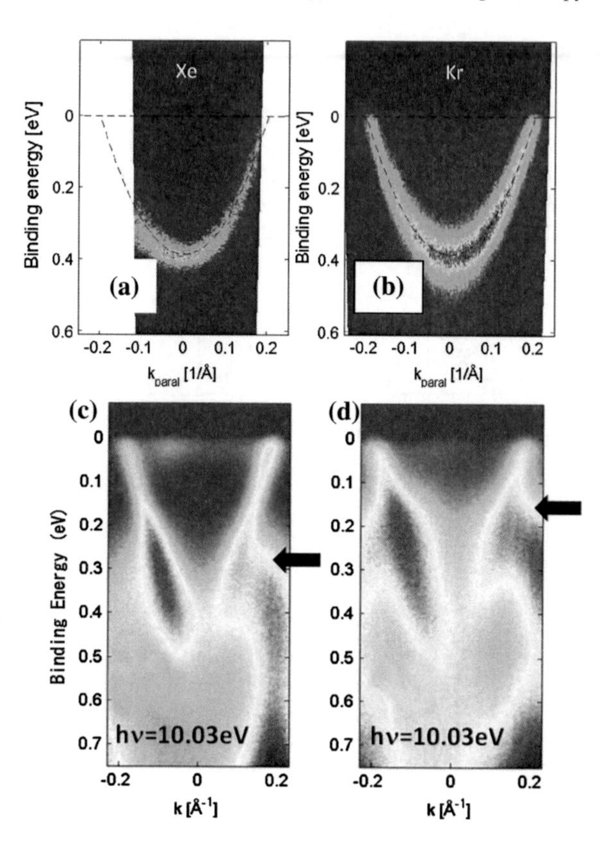

Fig. 9.7 ELEARPES of a Cu(111) clean surface measured at 90 K by Xe (8.4 eV) and Kr (10.0 eV) [37]. An analyzer pass voltage of 2 V and a 0.2 mm slit width were employed using a SCIENTA SES200 analyzer. A Xe spectrum took 5 min. A Kr spectrum took longer integration time because of the absorption of photons by the MgF_2 film coating on the Al toroidal mirror. **c** and **d** ELEARPES of 10 quintuple layer (QL) and 8.8 QL Bi_2Te_3 prepared by cycles of Ar-ion sputtering and annealing at 250 °C and measured at hν = 10.0 eV and 90 K [40]. The V-shaped dispersion represents the Dirac cone. The arrows show the shift of the possible second dispersive branch of the surface states

component, the probing depth or λ_{mp} at hν ~ 8 eV was estimated to be around 5–6 Å in Yb metal [38]. This value is almost 1/4 of the value theoretically predicted [39] and suggests that the ELEPES down to hν ~ 7.5 eV is not bulk sensitive in the case of Yb, even though λ_{mp} increases gradually with decreasing hν.

The next examples are Cu(111) and the topological insulator Bi_2Te_3. ELEARPES of a clean Cu(111) surface prepared by sputter-annealing cycles and measured at 90 K and hν = 8.4 and 10.0 eV are shown in Fig. 9.7(a) and (b) [37]. In this configuration and energy region, there are no bulk states accessible by the present ELEARPES. The high counting rate means the high surface sensitivity in this case. The results on MBE-grown Bi_2Te_3 films on n-type Si(111) wafers cleaned by cycles of Ar-ion sputtering at 500 eV and subsequent annealing at 250 °C were shown in Fig. 9.7(c) and (d) along the $\overline{\Gamma}\,\overline{K}$ axis [40]. The V-like dispersion is due to the Dirac cone characteristic of the surface of topological insulators. The bulk conduction band feature as well as the bulk valence band feature were observed in the range of film thicknesses between 16 and 6 Bi_2Te_3 quintuple layers (QL) by HeI excitation ARPES (hν = 21.2 eV) (Fig. 2 of [40]). However, E_B of the most pronounced bulk band feature around E_B ~ 1.25 eV was unchanged in this thickness region. Therefore, the shift of the hν = 10.0 eV

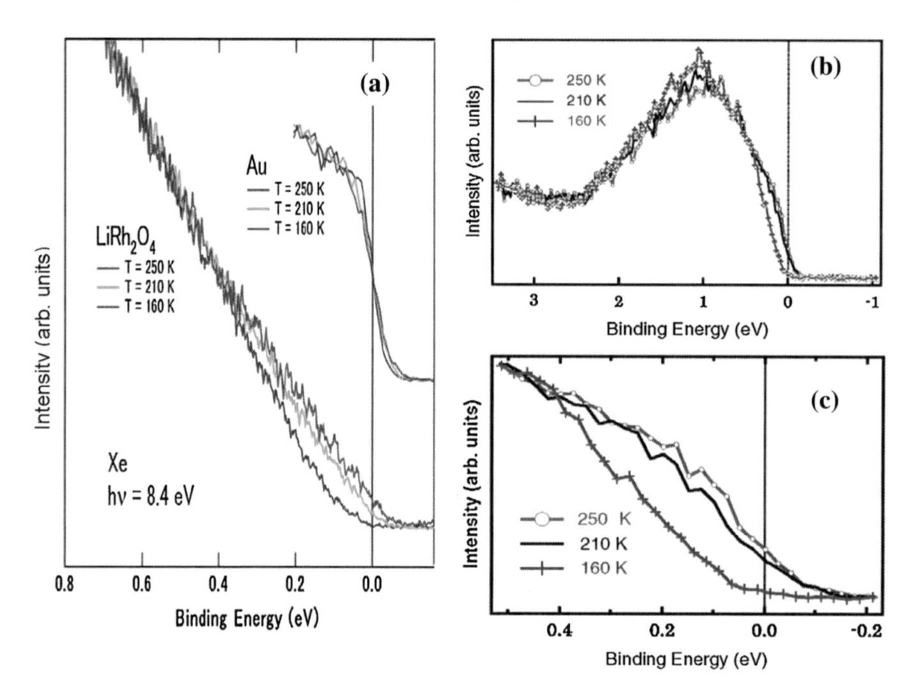

Fig. 9.8 ELEPES of $LiRh_2O_4$ at hν = 8.4 eV **a** [41]. HAXPES at hν = 8,180 eV **b** at 160, 210 and 250 K. Expanded spectra near E_F in **c**. The metal–insulator transition takes place at 170 K

ELEARPES feature at ~0.28 eV for 10 QL to ~0.15 eV for 8.8 QL film as indicated by horizontal arrows in Fig. 9.7(c), (d) was revealed not to be related to bulk bands but related to the second dispersive branch of the surface state. It was demonstrated in both cases of Cu(111) and Bi_2Te_3 that ELEARPES can be used as a surface sensitive technique if the bulk states are not accessible or not predominating under the given experimental conditions.

The fourth example is the spinel type compound $LiRh_2O_4$, which shows two step MITs via an intermediate phase. A transition from an orbital-disordered metal phase to an orbital-ordered metal phase occurs at 230 K and a charge-ordered valence bond solid (insulator phase) forms below 170 K. ELEPES at hν = 8.4 eV measured on an in situ fractured polycrystal is shown in Fig. 9.8(a) at 250 K (metal), 210 K (intermediate phase) and 160 K (insulating phase). It is clearly observed in (a) that the spectral intensity decreases on approaching E_F from E_B = 0.8 eV and the intensity just at E_F decreases through 250, 210 and 160 K, where no intensity is seen at E_F at 160 K, confirming the occurrence of the MIT transition between 210 and 160 K [41]. In the HAXPES in Fig. 9.8(b) and (c), recoil effects on the photoelectron emission were not observed in $LiRh_2O_4$. In HAXPES, Rh 4d states were predominating in the valence band because its PICS is two orders of magnitude larger than that of the O 2p state. A possible recoil effect of ~20 meV could be almost negligible in Fig. 9.8(b) and (c) even if it took place. The bulk sensitive results in Fig. 9.8(b) and (c) are consistent with the result

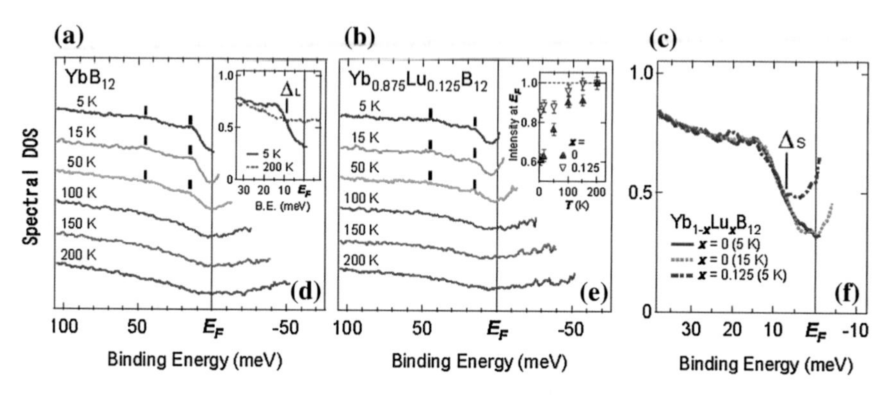

Fig. 9.9 Temperature dependence of DOS of $Yb_{1-x}Lu_xB_{12}$ derived from ELEPES performed at $h\nu = 8.4$ eV between 200 and 5 K on in situ fractured single crystals [42]

of ELEPES in Fig. 9.8(a), confirming the bulk sensitivity of ELEPES in $LiRh_2O_4$ with a rather low density of states near E_F.

The fifth example is on $Yb_{1-x}Lu_xB_{12}$ single crystals cleaved in situ in UHV [42]. YbB_{12} is a well known valence fluctuation material, which behaves as a metal at high T with localized f magnetic moments whereas a narrow gap opens at E_F below the characteristic temperature T* ~ 80 K showing semiconducting behavior. The gap of a YbB_{12} single crystal was so far discussed from bulk sensitive experiments such as transport, optical and inelastic neutron scattering. Two types of gaps with different magnitude were reported in this material from electrical resistivity measurements [43], where the large gap Δ_L and the small gap Δ_S were evaluated as ~12 meV for $15\ K < T < 40\ K$ and ~4 meV for $T < 15$ K. The small gap was interpreted to be due to an in-gap state within the large gap. In the HAXPES studies of $Yb_{1-x}Lu_xB_{12}$ discussed in Sect. 8.4, it was demonstrated that the Yb 4f lattice coherence collapses by Lu substitution already for x = 0.125 [44]. Optical studies independently showed the gap collapse for x = 0.125 [45]. ELEPES with high enough energy resolution was then employed to study the relation between the gap formation and the Yb 4f lattice coherence effect in $Yb_{1-x}Lu_xB_{12}$.

The ELEPES spectra were measured on both YbB_{12} and $Yb_{0.875}Lu_{0.125}B_{12}$ single crystals at T between 200 and 5 K with a total energy resolution of 6 meV. The observed EDC near E_F are dominated by the non-4f states, namely by the B 2sp and Yb 5d states hybridized with the Yb 4f states, where two peaks are observed at ~45 and ~15 meV for $T < 50$ K and the intensity at E_F decreases gradually with the decrease in T. By dividing the EDC spectra with the FD function convoluted with the instrumental resolution, DOS information becomes available as shown in Fig. 9.9. The gap in YbB_{12} is evaluated as ~10 meV in the inset of Fig. 9.9(a) in consistence with the result of Δ_L reported from electrical resistivity [43]. It was also noticed that the reduction of the intensity at E_F becomes steep below T* ~ 80 K for both x = 0 and 0.125. The degree of intensity decrease is, however, much less for x = 0.125 compared with that for x = 0 (inset of Fig. 9.9(b)). This situation is very clear in Fig. 9.9(c) as the intensity within ~7 meV from E_F at 5 K is much higher for x = 0.125. This energy scale may correspond to Δ_S, suggesting that the small gap collapsed in $Yb_{0.875}Lu_{0.125}B_{12}$ (x = 0.125) at low T below 50 K while the large gap remained still with the Lu doping of x = 0.125. The collapase of the 4f lattice coherence for x = 0.125 was already revealed by HAXPES. From these results, we could conclude that ELEPES is bulk sensitive in $Yb_{1-x}Lu_xB_{12}$, where the small gap collapses when the 4f lattice coherence is broken but the large gap survives because it is due to the 4f single-site effect [42].

9.4 Two Photon Excitation Photoelectron Spectroscopy

Most photoelectron spectroscopies are performed by one photon excitation with hν higher than the work function. Even when hν is lower than the work function, photoelectron excitation is possible by multiphoton excitation processes. Short pulse lasers with high photon flux can be used for such purposes. Studies of ultrafast dynamical processes and relaxation processes of charge carriers are feasible due to the short pulse width (often in <100 fs range) of such lasers. Time-resolved two-photon photoemission (TR-2PP) spectroscopy is one of several techniques for studying ultrafast charge carrier dynamics in the bulk, interface and surface regions. The time delay between the pump and probe laser pulses can be arbitrarily changed. In addition, the excited states which cannot be accessible by one photon excitation due to the dipole selection rules are accessible by two photon excitations. Here the application of two photon excitation PES is briefly overviewed. TR-2PP was performed for nearly 3 decades. Initially the time resolution was in the ps range and then came into the sub ps and fs range. Electron–phonon scattering in metals, dynamics of thermalized band edge carriers in semiconductors, unoccupied bands and surface states in metals as well as image potential states were so far extensively studied. The modification of the electronic states and carrier dynamics by chemisorption and/or physisorption was also widely studied. The coherence between the electric field of the excitation pulse and the excited electron–hole polarization was also an interesting subject.

An example of a TR-2PP experimental set up is illustrated in Fig. 9.10 [46]. A commercial Ti:sapphire laser provides high intensity light with a typical frequency of 80 MHz and a pulse width <100 fs at hν ~ 1.25–1.75 eV. By use of non-linear optical crystals such as β-BaB_2O_4 (abbreviated as BBO), the second harmonic generation (SHG) is realized. Then the second harmonic in hν = 2.5–3.5 eV becomes available. Pump and probe experiments are feasible using a Mach–Zehnder interferometer by splitting the SHG pulse with a beam splitter (BS) and recombining them by another BS at the exit of the interferometer after a variable delay. An extreme resolution down to 50 attoseconds (as) due to 15 nm accuracy using a piezoelectric drive is feasible in this system. The polarization of the incidence light onto the sample in the analyzer chamber can be controlled by use of a $\lambda/2$ or $\lambda/4$ phase plate. In the TR-2PP, the pump and probe light pulses with

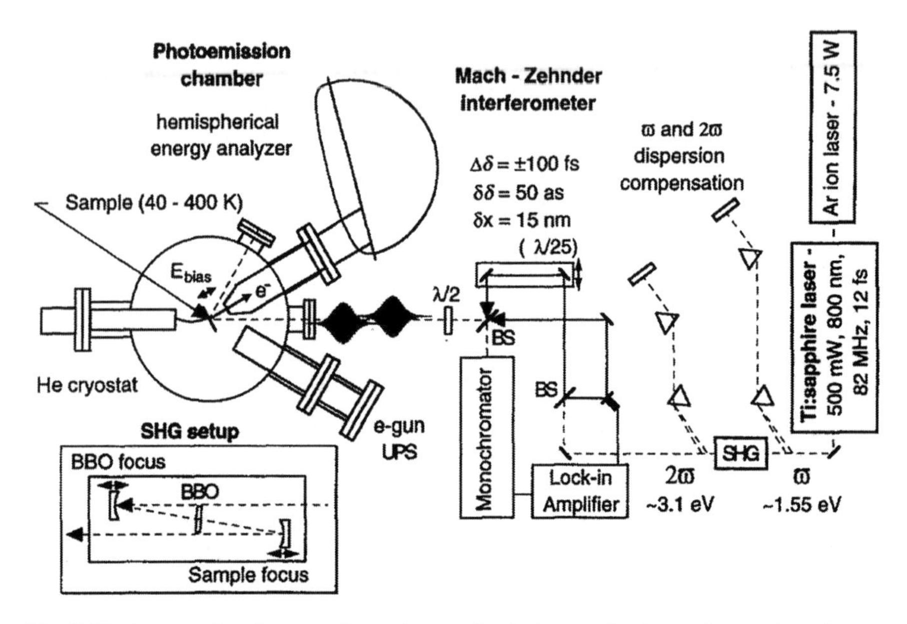

Fig. 9.10 An example of an experimental setup for fs time resolved two-photon photoelectron spectroscopy (TR-2PP) [46]. The inset (*left down*) shows the details of the second harmonic generator (SHG) with use of a 80 μm thick β-BaB_2O_4 (BBO) crystal

certain time delay can have different energies, different polarizations and different incidence directions but should be incident on the same area of the sample with controlled focusing, for which high skill is required.

The uncertainty principle provides the relation $\Delta E \cdot \Delta t \geq h/2\pi = 0.66$ meV·ps, where ΔE is the energy width and Δt the pulse width. So the energy resolution becomes 6.6 meV for a 100 fs pulse and 66 meV for a 10 fs pulse. This relation should be taken into account when one uses short pulses for a high energy resolution experiment, though it is not really necessary for ps pulses. The electron energy analysis is performed by a conventional hemispherical electron analyzer. The resolution of 100 meV is easily achieved by an analyzer with a radius of 100 mm. Although this type of analyzer is very convenient for high repetition excitation, the time of flight measurement is superior to this for low repetition excitation like 2PP spectroscopy. For low kinetic energy electrons, the time of flight experiment can realize better energy resolution. For a typical flight length of 30 cm, an energy resolution of 5 meV can easily be achieved by a time resolution of 0.1 ns for 1 eV electrons. By using the 2 dimensional detector, even angle resolved measurements are feasible.

As discussed in the next chapter (Chap. 10), the image potential state located below the vacuum level was mostly studied by inverse photoemission spectroscopy. However, it was also accessible with higher accuracy ($\sim$40 meV) by the 2PP spectroscopy. A frequency doubled dye laser pumped by an excimer laser or a frequency tripled Ti:sapphire laser pumped by an Ar laser was used for the excitation of the image potential states in W(110) as well as bcc Fe(110) films on W(110) and Cu(100) substrates [47]. The polarization of the light was easily controlled by Pockels cells or proper phase plates. By using a proper hν, even the n = 2 image potential state is clearly resolved in addition to the n = 1 state by this technique as shown in Fig. 9.11(a). When hν is reduced, the n = 2 state is no more observable and only the n = 1 state is detected. In the magnetic materials, the band structures are different for different spin states. The energies of the image potential states depend also on the spin orientations. Therefore, one could evaluate the exchange splitting of the image potential states in Fe/W(110) and Fe/Cu(100) as 57–85 meV.

Later many studies were reported on image-potential states and the dynamics on clean and adsorbate-covered metal surfaces. In the case of highly oriented pyrolitic graphite (HOPG) [48] and Cu (100) [49], image-potential states up to n = 3 were clearly observed by 2PP (Fig. 9.11(b) for HOPG) due to the improved energy resolution. In the case of an Ar layer on Cu(100), the excited electrons were attracted towards the metal by imaging forces while the projected band gap of Cu(100) hindered them to penetrate into the metal. Then states analogous to the image potential states on a Cu clean surface were formed in the interface region. This buried interface state under the Ar layer as thick as 100 Å could be detected by 2PP spectroscopy. After the interface states with $n' = 1$ and 2 were populated with electrons from the metal by the pump laser pulse, the probe laser pulse promoted the electrons into the Ar conduction band, from where the electrons escaped into the vacuum. The third harmonic of the Ti:sapphire laser at hν = 4.71 eV was used for pumping and the fundamental pulse at hν = 1.57 eV was used for the time delayed probe pulse [49]. First the lifetime of the interface state was found to increase with the Ar thickness up to 15 ML (45 Å) and then stays constant around ~100 fs above this thickness. 2PP spectra of Ar/Cu(100) interface thus measured for different emission angle or $k_{||}$ are shown in Fig. 9.11(c), where the inset shows the dispersion along $k_{||}$ for the interface state ($n' = 1$) in comparison with the n = 1 image potential state of clean Cu(100). The effective mass m_{eff} for the interface state was 0.61 in contrast to 0.98 of the image potential state of Cu(100). Since image-potential states on clean Cu(100) are mainly located in the vacuum side in front of the surface, they might disperse with the mass very close to the free-electron mass. On the other hand, the interface states are strongly influenced by the conduction band minimum of bulk Ar. The small m_{eff} inside the Ar film is a consequence of the atomic potential corrugation effect. Figure 9.11(d) shows the time resolved 2PP intensity of the $n' = 1$ interface state for 3 different $k_{||}$. In comparison with the life time of the n = 1 image potential state of 40 fs, the $n' = 1$ interface state had a much longer life time but it decreased with increasing $k_{||}$ (111, 100 and 84 fs at $k_{||} = 0$, 0.11 and 0.18 Å^{-1}). Information on the inter- and intra-band decay processes might be obtained from these results.

Although spin polarized PES is later discussed in Sect.11.3, spin polarized fs TR-2PP is introduced here on this occasion. It was applied to a 3 ML Fe film on Cu(100) with a total energy resolution of 90 meV, an angular resolution of ±2.5°

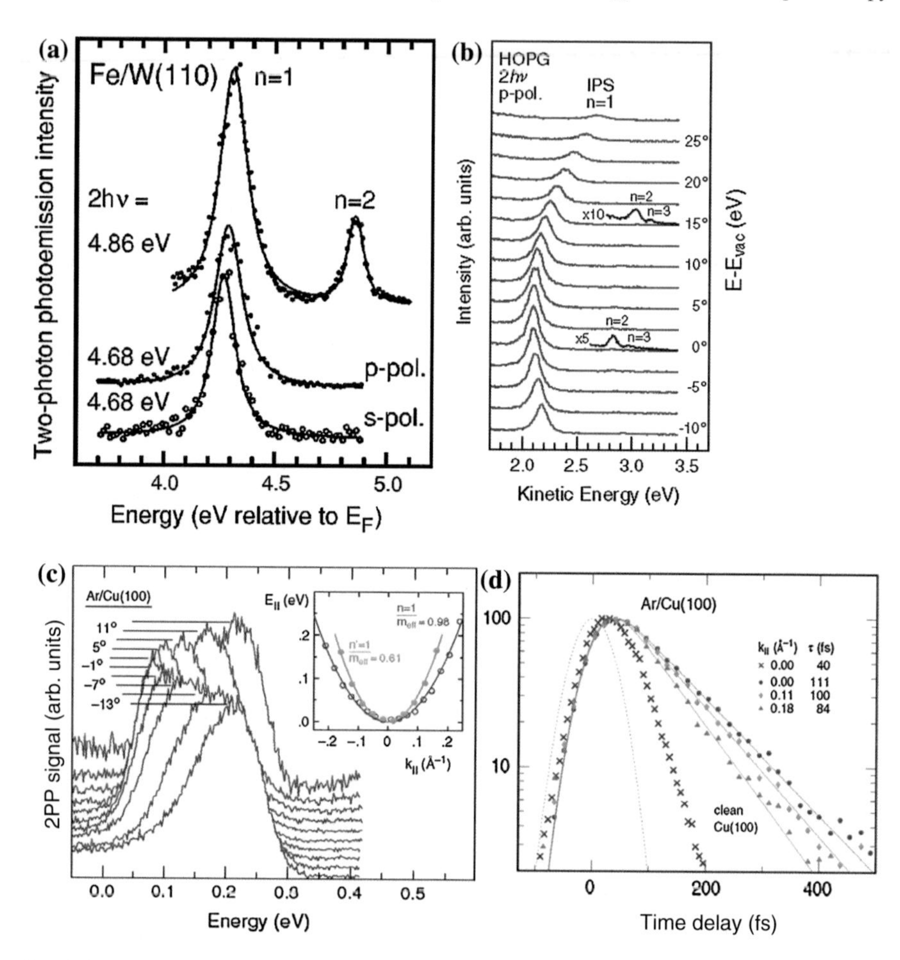

Fig. 9.11 Two-photon photoemission spectra of Fe(110) on a W(110) substrate excited by a frequency doubled dye laser with an energy of 2 h*ν* to populate the image-potential states **a** [47]. Image potential states in HOPG probed by 2PP **b** [48]. 2PP spectra of the Ar/Cu(100) interface measured at different emission angles are shown in **c** [49], where the inset shows the dispersion along $k_{||}$ for the interface state ($n' = 1$) and the $n = 1$ image potential state of clean Cu(100). $m_{eff}(n' = 1)$ was evaluated as 0.61 in contrast to $m_{eff}(n = 1) = 0.98$. **d** Time dependence of the $n' = 1$ Ar/Cu(100) interface state intensity (full marks) compared with that of the $n = 1$ image potential state of Cu(100) (*violet cross*) for different $k_{||}$

and a count rate of a few kHz [50]. 2PP was performed by use of a third harmonic Ti:sapphire laser pump (55 fs, 4.68 eV, 10^{18} photons/s) and fundamental laser probe (40 fs, 1.56 eV, 8×10^{15} photons/s). Pulses were focused to a spot of 80 μm diameter. Spin was detected by a W-LEED detector. Exchange splitting of both the $n = 1$ and $n = 2$ image-potential states was clearly resolved as seen in Fig. 9.12 to be 56 ± 10 and 7 ± 3 meV. In this experiment the delay between the pump and probe was set to zero. In addition, a strong dependence of the spin-polarization on the polarization of the pump laser was revealed. Spin polarization

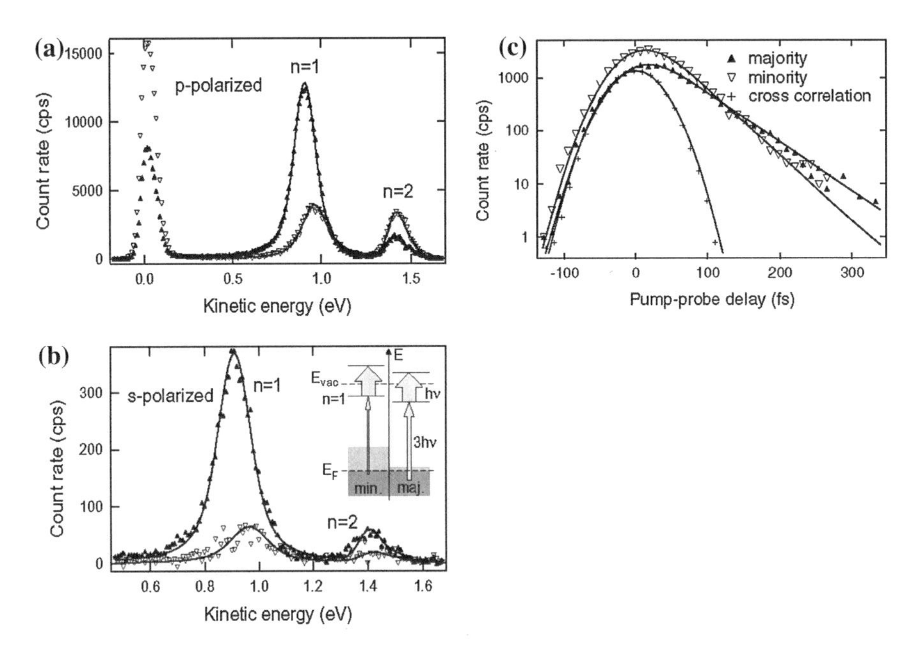

Fig. 9.12 Spin resolved 2PP spectra measured in normal emission pumped at 3 hν = 4.75 eV and probed at hν = 1.58 eV without mutual delay. Majority (minority) spin is shown by full (empty) *triangles*. **a** p-polarized pump and probe [50]. **b** s-polarized pump and p-polarized probe. **c** Time and spin resolved 2PP for the majority and minority spin components of the n = 2 image potential state

for the n = 1 and 2 image potential states in Fig. 9.12(a) was 47 % and −33 % and that in Fig. 9.12(b) was 66 and 45 %. The lifetime of the excited majority spin electrons was found to be 1.4 times longer than that of the minority spin electrons (16 ± 2 to 11 ± 2 fs). With increasing the delay between probe and pump laser, the intensity of the minority spin state dropped faster than the majority spin state as revealed in Fig. 9.12(c). The lifetime of the n = 2 image potential states was found to be 47 ± 3 (35 ± 3) fs for the majority (minority) spin electrons in this case. The major decay channel for image potential state electrons is the inelastic electron–electron scattering. The shorter lifetime of the minority spin electrons might be due to the larger number of unoccupied states. Spin, momentum and time resolved 2PP spectroscopy with high energy resolution and high count rate has a bright future in materials sciences.

A momentum microscope explained in Sect. 3.6.5 was effectively used for the $k_x - k_y$ simultaneously resolved multiphoton ARPES [51]. Stimulating results were shown for the occupied and unoccupied Shockley surface state of Cu(111) excited by multiphotons (two and three photons) using the second harmonic of a Ti:sapphire fs laser with a single photon energy of hν = 3.1 eV and a pulse width of ~20 fs. In the case of 2PP, excitation at 6.2 eV was feasible and the occupied surface state as well as the sp–sp bulk transition was observed. In the case of 3PP, the unoccupied Shockley surface state was observed in addition to the n = 1 image potential state. The dispersion obtained from these experiments was also compared with the dispersion obtained by Fourier transformed scanning tunneling spectroscopy (FT-STS) explained later in Chap. 14. The dispersion of the occupied and unoccupied Shockley surface states of Cu(111) deviated noticeably from the paradigmatic parabolic behavior of quasi-free electrons with increasing $k_{||}$ after crossing E_F and such a behavior was ascribed to the shift of the spectral weight of the surface state into the bulk resulting from the enhanced hybridization with increasing energy [51].

References

1. T. Kiss, T. Shimojima, K. Ishizaka, A. Chainani, T. Togashi, T. Kanai, X.-Y. Wang, C.-T. Chen, S. Watanabe, S. Shin, Rev. Sci. Instrum. **79**, 023106 (2008)
2. S. Tsuda, T. Yokoya, T. Kiss, T. Shimojima, S. Shin, T. Togashi, S. Watanabe, C.Q. Zhang, C.T. Chen, S. Lee, H. Uchiyama, S. Tajima, N. Nakai, K. Machida, Physica C **460–462**, 80 (2007)
3. H. Uchiyama, K.M. Shen, S. Lee, A. Damascelli, D.H. Lu, D.L. Feng, Z.-X. Shen, S. Tajima, Phys. Rev. Lett. **88**, 157002 (2002). 28 eV
4. S. Souma, Y. Machida, T. Sato, T. Takahashi, H. Matsui, S.-C. Wang, H. Ding, A. Kaminski, J.C. Campuzano, S. Sasaki, K. Kadowaki, Nature **423**, 65 (2003). 28 eV
5. S. Tsuda, T. Yokoya, T. Kiss, Y. Takano, K. Togano, H. Kito, H. Ihara, S. Shin, Phys. Rev. Lett. **87**, 177006 (2001)
6. A. Sekiyama, S. Suga, S. Ueda, S. Imada, K. Matsuda, T. Iwasaki, M. Hedo, E. Yamamoto, Y. Haga, Y. Onuki, Y. Saitoh, T. Matsushita, T. Nakatani, M. Kotsugi, S. Tanaka, H. Harima, Solid State Commun. **121**, 561 (2002)
7. A. Sekiyama, T. Iwasaki, K. Matsuda, Y. Saitoh, Y. Onuki, S. Suga, Nature **403**, 396 (2000)
8. T. Kiss, F. Kanetaka, T. Yokoya, T. Shimojima, K. Kanai, S. Shin, Y. Onuki, T. Togashi, C. Zhang, C.T. Chen, S. Watanabe, Phys. Rev. Lett. **94**, 057001 (2005)
9. M. Hedo, Y. Inada, E. Yamamoto, Y. Haga, Y. Ōnuki, Y. Aoki, T.D. Matsuda, H. Sato, S. Takahashi, J. Phys. Soc. Jpn. **67**, 272 (1998)
10. K. Ishida, H. Mukuda, Y. Kitaoka, K. Asayama and Y. Onuki, Z. Naturforsch. **51a**, 793 (1996)
11. H. Mukuda, K. Ishida, Y. Kitaoka, K. Asayama, J. Phys. Soc. Jpn. **67**, 2101 (1998)
12. W. Schmitt, G. Guntherodt, J. Magn. Magn. Mater. **47–48**, 542 (1985)
13. H. Sakata, N. Nishida, M. Hedo, K. Sakurai, Y. Inada, Y. Ōnuki, E. Yamamoto, Y. Haga, J. Phys. Soc. Jpn. **69**, 1970 (2000)
14. A. Shimoyamada, S. Tsuda, K. Ishizaka, T. Kiss, T. Shimojima, T. Togashi, S. Watanabe, C.Q. Zhang, C.T. Chen, Y. Matsushita, H. Ueda, Y. Ueda, S. Shin, Phys. Rev. Lett. **96**, 026403 (2006)
15. S. Suga, A. Sekiyama, H. Fujiwara, Y. Nakatsu, J. Yamaguchi, M. Kimura, K. Murakami, S. Niitaka, H. Takagi, M. Yabashi, K. Tamasaku, A. Higashiya, T. Ishikawa, I. Nekrasov, J. Phys. Soc. Jpn. **79**, 044711 (2010)
16. T. Yokoya, T. Kiss, A. Chainani, S. Shin, M. Nohara, H. Takagi, Science **294**, 2417 (2001)
17. T. Kiss, T. Yokoya, A. Chainani, S. Shin, T. Hanaguri, M. Nohara, H. Takagi, Nat. Phys. **3**, 720 (2007)
18. J.D. Koralek, J.F. Douglas, N.C. Plumb, Z. Sun, A.V. Fedorov, M.M. Murnane, H.C. Kapteyn, S.T. Cundiff, Y. Aiura, K. Oka, H. Eisaki, D.S. Dessau, Phys. Rev. Lett. **96**, 017005 (2006)

19. J.D. Koralek, J.F. Douglas, N.C. Plumb, J.D. Griffith, S.T. Cundiff, H.C. Kapteyn, M.M. Murnane, D.S. Dessau, Rev. Sci. Instrum. **78**, 053905 (2007)
20. T. Yamasaki, K. Yamazaki, A. Ino, M. Arita, H. Namatame, M. Taniguchi, A. Fujimori, Z.-X. Shen, M. Ishikado, S. Uchida, Phys. Rev. B **75**, 140513 (2007)
21. J. Corson, J. Orenstein, S. Oh, J. O'Donnell, J.N. Eckstein, Phys. Rev. Lett. **85**, 2569 (2000)
22. A. Lanzara, P.V. Bogdanov, X.J. Zhou, S.A. Kellar, D.L. Feng, E.D. Lu, T. Yoshida, H. Eisaki, A. Fujimori, K. Kishio, J.-I. Shimoyama, T. Noda, S. Uchida, Z. Hussain, Z.-X. Shen, Nature **412**, 510 (2001)
23. N.C. Plumb, T.J. Reber, J.D. Koralek, Z. Sun, J.F. Douglas, Y. Aiura, K. Oka, H. Eisaki, D.S. Dessau, Phys. Rev. Lett. **105**, 046402 (2010)
24. I.M. Vishik, W.S. Lee, F. Schmitt, B. Moritz, T. Sasagawa, S. Uchida, K. Fujita, S. Ishida, C. Zhang, T.P. Devereaux, Z.X. Shen, Phys. Rev. Lett. **104**, 207002 (2010)
25. W. Zhang, G. Liu, J. Meng, L. Zhao, H. Liu, X. Dong, W. Lu, J.S. Wen, Z.J. Xu, G.D. Gu, T. Sasagawa, G. Wang, Y. Zhu, H. Zhang, Y. Zhou, X. Wang, Z. Zhao, C. Chen, Z. Xu, X.J. Zhou, Phys. Rev. Lett. **101**, 017002 (2008)
26. T. Shimojima, F. Sakaguchi, K. Ishizaka, Y. Ishida, T. Kiss, M. Okawa, T. Togashi, C.-T. Chen, S. Watanabe, M. Arita, K. Shimada, H. Namatame, M. Taniguchi, K. Ohgushi, S. Kasahara, T. Terashima, T. Shibauchi, Y. Matsuda, A. Chainani, S. Shin, Science **332**, 564 (2011)
27. T. Yoshida, I. Nishi, S. Ideta, A. Fujimori, M. Kubota, K. Ono, S. Kasahara, T. Shibauchi, T. Terashima, Y. Matsuda, H. Ikeda, R. Arita, Phys. Rev. Lett. **106**, 117001 (2011)
28. R. Yoshida, Y. Nakamura, M. Fukui, Y. Haga, E. Yamamoto, Y. Ōnuki, M. Okawa, S. Shin, M. Hirai, Y. Muraoka, T. Yokoya, Phys. Rev. **B82**, 205108 (2010)
29. T. Valla, A.V. Fedorov, P.D. Johnson, B.O. Wells, S.L. Hulbert, Q. Li, G.D. Gu, N. Koshizuka, Science **285**, 2110 (1999)
30. F. Rodolakis, B. Mansart, E. Papalazarou, S. Gorovikov, P. Vilmercati, L. Petaccia, A. Goldoni, J.P. Rueff, S. Lupi, P. Metcalf, M. Marsi, Phys. Rev. Lett. **102**, 066805 (2009)
31. H. Fujiwara, A. Sekiyama, S.-K. Mo, J.W. Allen, J. Yamaguchi, G. Funabashi, S. Imada, P. Metcalf, A. Higashiya, M. Yabashi, K. Tamasaku, T. Ishikawa, S. Suga, Phys. Rev. **B84**, 075117 (2011)
32. S.-K. Mo, H.-D. Kim, J.D. Denlinger, J.W. Allen, J.-H. Park, A. Sekiyama, A. Yamasaki, S. Suga, Y. Saitoh, T. Muro, P. Metcalf, Phys. Rev. B **74**, 165101 (2006)
33. M. Kimura, H. Fujiwara, A. Sekiyama, J. Yamaguchi, K. Kishimoto, H. Sugiyama, G. Funabashi, S. Imada, S. Iguchi, Y. Tokura, A. Higashiya, M. Yabashi, K. Tamasaku, T. Ishikawa, T. Ito, S. Kimura, S. Suga, J. Phys. Soc. Jpn. **79**, 064710 (2010)
34. D. Schrupp, M. Sing, M. Tsunekawa, H. Fujiwara, S. Kasai, A. Sekiyama, S. Suga, T. Muro, V.A.M. Brabers, R. Classen, Europhys. Lett. **70**, 789 (2005)
35. S. Souma, T. Sato, T. Takahashi, P. Baltzer, Rev. Sci. Instrum. **78**, 123104 (2007)
36. G. Funabashi, H. Fujiwara, A. Sekiyama, M. Hasumoto, T. Itoh, S. Kimura, P. Baltzer, S. Suga, Jpn. J. Appl. Phys. **47**, 2265 (2008)
37. S. Suga, A. Sekiyama, G. Funabashi, J. Yamaguchi, M. Kimura, M. Tsujibayashi, T. Uyama, H. Sugiyama, Y. Tomida, G. Kuwahara, S. Kitayama, K. Fukushima, K. Kimura, T. Yokoi, K. Murakami, H. Fujiwara, L. Plucinski, C.M. Schneider, Rev. Sci. Instrum. **81**, 105111 (2010)
38. F. Offi, S. Iacobucci, L. Petaccia, S. Gorovikov, P. Vilmercati, A. Rizzo, A. Ruocco, A. Goldoni, G. Stefani, G. Panaccione, J. Phys.: Condens. Matter **22**, 305002 (2010)
39. M.P. Seah, W.A. Dench, Surf. Interface Anal. **1**, 2 (1979)
40. L. Plucinski, G. Mussler, J. Krumrain, A. Herdt, S. Suga, D. Grützmacher, C.M. Schneider, Appl. Phys. Lett. **98**, 222503 (2011)
41. Y. Nakatsu, A. Sekiyama, S. Imada, Y. Okamoto, S. Niitaka, H. Takagi, A. Higashiya, M. Yabashi, K. Tamasaku, T. Ishikawa, S. Suga, Phys. Rev. B **83**, 115120 (2011)
42. J. Yamaguchi, A. Sekiyama, M. Y. Kimura, H. Sugiyama,Y. Tomida, G. Funabashi, S. Komori, T. Balashov, W. Wulfhekel, T. Ito, S. Kimura, A. Higashiya, K. Tamasaku, M.

Yabashi, T. Ishikawa, S. Yeo, S.-I. Lee, F. Iga, T. Takabatake, S. Suga, New J. Phys. **15**, 043042 (2013)
43. F. Iga, N. Shimizu, T. Takabatake, J. Magn. Magn. Mater. **177–181**, 337 (1998)
44. J. Yamaguchi, A. Sekiyama, S. Imada, H. Fujiwara, M. Yano, T. Miyamachi, G. Funabashi, A. Higashiya, K. Tamasaku, M. Yabashi, T. Ishikawa, F. Iga, T. Takabatake, S. Suga, Phys. Rev. B **79**, 125121 (2009)
45. H. Okamura, M. Matsunami, T. Inaoka, T. Nanba, S. Kimura, F. Iga, S. Hiura, J. Klijn, T. Takabatake, Phys. Rev. **B62**, R13265 (2000)
46. H. Petek, S. Ogawa, Prog. Surf. Sci. **56**, 239 (1997)
47. U. Thomann, Ch. Reuß, Th. Fauster, F. Passek, M. Donath, Phys. Rev. **B61**, 16163 (2000)
48. K. Takahashi, J. Azuma, M. Kamada, Phys. Rev. **B85**, 075325 (2012)
49. M. Rohleder, K. Duncker, W. Berthold, J. Güdde, U Höfer, New J. Phys. **7**, 103 (2005)
50. M. Pickel, A.B. Schmidt, M. Donath, M. Weinelt, Surf. Sci. **600**, 4176 (2006)
51. A.A. Ünal, C. Tusche, S. Ouazi, S. Wedekind, C.-T. Chiang, A. Winkelmann, D. Sander, J. Henk, J. Kirschner, Phys. Rev. **B84**, 073107 (2011)

Chapter 10
Inverse Photoemission

Occupied electronic states are effectively probed by photoelectron spectroscopy as already explained. On the other hand, unoccupied electronic states can be probed either by the absorption spectroscopy (as explained later in Sect. 13.1) or by the inverse photoemission spectroscopy (IPES) to be explained in this chapter except for the case of multiphoton PES. Since the photon absorption is related to the joint density of both unoccupied and occupied states and the Coulomb interaction between the core hole and electron(s) cannot be neglected, the IPES is superior to the absorption spectroscopy to obtain direct information on unoccupied states [1–6]. Even momentum information can be obtained in the case of angle resolved IPES.

There are two detection methods in IPES. One at a constant hν and another by means of photon spectroscopy. Some examples of comparison between PES and IPES are then introduced. Resonance IPES is also possible by use of the spectroscopy mode. Angle resolved IPES is shown to be also useful for probing dispersions of unoccupied states. Finally image potential states and QWS detected by angle revolved IPES are discussed.

10.1 General Concept

In the IPES, monochromatic electrons are impinged onto the solid and accommodated in the high energy part of the unoccupied electronic states. These electrons then relax to the lower energy part of the unoccupied states emitting photons besides several other non-radiative processes. Since the electron-in and photon-out process (Fig. 10.1 [5]) is the reverse of the photon-in and electron-out process in the photoelectron excitation, this is called inverse photoemission process. The transition can occur to the final states above the Fermi level and one can even probe the electronic states between the vacuum level E_V and the Fermi level E_F, which are not accessible by conventional PES. IPES can be measured in a wide energy range from X-rays down to vacuum ultraviolet. There are two different detection methods for the emitted photons. One method is to use a band pass

S. Suga and A. Sekiyama, *Photoelectron Spectroscopy*,
Springer Series in Optical Sciences 176, DOI: 10.1007/978-3-642-37530-9_10,

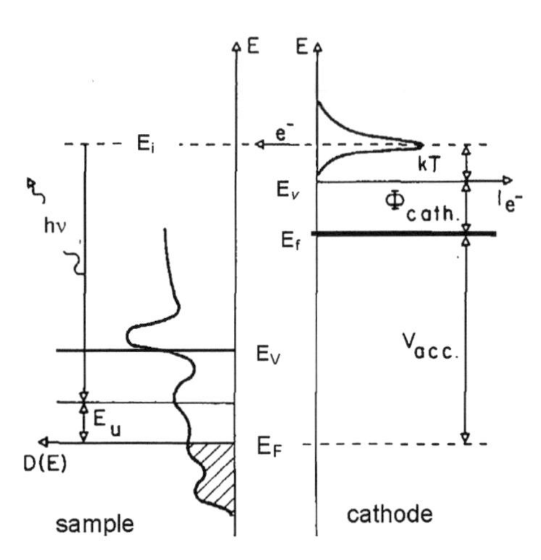

Fig. 10.1 Energy diagram and principle of inverse photoemission spectroscopy (IPES). Fermi level and vacuum level of the sample are indicated by E_F and E_v. E_i is the incident electron energy onto the sample and E_u is the energy of the unoccupied state in the sample corresponding to the final state of the IPES process measured from E_F. Then photons with $h\nu = E_i - E_u$ can be observed. E_f and E_v in the right panel are the Fermi level and vacuum level of the cathode shifted by the acceleration voltage V_{acc}. The intensity distribution of the emitted electrons (thermionic emission) from the cathode is schematically shown by I_{e^-}

detector tuned to one particular hν while scanning the excitation electron energy. This method is called isochromat spectroscopy and often called X-ray Bremsstrahlung Isochromat Spectroscopy (X-ray BIS) when used in the X-ray region. The other method is to detect the spectrum of emitted photons at a certain incident electron energy. In the latter case, a position sensitive detector can be used in combination with a grating monochromator. Angle resolved IPES was often performed in the case of low energy IPES in both detection methods. By utilizing spin-polarized electrons for excitation, spin-polarized IPES (SP-IPES) is also feasible.

10.2 Isochromat IPES

In the case of high energy IPES as the X-ray BIS, the angular divergence of the electron beam provides substantial momentum broadening making the angle resolved excitation impractical. Then one can only probe the density of unoccupied states. Since the count rate of IPES is generally much lower than in typical PES, high sensitivity detectors and a high excitation current are always required. For the monochromatic energy electron excitation, specially designed electron guns are used to provide an electron beam with small beam size and high flux. Indirectly heated BaO is often used as a thermal cathode instead of the direct

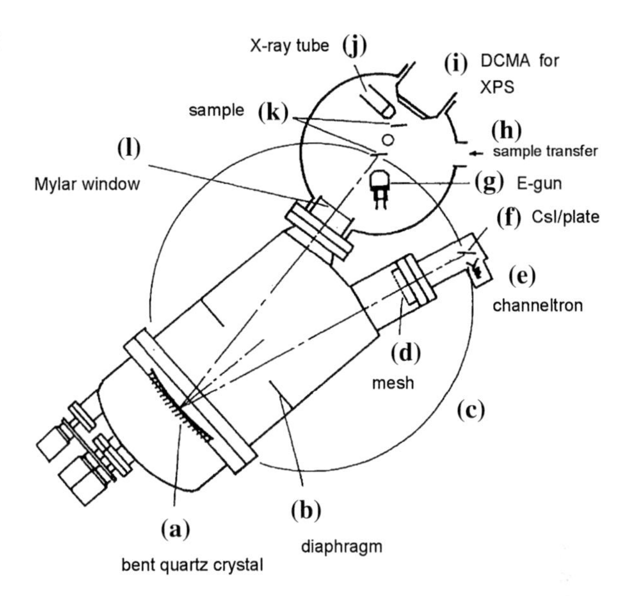

Fig. 10.2 Schematic view of an X-ray BIS-XPS combined apparatus. (c) is a Rowland circle with a diameter of 700 mm

heating type W cathode. The work function of the BaO cathode is ~2 eV and a relatively low temperature of 800–1,000 °C is enough to get thermal electrons, providing a resolution of ~0.2 eV. In the case of the X ray-BIS, for example, a few hundred μA beams are typically used. Then the damage of the surface by electrons is very serious and the surface must be cleaned from time to time during the measurement. Crystal monochromators are employed for the X ray-BIS measurements. Typically a Bragg reflection by means of the SiO_2 (1010) plane is often employed for detecting 1,486.6 eV photons at a Bragg angle ~78°. The resolution of X ray-BIS is mostly limited by the band pass of the crystal monochromator, which is often of the order of 1 eV. Figure 10.2 schematically shows a conventional X-ray BIS instrument combined with XPS [7].

In general cases, better understanding is obtained if IPES is compared with PES as shown in Fig. 10.3 for the case of NiO (100) [8], which is an antiferromagnetic insulator with $T_N = 520$ K. From this comparison the gap energy was evaluated as 4.3 eV, which is much larger than the prediction by the band calculations within the local density approximation [8, 9]. In the upper part are shown the electronic structures predicted by a local cluster model calculation for $(NiO_6)^{10-}$, where the hybridization between the Ni 3d and O 2p states was taken into account. Denoting the ligand hole as $\underline{L}$, the ground state is represented by a linear combination of $|d^8\rangle$ and $|d^9\underline{L}\rangle$ configurations. The photoelectron excited states are given as linear combinations of $|d^7\rangle$, $|d^8\underline{L}\rangle$ and $|d^9\underline{L}^2\rangle$ and the X-BIS final states are given by $|d^9\rangle$, and $|d^{10}L\rangle$ [9].

The energy required to move one electron from the ligand atom to the Ni atom is defined as Δ, which is called the charge-transfer energy. Due to the hybridization between the Ni 3d and O 2p orbitals, the level energies shift appreciably. The parameters employed in the calculation in Fig. 10.3 [8, 9] are $E_L = E(|d^n\underline{L}\rangle) -$

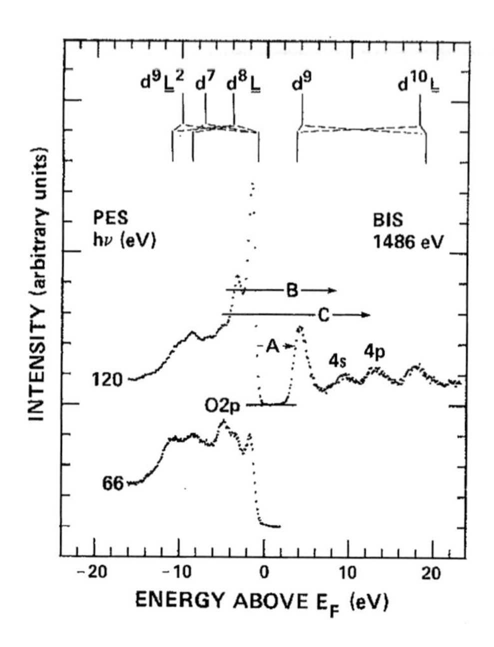

Fig. 10.3 Valence-band structures of cleaved NiO probed by PES and X-BIS [8–10]. The bar diagram shows the assignments of the peaks according to the local-cluster calculation. The unoccupied 4s and 4p bands are also assigned. The arrows indicate the expected energies of the non-excitonic optical transitions

$E(|d^n\rangle) = 3.5\,eV, E_d = E(|d^9\rangle) - E(|d^8\rangle) = 3\,eV, U = 9\,eV$ and the level shift for $|d^8\rangle$ is 1 eV. Therefore it was noticed that the energy separation between the bare $|d^7\rangle$ and $|d^9\rangle$ states is not equal to U because the ground state hybridization shift was included. Δ was evaluated as 6.5 eV in this case. The observed large band gap does not directly correspond to U and is related to Δ. While the concept of a Mott–Hubbard insulator holds for $U < \Delta$, the concept of a charge-transfer insulator is used for the cases of $U > \Delta$. It was revealed that NiO belonged to the charge transfer insulator regime according to the Zaanen-Sawatzky-Allen diagram [10].

For a low energy IPES, a much lower electron beam current of a few to tens μA was used from a BaO cathode or a negative electron affinity (NEA) photocathode for spin-polarized IPES as explained later in Sect.11.5. In the case of isochromat ultraviolet (UV) IPES, a high hν pass detector was used in combination with a low hν pass filter. As an example, a Dose detector is schematically shown in Fig. 10.4, based on the Geiger-Müller (GM) counter [11]. The transmittance of ionic crystals decreases suddenly above a certain hν corresponding to the exciton or interband absorption threshold and the ionic crystals work as low hν pass filters. On the other hand, ionization of I_2 gas takes place above a certain hν realizing a high pass filter. The combination of a CaF_2 window and I_2 gas can provide a band pass of FWHM = 0.8 eV near hν = 9.7 eV. Later, the combination of a SrF_2 window and I_2 gas provided FWHM = 0.35 eV around hν = 9.5 eV [11]. By slightly increasing the temperature of the window material, the resolution can be further improved [12].

In addition to GM type detectors, a solid state detector combining a Cu-BeO cathode photomultiplier and a CaF_2 or SrF_2 window provides a band pass near 9.8 or 9.4 eV. By depositing a KBr or KCl thin film (~1,000 Å) onto the Cu-BeO

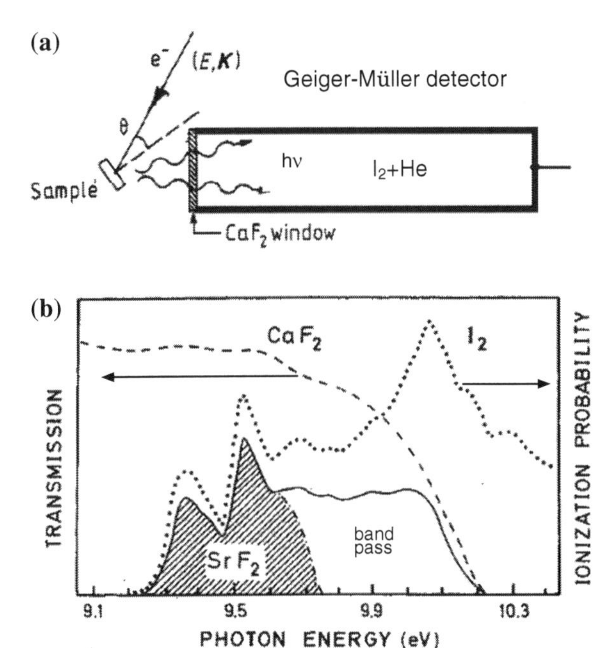

Fig. 10.4 Schematic drawing of the ultraviolet IPES in the isochromat mode. The upper diagram (**a**) represents a typical experimental arrangement with a Geiger–Müller type band pass detector. If the incident electron energy E is swept while holding $h\nu$ constant, the measured photon counting rate will replicate the unoccupied density of states. The transmittance of an ionic crystal CaF_2 and photoionization probability of the I_2 gas are shown in (**b**), where the characteristic of the detector as a band pass filter is shown for the cases of CaF_2 and SrF_2 windows [11]

photocathode, the sensitivity increases by a factor of 3–5. Even the FWHM resolution can be improved down to 0.47 eV from ~0.6 eV by the combination a SrF_2 window and a KCl coated Cu-BeO photocathode [13]. It was also proposed to use the SrF_2 window or filter at different T and mutually subtract the two IPES spectra detected by this photocathode to obtain a higher resolution spectrum [14]. By taking the difference of the two IPES spectra measured with this filter at room temperature (R.T.) and 100 K above, one expected to realize a FWHM of 0.23 eV. The details are shown in Fig. 10.5(a) and (b). This is because the energy of the exciton absorption edge depends on T according to the Urbach tail behavior [15]. The resolution of the detector can further be improved by cooling the filter material. Only the high energy cut-off part changes in the two spectra and their difference stands for the spectrum probed by the high resolution band pass between the two high energy cut-offs.

The unoccupied DOS can be probed by these UV IPES for polycrystalline samples and a wide angular divergence of the electron beam. There were so far many reports, which compared angle-integrated PES and IPES in the UV region. One example is shown in Fig. 10.6 for Cu_2Sb type intermetallic compounds (Cr_2As, Fe_2As and Mn_2As are antiferromagnetic, Mn_2Sb is ferrimagnetic and MnAlGe is ferromagnetic with T_N = 393, 353, 573 K and T_C = 550, 518 K) [16]. There are two different metal sites M(I) and M(II) at tetrahedral and octahedral sites. In the case of MnAlGe, non-magnetic Al atoms preferentially occupy the M(II) site. Through all these compounds except for MnAlGe, the M(I) moment was known to be much smaller than the M(II) moment. It is clearly seen that the

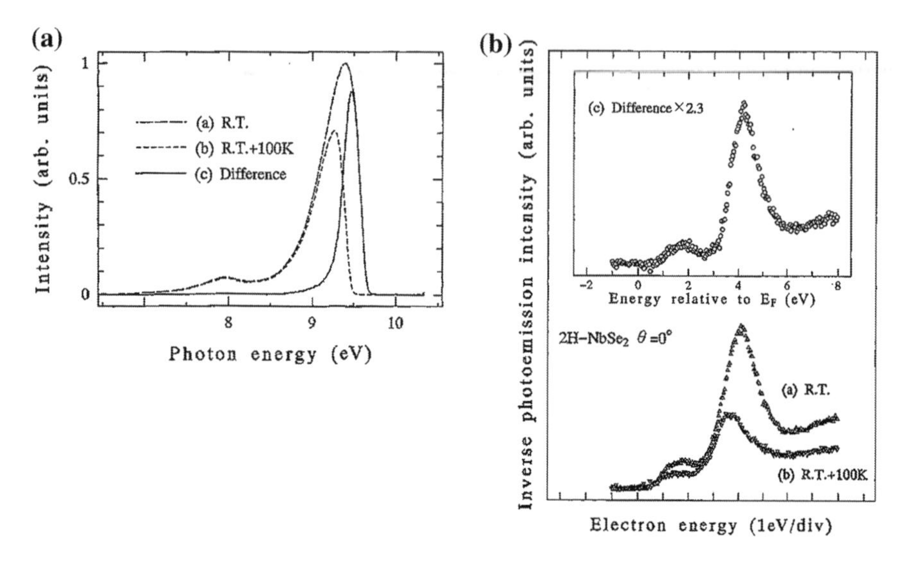

Fig. 10.5 Principle of a high resolution IPES solid-state band pass detector composed of a 20 stages photomultiplier with KCl coated Cu-BeO photocathode and a 2 mm thick SrF_2 crystal. The high energy cut-off can be tuned by changing the T of SrF_2. (**a**) Detector sensitivity spectra at two T of SrF_2, namely at room temperature and at 100 K above room temperature. The difference of the sensitivity spectra realizes a band pass of 0.23 eV. (**b**) An example applied to IPES of 2H-$NbSe_2$ [14]

prominent peak in the IPES is missing in MnAlGe. This result reveals that the IPES peak at 1.7–2.0 eV is mainly due to the Mn(II) 3d state and that the Mn(I) state is rather spread in a wide energy region mixed with the Al 3p and Ge 4p states. In UPS, on the other hand, Mn 3d states are mainly detected because the PICS are much smaller for the As 4p, Ge 4p, Sb 5p and Al 3p states [17]. It is seen in Fig. 10.6 that the occupied Mn 3d states show a prominent peak at $E_B \sim 3$ eV in Mn_2As and Mn_2Sb, whereas Cr_2As and Fe_2As show a prominent feature at $E_B \sim 1$ eV. Since MnAlGe show a rather spread M(I) 3d feature, the $\sim$3 eV feature in Mn_2As and Mn_2Sb can be understood to result from the M(II) states. The exchange splitting in the Mn(II) site was then evaluated as 5.0 eV in Mn_2As, which is close to but slightly larger than that in Mn_2Sb (4.7 eV) [16].

Though the counting efficiency is much higher in UV IPES than in X-ray BIS, still high efficiency detectors are desirable to reduce the radiation damage and surface degradation during the measurement. Therefore the collection angle restricted by the size of the window material and the detector collection angle should be set as wide as possible. In the case of solid state detectors, a large area Cu-BeO plate and either a microchannel plate or a large detection area channeltron would facilitate the measurement. It would also be possible to accept the emitted photons by a mirror, which focuses the light and guide it to a detector element. Except for the possible shadowing by the sample and the sample holder, the

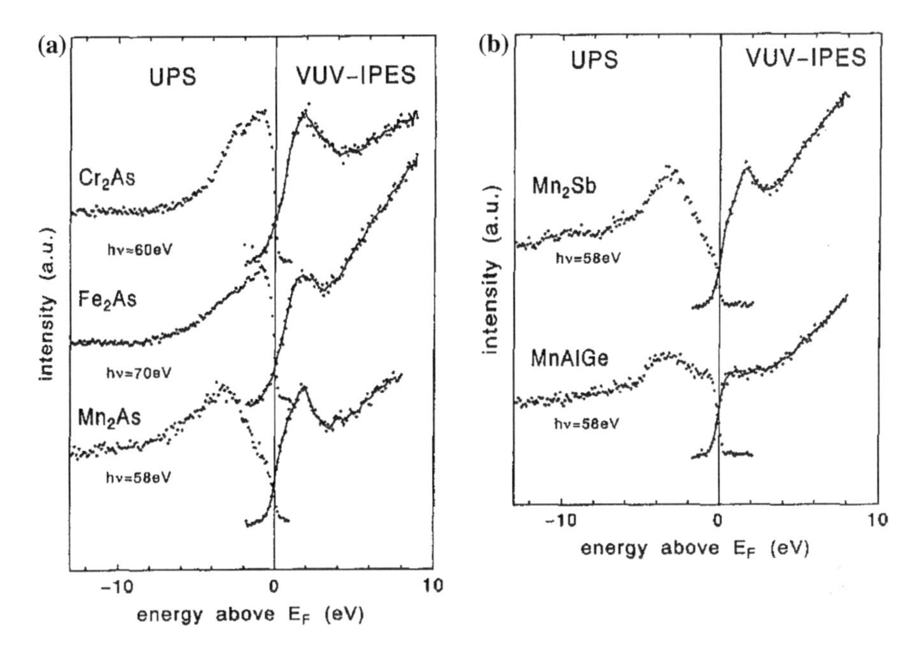

Fig. 10.6 Comparison of IPES of Cr_2As, Fe_2As, Mn_2As, Mn_2Sb and MnAlGe measured at room temperature with conventional UPS [16]

emitted photons can be efficiently detected by use of a focusing mirror with high reflectivity in this low hν region.

The development of higher efficiency band pass detectors continues up to now [18–20]. A band pass filter made of a $Sr_{0.7}Ca_{0.3}F_2$ entrance window with a transmission cut-off at 9.85 eV and acetone as detection gas with 9.7 eV photo-ionization threshold was very recently reported. An energy resolution of FWHM 82 $\pm$ 2 meV was achieved [20]. High resolution IPES is expected to strongly promote the frontier of new physics and materials sciences.

10.3 Angle-Resolved IPES

Although the angular spread of the electron beam prohibits momentum resolution in the X-ray BIS, the momentum spread in the case of the low energy IPES can be confined to a certain level even after the weak focusing onto the sample. By rotating the sample, the $k_{||}$ can be tuned as in ARPES. Since the low hν photons have marginal momentum **q** relative to the reciprocal lattice vector, the relation $E(k_{||})$ can be probed by this technique, providing the dispersion of the unoccupied electronic states (angle resolved IPES for band mapping).

Some angle-resolved IPES results measured on Ni (001) surface are shown in Fig. 10.7(a) and (b) [11] as a function of the incidence polar angle θ and $k_{||}$ in the

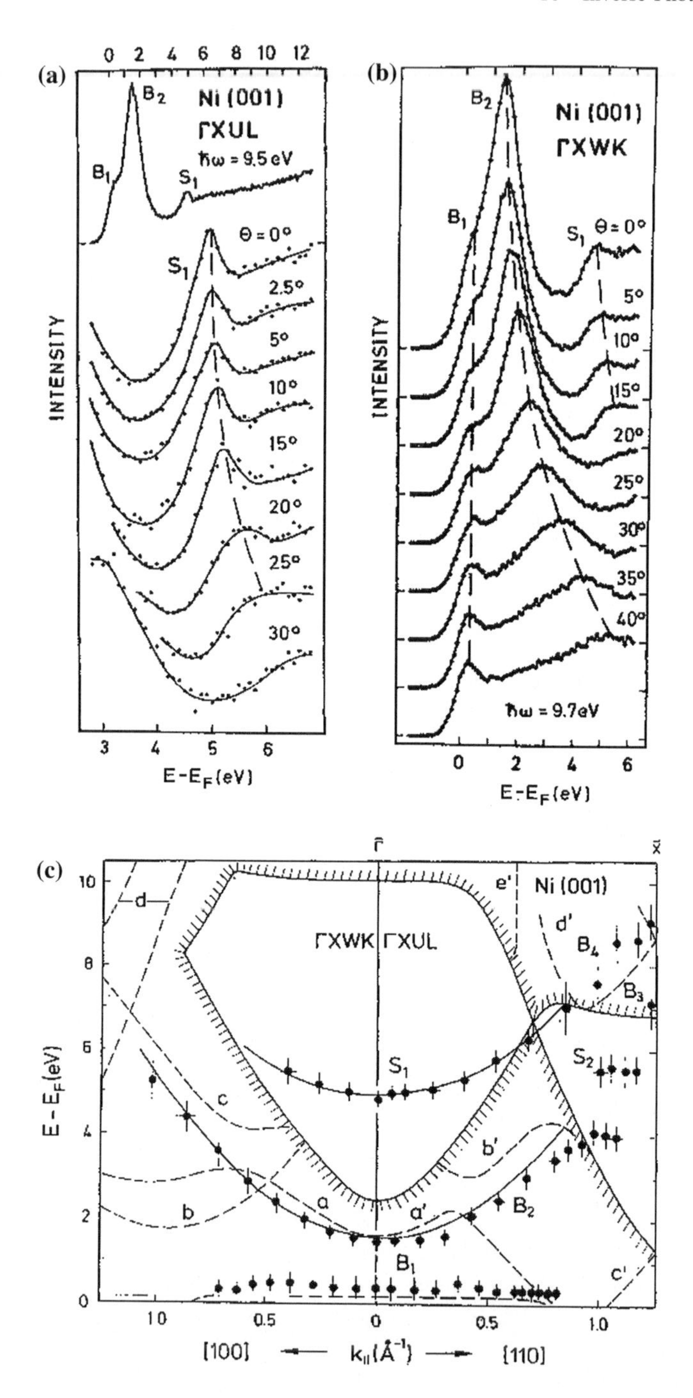
(a)
Ni (001)
ΓXUL
ħω = 9.5 eV
B1
B2
S1
Θ = 0°
2.5°
5°
10°
15°
20°
25°
30°
INTENSITY
E - EF (eV)
(b)
Ni (001)
ΓXWK
B1
B2
S1
Θ = 0°
5°
10°
15°
20°
25°
30°
35°
40°
ħω = 9.7 eV
INTENSITY
E - EF (eV)
(c)
Ni (001)
ΓXWK ΓXUL
S1
S2
B1
B2
B3
B4
E - EF (eV)
[100]
k∥ (Å⁻¹)
[110]

◀ **Fig. 10.7** Angle resolved IPES of Ni (001) measured in isochromat mode for electrons incident at different polar angles θ in the ΓXUL bulk mirror plane. The uppermost spectrum in (**a**) is for normal incidence ($\theta = 0$) shown in a wide energy range (*upper scale*). Other spectra in (**a**) show the variation of the peak S_1 with θ. All spectra were measured at $h\nu = 9.5$ eV with a resolution of ~0.4 eV. (**b**) Isochromat spectra for the ΓXWK bulk mirror plane measured at $h\nu = 9.7$ eV with a resolution of ~0.8 eV. Final state energies as a function of $k_{||}$ are shown in (**c**), where the *left* (*right*) side corresponds to the ΓXWK (ΓXUL) bulk mirror plane and the unshaded areas correspond to the gap of the projected bulk band structures. The *dashed lines* show the kinematically allowed bulk band transitions [11]

ΓXUL and ΓXWK bulk mirror planes up to 12 eV above E_F. In Fig. 10.7(b), the feature B_1 at $\theta = 0°$ remains at the same energy with the change of θ, while it develops into a peak for $\theta \geq 20°$. The peak B_2 being prominent at $\theta = 0°$ decreases its intensity and shifts towards higher energies with the increase of θ. In addition, a small peak S_1 is observed which also shows a noticeable dispersion. S_1 is also observed in Fig. 10.7(a) with a noticeable dispersion. The whole results are plotted in (c) together with the results of band structure calculations. In angle-resolved IPES, the impinging electrons settle in the Bloch band state at a certain wave vector **k** in the three dimensional **k** space and then decay through a momentum conserving radiative transition. The feature B_1 is interpreted as a transition to the d band just above E_F. Its intensity increases with θ. This result was interpreted in terms of a relaxation of the selection rules and the simultaneous increase in the matrix element. The structure B_2 behaves as the bands a and c along ΓXWK (left) and the bands a′ and b′ along ΓXUL (right) in Fig. 10.7(c).

The unhatched region shows the gap of the projected bulk band structure. A clear structure S_1 observed in this region was interpreted as an image potential state or barrier-induced state whose amplitude is largest in vacuum far from the surface compared with the crystal-induced surface state well confined to the region near the surface. If there is no Bloch state which could accommodate the incident electron, the so-called image charge is induced within the crystal. Then the incident electron is bound with the image charge forming a Rydberg series. Then the energy is given by

$$E(k_{||}) = \hbar^2 k_{||}^2 / 2m^* - \varepsilon_n + \phi,$$

where m* is the electron's effective mass, ϕ is the work function and ε_n is the energy of the bound state with the quantum number n. ε_n is thought to be inversely proportional to n^2. However, $n > 1$ components were not resolved under the present experimental resolution. The image potential state was so far observed in Ni, Cu, Au and so on. The typical value of $\varepsilon_{n=1}$ is ~0.6–0.7 eV at $k_{||} = 0$ for the (001) surface. The image potential states were also observed on Ni(111) and Ni(110) surfaces [11]. It was then found that m* depends upon the Miller indices of the surface in Ni as $m^*(001) = 1.2 \pm 0.2$, $m^*(111) = 1.6 \pm 0.2$ and $m^*(110) = 1.7 \pm 0.3$. The deviation of m* from 1 shows that the lateral motion of the electron is no more free electron like and modified by the crystal potential [11].

10.4 IPES with a Fixed Incident Electron Energy

It is then described how to use a spectrometer to simultaneously detect the IPES luminescence spectrum in a certain energy range with use of 2D detectors for a fixed energy of the incident electrons (Fig. 10.8 [21]). Cases for low energies with $h\nu < 30$–44 eV and wider applications in the soft X-ray region as well as resonance IPES are described in Refs. [21–23] and Refs. [24–27], respectively. The luminescence spectrum could be measured by use of a 2D detector such as a CsI coated microchannel plate and a resistive anode, where the signals are picked up at four corners of the resistive anode [21]. Since CsI is hygroscopic, the detector should not be exposed to air. The efficiency of this experiment is much higher than the BIS type detection. In the spectrometer type measurement, focusing of the electron beam into a narrow line corresponding to the entrance slit of the spectrometer is preferred to realize rather effective detection.

This method is very useful to study the resonance IPES process, where the behavior of the fluorescence spectrum must be measured as a function of the excitation electron energy. An example in the low $h\nu$ region is shown in Fig. 10.9(a) for the case of a Sb thin film (20 Å) on InP [28], where the incident electron energy E_i was scanned from 11.6 eV up to 28.6 eV with 0.5 eV steps while the fluorescence spectra were measured by the spectrograph with a position-sensitive detector [21]. A channel with almost constant $h\nu$ slightly above 16 eV was noticed in Fig.10.9(a). This energy is almost equal to the plasmon energy of Sb which is known to be 16.4 eV. For each E_i the highest $h\nu$ corresponds to the transition to empty electronic states just above E_F as shown by the top diagonal line marked by E_F. For the Sb film on InP(110) there are several empty electronic states at 0.25, 0.95 as well as 3.9 eV above E_F as recognized by ridge shapes. If the $h\nu$ of the IPES process corresponding to these states crosses the plasmon energy,

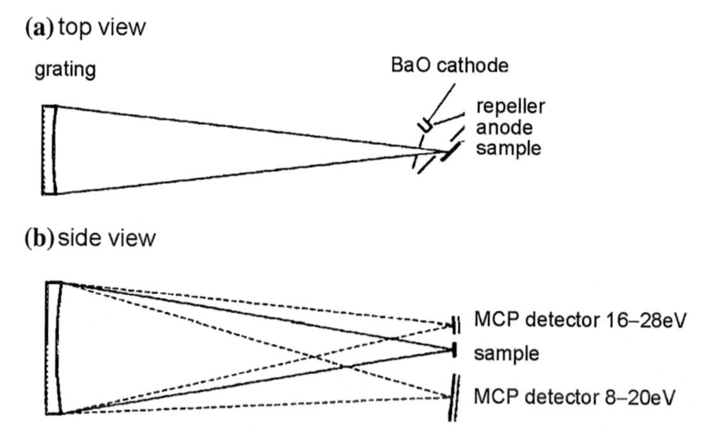

Fig. 10.8 Schematic view of the inverse photoemission spectrograph. (**a**) shows the top view with the BaO electron gun and the sample. (**b**) shows the side view of the MCP detectors together with the sample [21]

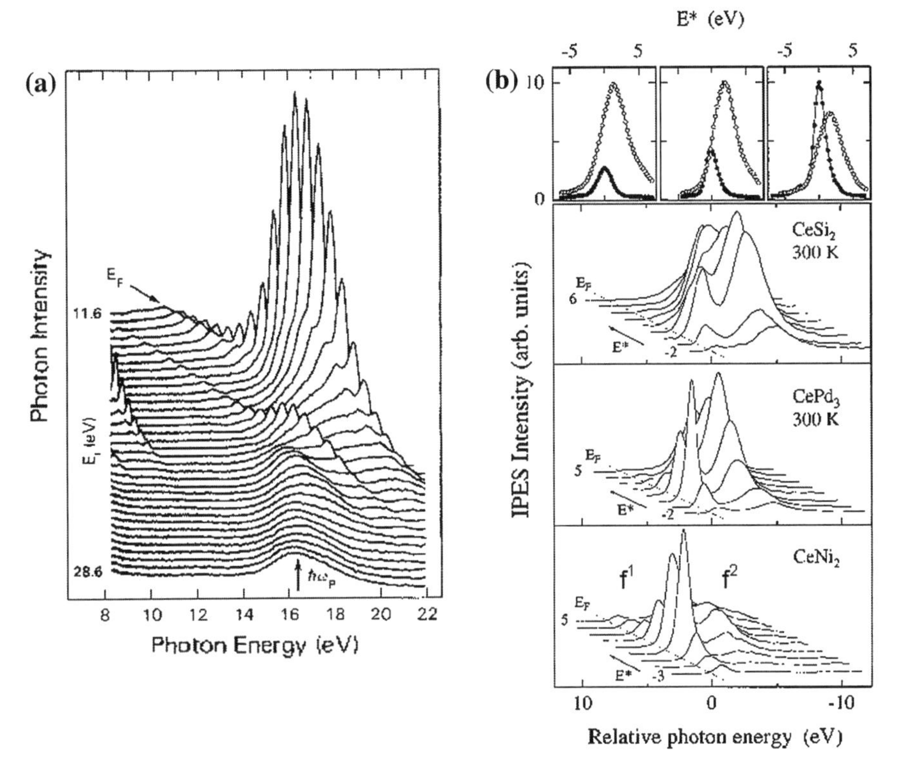

Fig. 10.9 Resonance low energy IPES of a thin Sb film (20 Å) on a cleaved InP(110) for various energies E_i of the incoming electrons at $k_{\|} = 0$ (**a**) [28]. E_i relative to E_F varies from 11.6 eV for the *top curve* to 28.6 eV for the *bottom curve* in steps of 0.5 eV. The plasmon energy is shown by the *upward arrow*. (**b**) Resonance soft X-ray IPES spectra of $CeSi_2$, $CePd_3$ and $CeNi_2$ [26]. The incident electron energies increase towards the up-left direction. E* means the relative electron energy. The *top panel* shows the corresponding CFS curves for these three materials as a function of the electron incidence energy: $CeSi_2$ (*left*), $CePd_3$ (*middle*) and $CeNi_2$ (*right*). Full (*empty*) dots stand for the f^1 (f^2) final states. The intensity of the f^1 resonance increases with the strength of the 4f–conduction band hybridization

resonance enhancement of the IPES took place. An enhancement factor of ~30 and a FWHM of 3.0 eV were observed for the $E_F + 0.25$ eV final state. The interference and the resulting resonance was thought to be taking place in this material between the direct IPES radiative decay and the non-radiative decay by emitting bulk plasmons [28].

A certain electronic state can be strongly excited when the excitation is tuned to the core excitation threshold. In the case of a RE $|4f^N\rangle$ electron system, for example, the following two processes can interfere on the incidence of the excitation electron. The first process is the usual radiative inverse photoemission process described as $e^- + |4f^N\rangle \rightarrow |4f^{N+1}\rangle + h\nu$ (direct process). The second process is due to the non-radiative core excitation process $e^- + |4f^N\rangle \rightarrow |\underline{c}4f^{N+2}\rangle$

followed by the core decay fluorescence process $|4f^{N+1}\rangle$ + hν (indirect process), where the core hole is represented by $\underline{c}$. Since both initial and final states are the same in the two processes, interference takes place between these two processes and resonance IPES behavior is expected for a certain incident electron energy. If the system is in a valence-mixing or valence-fluctuation condition, the resonance behavior will be different for the $|4f^{N}\rangle$ and $|4f^{N+1}\rangle$ states. Extensive studies were performed [26] and some examples are shown in Fig. 10.9(b) for $CeSi_2$, $CePd_3$ and $CeNi_2$, which have Kondo temperatures around 50–100, 240–350 and >1,000 K, with increasing the 4f hybridization with the conduction band. The experiments were performed for the Ce M_5 ($3d_{5/2}$) absorption edge with use of an indirectly heated BaO cathode with the energy resolution of 0.3–0.5 eV and a beryl $(10\bar{1}0)$ crystal. Since the actual resolution depends on the entrance slit of the monochromator or the illuminated area on the sample, a total energy resolution of 0.6–0.8 eV was achieved [26].

In the X-ray BIS spectrum of $CePd_3$ with $T_K \sim$ 240–350 K measured at hν = 1,486.6 eV, three structures were observed from E_F up to 10 eV above E_F, which were assigned to the Ce $4f^1$, 5d and $4f^2$ final states, respectively. Near E_F was located the $|4f^1\rangle$ Kondo resonance peak and near 4 eV was located the $|4f^2\rangle$ multiplet, whereas the Ce 5d structure was observed between 2 and 3 eV (Fig. 11 in [23]). It was recognized in the M_5 resonance IPES that the background and the Ce 5d band structures were almost negligible and one could clearly see the resonance behavior of both $|4f^1\rangle$ and $|4f^2\rangle$ final states. In the top panel of Fig. 10.9(b), the maximum of the $|f^1\rangle$ curve (full dots) was taken as the origin of the relative electron energy scale (E*) used below. The $|4f^1\rangle$ and $|4f^2\rangle$ peaks show different behaviors of the resonance enhancement as clearly shown in the upper panel. The resonance enhancement of the $|4f^1\rangle$ peak relative to that of the $|4f^2\rangle$ peak becomes more prominent through $CeSi_2$, $CePd_3$ and $CeNi_2$ with the increase in the Kondo temperature T_K. The $|4f^2\rangle$ resonance takes place at around E* = 2 eV. The results were also compared with the results of CeSb and $CeRh_3$. In CeSb, which has a magnetic ground state with a large Ce moment and low T_K, the hybridization is very small and the resonance IPES is dominated by the enhancement of the $|4f^2\rangle$ feature [26]. On the contrary, the resonance IPES spectra in $CeRh_3$, which has a large hybridization and high T_K ($\sim$1,350 K), are dominated by the $|4f^1\rangle$ Kondo resonance feature and the $|4f^2\rangle$ feature is very weak. Ce 4d–4f resonance IPES spectra were later reported on similar Ce systems [29], where some differences from the 3d–4f resonance IPES were recognized.

The advantage of the higher energy soft X-ray resonance IPES technique is its rather high bulk sensitivity, demonstrating the validity of scaling the spectra by the Kondo temperature as in the case of soft X-ray 3d–4f RPES. The strong enhancement of the 3d resonance IPES requires only a short measuring time (in some cases a few minutes) suppressing possible surface damage. The changes of the spectral shape with the excitation energy in resonance IPES thus facilitated detailed analyses.

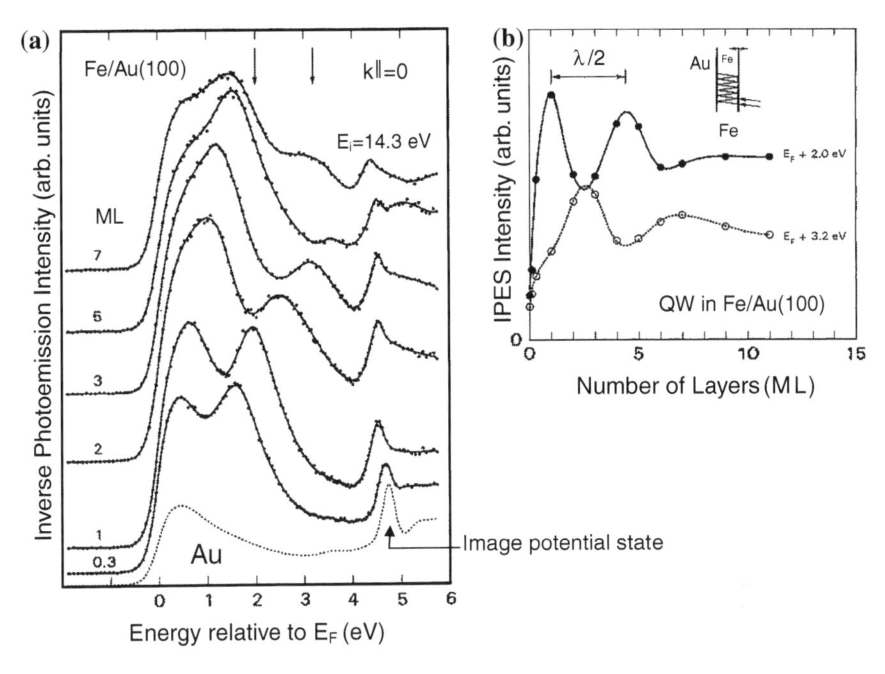

Fig. 10.10 Inverse photoemission spectra of quantum well states (QWS) of Fe/Au(100) measured at normal incidence excitation (**a**). QWS in the Fe layer vary with the film layer thickness. The intensity at the energy indicated by the *arrows* in (**a**) is plotted in (**b**), where the period of the oscillation corresponds to $\lambda/2$ (λ is the wavelength of the QWS) [30]

10.5 IPES of Quantum Well States

Returning to the non-resonance IPES, the study of the unoccupied part of the quantum well states (QWS) is briefly introduced in the case of an Fe film on Au(100) [30], which attracted attention from the viewpoint of magnetic properties. The measurement was performed by use of the spectrograph. A good lattice match of the Fe/Au(100) system within 1 % (lattice constant of fcc Au is a factor $\sqrt{2}$ larger than the lattice constant of bcc Fe) is suitable for narrow QWS. A large band offset and a large well depth of 9 eV were expected for this system. The layer-by-layer growth of Fe was known on Au (100) with one segregated Au layer on top of the growing Fe surface. This Au overlayer acts as a surfactant and lowers the surface energy of the growing film, preventing the island formation. Fe was capped by Au on both sides and one could then obtain an almost symmetric QWS.

Figure 10.10(a) shows the behavior of the newly emerging unoccupied states for the submonolayer to a few monolayers of Fe in the normal incidence configuration. Rather prominent changes are observed with the Fe layer thickness. Figure 10.10(b) shows that the intensity at certain energies indicated by the downward arrows in Fig. 10.10(a) oscillates with the number of Fe layers. The period of these oscillations corresponds to half the wavelength (λ) of a QWS like the interference fringes of a Fabry–Perot interferometer as shown in the upper right of Fig. 10.10(b). For the energy at 2 eV above E_F, for example, $\lambda/2$ was evaluated as 3.4 layers of Fe and λ was evaluated as 6.8 layers, which is 9.5 Å judging from a layer thickness of d = 1.4 Å. Then $k_\perp$ of the QWS was estimated as $k_\perp = 2\pi/\lambda = 0.66$ Å^{-1}. On the other hand, the Brillouin zone wave number k_{BZ} is given by $\pi/d = 2.24$ Å^{-1}, which is the momentum at the boundary H of the bulk BZ in the [100] direction. The above $k_\perp$ is equal to k_{BZ} divided by 3.4. At higher energies, λ increases ($\lambda = 11$ and 14 Å at $E_F + 3.2$ and $E_F + 5.7$ eV) and the phase of the oscillations changes. The $k_\perp$ represents the momentum of the so-called envelope function, which modulates a rapidly oscillating Bloch function derived from band edge states in the quantum well. In the present case, the band edge of the Fe quantum well is formed by the H_{15} point at the BZ boundary. The total momentum is then obtained by combining the momentum $k_{0\perp}$ of the band edge states with the $\pm k_\perp$ of the envelope function [30]. In accord with the development of the spin polarized electron sources, spin polarized IPES (SP-IPES) became quite popular later. Even the spin polarized QWS were studied by angle resolved SP-IPES as later explained in Sect. 11.5.

References

1. D.P. Woodruff, N.V. Smith, Phys. Rev. Lett. **48**, 283 (1982)
2. P.D. Johnson, N.V. Smith, Phys. Rev. **B27**, 2527 (1983)
3. N.V. Smith, Vacuum **33**, 803 (1983)
4. N.V. Smith, Rep. Prog. Phys. **51**, 1227 (1988)
5. J.K. Lang, Y. Baer, Rev. Sci. Instrum. **50**, 221 (1979)
6. F.U. Hillebrecht, J.C. Fuggle, G.A. Sawatzky, M. Campagna, Q. Gunnarsson, K. Schonhammer, Phys. Rev. **B30**, 1777 (1984)
7. S. Ogawa, S. Suga, A.E. Bocquet, F. Iga, M. Kasaya, T. Kasuya, A. Fujimori, J. Phys. Soc. Jpn. **62**, 3575 (1993)
8. G.A. Sawatzky, J.W. Allen, Phys. Rev. Lett. **53**, 2339 (1984)
9. J.W. Allen, in *Synchrotron Radiation Research: Advances in Surface and Interface Science, Techniques*, vol. 1, ed. by Z. Bachrach (Prenum Press, New York, 1992)
10. J. Zaanen, G.A. Sawatzky, J.W. Allen, Phys. Rev. Lett. **55**, 418 (1985)
11. A. Goldmann, M. Donath, W. Altmann, V. Dose, Phys. Rev. **B32**, 837 (1985)
12. D.H. Yu, M. Donath, J. Braun, G. Rangelov, Phys. Rev. **B68**, 155415 (2003)
13. K. Yokoyama, K. Nishihara, K. Mimura, Y. Hari, M. Taniguchi, Y. Ueda, M. Fujisawa, Rev. Sci. Instrum. **64**, 87 (1993)
14. S. Suga, T. Matsushita, H. Shigeoka, A. Kimura, H. Namatame, Jpn. J. Appl. Phys. **32**, L1841 (1993)
15. H. Sumi, Y. Toyozawa, J. Phys. Soc. Jpn. **31**, 342 (1971)
16. A. Kimura, S. Suga, T. Matsushita, T. Kaneko, T. Kanomata, Solid State Commun. **85**, 901 (1993)
17. J.J. Yeh, I. Lindau, At. Data and Nucl. Data Tables **32**, 1 (1985)
18. R. Stiepel, R. Ostendorf, C. Benesch, H. Zacharias, Rev. Sci. Instrum. **76**, 063109 (2005)
19. M. Budke, V. Renken, H. Liebl, G. Rangelov, M. Donath, Rev. Sci. Instrum. **78**, 083903 (2007)

20. M. Maniraj, S.W. D'Souza, J. Nayak, A. Rai, S. Singh, B.N.R. Sekhar, S.R. Barman, Rev. Sci. Instrum. **82**, 093901 (2011)
21. Th Fauster, D. Straub, J.J. Donelon, D. Grimm, A. Marx, F.J. Himpsel, Rev. Sci. Instrum. **56**, 1212 (1985)
22. P.D. Johnson, S.L. Hulbert, R.F. Garrett, M.R. Howells, Rev. Sci. Instrum. **57**, 1324 (1986)
23. Y. Gao, M. Grioni, B. Smandek, J.H. Weaver, T. Tyrie, J. Phys. **E21**, 489 (1988)
24. P. Weibel, M. Grioni, D. Malterre, B. Dardel, Y. Baer, Phys. Rev. Lett. **72**, 1252 (1994)
25. P. Weibel, M. Grioni, D. Malterre, O. Manzardo, Y. Baer, G.L. Olcese, Europhys. Lett. **29**, 629 (1995)
26. M. Grioni, P. Weibel, D. Malterre, Y. Baer, L. Duo, Phys. Rev. **B55**, 2056 (1997)
27. M. Grioni, P. Weibel, M. Hengsberger, Y. Baer, J. Spectrosc. Rel. Phenom. **101–103**, 713 (1999)
28. W. Drube, F.J. Himpsel, Phys. Rev. Lett. **60**, 140 (1988)
29. K. Kanai, Y. Tezuka, T. Terashima, Y. Muro, M. Ishikawa, T. Uozumi, A. Kotani, G. Schmerber, J.P. Kappler, J.C. Parlebas, S. Shin, Phys. Rev. **B60**, 5244 (1999)
30. F.J. Himpsel, Phys. Rev. **B44**, 5966 (1991)

Chapter 11
Magnetic Dichroism and Spin Polarization in Photoelectron Spectroscopy

It is well know that each emitted photoelectron has its own spin. In the case of PES, however, many electrons in an instrumental acceptance angle and energy window are detected simultaneously. Then the spin information is often averaged out in the case of PES from non-magnetic materials even when the spin detector is used after the energy analysis. In the case of photoelectrons from remanently magnetized ferromagnetic materials, however, the spin of the photoelectrons are oriented to a certain direction and could be detected by a spin detector, irrespective of the mode of PES or ARPES. Even in the case of non-magnetic materials, the photoelectrons excited by a circularly polarized light can be spin polarized due to the spin–orbit interaction. This concept is fully utilized for realizing the spin-polarized electron sources. In the case of ARPES of magnetic materials, the dichroism of the photoelectron intensity depending on the helicity of the circularly polarized light and the direction of the linearly polarized light excitation is observed as magnetic circular and magnetic linear dichroisms called MCD and MLD. So MCD and MLD are first introduced. Then the principle and instrumentation for spin detection of photoelectrons are described including the up-to-date techniques. Spin polarized photoelectron spectroscopy in ARPES (SP-ARPES) on some typical materials are presented before introducing the SP-ARPES of topological insulators and Rashba effects. SP-ELEARPES is also very useful for such studies. For magnetic materials only typical results of SP-PES and SP-ARPES are presented. Finally spin polarized inverse photoemission (SP-IPES) is shown for QWS including the technique to provide the spin polarized electron sources.

11.1 Magnetic Circular and Linear Dichroism in Photoelectron Spectroscopy

It is well known that magnetic circular dichroism (MCD) of core absorption spectra (XAS) provides fruitful information on the electronic and magnetic structure in magnetic materials. As explained in Sect. 3.6.4, the real space imaging

S. Suga and A. Sekiyama, *Photoelectron Spectroscopy*,
Springer Series in Optical Sciences 176, DOI: 10.1007/978-3-642-37530-9_11,

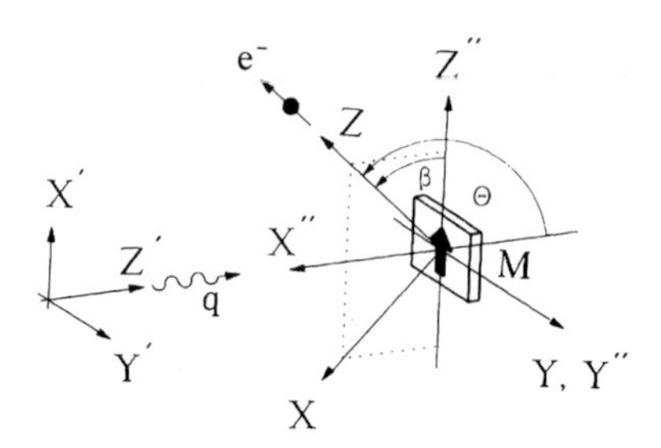

Fig. 11.1 Geometry of MCD experiments in core level photoemission from ferromagnets [2]. X″, Y″ and Z″ are the coordinates fixed to the sample. The in-plane magnetization of the sample is shown by the *thick arrow* **M**||Z″. The Z axis in the X, Y, Z coordinate system corresponds to the direction of photoelectrons to be detected. The Z′ axis in the X′, Y′, Z′ coordinate system for the incident light is along the light wave vector **q**||Z′, incident onto the sample. θ is the angle between the axes Z′ and Z. β is the angle between the axes Z and Z″. In the case of grazing incidence, $\theta-\beta$ becomes rather small [1]

was also facilitated by XAS-MCD PEEM. In MCD the spin–orbit and magnetic exchange interactions play important roles besides the dipole selection rule.

In photoelectron spectroscopy, magnetic circular as well as magnetic linear dichroism (MLD) was observed as explained below. Figure 11.1 shows the geometry employed in MCD measurements in photoelectron spectroscopy [1, 2]. The first experiment was performed on an Fe (110) surface for grazing incidence with $\theta - \beta = 10°$ (θ is the angle between the axes Z′ and Z) and $\beta = 55°$, where the emission angle was almost parallel to [111] and the photon wave vector **q** had a large component parallel to the remanent magnetization **M** [1]. Two spectra for the sample remanent magnetization **M** parallel and antiparallel to the fixed photon spin (helicity) were measured and their sum is shown in the upper panels of Fig. 11.2(b). Defining the photoemission intensity for the above-mentioned parallel (antiparallel) configuration as I^+ (I^-), the dichroic asymmetry A is here defined as $(I^+ - I^-)/(I^+ + I^-)$ [1]. The lower panel in Fig. 11.2(b) shows the asymmetry spectrum thus obtained. In the case of Fig. 11.2(a), the light was normally incident on the (110) crystal surface and the asymmetry is obtained by taking the difference between the two spectra for ±**M** and dividing by their sum [2].

Differential spectral shapes are observed in the asymmetry spectra in the Fe $2p_{3/2}$ and $2p_{1/2}$ peak regions, suggesting that the excitation energy is slightly shifted for opposite **M**. In the case of Fig. 11.2(b), where **q** has a large component along **M** as in most other MCD experiments, the peak splitting was roughly evaluated as 0.3 ± 0.2 and 0.5 ± 0.2 eV for the $2p_{1/2}$ and $2p_{3/2}$ peaks [1]. Such effects were expected under the coexistence of the spin–orbit interaction of the Fe 2p core hole and the exchange interaction between the core state and the 3d electrons. Namely, the photoexcitation follows the dipole selection rule from the initial state split into fourfold degenerate $|3/2, m\rangle$ (m = 3/2, 1/2, −1/2, −3/2) and twofold degenerate $|1/2, m\rangle$ (m = 1/2, −1/2) states to the unoccupied εd states. In the experiment in Fig. 11.2(b), no significant change of the asymmetry behavior was observed when

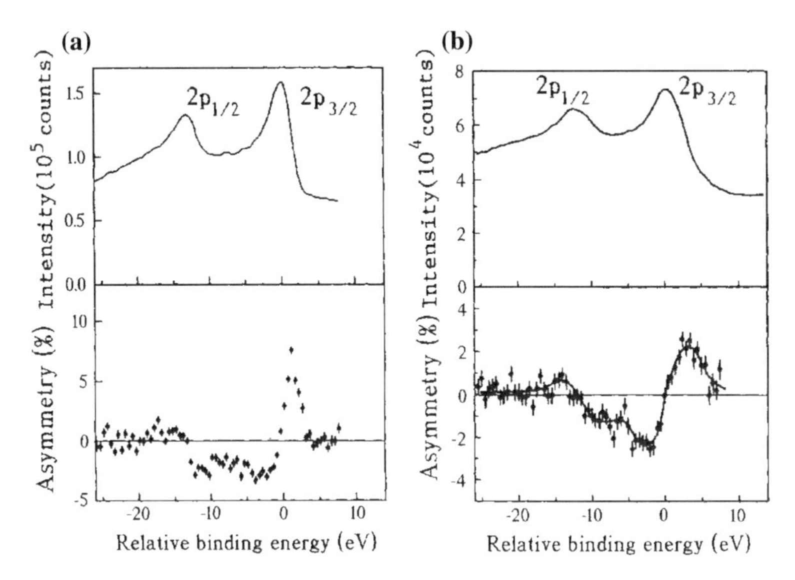

Fig. 11.2 MCD of an Fe (110) surface in the Fe 2p core excitation region for two different geometries. Circularly polarized light was normally incident onto the (110) surface in (**a**), where the vector **q** was perpendicular to the sample magnetization **M** along the [001] direction. Photoelectrons were collected at 45° to the surface normal and in the plane defined by **q** and **M**. The sum of the two spectra measured at $h\nu = 886$ eV for opposite **M** directions and the asymmetry are shown in the *upper* and *lower parts*, respectively [2]. The result at $h\nu = 850$ eV for grazing incidence [1] with nearly the same electron emission direction as (**a**) is shown in Fig. (**b**)

E_K was changed from 80 to 180 eV. So the final state effect was thought to be not playing a major role [1]. In fact, the matrix element for various absorption channels might have a different energy dependence and some final state effect might be observed as a function of E_K even in this case [3].

In the case of MCD, $\mathbf{q} \cdot \mathbf{M} \neq 0$ was thought to be a necessary condition. When the photon spin (light helicity) and the quantization direction defined by the magnetization or the exchange field are orthogonal, circular dichroism was expected to disappear as in the case of XAS-MCD. In this "forbidden" configuration with $\mathbf{q} \cdot \mathbf{M} = 0$, however, the magnetic circular dichroism was still observed in photoelectron spectra as shown in Fig. 11.2(a) [2], where $\theta - \beta = 90°$. Photoelectrons were collected along $\beta = 45°$ from **M** in the plane defined by **q** and **M**. The spectra were measured by reversing the remanent magnetization. The magnitude of the asymmetry (peak-to-peak asymmetry) was found to be up to 10 % near the $2p_{3/2}$ peak in Fig. 11.2(a), which was more than twice larger than that observed in Fig. 11.2(b). The present experiments were performed with some angle resolution in contrast to the XAS-MCD without angle resolution.

It was derived that the dichroic asymmetry for any linear combination of the final states derived from d wave functions contain two terms in the numerator of the expression for the before mentioned dichroic asymmetry, one proportional to $\mathbf{q} \cdot \mathbf{M}$ and the other proportional to $(\mathbf{q} \cdot \mathbf{z})(\mathbf{z} \cdot \mathbf{M})$, where **z** is the polar axis along

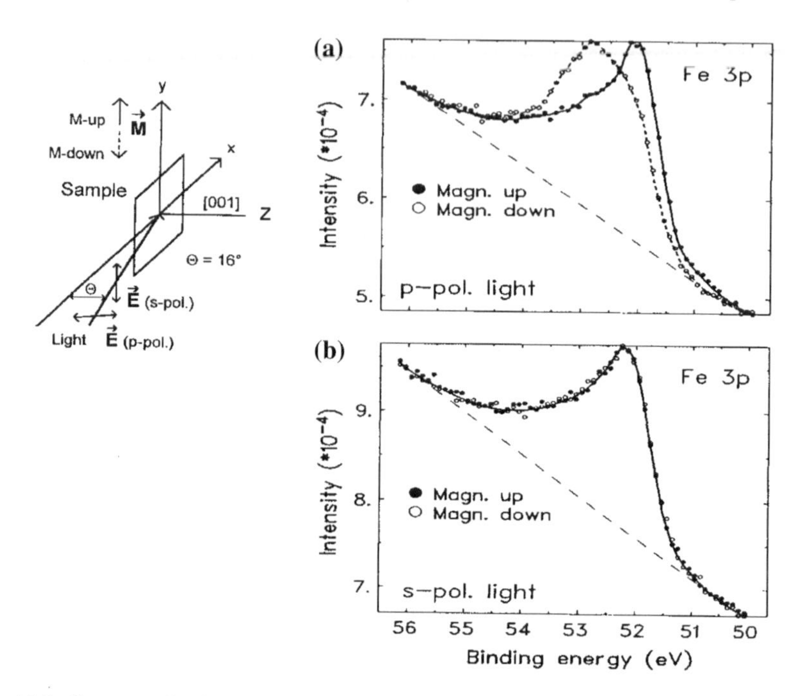

Fig. 11.3 Geometry for the measurement of MLD of the Fe 3p core level ARPES is shown on the *left*. At grazing incidence of $\theta = 16°$ p- or s-polarized light impinged onto the Fe (001) surface, nearly parallel to the [100] direction. Normal emission photoelectrons were collected within $\pm 4°$. The sample was remanently magnetized parallel or antiparallel to the y axis, as denoted by *Magn. up* and *Magn. down* in Figs. (**a**) and (**b**) [5]. Spin-integrated spectra excited at $h\nu = 90$ eV with p–polarized SR are shown in (**a**) and those with s-polarized SR in (**b**)

the electron emission direction [2–4]. In the latter term, the emission direction plays an importance role for providing a finite value even in the case of $\mathbf{q} \cdot \mathbf{M} = 0$, since both $(\mathbf{q} \cdot \mathbf{z})$ and $(\mathbf{z} \cdot \mathbf{M})$ can be non zero in ARPES. In an angle integrated experiment such as absorption or total yield measurement, the sum of the latter asymmetry term becomes zero and the asymmetry proportional to $\mathbf{q} \cdot \mathbf{M}$ is recovered. It is semi-quantitatively predicted that the magnitude of the asymmetry is approximately 50 % larger in the configuration of Fig. 11.2(a) compared to Fig. 11.2(b) [2]. ARPES intensity and the MCD for normal incidence would be thus useful to determine the magnetization direction in the crystal plane without sample rotation.

Even in the case of linearly polarized SR excitation, ARPES of ferromagnetic materials shows magnetic linear dichroism (MLD). The left panel in Fig. 11.3 shows the experimental geometry employed for the study of the 3p core level ARPES of Fe [5]. The measurement was performed at $h\nu = 90$ eV on a thin Fe (001) film epitaxially grown on Ag (001) or Au (001). The remanent magnetization $\mathbf{M}$ was reversed between the parallel and antiparallel direction to the y axis in the (001) sample plane. Normal emission spectra around the z axis were measured.

Figure 11.3(a) shows the ARPES spectra for the opposite magnetization directions under the p-polarized SR excitation. Above the background from inelastically scattered electrons, the 3p core level ARPES showed a clear difference between the up and down magnetizations. Namely the peak energies and spectral shapes are much different in between. In contrast, the spectra for the up and down magnetization were almost degenerate under the s-polarized SR excitation as shown in Fig. 11.3(b). In both measurements, electrons emitted to the [001] direction were detected. Under the presence of the spin–orbit interaction, excitation by p-polarized light (with the electric field vector **E**) leads to spin polarization with a quantization axis parallel to $\mathbf{E} \times \mathbf{p}$ (**p** is the photoelectron momentum) [6], namely along the y direction in the present case. The quantization axis for the exchange interaction is the long-range magnetization direction, which is either parallel or antiparallel to the y direction. In other words, two quantization axes are collinear in the present case for p-polarization. The intensity of the peak with spin polarization originating from the exchange interaction is either enhanced or suppressed depending upon whether exchange and spin–orbit induced spin polarizations are parallel or antiparallel. With s-polarized light, however, spin polarization due to the spin–orbit interaction cannot occur for symmetry reasons [6]. Even for p-polarized light, the spin polarization due to spin–orbit interaction vanishes upon full emission angle integration. The Fe 3p spectral shapes should not depend upon the magnetization direction in angle integrated PES.

Detailed incidence angle θ dependence of the MLD in core level photoemission was measured for the Co 3p core level of Co/Cu(100) ultrathin films [7] upon reversal of the remanent magnetization. Normal emission spectra were measured under variation of both the polar and azimuthal angles (θ, Φ) of the incident SR at $h\nu = 125$ eV. This film had an in-plane easy magnetization axis along the $\langle 110 \rangle$ azimuth with a Curie temperature well above room temperature for a film thickness higher than 2 ML. The right panel in Fig. 11.4 shows a series of asymmetry spectra of a 6 ML film as a function of θ at a fixed $\Phi = 0°$. In this case, the remanent **M** was perpendicular to the plane defined by the photon wave vector **q** and the sample surface normal **n** and reversed between the x and $-$x directions during the measurement. It was noticed that the spectral shape, a pronounced peak around $E_B = 58.5$ eV and a broader feature with the opposite asymmetry around $E_B = 60$ eV, did not change much with the change of θ. Only the amplitude and sign of the asymmetry changed with θ. If the θ dependence of the total asymmetry was plotted at $\Phi = 0°$, a $\sin 2\theta$ behavior was recognized, whereas the Φ dependence at $\theta = 36°$ showed roughly a $\cos\Phi$ dependence. According to a theory [8], an angular dependence of the dichroism for normal electron emission along a high symmetry direction was predicted as $4\sin\theta \cdot \cos\theta \cdot \cos\Phi \cdot \mathrm{Re}[\chi_{zy}]$ for linearly polarized light [7], where the $\sin\theta \cdot \cos\theta \cdot \cos\Phi$ dependence results from the projection of the polarization vector onto the z and y axes of the crystal. The matrix element χ_{zy} is given by $\langle \Psi_i(\mathbf{M})|z|\Psi_f(\mathbf{k})\rangle\langle \Psi_f(\mathbf{k})|y|\Psi_i(\mathbf{M})\rangle$, where $\Psi_i(\mathbf{M})$ and $\Psi_f(\mathbf{k})$ are the initial and final state wave functions, and does not exhibit an angular dependence. From the qualitative agreement of the angular dependence between the theory and experiments [7, 8], the angular functions in front of a fixed matrix

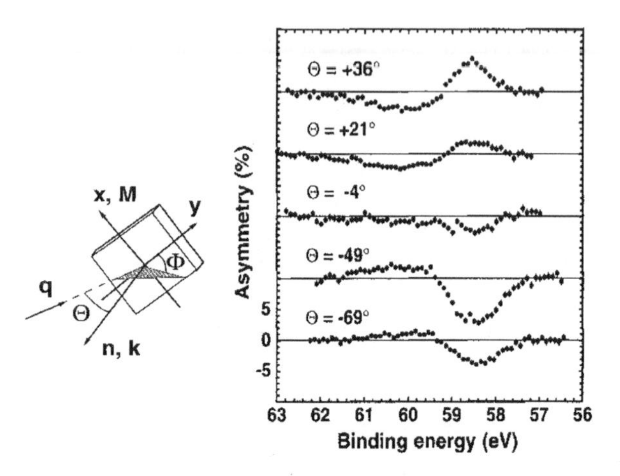

Fig. 11.4 MLD measurement in normal emission from the Co 3p core level in an ultrathin Co/Cu(100) film [7]. The experimental geometry is shown on the *left*. x and y are parallel to the ⟨110⟩ symmetry direction within the sample surface and **n** shows the surface normal direction. The electron momentum **k** was parallel to **n** in this case. Remanent magnetization **M** was aligned parallel or antiparallel to the x axis. Photons with momentum **q** were incident at a polar angle θ from **n** and at an azimuthal angle Φ from the y axis. The linear polarization **E** of the photons was in the plane defined by **q** and **n**. *The right panel* shows the θ dependence of the asymmetry spectra of the 6 ML Co/Cu(100) film measured at $\Phi = 0°$ and $h\nu = 125$ eV

element were thought to be playing a major role for the angular dependence. Such an effect as observed here in angle resolved measurements was not expected in angle integrated measurements.

In the case of unpolarized X ray excitation of an Fe (001) surface by a conventional laboratory X-ray source, magnetic dichroism was also observed around the low-index forward scattering direction [9] in the crystal for the Fe 2p core level photoemission [10], exhibiting antisymmetry with respect to the emission angle and E_B that was centered on each forward scattering peak. Photoelectron diffraction theory provided a quantitative explanation for this effect.

11.2 Principle and Instrumentation for Spin Polarized Photoelectron Spectroscopy

There are two major cases of application of spin polarized photoelectron spectroscopy (SP-PES). One is the application to nonmagnetic materials with utilizing the selection rules for circularly polarized light excitation based on the spin–orbit interaction. Another is the application to magnetic materials or spin polarized states even with unpolarized light. It was also revealed recently that partial spin polarization was induced on the surface of originally non-magnetic materials.

In all cases the spin of the emitted electrons should be analyzed by spin detectors. The spin polarization P_s can be defined as

$$P_s = (n\uparrow - n\downarrow)/(n\uparrow + n\downarrow), \tag{11.1}$$

where n↑ and n↓ represent the numbers of photoelectrons with up and down spins. In order to evaluate this quantity, various spin detectors can be employed. There are a few major spin detectors employed so far for experiments. The first one, which was most widely employed in the past, is the Mott detector [11]. While 100 kV detectors with a Au target had been traditionally used [12], low energy Mott detectors (called mini Mott detector) working in the range of 20–40 kV became more popular [13]. The working principle can be explained by the Dirac equation, which explicitly takes into account the spin and the spin–orbit interaction as below:

$$\left[(\mathbf{p} + e\mathbf{A})^2/(2m) - e\Phi + e\hbar\sigma\cdot\mathbf{B}/(2m) - ie\hbar\mathbf{E}\cdot\mathbf{p}/\left(4m^2c^2\right) + e\hbar\sigma\cdot(\mathbf{E}\times\mathbf{p})/\left(4m^2c^2\right)\right]\psi = W\psi, \tag{11.2}$$

where W + mc^2 is the total energy. **A** is the vector potential, **p** is the electron momentum, Φ is the scalar potential, σ is the electron spin and **B** is the magnetic field. The third term is the Zeeman energy and the fourth term is a relativistic correction. The last term stands for the spin–orbit energy, which plays an important role for the right-left scattering asymmetry. In this last term, the electric field **E** near the target atom can be proportional to dV/dr·**r**/r, where V represents the central field potential and r represents the distance. Then the vector product between **E** and **p** is proportional to **r** × **p**. This energy term then depends upon the inner product between this vector product and the spin. The energy for the ↑ and ↓ spin is therefore different as understood in Fig. 11.5(a), where the solid curve shows the energy before the spin-orbit interaction term is considered. In this case the energy is lower when the ↑ electrons pass on the left hand side of the target atom. For the ↓ spin electrons the energy becomes higher than that. The left–right asymmetry of the scattered intensity,

$$A(\theta) = (N_L - N_R)/(N_L + N_R) \tag{11.3}$$

of the scattered spin-polarized electrons, where θ is the scattering angle [11–13], is thus induced by the spin–orbit interaction [11–13]. Since the magnitude of the spin–orbit interaction is larger for heavier atoms, such materials as W, Au, Hg, Pt are used as target materials for scattering of spin polarized electrons. The P_s can be indirectly estimated by the experimentally obtained left–right asymmetry A(θ) as

$$P_s = A(\theta)/S(\theta), \tag{11.4}$$

where S(θ) is the Sherman function depending upon the scattering angle θ. Even when completely spin polarized electrons with P_s = 1.0 are incident onto the spin polarimeter, A(θ) shows the value of S(θ), which is much smaller than 1.0. S(θ)

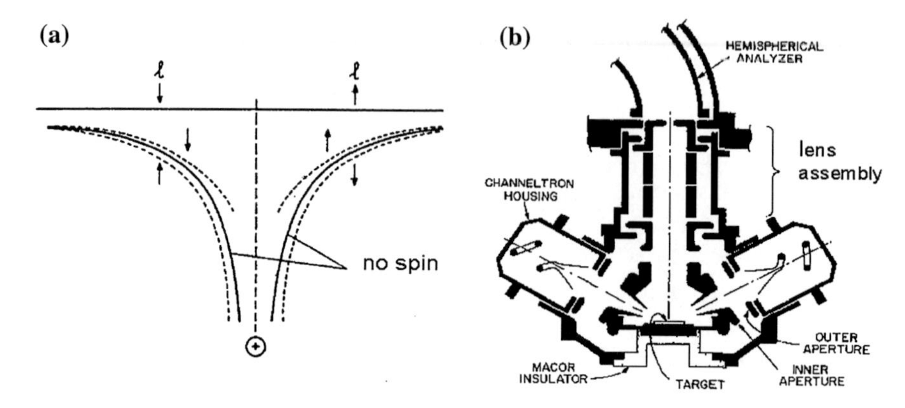

Fig. 11.5 **a** Schematic energy diagram of an electron passing on the *left* and *right hand* side of a positive ion. *The solid line* shows the potential without considering the spin–orbit interaction. *The dashed line* shows the energy when the spin–orbit interaction is taken into account. ↑ and ↓ show the incident electron spin perpendicular to the scattering plane. The intensity of the scattered electrons into the left and right directions (N_L and N_R) in regard to the spin direction is measured [11]. **b** Compact retardation-potential polarimeter operated at ~20 kV [13]. In this case the spin perpendicular to the paper plane is resolved. The in-plane spin can be resolved by the other two detectors located back and forth of the paper plane

depends upon the employed target material, the electron collision energy and the range of detection angle. In order to evaluate the absolute spin polarization P_s of the incident electrons, a careful calibration of the spin detectors is therefore required.

For Mott detectors, Au thin foils are often used as the target. Since high voltage (100–120 kV) Mott detectors [12] require high security and occupy a big volume, lower energy (20–40 kV) mini-Mott detectors became more popular. Figure 11.5(b) shows an analyzer operated at ~20 kV. The effective Sherman function S_{eff} of Mott type polarimeters lies in the range of 0.1–0.4. The figure of merit or the efficiency defined by $(I/I_0)S_{eff}^2$, where I_0 is the electron current entering the polarimeter and I is the total scattered current measured by the left and right detectors, lies in the range of $\sim 10^{-7}$–2×10^{-4} [13]. In a compact 25 kV mini-Mott spin polarimeter, for example, $S_{eff} = 0.17$ and a figure of merit of 1.4×10^{-4} were reported [14].

As explained in Chap. 9, ELEPES became very popular in both angle-integrated and -resolved modes. Then SP-PES with much higher energy resolution was desired. Figure 11.6 illustrates a mini Mott detector combined with a hemispherical analyzer. A very bright Xe lamp combined with a cylindrical mirror and a spherical grating monochromator, all designed and manufactured by MB Scientifc AB (MBS) are used as a monochromatic light source, which can be focused down to 3 mm Φ onto the sample surface [15]. Both mirror and grating surfaces were coated with thin MgF_2 on Al. The MCP was displaced by 1 cm towards the inner sphere of the analyzer and a 4 mm Φ hole was introduced between the MCP and the outer sphere to guide the photoelectrons to the 90° deflector and a mini

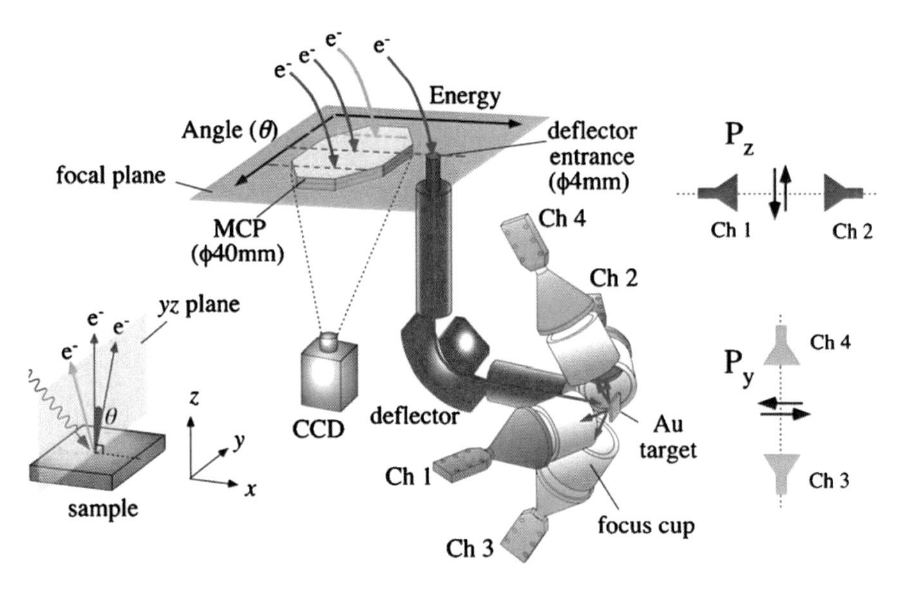

Fig. 11.6 Schematic drawing of a very high resolution 25 kV mini Mott detector, MUSPIN, manufactured by MB Scientific AB and used in SP-PES with a Xe lamp [15]. In the focal plane between the inner and outer spheres of the MBS A-1 analyzer, a MCP and an entrance hole of the 90° deflector to spin detectors were located closer to the inner and outer spheres. The size of the entrance hole was 4 mmΦ. By rotating the sample by 90° in the focal plane all three spin components can be detected

Mott detector. Best energy and angular resolutions in the spin-integrated mode were 0.9 meV and 0.2° [15]. This 25 keV mini Mott detector was used for spin analysis in this configuration. In the spin polarized mode, the angle over the 4 mm Φ hole was integrated, resulting in an angular acceptance of $\sim\pm0.6°$. Electrons were accelerated to 25 keV and backscattered after hitting a Au target. Scattered electrons within a $\pm15°$ cone were collected at $\pm120°$ from the normal of the Au surface. Inelastically scattered electrons were suppressed by the retarding field of 24 kV at the channeltron. Three spin components were detected by rotating the sample by 90° in the x–y plane. Very high energy resolutions of 8 meV were achieved for spin-polarized measurements.

The low-energy-electron-diffraction (LEED) spin detector based on a W single crystal (W SPLEED detector) was also used for spin analysis [16]. As shown in Fig. 11.7(a), the electrons pass through the center-holed resistive-anode and the MCP and hit the W (001) surface in normal incidence. Among the LEED spots excited at 104.5 eV, the four (2,0) LEED spots (Fig. 11.7(b)) are used to evaluate the spin polarization parallel to the W(001) surface. Although a compact design was possible owing to the use of low voltages, the W (001) single crystal surface should be cleaned to keep the high value of the Sherman function of about 0.27 ± 0.02. The cleaning could be done by flashing every 20–30 min in a vacuum of $1\text{–}2 \times 10^{-8}$ Pa. The figure of merit is about 1.6×10^{-4}.

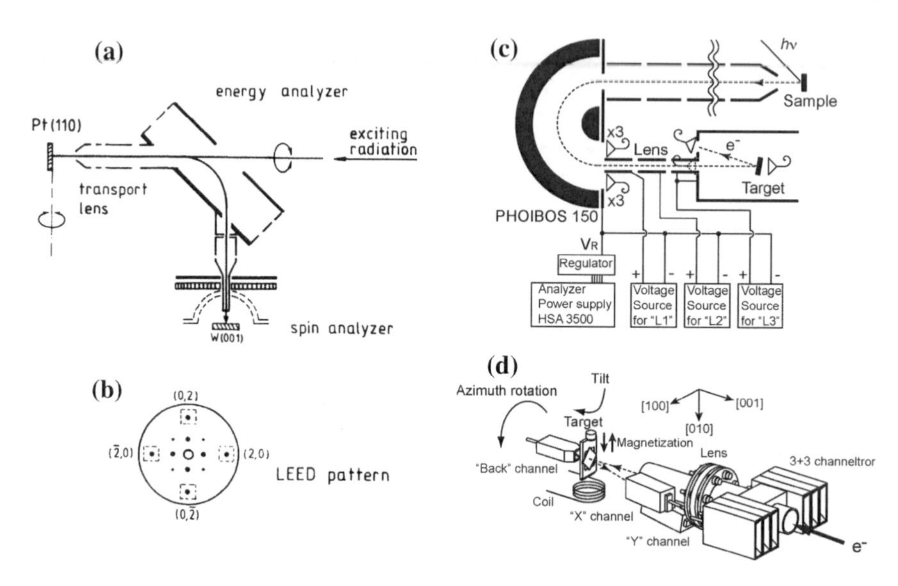

Fig. 11.7 W (001) spin LEED detector employed for the spin polarized electron detection excited by circularly polarized SR (**a**) [16]. **b** Intensity of the (2, 0) LEED spots is used to estimate the spin polarization. For example, (0, 2) and (0, $\bar{2}$) beams measure the spin polarization component along the surface normal. Although not necessary in the present case, the surface parallel spin polarization can be measured by the ($\bar{2}$, 0) and (2, 0) spots. In addition, both the electron energy analyzer and the spin analyzer are rotatable around the horizontal axis. A very-low-energy-electron-diffraction (VLEED) spin detection system is shown in (**c**) [18]. Details near the Fe (001) target are shown in (**d**), where the electron comes from *the right bottom* onto the magnetized Fe (001)−p(1 × 1)−O surface grown on a MgO (001) substrate

In Fig. 11.7(c) is shown a very-low-energy-electron-diffraction (VLEED) spin detector using an Fe (001) surface [17, 18]. The spin exchange interaction is used to measure the spin polarization of the electrons incident onto the magnetized ferromagnetic target. In the original design, the Fe (001) surface was prepared by depositing Fe onto a Ag(100) surface. However, cleaning of the surface after several hours was required to keep the high spin sensitivity. Later an oxygen passivated Fe (001) − p(1 × 1) − O surface was proposed as a target of the VLEED detector [18]. By tuning the electron energy to a critical point of the target's band structure, a large spin asymmetry was realized for the specular reflection. When the electrons impinged onto the target surface, they were absorbed if any matching band exists. If the band structure showed a gap in this energy region, they were not absorbed and efficiently reflected back. Since the band structure depends upon the spin in ferromagnetic materials, the absorption and reflection depended critically upon the incidence energy. The spin sensitivity was expected larger for a ferromagnet with larger spin exchange splitting. In the case of this target, the maximum figure of merit was realized at $E_K = 6$ eV. When a ferromagnetic target is employed, the magnetization of the target should be reversed and two sequential measurements are required to evaluate the electron spin polarization. Spin polarized electrons even from non-magnetic materials could be reliably measured by this spin detector. An effective Sherman function S_{eff} of about 0.40 ± 0.02 and a figure of merit of about (1.9 ± 0.2) × 10^{-2} were obtained [18]. By the set up in Fig. 11.7(c), only the spin polarization perpendicular to the paper plane could be evaluated. In the geometry of Fig. 11.7(d), spins along the [010] direction could be measured by the channeltron X. By rotating the target and the coil by 90°, spins along the [100] direction could be measured by the channeltron Y. The stray magnetic field was minimized by the use of the Fe film. Quite high figure of merit and stable performance over weeks are the merit of this type of VLEED spin detector.

The low efficiency of single-channel electron spin polarimeters has so far hindered this technique to be widely applied to various fundamental experiments. By using the concept of multichannel detection, however, the detection efficiency can be improved by orders of magnitude, opening the possibility of much wider application to studies such as spin polarized HAXPES with high bulk sensitivity for the valence band, dynamical processes, microscopy and so on [19, 20]. 2D multichannel spin detection is based on the idea of preserving the 2D electron distribution after a hemispherical analyzer in the next process of spin-polarized low-energy electron diffraction [19]. Figure 11.8 schematically describes the setup, where a W (100) crystal is used as the spin filter. The incident electrons are specularly deflected and the electron momentum parallel to the W crystal surface is conserved due to the (0, 0) beam diffraction. In this case, the dispersive direction corresponds to the electron energy (E_i) and the nondispersive direction corresponds to the emission angle (θ_j) in the exit field of a modern electron spectrometer. The polarization perpendicular to the scattering plane, namely P↑ and P↓, could be analyzed via diffraction at an electron kinetic energy of about 26–27 eV, where the optimal efficiency is realized with the Sherman function S ~0.43 and

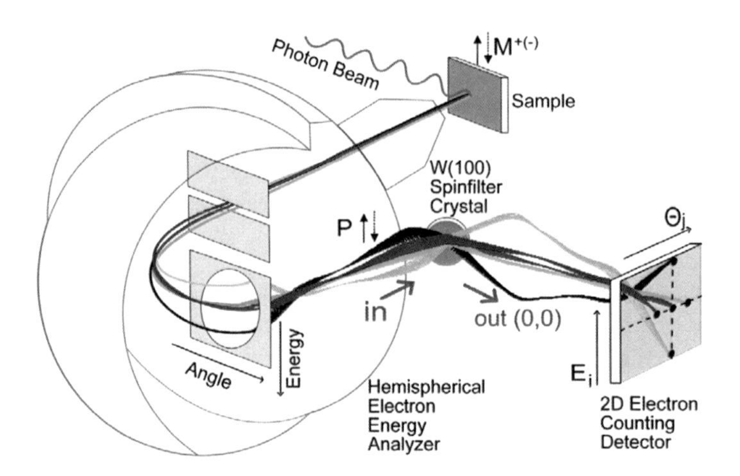

Fig. 11.8 Schematic view of the multichannel spin polarimeter [19]. Angle dependence and energy dependence are illustrated

the reflectivity I/I_0 of about 0.012 [19]. A figure-of-merit of about 2.1×10^{-3} was then obtained, which is higher than most other single-channel spin-detectors [21] except fo VLEED [18]. The spin filtered image was then recorded by a 2D electron counting detector such as a delay line detector with a lateral resolution of $\sim$50 μm and a lateral size of 40 mm [19]. For 64 resolvable angular slices and 17 resolvable energy points, for example, 1,044 data points were simultaneously recorded. The 2D spin polarization pattern $P_{i,\,j}$ was obtained via pixel (i, j) based processing of two electron-distribution patterns measured for two sample magnetizations.

Compared to the single channel spin detectors for electron spectroscopy, this type of multichannel spin polarimeter has the advantage of a greatly increased efficiency due to the multiplication of the number of resolved data points and the figure of merit. The resulting 2D effective figure-of-merit can be $\sim 10^4$ times larger than the typical value for a single channel spin detector. This technique is very useful for both angle-integrated and -resolved photoelectron spectroscopy with spin analysis. By using an optimized spectrometer, the numbers of resolvable points may be remarkably increased and a 2D figure-of-merit being $\sim 10^5$ times larger than the typical single channel spin detector will be feasible. In the typical case of a W (100) single crystal, an ideal parallel incident beam will be diffracted into a cone of around 0.1° by the mosaic spread of the crystal inducing a spatial deviation of sub mm in the imaging plane [19]. In spite of this limitation, the above mentioned numbers will be attainable.

Meanwhile, a momentum microscope composed of a PEEM column and an aberration corrected electrostatic energy analyzer was developed as schematically shown already in Fig. 3.28(d) [22]. Together with an ultrafast laser light source and an imaging spin filter, it is possible to combine spatial, momentum, energy, time and spin resolutions based on this instrument. The extraction voltage for photoelectrons from the sample can reach up to 15 kV. Either spatial resolved imaging or angular resolved k-space imaging is possible at a certain E_K by switching between the spatial and momentum resolving modes while keeping the position of the first real image in the field aperture and the position of the k-space image in the contrast aperture. A multichannel spin-polarization analysis was combined with this technique, where the spin filter was installed directly after the imaging double energy analyzer (IDEA) [20] (Fig. 3.28(d)), resulting in the unique combination of a 2D spin filter and a momentum microscope. In the case of the spatially resolved imaging (Figs.11.9 and 3.28(c)), electrons after the energy analyzer are decelerated from the pass energy of the energy analyzer ($\sim$100 eV) to the scattering energy ($\sim$30 eV) by an electrostatic retarding lens and the electrons originating from the same point in the spatial image are incident on the W(100) crystal as a parallel beam (Fig. 11.9). The angle of incidence is conserved upon the specular reflection by W(100). Then the spatial image is recovered after the second symmetrical retarding lens with spin contrast [20]. An instrumental image resolution <300 nm was achieved. For scattering energies of 15–90 eV, the reflected intensity from a W(100) crystal varies over two orders of magnitude and the asymmetry shows an oscillatory behavior with sign changes. A scattering energy near 27 eV results in optimal efficiency. Likewise, the spin polarization at

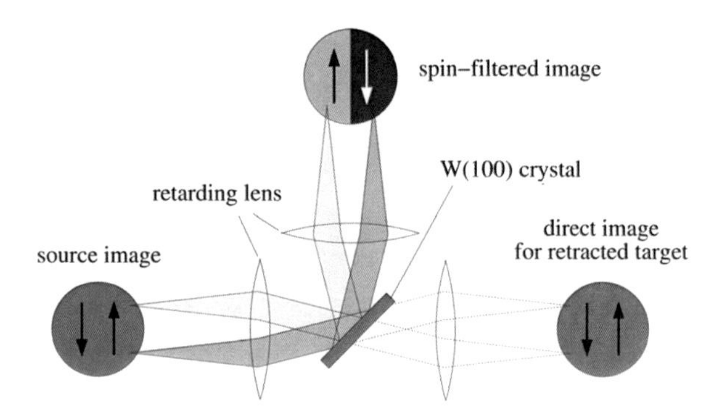

Fig. 11.9 Illustration of spin-filtering of two dimensional images by low energy specular reflection of electrons at a W (100) surface [20]. The W spin filter is inserted into the electron optical path directly after the IDEA energy analyzer shown in Fig. 3.28(c)

(k_x, k_y) can also be detected in the k-space imaging mode. This novel concept of multichannel spin-polarization analysis shows great potential for application in various new fields ranging from real-space imaging of nanomagnetism to k-resolved studies of the spin-dependent electronic structure of materials.

11.3 Spin Polarized Photoelectron Spectroscopy for Non-Magnetic Materials

Let us move to the discussion of several experimental results of spin polarization of photoelectrons from nonmagnetic materials. SP-PES is a powerful means to study the occupied spin states. By use of circularly polarized light, one can preferentially excite the spin polarized electrons even in non-magnetic materials by the dipole selection rule due to the spin–orbit interaction as schematically shown in Fig. 11.10 [23] for the p-like initial state. A similar approach is feasible in the case of a d-like initial state. The spin polarization of angle-resolved photoelectrons from nonmagnetic materials excited by circularly polarized light is thus generally induced by the spin-orbit interaction and dipole selection rule in contrast to the case of magnetic materials, where the spin polarization is primarily attributed to the initial state.

11.3.1 Pt

SP-ARPES experiments were performed for the Pt (111) unreconstructed surface by use of a Mott detector in the region of $h\nu = 6.5-24$ eV [24] and by a W SPLEED detector [25] in the normal incidence and normal emission configuration. The circular polarization of the incident SR light was estimated as $\sim\pm92 \pm 1$ % [12]. The spin polarized photoelectron spectra on Pt (111) measured with good statistics by a W SPLEED detector at $h\nu = 13$ eV are reproduced in Fig. 11.11(a–c) [25]. The SP-ARPES spectra of (1 $\times$ 2) reconstructed Pt(110) surface measured by the

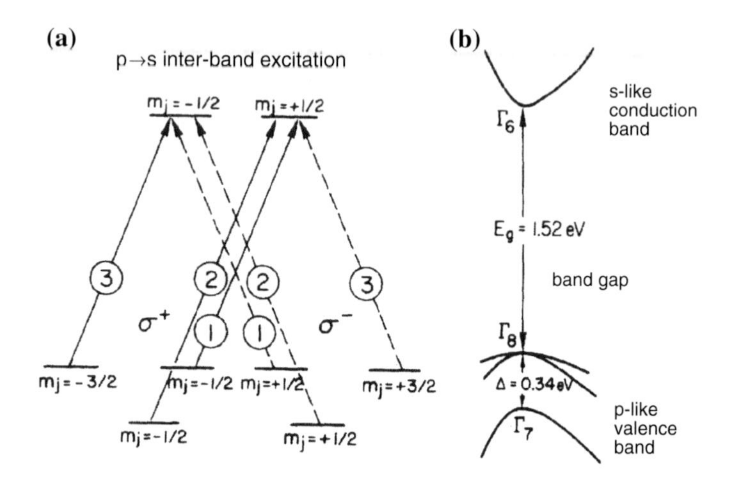

Fig. 11.10 Dipole selection rule and the transition probability (numbers encircled) from the p state split by the spin–orbit interaction to an s state for the left helicity (or *left handed screw*) light (σ^+) and the right helicity light σ^- (**a**) [23]. Schematic energy diagram of GaAs employed for the spin polarized electron source (**b**) as later explained in Sect. 11.5

same SPLEED detector are shown in Fig. 11.11(e) and (f) [26]. Polarization spectra correspond to the spin asymmetry and the intensity spectra (c) and (f) show the spin polarized spectra (Fig. 11.11). In all these cases the wave vector $k_{||}$ (parallel to the surface) was kept 0 and the wave vector $k_{\perp}$ (perpendicular to the surface) (parallel to $\langle 111 \rangle$ or $\langle 110 \rangle$) was changed as a function of hν. Clear spin polarization up to 20–40 % is observed in some E_B regions depending upon hν. In Fig. 11.11(c), the spin polarized spectra are derived from the total intensity I and the spin polarization P as $I^+ = I(1 + P)/2$ and $I^- = I(1-P)/2$, where I^+ and I^- stand for the numbers of photoelectrons with the spin parallel and antiparallel to the surface normal. It was straightforward recognized in Fig. 11.11(c) that the peaks A and B (C and D) in the total intensity spectrum correspond to peaks in different spin polarization spectra. Since these states share the same final band with nearly free-electron character [25], the peak B and C were thought to have the same λ_6 symmetry initial states as shown in (Fig. 11.11(d)). The peaks A and D belong to the $\lambda_4\lambda_5$ states. The broad feature C has two structures in the I^- spectrum (c) as indicated by the downward arrows. In comparison with theoretical band-structure calculations, the initial states of these two structures must have the same symmetry, suggesting that these two bands are not crossing each other but show anticrossing behavior. In wide energy regions in the valence band, band crossing and anticrossing take place occasionally. It was rather difficult by conventional ARPES to discriminate these differences. By virtue of the spin polarization behavior, one could rather easily discriminate these two cases as in this case of SP-ARPES of Pt (111) [25]. A similar experiment was performed for the (1 × 2) reconstructed Pt (110) surface, the metastable unreconstructed Pt (110) surface and the CO adsorbed (2 × 1) p1g1 surface [26]. SP-ARPES on the (1 × 2) reconstructed Pt (110) surface is shown in Fig. 11.11(e) and (f). Although a

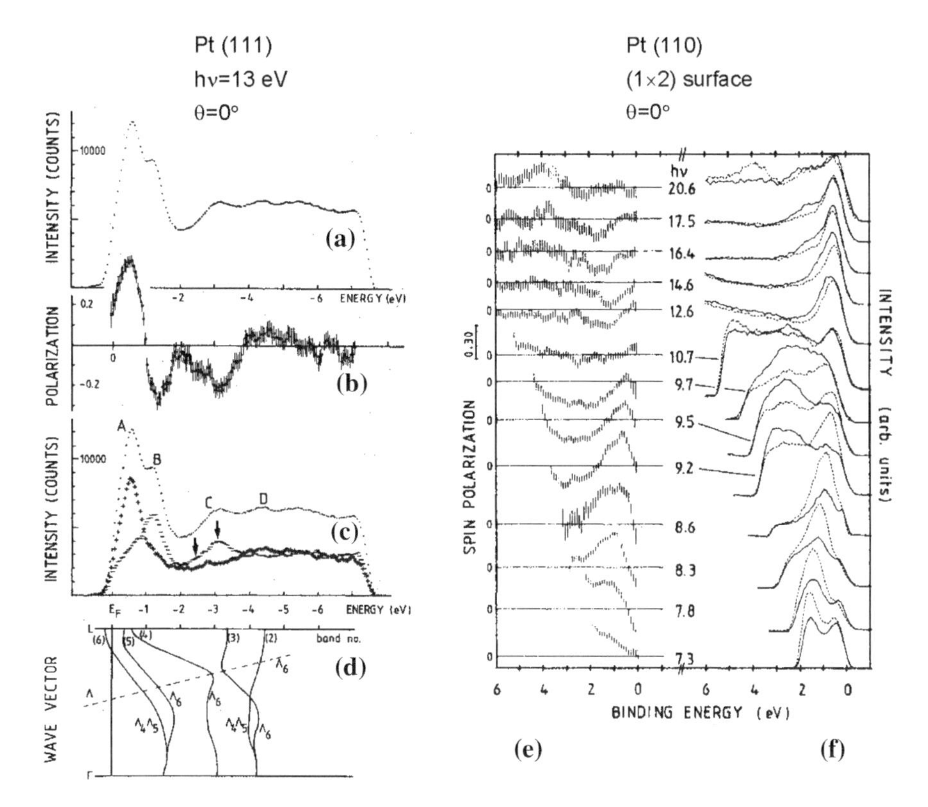

Fig. 11.11 SP-ARPES in the normal incidence and normal emission configuration for the Pt (111) surface in Figs. (**a**)–(**d**) [25] and for the (1 × 2) reconstructed Pt(110) surface in Figs. (**e**) and (**f**) [26]. *The dashed line* in (**d**) corresponds to the final-state band shifted by 13 eV. All spectra were measured by the apparatus shown in Fig. 11.7(a). In order to suppress secondary electrons, the sample was, for example, biased by –10 V relative to the ground. The angular acceptance was estimated to be about ±5° and the total energy resolution was set to about 0.3–0.4 eV

straightforward theoretical interpretation was difficult regarding the $\langle 110 \rangle$ direction, the results experimentally clarified the spin dependent hybridization of the initial state along the Σ ($\langle 110 \rangle$) direction in the BZ.

11.3.2 High-T_c Cuprate

The circularly polarized soft X-ray excitation for spin analysis was applied to the study of the valence bands of the high T_c cuprate $Bi_2Sr_2CaCu_2O_{8+\delta}$ (Bi2212) [27]. The hν was tuned to the Cu $2p_{3/2} \rightarrow 3d$ absorption edge to induce resonance enhancement of the Cu 3d states. The resonance is a consequence of interference between the $|3d^9\rangle + h\nu \rightarrow |3d^8\rangle + \varepsilon$ direct photoemission and the core level absorption $|2p^63d^9\rangle + h\nu \rightarrow |2p^53d^{10}\rangle$ followed by the direct recombination process as $|2p^53d^{10}\rangle \rightarrow |2p^63d^8\rangle + \varepsilon$, where ε represents the photoelectron. Due to this resonance effect the $|3d^8\rangle$ final state PES intensity is increased by two orders of magnitude, facilitating the spin measurement. In the experiment at ID12B beam line of ESRF, optimally doped Bi2212 crystals were cleaved in UHV and excited by a circularly polarized light with a degree of polarization of ~92 %. The spectra obtained by measurements at room temperature with an angular acceptance of ±20° by use of a 25 kV Mott detector with a Sherman function of 0.17 and with a total energy resolution of 0.75 eV is shown in Fig. 11.12 [27]. The measurements were made for a certain helicity and repeated for the opposite helicity to remove systematic errors. Both spin integrated and spin polarization spectra were measured. The resonance spectrum mainly represents the $3d^8$ final states. The prominent peak at around 12 eV with the spin polarization of ~80 % was ascribed to the ^1G spin singlet state. If the $|3d^9\rangle$ initial state has a hole in the x^2-y^2 orbital and the **E** vector of the excitation light is in the x–y plane, a spin polarization of 83.3 % (−27.8 %) is expected for the pure spin singlet (triplet) state [27, 28]. The strong dip in the spin polarization around 9.3 eV was then interpreted as due to the contribution of the spin triplet states.

Assuming the above mentioned spin polarization for pure spin states, spectral contributions of the singlet and triplet components were estimated as shown by thick marks for spin singlet and triplet in Fig. 11.12(a). The increase of the spin polarization near E_F (Fig. 11.12(b)) suggests the contribution of the spin singlet state. The difference between the singlet and triplet components is very clearly seen in Fig. 11.12(c) near E_F with the expanded ordinate. The mutual energy

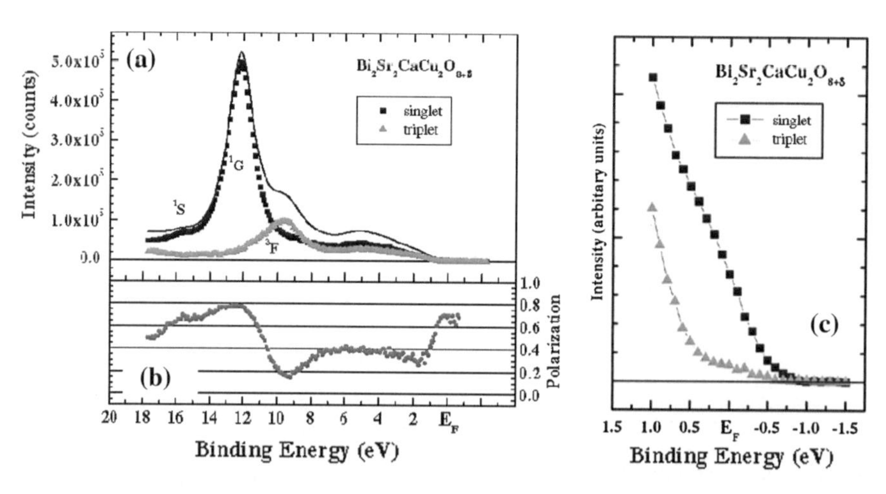

Fig. 11.12 Spin polarized photoelectron spectra from $Bi_2Sr_2CaCu_2O_{8+\delta}$ [27]. *The full thin line* shows the spin integrated resonant PES excited at the Cu $2p_{3/2}$ absorption edge. *The full black rectangles* and *grey triangles* show the spin singlet and spin triplet components (**a**). The spin polarization is shown in (**b**). Spin polarized photoelectron spectra near the Fermi level are shown in (**c**), where the spin singlet component is well separated from the spin triplet component

separation was found to be close to 1 eV. The behavior below the superconducting transition temperature was found to be essentially the same. It was thus found that the spin singlet state is very stable even at room temperature. The singlet states are mostly of $|3d^9\underline{L}\rangle$ character ($\underline{L}$ represents the hole on the O sites) [29, 30] and these states are probed through the hybridization with the $|3d^8\rangle$ states which are strongly enhanced by the resonance in this experiment. The robustness of the singlet character against the hole doping was thus confirmed in high T_c cuprates.

11.3.3 Rashba Effect and Topological Insulators

Spin-polarized (or resolved) and angle-resolved PES (SP-ARPES) can also be applied without employing circularly polarized light to the spin measurements of non-magnetic materials. Some examples are shown in the following, first for the Rashba effect and then for topological insulators, which were already explained to some extent in Chap. 6. The Rashba effect [31] is a low dimensional physical property to produce a spin-polarized 2D electron gas even for nonmagnetic materials. While the opposite spin states are degenerate in nonmagnetic bulk materials (Kramers degeneracy) due to the presence of both time-reversal and space-inversion symmetries, this degeneracy is lifted by the spin–orbit interaction when the space-inversion symmetry is broken in a 2D system such as the crystal surface. The splitting thus induced is called Rashba splitting and was observed on clean surfaces of noble metals and heavy group V elements. It was gradually recognized that the splitting is enhanced when heavy element atoms are adsorbed on the substrate of lighter elements such as the Bi adsorbed Ag (111) surface (long-range ordered surface Bi/Ag(111)) [32]. The Rashba spin-split bands observed in most systems show rather common features such as pairs of split bands and the spin lying almost in the surface plane.

Beside these normal Rashba spins, a peculiar Rashba effect has been reported on a 1 ML Tl adsorbed Si (111) surface (Tl/Si(111)) with (1 × 1) structure [33]. In this study, a SP-ARPES experiment was performed along the $\overline{\Gamma}$–$\overline{K}$ direction with two different setups of the spin detector. The SP-ARPES spectra are shown for the in-plane spin (perpendicular to the $\overline{\Gamma}$–$\overline{K}$ axis) and out-of-plane spin in Fig. 11.13(a) and (b), respectively. At the emission angle $\theta_e = 0°$, or the $\overline{\Gamma}$ point, the two spin-polarized spectra become degenerate in both cases of in-plane and out-of-plane spins. In the case of in-plane spin polarized spectra, E_{Bs} of the peaks for opposite spin polarizations are much different for any θ_e between 10 and 26° while they became close above 28°. On the other hand, E_{Bs} of the peaks for opposite spin polarizations perpendicular to the surface were much different for θ_e only above 28°. It was thus revealed that the spin-polarization of the surface state band of this system rotates very rapidly from the in-plane direction to the out-of-plane direction across $\theta_e = 28°$. The maximum spin polarization in Fig. 11.13(b), observed near the $\overline{K}$ point between $\theta_e = 32$ and 34° was 100 %. These results are

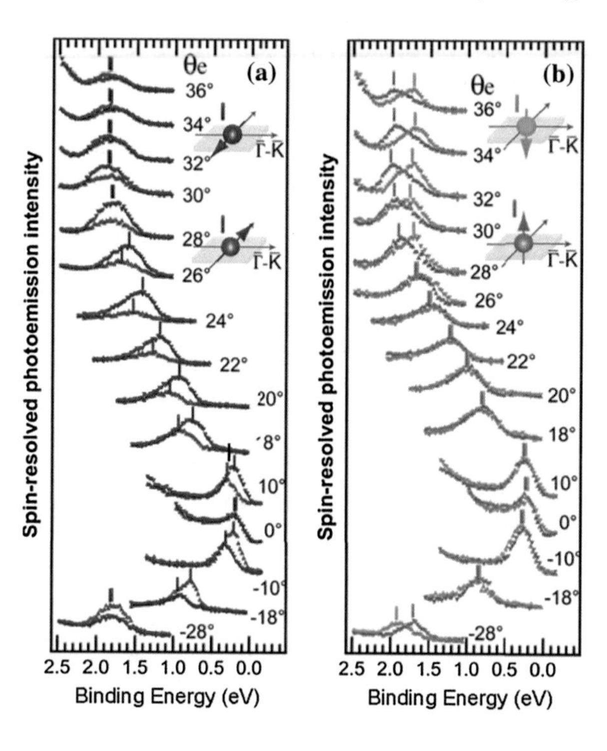

Fig. 11.13 Spin-ARPES spectra of a Tl/Si(111)–(1 × 1) surface measured at $h\nu = 21.2$ eV by use of a 25 kV mini Mott detector [33]. **a** Spin polarization parallel to the surface and perpendicular to the $\overline{\Gamma}$–$\overline{K}$ direction. **b** Spin polarization perpendicular to the surface. $\theta_e = 0°$ corresponds to $\overline{\Gamma}$, and $\overline{K}$ is located between $\theta_e = 32$ and 34°. Detected spin directions are schematically shown in the *insets*

in strong contrast to the cases of ~10 % spin polarization along the surface normal direction of Bi/Ag(111) [32]. The abrupt rotation of the Rashba spin and the curious splitting around the $\overline{K}$ cannot be explained by the simple Rashba effect in an ideal 2D system. In the real 2D symmetry of the hexagonal structure, a Zeeman term appears and acts as an effective magnetic field to the band spin magnetization. The theoretically calculated results provided good agreement with the experimental results clarifying the origin of the spin splitting at $\overline{K}$ in accordance with the C_3 symmetry [33].

A peculiar Rashba effect was also found in the β-Bi/Si(111) – ($\sqrt{3} \times \sqrt{3}$) surface [34], where the symmetry of the $\overline{K}$ point is different from that in Tl/Si(111). Along the $\overline{\Gamma}$–$\overline{K}$–$\overline{\Gamma}$ direction, two surface states S1 (with the lowest (smallest) E_B) and S2 (with slightly higher E_B) show Rashba splitting at $\overline{M}$ and the third one S3 (with the highest E_B) shows the Rashba splitting at $\overline{\Gamma}$. The three surface states were also observed along the $\overline{\Gamma}$–$\overline{K}$–$\overline{M}$ direction. The spin polarization was then experimentally measured along the $\overline{\Gamma}$–$\overline{M}$–$\overline{\Gamma}$ and $\overline{\Gamma}$–$\overline{K}$–$\overline{M}$ directions. Opposite spin polarization was observed for the S1 band with respect to the $\overline{M}$ and $\overline{K}$ points and for the S3 band with respect to the $\overline{\Gamma}$ point. Although the spin splitting along the $\overline{\Gamma}$–$\overline{M}$–$\overline{\Gamma}$ direction was ascribed to the normal Rashba effect, the spin splitting around the $\overline{K}$ point was unusual since this point has no time reversal symmetry, a symmetry that is thought to be essential for the Rashba effect. However, these experimental results of peculiar Rashba splitting were in a close

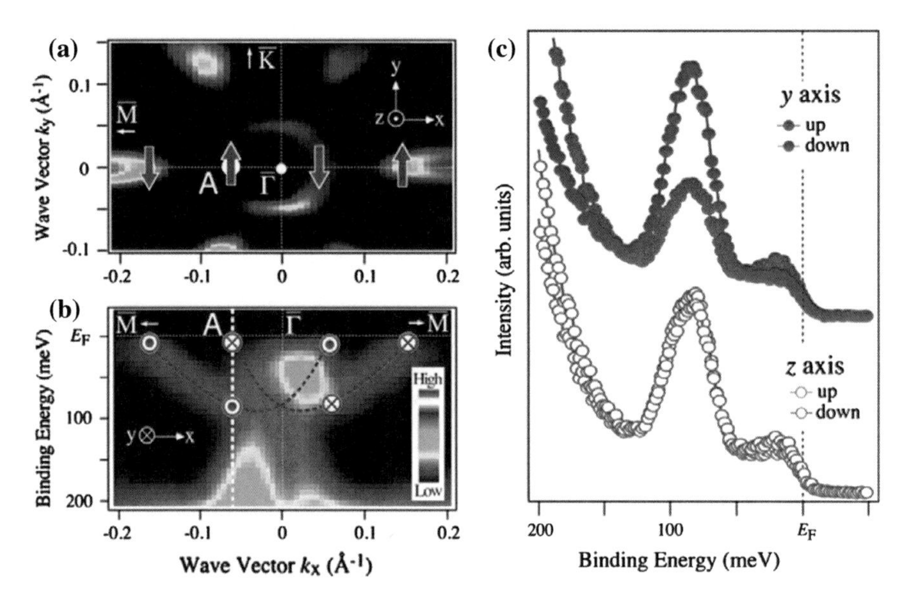

Fig. 11.14 ELEARPES and spin ELEPES of a ~80 monolayers single crystal Bi thin film grown on a Si(111) 7 × 7 surface [15]. ARPES intensity at E_F (**a**) with the high symmetry points in the SBZ. Measurement was made at 30 K with use of the Xe 8.4 eV line. The energy resolution was set to 0.9 and 8 meV for spin integrated and SP-ARPES. The angular acceptance for the SP-ARPES was set to ~4°. Band dispersions along the $\overline{\Gamma}$–$\overline{M}$ axis are shown in (**b**). Spin polarized ELEPES around the point A in (**a**) are shown in (**c**)

agreement with the results of first principle band-structure calculations. The peculiar Rashba splitting at the $\overline{K}$ point without time reversal symmetry suggested the necessity of detailed knowledge on the 2D symmetry of the system for a thorough understanding of the Rashba effects [34].

SP-ELEPES with much higher energy resolution became available by use of such a high intensity light source as a microwave excited rare gas lamp and a high performance mini Mott detector as already explained. Figure 11.14 shows the spin integrated ELEARPES (a, b) and spin polarized ELEPES (SP-ELEPES) (c) of Bi(111)/Si(111) with a Bi film thickness of ~80 monolayers measured at 30 K by use of T-1 lamp with Xe gas, a TM-1 monochromator, an A-1 analyzer and a MUSPIN mini Mott detector manufactured by MB Scientific [15]. Rashba splitting and the in-plane spin polarization along the y direction were clearly observed at the point A (Fig. 11.14(b, c)), whereas the spin polarization along the z direction perpendicular to the surface was found to be much smaller (Fig. 11.14(c)). The integration time for the spin polarized spectra (11.14(c)) was ~6 h. The Bi(111)/Si(111) was further studied at 30 K with energy resolutions of 40 and 6 meV for the spin-polarized and spin-integrated ELEARPES in wider SBZ including the $\overline{M}$ point [35] ($\overline{M}$ point is far outside of the area shown in Fig. 11.14(a). Measurements were performed in the region along the $\overline{\Gamma}$–$\overline{M}$ line. In addition to a small electron

FS (S_1) centered around the $\overline{\Gamma}$ point, 6 elongated hole pocket FSs (S_2) were observed. In addition, an ellipsoidal electron FS was observed near the $\overline{M}$ point (not shown). All these states were ascribed to the surface states, while quantum well states (QWSs) were observed in $E_B > 0.2$ eV. Then the spin polarized spectra were measured with a k resolution of ~ 0.06 Å^{-1}. Intensive efforts were made for the measurements of in-plane (y and x) and out-of-plane (z) spin polarizations of the S_2 hole pockets (Figs.1 and 2 of [35]). It was found that the magnitude of the in-plane spin polarization was asymmetric between the two S_2 hole pockets located on the opposite side of the $\overline{\Gamma}$ point. In addition, quite strong out-of-plane spin polarization was observed as in the case of Tl/Si(111) [33] with changing its sign across the $\overline{\Gamma M}$ line. The out-of-plane spin polarization switched its sign by every 60° in contrast to the general Rashba effect. The threefold symmetry of this spin polarization might be due to the threefold crystal symmetry due to the presence of the second Bi layer [35].

Spin splitting was also observed on the Si(557)-Au surface prepared by deposition of 0.2 ML Au onto an atomically clean well ordered Si(557) surface at 650 °C with a post annealing at 850 °C. Au chains are known to grow on vicinal Si surface as Si(335), Si(553) and so on. A 1D metallic character with double parabolic band dispersions along the chain direction and an almost straight FS line perpendicular to the chain were recognized. Later, a Rashba type splitting scenario was proposed. SP-ARPES measurements were performed at $h\nu = 34$ eV along the direction parallel to the Au chain with an energy and angular resolution of ~ 100 meV and $\pm 1°$ [18, 36]. The unambiguous spin splitting was consistent with first-principle calculations including the spin–orbit splitting of Au atoms in the Rashba scenario. The observed spin polarization was along a direction with a certain out-of-plane component in accordance with the anisotropic spatial distribution of the Au wave function. 1D Rashba spin splitting surface states were thus observed due to the highly anisotropic wave function of the Au chains [36].

In contrast to a trivial insulator, in which the surface states show a gap, the nontrivial topological insulators show spin-polarized metallic edge states due to the strong spin–orbit interaction and space inversion asymmetry [37]. According to the Kramers theorem, however, the spin must still be degenerate at special symmetry points with time-reversal invariant momenta. In the case of the (111) surface of the $Bi_{1-x}Sb_x$ topological insulator system, such momenta correspond to the $\overline{\Gamma}$ and $\overline{M}$ points in the surface Brillouin zone (SBZ) [37]. In addition to the in-plane Fermi surface topology of the surface states probed by the ARPES, spin-polarized bands were detected by SP-ARPES performed by a 40 kV Mott detector [37] and by a VLEED detector [38]. Figure 11.15 shows the latter result obtained along the $\overline{\Gamma}$–$\overline{M}$ direction. It was noticed that the Σ_2 electron pocket band centered at the $\overline{\Gamma}$ point with the Fermi wave vector of about 0.1 Å^{-1} had an opposite spin polarization vector with respect to $\overline{\Gamma}$. The Σ_2 band again crosses E_F at around 0.5 Å^{-1} and shows another electron pocket around $\overline{M}$. The Σ_1' band is located very close to this part of the Σ_2 band crossing E_F near 0.6 Å^{-1} but could still be discriminated by the high resolution SP-ARPES owing to its opposite spin polarization. It was noticed

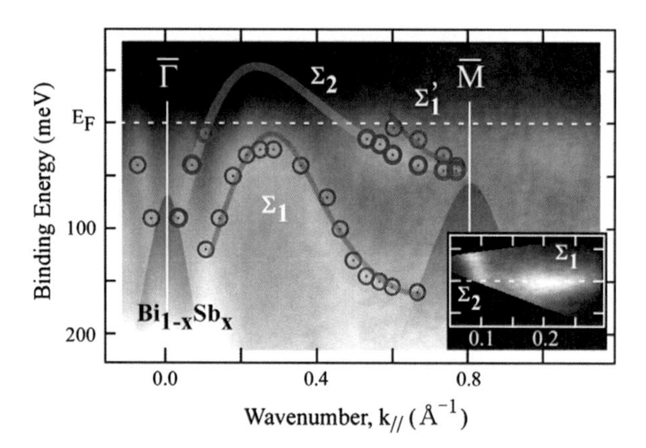

Fig. 11.15 Spin-polarized band dispersions of the surface states of $Bi_{1-x}Sb_x$ (x = 0.12–0.13) [38]. The spin-resolved structures are plotted with *thin* and *thick open circles* for opposite spin polarizations. *The right bottom inset* shows the two-dimensional k-resolved E_F intensity map, where the $\bar{\Gamma}$–$\bar{M}$ axis is shown by *the dashed line*

that the Σ_1' band and the Σ_2 band merge at $\bar{M}$ as predicted before. Besides, the Σ_1 band does not show any E_F crossing in contrast to the result for x = 0.10 [39] possibly because of the chemical potential shift for x=0.12-0.13. These observed surface states were ascribed to the edge states of the 3D quantum spin Hall phase [37, 38]. In this way, direct evidence for the quantum spin Hall phase in 3D topological insulators was provided.

Bi_2X_3 (X = Se, Te) were proposed to be more promising 3D topological insulators to be applied in spintronics devices and quantum computing technologies possibly at room temperature. Using the double Mott detector setup, three spin components were detected by means of SP-ARPES performed at $h\nu = 20-22$ eV with energy and momentum resolutions of 80 meV and 3 % of the SBZ [40]. At $E_B = 20$ meV along the $k_x(\|\bar{\Gamma}-\bar{M})$ direction in Bi_2Te_3, spin polarization along the y direction (P_y) was clearly resolved with opposite signs across $k_x = 0$ in contrast to the unclear polarization along the other two directions (P_x and P_z where z is the out-of-plane direction). Then the surface electrons of Bi_2Te_3 were confirmed to be helical Dirac fermions of Z_2 topological-order origin [40]. For the application purpose, the electronic structure must be in the topological transport regime, where there is zero charge fermion density. This regime is achieved when E_F lies in the gap between the bulk VB maximum and the bulk CB minimum, and exactly at the surface or edge Dirac point. Unfortunately, this is not the case for undoped Bi_2Te_3 and Bi_2Se_3 [40]. In order to lower the E_F into the bulk band gap in Bi_2Se_3, a low amount of Ca^{2+} was substituted for Bi^{3+}. As the Ca concentration increased from 0 to 0.5 %, low T resistivity showed a sharp peak at 0.25 %, suggesting the metal to insulator to metal transition. At the concentration of 0.25 %, a change in the sign of the Hall carrier density was also observed. ARPES was measured for $Bi_{2-\delta}Ca_{\delta}Se_3$ and the δ dependence of the electronic structure was studied. For $\delta = 0$, E_F was ~0.3 eV above the Dirac node while E_F was lowered to lie near the Dirac node for $\delta = 0.25$ %. For $\delta = 1$ %, E_F was expected to intersect the hole-like bulk VB. However, the surface was not stable and the E_F rose back to the $\delta = 0$ position on a typical time scale of 18 h as in Bi_2Te_3, possibly due to the surface band bending effect. Upon dosing with NO_2 molecules, the E_B of the surface Dirac

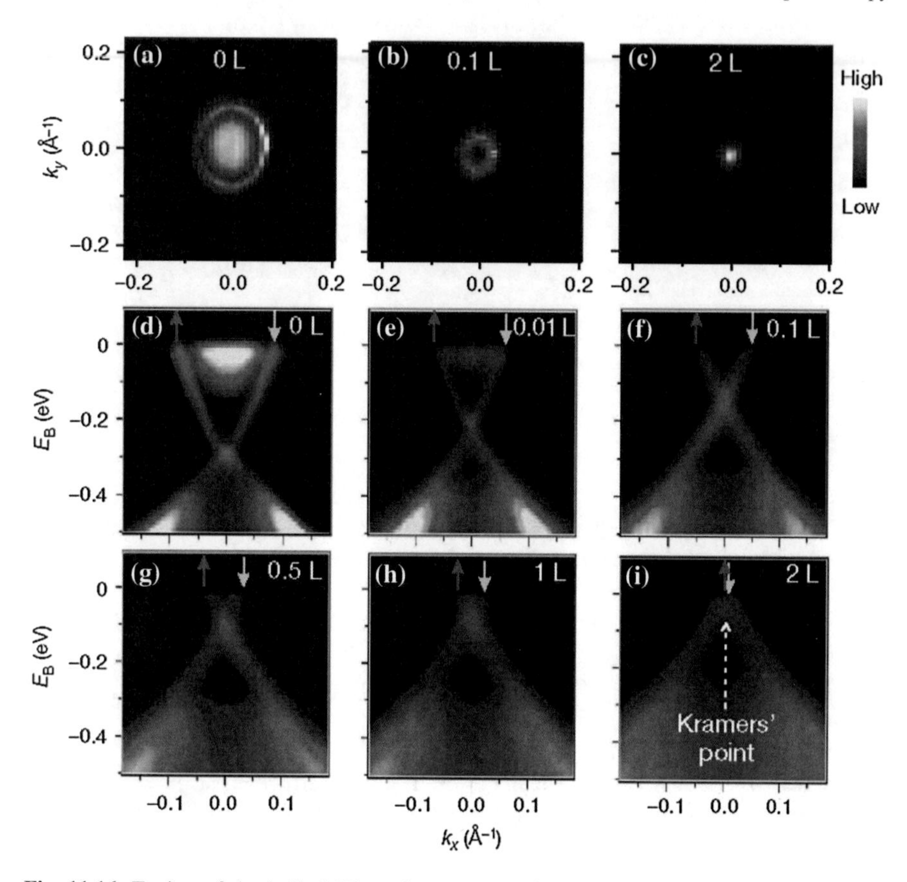

Fig. 11.16 Tuning of the helical Dirac electron states in $Bi_{2-\delta}Ca_{\delta}Se_3$ to the spin degenerate Kramers point and topological transport condition [40]. Topological surface FS near $\overline{\Gamma}$ of $Bi_{2-\delta}Ca_{\delta}Se_3$ (111) surface (**a**) and 0.1 L NO_2 dosed surface (**b**) and after 2 L dosage (**c**). **d–i** ARPES through $\overline{\Gamma}$ after NO_2 dosage as indicated in each figure

point was found to monotonously rise towards E_F as shown in Fig. 11.16 [40]. It reached the charge neutrality point ($E_B = 0$ eV) at NO_2 dosing of 2L (Langmuir) and no further changes of the chemical potential were observed with higher dosages. It should be noticed that surface dosing did not affect the carrier density in the bulk. T dependent ARPES demonstrated that the charge neutral point-like FS was robust up to T = 300 K over days. The low energy properties of stoichiometric Bi_2Se_3 or the $\delta = 0.0025$ Ca doped sample with proper NO_2 dosing were dominated by a novel topological ground state with massless helical Dirac fermion features with nearly 100 % spin polarization. The topological order found at room temperature might open up the possibility of using quantum Hall like phenomena and spin-polarized protected edge channels for spintronics applications without high magnetic field or delicate cryogenics [40].

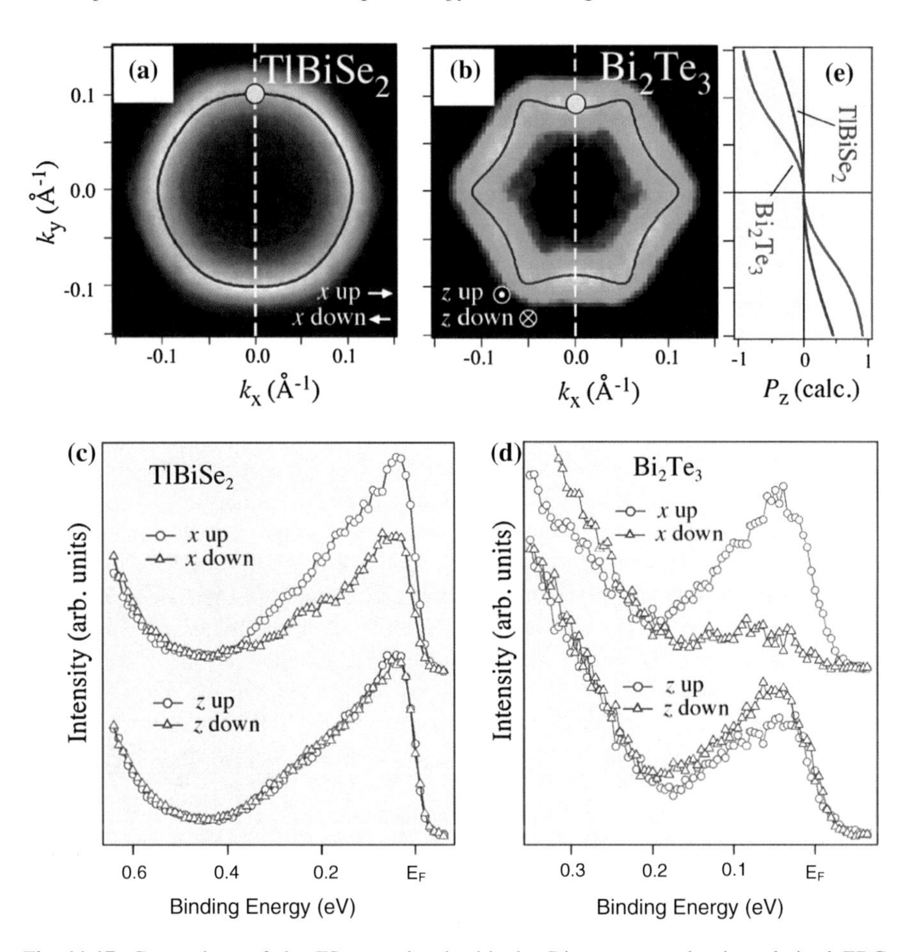

Fig. 11.17 Comparison of the FSs associated with the Dirac cone and spin polarized EDC between $TlBiSe_2$ (**a**), (**c**) and Bi_2Te_3 (**b**), (**d**) [41]. (**c**) and (**d**) are measured along the $\overline{\Gamma}$–$\overline{K}$ line. **e** Calculated P_z at $k_x = 0$ as a function of k_y

SP-ARPES was performed on Bi_2Te_3 with a total energy resolution of 40 meV at $h\nu = 8.4$ eV by use of Xe I photons with high S/N ratio. In contrast to the rather circular FS of $TlBiSe_2$ (Fig. 11.17(a)), a hexagonally distorted FS was observed in Bi_2Te_3 (Fig. 11.17(b)). An out-of-plane spin component in Bi_2Te_3 was unambiguously clarified in Fig. 11.17(d) [41] in contrast to $TlBiSe_2$ shown in Fig. 11.17(c) and many other topological insulators so far reported. On the hexagonally deformed FS, the out-of-plane spin polarization was up to 25 % of the in-plane spin polarization. Although hexagonal deformation of the FS was proposed to be responsible for such a deviation from the ideal helical spin configuration [41], the observed magnitude of the out-of-plane spin polarization was much reduced compared to the theoretical prediction. Such effects as interband scattering between the surface band to the non-spin-polarized bulk band and/or many-body effects such as a spin density wave might be contributing to the reduced out-of-plane spin polarization in Bi_2Te_3 [41].

11.4 Spin Polarized Photoelectron Spectroscopy of Magnetic Materials

Now we move to ferromagnetic materials for which a tremendous amount of work was accumulated in regard to the bulk and surface magnetism. In order not to disturb the direction of the emitted photoelectrons, the samples should be remanently magnetized. The polarization of the incident light is not restricted in most cases except for some special experiments. SP-ARPES was extensively applied to Fe, Co and Ni systems [43–49]. Among many interesting results, the rather contrasting behavior between Fe (100) and Ni (110) as a function of temperature (Fig. 11.18) attracted much attention [47, 49]. For example, the exchange splitting of the Fe 3d spectral density at Γ_{25}' remains almost constant (2.3 $\pm$ 0.2 eV) up to T_C, though the intensity difference between the $\uparrow$ and $\downarrow$ states decreases with increasing T. The $\Gamma_{12}\uparrow$ state is formally forbidden but actually observed due to the incomplete light polarization and finite angular resolution [47]. Complicated behavior was observed as a functions of hν and T [48]. On the other hand, the peaks of minority and majority spin character in Ni (110) probed near the X point with an exchange splitting of $\sim$0.17 $\pm$ 0.01 eV at room temperature, merged into a single peak at T_C [49].

There were two approaches to tackle the problem at finite T near T_C [46]: (1) A disordered-local-moment picture (no short-range order) and (2) a local-band picture (massive short-range order). They are two limiting cases. The results of Ni (110) [49] were successfully interpreted with the assumption of substantial short-range order [50]. Although the hν and T dependence of the SP-ARPES for Fe (100) were, on the contrary, thought to be in qualitative agreement with the disordered-local moment picture [47], there still remained some disagreement [46]. Therefore, a more comprehensive theory to fill the gap between these two extreme pictures was required for Fe (100). SP-ARPES was calculated in a bulk tight-binding formalism for bcc Fe clusters with varying local exchange fields. From a comparison with experimental results, it was found that Fe has a short-range magnetic order of at least 4 Å near T_C, ruling out the disordered-local-moment picture [46]. It was also demonstrated in the coherent potential approximation (CPA) that the spectral density for a given average magnetization could be predicted by three parameters: (1) the exchange splitting, (2) the effective width of the band and (3) the energy position of the level in the band even in the presence of short range order [51, 52]. So far the calculations with the disordered-local-moment model simulating Fe showed that the T dependence of the spectral density varies strongly with the wave vector and spin. It was shown in the calculation in [51, 52] that the spectral density of the level near the center of the band shows a double peak

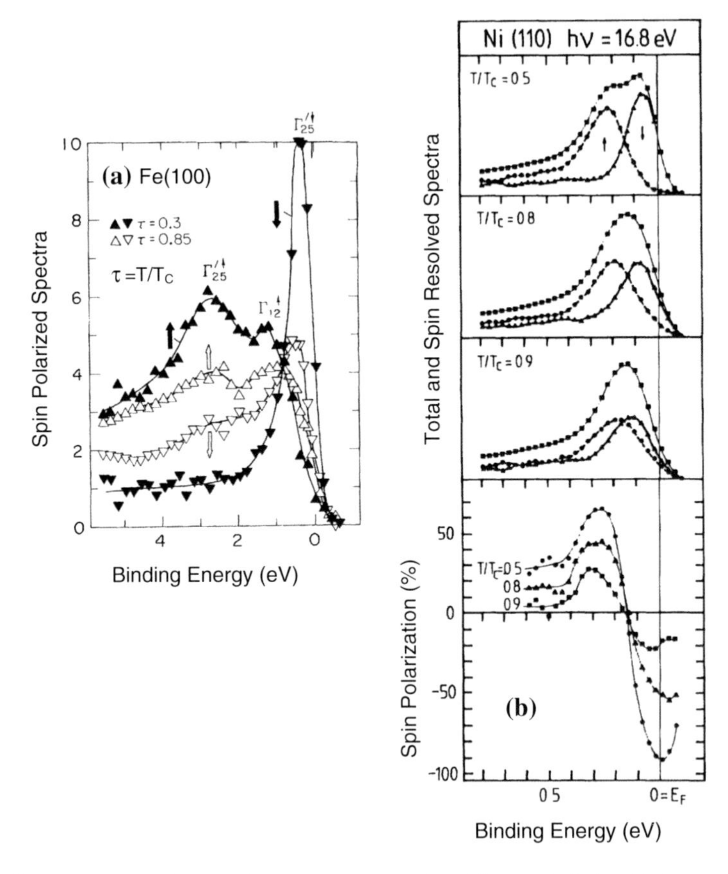

Fig. 11.18 Spin-polarized and angle-resolved valence band PES of Fe(100) [47] (**a**) and Ni(110) [49] (**b**). In (**a**) are shown the results obtained at hν = 60 eV with s-polarized light for normal emission from Fe(100), where $\tau = T/T_C$ (T_C is the Curie temperature). Spin resolved spectra at two temperatures are shown. In (**b**) are shown the results obtained at hν = 16.85 eV with the light polartization $\|$ [110] for normal emission by use of the Ne I line of a resonance lamp for Ni(110), where the sum of the $\uparrow$ and $\downarrow$ spectra and also their differences are shown in addition to the spin resolved spectra. Only the bands with S_4 symmetry at the X point are thought to contribute to the observed emission

structure even at T_C, if the exchange splitting is large enough compared with the effective width. On the other hand, the minority spin level in the lower half of the band or the majority spin level in the upper half of the band have a tendency to join its opposite spin counterpart with increasing T, shifting towards it as a single peak if the level position is remote enough from the band center and the exchange splitting is relatively small. In addition, the growth of a secondary peak at the position of opposite spin near T_C was predicted if the level is closer to the band center and the exchange splitting is large. Almost all features discussed above for Fe (100) and Ni (110) seemed to be reasonably explained by this calculation [51, 52].

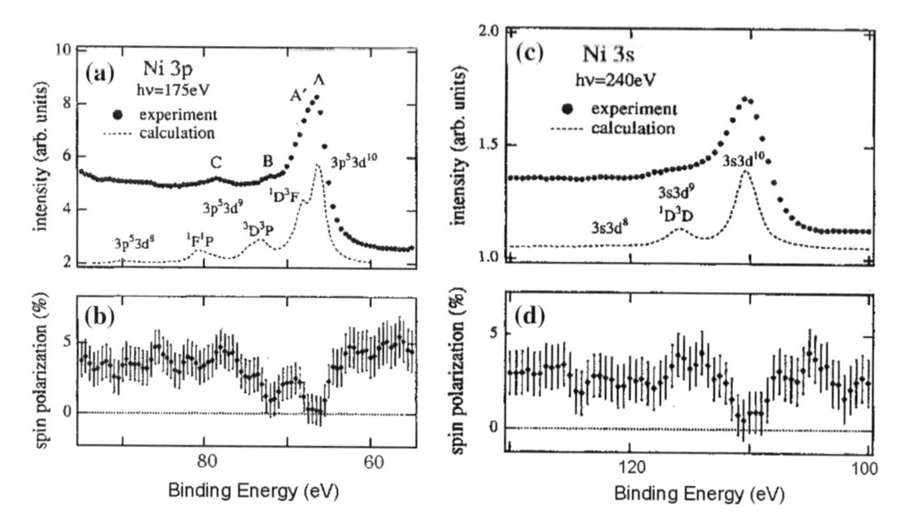

Fig. 11.19 Spin-PES of a Ni(110) single crystal surface. Ni 3p core level spectra (**a**) and (**b**) were measured at hν = 175 eV and Ni 3s core level spectra (**c**) and (**d**) were measured at hν = 240 eV [56]. *The dashed curves* in (**a**) and (**c**) are calculated results [57, 58]

SP-PES was also applied to various core levels [43, 53, 54]. In addition to the MLD in angle-resolved and spin-integrated photoemission, SP-PES was performed to the Fe 3p core level of an Fe (001) film grown on Ag (001) or Au (001) [55]. Meanwhile the SP-PES results on a Ni (110) single crystal are reproduced in Fig. 11.19, where the left (right) panel represents the results for 3p (3s) core level photoemission [56]. The picture frame shaped sample with each edge parallel to ⟨111⟩ was remanently magnetized by a pulse current through a coil wound around one edge of the sample. The 3p spectra were measured at hν = 175 eV, where the structures A, B and C are resolved in the spin-averaged spectrum (a) and spin-polarization is shown in (b). In the core level region, the background is often noticeably spin polarized and the ↑ and ↓ components derived by combining (a) and (b) show only a small difference. Therefore a proper subtraction of the background spin polarization would be required. According to the SIAM calculation [57], the main peak has primarily $|3p^53d^{10}\rangle$ character and the satellite structures have mainly $|3p^53d^9\rangle$ character. The $|3p^53d^9\rangle$ final states are split into three peaks A′, B and C corresponding to $^3F + {}^1D$, $^3P + {}^3D$, $^1P + {}^1F$ states, respectively. The calculation qualitatively reproduces the satellite structures and the shoulder structure of the main peak. If the energy scale of the calculated results is scaled down to 80 %, quite good correspondence was noticed with the experimental results. It is rather clear that the $^1P + {}^1F$ states have positive spin polarization and the $^3P + {}^3D$ states have negative spin polarization in the satellites region. The negative spin polarization in the main peak region was therefore ascribed to the ^3F contribution hybridized with the $|3p^53d^{10}\rangle$ component.

The results of the 3s core level spectrum measured at hν = 240 eV are shown in the right panel. The main peak has a predominantly negative spin polarization and the satellite shows a positive spin polarization (Fig. 11.19(d)). Since no spin polarization was expected for the $|3s3d^{10}\rangle$ state, the experimental results suggest an appreciable hybridization of the spin triplet ^3D component of the $|3s3d^9\rangle$ state with the main peak. On the other hand, the ^1D component is strongly contributing to the positive spin polarization of the satellite [56, 58].

11.5 Spin Polarized Inverse Photoemission Spectroscopy (SP-IPES)

11.5.1 Principle and Instrumentation

As a spin polarized electron source, the negative electron affinity (NEA) GaAs photocathode excited by circularly polarized laser is often used [23, 59 and references therein]. Cs is deposited onto a p-type GaAs to reduce the work function and realize NEA, where the vacuum level at the surface is located below the bottom of the conduction band in the bulk. This does not necessarily mean that the vacuum level is below the bottom of the conduction band at the surface which is subjected to a noticeable amount of surface band bending. This NEA surface is stabilized by slight adsorption of O_2. When the electrons are excited in the conduction band by laser photon excitation, they can easily escape into the vacuum. If the energy of the photon is only slightly beyond the band gap energy, electrons with a small energy distribution are emitted from the NEA cathode. This electron source can then be used as a high energy resolution electron source.

If circularly polarized laser light (σ^+ or σ^-) is incident on a GaAs surface in the normal incidence ($\mathbf{q}\|\mathbf{z}$) configuration, the quantization axis is the z-axis and the spin of the emitted electron can be defined as either $\uparrow$ (parallel) or $\downarrow$ (antiparallel) with respect to the z-axis. As already discussed in Sect. 11.2, the emitted photoelectrons can be partially spin-polarized according to the dipole selection rules for circularly polarized light. The maximum obtainable polarization for the σ^+ excitation can be -50 % (in the case described in Fig. 11.10). In practice much less polarization is realized in conventional sources. The longitudinally polarized electrons can be emitted to the surface-normal direction. Transversly polarized electron beams can be obtained by using a 90° deflector in front of the GaAs source through which the excitation laser passes. At the kinetic energies of $\sim$10 eV, for example, a current of the order of 10–20 μA and a spin polarization <40 % were obtained by conventional GaAs sources. The practically available current depends on the energy width and the lifetime of the source. A few μA current with a lifetime of more than a few hours is a typical example. By using strained GaAs or strained layer InGaP- or GaAsP-cathodes, a higher polarization even above 80 % was achieved at the sacrifice of the quantum efficiency [60, 61].

A He–Ne laser was used in combination with GaAsP to do momentum and spin polarized IPES (SP-IPES) of Fe (110) (Fig. 11.20) [62]. A solid state GaAlAs laser

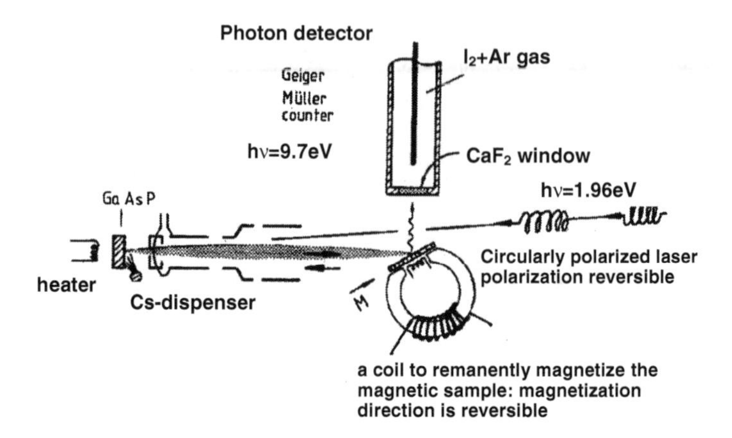

Fig. 11.20 An example of a spin polarized inverse photoemission (SP-IPES) system [62]. Circularly polarized light from a He–Ne laser through a $\lambda/4$ plate is incident onto a NEA GaAsP photocathode activated by Cs and O. The spin polarized photoelectrons from GaAsP are accelerated and focused onto the sample. The emitted photons from the sample are detected by a band pass photon detector. In this case, a Geiger Müller counter with I_2 + Ar gas with a CaF_2 window was employed. Solid state detectors with solid photocathodes were also widely available

diode could also be used in combination with GaAs, where the available hν could be properly selected by the Al concentration [63]. A Geiger-Müller counter with a CaF_2 window works as a band pass detector for IPES. Stimulated by the importance of SP-IPES for studying magnetism, various instruments were so far developed [64–67]. For example, a bolt-on type portable spin-polarized electron gun was developed [66], where the distance between the last element of the gun and the sample could be about 150 mm. For the electron energies from 7 to 20 eV, the transmission efficiency of the electron optics was >70 %. The FWHM resolution of the electron beam was better than 0.27 eV with an absolute value of the polarization of 25 $\pm$ 5 % in this gun. The FWHM of the electron beam energy increased when higher current was extracted. The increase of the collection angle was also quite important to realize good statistics for spin polarized spectra. BIS type detectors with various combinations of solid photocathodes and filter windows were developed (such as KCl coated photocathode + CaF_2 window and NaCl coated photocathode + SrF_2 window, with an energy resolution of 0.73 eV and 0.42 eV). Higher counting rate detectors were strongly desired over decades in SP-IPES. An efficient compact detector with an acceptance angle as wide as 2.7 steradian was developed for this purpose [67]. This detector had a cylindrical symmetry of the KBr coated photocathode and a SrF_2 window coaxial with the electron gun. A total energy resolution of 0.75 eV and 3000 cps/μA was realized for the unoccupied states of Ag (001).

The improvement of spin polarized electron sources progressed further in parallel. The GaAs-GaAsP strained-layer superlattice photocathode achieved the maximum spin-polarization of ~92 % with a quantum efficiency of 0.5 % by means of the conventional reflection type photocathode [68]. While the surface

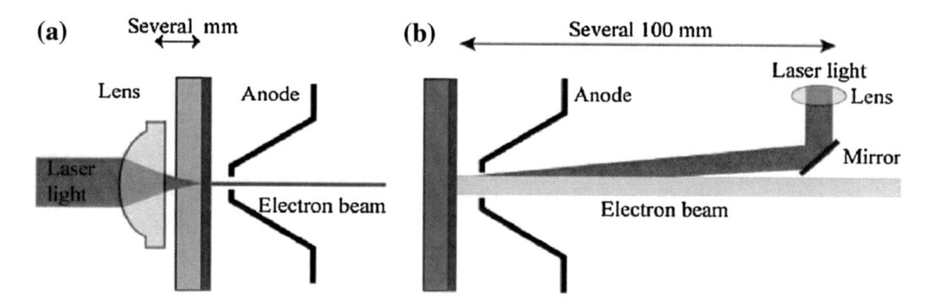

Fig. 11.21 New transmission type photocathode (**a**), compared with the conventional reflection type photocathode (**b**). Sharp focus of the laser light is possible in (**a**) [69, 70]

layer was heavily doped to achieve large band bending at the NEA surface, superlattice layers (alternate GaAs well layer and $GaAs_{1-x}P_x$ barrier layer) were medium doped to avoid the spin-flip depolarization. Such a modulation doped superlattice photocathode could solve the problem of the surface charge limit. Zn was used as p-type dopant. The thicknesses of the wells and barrier layers were chosen to maximize the heavy hole-light hole splitting and the total thickness was ~100 nm. Commercial GaAs with high Zn doping was used as a substrate and a strain relaxed GaAsP buffer layer (2 μm thick) was grown on it to induce large strains in the GaAs well layer (a few nm thick) to be grown later in situ by MOCVD (metal-oxide chemical vapor deposition) [68]. In the conventional reflection type photocathode, however, the interference between the laser excitation system and the electron beam line was a serious problem for the improvement of the overall performance of the spin polarized electron source.

Then the transmission type photocathode was developed as illustrated in Fig. 11.21 [69, 70], where the substrate was changed from a GaAs to a GaP wafer (with high Zn doping) to transmit the laser light at ~780 nm. Laser light was incident from the backside of the photocathode through a focus lens. Since the distance between the lens and the front side of the photocathode with the NEA surface could be as short as a few mm, the laser spot diameter in the active GaAs-GaAsP superlattice layer (for example 12 pairs of GaAs and GaAsP layers) could be as small as a few μm. The superlattice layers were terminated by a heavily Zn doped GaAs surface layer to realize a thin region of band bending. Then the brightness of the polarized electron beam became very high due to the shrinkage of the area of the electron generation (for example, 3 orders of magnitude higher than that of the SPLEEM gun). A long lifetime of the photocathode above several tens of hours was realized. Then spin polarization of ~77 % was achieved. In an improved design, a GaAs layer was inserted between the GaP substrate and the GaAsP buffer layer to control the strain relaxation process in the GaAsP buffer layer. Then high polarization up to ~90 % and high brightness of $\sim 1.3 \times 10^7$ A cm^{-2} sr^{-1} were realized even in the transmission type photocathode [70]. Now wide application of the spin polarized electron source is expected in materials sciences.

SP-IPES was applied to various ferromagnetic materials. The samples should be remanently magnetized not to influence the incident electrons' motion. The detection system could be the same as used for conventional IPES. In contrast to the spin polarized EELS (electron energy loss spectroscopy), where the spins of the scattered electrons are also analyzed [71], the polarization of the IPES photons is usually not discriminated in the SP-IPES, though the spin polarized electrons are used for excitation. By modulating the helicity of the excitation laser light by a Pockels cell, the preferential spin of the electrons incident onto the sample can be modulated between ↑ and ↓ states, which should be either parallel or antiparallel to the magnetization of the sample. In the BIS type detection, both the difference and the average (or sum) of the IPES intensity are simultaneously measured for two helicities at each excitation energy and the ↑ and ↓ IPES spectra are obtained after normalization by the degree of spin polarization of the electron beam. In order to eliminate the systematic error, the measurement is repeated by magnetizing the sample to the opposite direction.

11.5.2 Several SP-IPES Studies

Angle resolved SP-IPES was measured for the case of an Fe (100) surface with a total energy resolution of 0.8 eV [72] at various T normalized by the Curie temperature T_C. Figure 11.22(a) shows the SP-IPES at the angle of incidence $\theta = 75°$ and 11.22(b) spin integrated spectra at $\theta = 15°$. It was found that the T dependence is rather different depending on the region in the BZ. Both results probing different k regions in the BZ showed rather contrasting behavior. Namely Fig. 11.22(a) showed a broad peak ascribable to the minority spin band and the majority spin band just above E_F. What is observed with increasing T is the decrease of the intensity of the minority spin band while keeping its location. The spin-exchange splitting is almost negligible at $\theta = 75°$ at T below T_C. On the other hand, Fig. 11.22(b) shows two peaks ascribable to the majority (closer to E_F) and minority spin bands. The exchange splitting is 1.6 eV at room temperature but it decreases with increasing T towards T_C as expected from the Stoner model. Such non-collapsing and collapsing behavior seems to be correlated with the observation discussed in Refs. [51, 52] in SP-PES.

Angle resolved SP-IPES could be applied to the study of quantum well states (QWS). QWS of magnetic materials were a subject of intense experimental and theoretical studies because of their significance for applications in spintronics. Co/Cu structures showed layer-by-layer growth and were prototype systems. The electronic structure of Cu was well understood by PES and IPES. Then the electronic structure associated with the quantum confinement in ultrathin Co layers was studied. So far ultrathin Co films deposited on Cu (100) were studied by SP-PES but no clear evidence of QWS was confirmed [73, 74]. However, unoccupied QWS were observed by conventional IPES [75]. Depending upon whether its energy overlapped with the substrate continuum or not, we call it quantum well

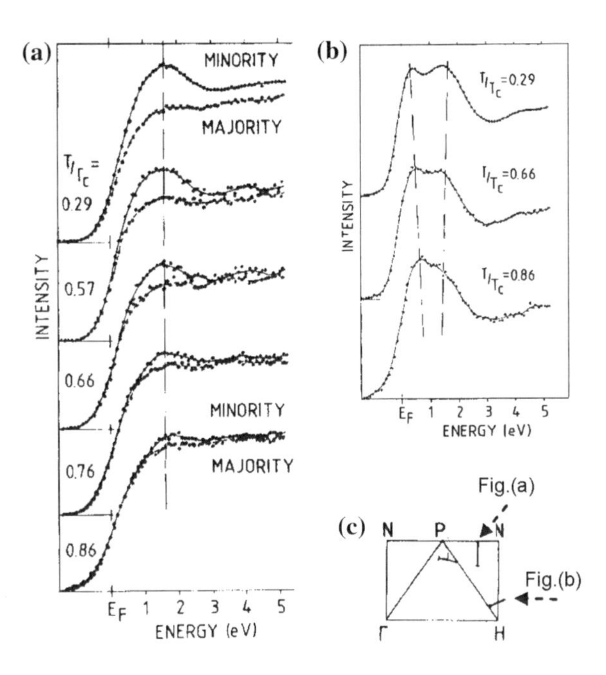

Fig. 11.22 SP-IPES spectra of Fe(100) at various reduced temperatures T/T_c [72]. Angle of incidence $\theta = 75°$ for (**a**) and $\theta = 15°$ for (**b**). **b** shows a spin integrated spectra. The wave vector region in the BZ during the isochromat energy scan from 0 to ~5 eV above E_F is shown by solid lines in (**c**) for both cases of (**a**) and (**b**). The region near the P point is for $\theta = 60°$ results on Fe(110)

resonance (QWR) or quantum well state (QWS). The Co films were remanently magnetized to saturation along the [110] easy magnetization direction. Clear QWS were observed in this system by SP-IPES [76]. We compare the band structures of fcc Cu and Co↓ and Co↑ bands along the [100] direction Γ-X in Fig. 11.23(a) and (b). We notice that the top of the p-like X_4' band for Co↓ is 0.9 eV and that for Co↑ is 0.7 eV higher in energy than that of Cu, while the top of the X_4' band in Cu is located at 1.8 eV above E_F as shown in Fig. 11.23. Therefore, the QWS were expected between 2.7 and 1.8 eV for ↓ minority state and 2.5 and 1.8 eV for ↑ majority state above E_F. QWR was expected below 1.8 eV. Here the bulk-like direct optical transition with $h\nu = 9.4$ eV (defined by the band pass detector) was considered for normal incidence. Although no spin dependence was observed for Cu (interband transition between Cu sp states, Fig. 11.23(c)), the 5 ML and 12 ML Co on Cu(100) (Co/Cu(100)) showed a clear spin dependence as reproduced in Fig. 11.23(d) and (e). In the 12 ML Co films (11.23(e)), two prominent structures were observed in the minority spin spectrum, whereas one structure was prominent in the majority spin spectrum. The spectral features around 2 eV above E_F are due to the bulk-like Co sp band transitions which show a clear spin splitting of 0.4 eV due to the ferromagnetism. In this case, the observed spin splitting is not due to the exchange splitting of the final band alone but is ascribed to the combined exchange splitting of both the initial and final bands. The initial state near the X_1 point has strong s-d hybridization in contrast to the p-like final state near the X_4' final band. Therefore the initial state splitting was thought to be larger than that of the final state. According to the band-structure calculation, no d hole is present for Cu and Co↑ bands. Only the Co↓ band has a d hole. However, the direct transition to this

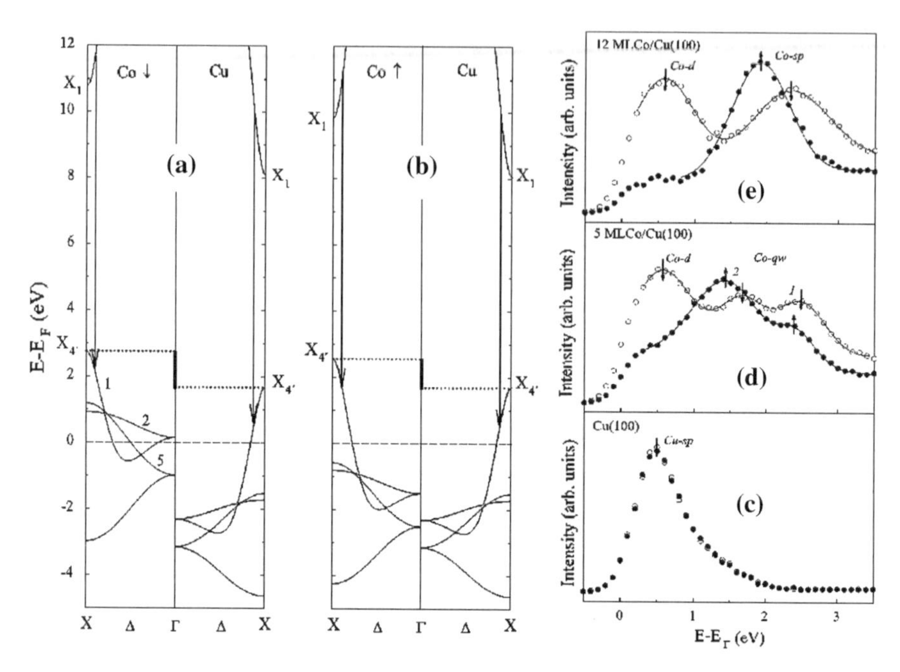

Fig. 11.23 Comparison of the energy bands E(k) along the [100] direction Γ−X for $Co_{\uparrow,\downarrow}$ and Cu (**a**) and (**b**). The X_4' point of Cu is 0.9 and 0.7 eV lower than that of $Co_{\uparrow,\downarrow}$. *Arrows* indicate direct optical transitions with hν = 9.4 eV. Spin polarized IPES for normal electron incidence on (**c**) Cu (100), (**d**) 5ML Co on Cu (100) and (**e**) 12 ML Co on Cu (100). *Open* and *closed circles* represent the spectra for the minority (↓) and majority (↑) spin [76]

Co↓ d hole state is not allowed along the Γ–X direction because of the energy conservation (9.4 eV) and selection rules [76]. The peak around 0.5 eV just above E_F marked as Co-d ↓ character was then interpreted as due to the indirect or non-k-conserving transitions induced by the high density of d holes close to E_F.

The results for the 5 ML Co/Cu(100) (11.23(d)) showed similar features near E_F with the Co-d ↓ character but with much reduced spin asymmetry possibly due to the contribution of the Cu substrate. In the region corresponding to the Co sp state in 12 ML Co/Cu(100), two features 1 and 2 were observed in a 5 ML Co/Cu(100) sample. Both features 1 and 2 showed clear spin splitting of 0.1 and 0.2 eV, respectively, which was appreciably smaller than the spin splitting of the bulk Co sp bands. Judging from its energy position, the feature 1 (2) was ascribed to the QWS (QWR). A full set of results was obtained for a Co thickness from 0 to 12 ML [76]. The first QWS appeared after the deposition of 1 or 2 ML Co. The second feature, namely QWR, appeared after deposition of about 4 ML. Even a 3rd feature, QWR, appeared for 10 ML Co. Both features 1 and 2 moved towards higher energies with the increase of the Co film thickness.

The phase-accumulation model [77] was used to interpret these results. The electron would be confined in the thin Co film when the following phase relation is satisfied:

$$\Phi_C + \Phi_B + m2ka = 2\pi n, \tag{11.5}$$

where Φ_C and Φ_B are the phase changes at the Co/Cu and Co/vacuum interfaces and m and n are integers. In the third term, a is the thickness of one ML Co (1.8 Å) and m the number of ML of the Co film, k the electron wave vector perpendicular to the film plane. Φ_B is given by

$$\Phi_B = \pi[3.4(\mathrm{eV})/(E_V - E)]^{1/2} - \pi, \tag{11.6}$$

where E_V is the vacuum level energy of Co relative to E_F [77] (5 eV is employed in [76]). On the other hand, Φ_C at the Co/Cu interface is empirically expressed as

$$\Phi_C = 2\arcsin[(E - E_L)/(E_U - E_L)]^{1/2} - \pi, \tag{11.7}$$

where E_U and E_L are the upper and lower band-gap boundary energies [77]. In the present case were employed $E_L = 1.8$ eV and $E_U\uparrow = 2.5$ eV and $E_U\downarrow = 2.7$ eV. The comparison between the calculated and experimental results for ↓ and ↑ QWS as a function of the Co film thickness showed good agreement due to the complete confinement. Slight disagreement was seen for the QWR states. The potential of the SP-IPES for the study of quantum well states in thin magnetic materials was thus demonstrated and such studies would be further advanced with employing brighter, very stable and long lifetime spin polarized electron sources.

References

1. L. Baumgarten, C.M. Schneider, H. Petersen, F. Scäfers, J. Kirschner, Phys. Rev. Lett. **65**, 492 (1990)
2. C.M. Schneider, D. Venus, J. Kirschner, Phys. Rev. **B45**, 5041 (1992)
3. H. Ebert, L. Baumgarten, C.M. Schneider, J. Kirschner, Phys. Rev. **B44**, 4406 (1991)
4. D. Venus, L. Baumgarten, C.M. Schneider, C. Boeglin, J. Kirschner, J. Phys.: Condens. Matter **5**, 1239 (1993)
5. Ch. Roth, F.U. Hillebrecht, H.B. Rose, E. Kisker, Phys. Rev. Lett. **70**, 3479 (1993)
6. E. Tanuma, R. Feder, Europhys. Lett. **16**, 695 (1991)
7. W. Kuch, M.-T. Lin, W. Steinhögl, C.M. Schneider, D. Venus, J. Kirschner, Phys. Rev. **B51**, 609 (1995)
8. D. Venus, Phys. Rev. **B49**, 8821 (1994)
9. Forward scattering and photoelectron diffraction is handled in Ch. 12.
10. A. Fanelsa, R. Schellenberg, F.U. Hillebrecht, E. Kisker, J.G. Menchero, A.P. Kaduwela, C.S. Fadley, M.A. Van Hove, Phys. Rev. **B54**, 17962 (1996)
11. J. Kessler, *Polarized Electrons*, Springer Series on Atoms and Plasmas (Springer, Berlin, 1985)
12. A. Eyers, F. Schäfers, G. Schönhense, U. Heinzmann, H.P. Oepen, K. Hünlich, J. Kirschner, G. Borstel, Phys. Rev. Lett. **52**, 1559 (1984)
13. F.B. Dunning, Nucl. Instrum. Meth. Phys. Res. **A347**, 152 (1994)
14. G. Ghiringhelli, K. Larsson, N.B. Brookes, Rev. Sci. Instrum. **70**, 4225 (1999)
15. S. Souma, A. Takayama, K. Sugawara, T. Sato, T. Takahashi, Rev. Sci. Instrum. **81**, 095101 (2010)

16. J. Kirschner, Polarized Electrons at Surfaces, Springer Tracts in Modern Physics, vol. 106 (Springer, Berlin, 1985)
17. R. Jungblut, Ch. Roth, F.U. Hillebrecht, E. Kisker, Surf. Sci. **269/270**, 615 (1992)
18. T. Okuda, Y. Takeichi, Y. Maeda, A. Harasawa, I. Matsuda, T. Kinoshita, A. Kakizaki, Rev. Sci. Instrum. **79**, 123117 (2008)
19. M. Kolbe, P. Lushchyk, B. Petereit, H.J. Elmers, G. Schönhense, A. Oelsner, C. Tusche, J. Kirschner, Phys. Rev. Lett. **107**, 207601 (2011)
20. C. Tusche, M. Ellguth, A.A. Ünal, C.-T. Chiang, A. Winkelmann, A. Krasyuk, M. Hahn, G. Schönhense, J. Kirschner, Appl. Phys. Lett. **99**, 032505 (2011)
21. M. Kolbe, P. Lushchyk, B. Petereit, H. J. Elmers, G. Schönhense, A. Oelsner, C. Tusche, J. Kirschner, Supplemental material at http://link.aps.org/supplemental/10.1103/PhysRevLett.107.207601
22. B. Krömker, M. Escher, D. Funnemann, D. Hartung, H. Engelhard, J. Kirschner, Rev. Sci. Instrum. **79**, 053702 (2008)
23. D.T. Pierce, F. Meier, Phys. Rev. **B13**, 5484 (1976)
24. A. Eyers, F. Schäfers, G. Schönhense, U. Heinzmann, H. P. Oepen, K. Hünlich, J. Kirschner and G. Borstel, Phys. Rev. Lett. **52**, 1559 (1984)
25. H.P. Oepen, K. Hünlich, J. Kirschner, A. Eyers, F. Schäfers, G. Schönhense, U. Heinzmann, Phys. Rev. **B31**, 6846 (1985)
26. D. Venus, J. Garbe, S. Suga, C. Schneider, J. Kirschner, Phys. Rev. **B34**, 8435 (1986)
27. N.B. Brookes, G. Ghiringhelli, O. Tjernberg, L.H. Tjeng, T. Mizokawa, T.W. Li, A.A. Menovsky, Phys. Rev. Lett. **87**, 237003 (2001)
28. L.H. Tjeng, C.T. Cheng, S.-W. Cheong, Phys. Rev. **B45**, 8205 (1992)
29. F.C. Zhang, T. Rice, Phys. Rev. **B37**, 3759 (1988)
30. H. Eskes, G.A. Sawatzky, Phys. Rev. Lett. **61**, 1415 (1988)
31. Y.A. Bychkov, E.I. Rashba, JETP Lett. **39**, 78 (1984)
32. C.R. Ast, J. Henk, A. Ernst, L. Moreschini, M.C. Falub, D. Pacile, P. Bruno, K. Kern, M. Grioni, Phys. Rev. Lett. **98**, 186807 (2007)
33. K. Sakamoto, T. Oda, A. Kimura, K. Miyamoto, M. Tsujikawa, A. Imai, N. Ueno, H. Namatame, M. Taniguchi, P.E.J. Eriksson, R.I.G. Uhrberg, Phys. Rev. Lett. **102**, 096805 (2009)
34. K. Sakamoto, H. Kakuta, K. Sugawara, K. Miyamoto, A. Kimura, T. Kuzumaki, N. Ueno, E. Annese, J. Fujii, A. Kodama, T. Shishidou, H. Namatame, M. Taniguchi, T. Sato, T. Takahashi, T. Oguchi, Phys. Rev. Lett. **103**, 156801 (2009)
35. A. Takayama, T. Sato, S. Souma, T. Takahashi, Phys. Rev. Lett. **106**, 166401 (2011)
36. T. Okuda, K. Miyamaoto, Y. Takeichi, H. Miyahara, M. Ogawa, A. Harasawa, A. Kimura, I. Matsuda, A. Kakizaki, T. Shishidou, T. Oguchi, Phys. Rev. **B82**, 161410 (R) (2010)
37. D. Hsieh, Y. Xia, L. Wray, D. Qian, A. Pal, J.H. Dil, J. Osterwalder, F. Meier, G. Bihlmayer, C.L. Kane, Y.S. Hor, R.J. Cava, M.Z. Hasan, Science **323**, 919 (2009)
38. A. Nishide, A.A. Taskin, Y. Takeichi, T. Okuda, A. Kakizaki, T. Hirahara, K. Nakatsuji, F. Komori, Y. Ando, I. Matsuda, Phys. Rev. **B81**, 041309 (R) (2010)
39. D. Hsieh, D. Qia, L. Wray, Y. Xia, Y.S. Hor, R.J. Cava, M.Z. Hasan, Nature **452**, 970 (2008)
40. D. Hsieh, Y. Xia, D. Qian, L. Wray, J.H. Dil, F. Meier, J. Osterwalder, L. Patthey, J.G. Checkelsky, N.P. Ong, A.V. Fedorov, H. Lin, A. Bansil, D. Grauer, Y.S. Hor, R.J. Cava, M.Z. Hasan, Nature **460**, 1101 (2009)
41. S. Souma, K. Kosaka, T. Sato, M. Komatsu, A. Takayama, T. Takahashi, M. Kriener, K. Segawa, Y. Ando, Phys. Rev. Lett. **106**, 216803 (2011)
42. L. Fu, Phys. Rev. Lett. **103**, 266801 (2009)
43. S. Hüfner, *Photoelectron Spectroscopy, Principles and Application* (Springer, Berlin, 2003)
44. R. Feder (ed.), *Polarized Electrons in Surface Physics* (World Scientific, Singapore, 1985)
45. J. Kirschner, Polarized Electrons at Surfaces, Springer Tracts Mod. Phys. **106** (Springer, Berlin, Heidelberg 1976)
46. E.M. Haines, R. Clauberg, R. Feder, Phys. Rev. Lett. **54**, 932 (1985)
47. E. Kisker, K. Schroeder, M. Campagna, W. Gudat, Phys. Rev. Lett. **52**, 2285 (1984)

48. E. Kisker, K. Schroeder, W. Gudat, M. Campagna, Phys. Rev. **B31**, 329 (1985)
49. H. Hopster, R. Raue, G. Guentherodt, E. Kisker, R. Clauberg, M. Campagna, Phys. Rev. Lett. **51**, 829 (1983)
50. V. Korenman, R.E. Prange, Phys. Rev. Lett. **53**, 186 (1984)
51. J. Kanamori, S. Imada, J. Phys. **49 (C-8)**, 97–98 (1988)
52. J. Kanamori, in *Core-level Spectroscopy in Condensed Matter*, eds. J. Kanamori, A. Kotani, Solid-State Sci. (Springer) **81**, 160 (1988)
53. C. Carbone, E. Kisker, Solid State Commun. **65**, 1107 (1988)
54. G. Rossi, F. Sirotti, N.A. Cherepkov, F. Combet, F.G. Panaccione, Solid State Commun. **90**, 557 (1994)
55. C. Roth, F. U. Hillebrecht, H. B. Rose and E. Kisker, Phys. Rev. Lett. *70*, 3479 (1993)
56. Y. Saitoh, T. Matsushita, S. Imada, H. Daimon, S. Suga, J. Fujii, K. Shimada, A. Kakizaki, K. Ono, M. Fujisawa, T. Kinoshita, T. Ishii, Phys. Rev. **B52**, R11549 (1995)
57. B.T. Thole, G. van der Laan, Phys. Rev. Lett. **67**, 3306 (1991)
58. G. van der Laan, B.T. Thole, H. Ogasawara, Y. Seino, A. Kotani, Phys. Rev. **B46**, 7221 (1992)
59. D.T. Pierce, R.J. Celotta, G.C. Wang, W.N. Unertl, A. Galejs, C.E. Kuyatt, S.R. Mielczarek, Rev. Sci. Instrum. **51**, 478 (1980)
60. R. Alley, H. Aoyagi, J. Clendenin, J. Frisch, C. Garden, E. Hoyt, R. Kirby, L. Klaisner, A. Kulikov, R. Miller, G. Mulhollan, C. Prescott, P. Sgez, D. Schultz, H. Tang, J. Turner, K. Witte, M. Woods, A.D. Yeremian, M. Zolotorev, Nucl. Instrum. Meth. Phys. Res. **A365**, 1 (1995)
61. K. Aulenbacher, Ch. Nachtigall, H.G. Andresen, J. Bermuth, Th Dombo, P. Drescher, H. Euteneuer, H. Fischer, D.V. Harrach, P. Hartmann, J. Hoffmann, P. Jennewein, K.H. Kaiser, S. Köbis, H.J. Kreidel, J. Langbein, M. Petri, S. Plützer, E. Reichert, M. Schemies, H.-J. Schöpe, K.-H. Steffens, M. Steigerwald, H. Trautner, Th Weis, Nucl. Instrum. Meth. Phys. Res. **A391**, 498 (1997)
62. H. Scheidt, M. Glöbl, V. Dose, J. Kirschner, Phys. Rev. Lett. **51**, 1688 (1983)
63. W. Grentz, M. Tschudy, B. Reihl, G. Kaindl, Rev. Sci. Instrum. **61**, 2528 (1990)
64. U. Kolac, M. Donath, K. Ertl, H. Liebl, V. Dose, Rev. Sci. Instrum. **59**, 1933 (1988)
65. F. Ciccacci, E. Vescovo, G. Chiaia, D. De Rossi, M. Tosca, Rev. Sci. Instrum. **63**, 3333 (1992)
66. F. Schedin, R. Warburton, G. Thornton, Rev. Sci. Instrum. **69**, 2297 (1998)
67. M. Cantoni, R. Bertacco, Rev. Sci. Instrum. **75**, 2387 (2004)
68. T. Nishitani, T. Nakanishi, M. Yamamoto, S. Okumi, F. Furuta, M. Miyamoto, M. Kuwahara, N. Yamamoto, K. Naniwa, O. Watanabe, Y. Takeda, H. Kobayakawa, Y. Takashima, H. Horinaka, T. Matsuyama, K. Togawa, T. Saka, M. Tawada, T. Omori, Y. Kurihara, M. Yoshioka, K. Kato, T. Baba, J. Appl. Phys. **97**, 094907 (2005)
69. N. Yamamoto, T. Nakanishi, A. Mano, Y. Nakagawa, S. Okumi, M. Yamamoto, T. Konomi, X. Jin, T. Ujihara, Y. Takeda, T. Ohshima, T. Saka, T. Kato, H. Horinaka, T. Yasue, T. Koshikawa, M. Kuwahara, J. Appl. Phys. **103**, 064905 (2008)
70. X. Jin, N. Yamamoto, Y. Nakagawa, A. Mano, T. Kato, M. Tanioku, T. Ujihara, Y. Takeda, S. Okumi, M. Yamamoto, T. Nakanishi, T. Saka, H. Horinaka, T. Kato, T. Yasue, T. Koshikawa, Appl. Phys. Express **1**, 045002 (2008)
71. J. Kirschner, S. Suga, Solid State Commun. **64**, 997 (1987)
72. J. Kirschner, M. Glöbl, V. Dose, H. Scheidt, Phys. Rev. Lett. **53**, 612 (1984)
73. C.M. Schneider, P. Schunster, M. Hammond, H. Ebert, J. Noffke, J. Kirschner, J. Phys. Condens. Matter **3**, 4349 (1991)
74. W. Clemens, T. Kachel, O. Rader, E. Vescovo, S. Blügel, C. Carbone, W. Eberhardt, Solid State Commun. **81**, 739 (1992)
75. J.E. Ortega, F.J. Himpsel, G.J. Mankey, R.F. Willis, Phys. Rev. **B47**, 1540 (1993)
76. D.H. Yu, M. Donath, J. Braun, G. Rangelov, Phys. Rev. **B68**, 155415 (2003)
77. N.V. Smith, N.B. Brookes, Y. Chang, P.D. Johnson, Phys. Rev. **B49**, 332 (1994)

Chapter 12
Photoelectron Diffraction and Photoelectron Holography

Photoelectron spectroscopy is mostly used for the analysis of electronic structures. ARPES is a powerful tool for band mapping and Fermiology as explained already. It is also widely known that the photoelectrons are diffracted by neighboring atoms and provide characteristic diffraction patterns [1]. From the analyses of diffraction patterns the atomic structures of surfaces were expected to be revealed. Then adsorbed surfaces as well as clean surfaces were intensively studied. The photoelectron diffraction (PED) occurs when a photo-excited electron finds more than one path into a detector. Direct propagation to the detector provides one path and the other path available is by the elastic scattering of a photoelectron by nearby atoms as illustrated in Fig. 12.1. In other words, direct photoemission plays the role of the reference wave and the scattered waves are the objective waves. The interference pattern is thought to contain the information on the geometry of atoms surrounding the photoelectron- emitting atom. In comparison with LEED for surface structural analysis, PED can also be applied to adsorbed atoms on the surface, which do not have lateral periodicity. In addition, adsorbed-atom specific information can be obtained because one can tune the detection E_K to a certain core level of the adsorbed atoms. A complete 3D image of the surface structure surrounding the emitter can be reconstructed by Fourier transformation with a sophisticated treatment [2], which is called photoelectron holography. Numerical simulations were also used to find out the realistic geometry. Since then many studies of PED and photoelectron holography have been integrated as reviewed in [3–5]. A more recent review is given for example in [6].

Photoelectron holography suffers from serious image aberrations caused by the strong forward–scattering peaks. The photoelectron atomic scattering factor is highly anisotropic with respect to the scattering angle and depends strongly on E_K above a few hundred eV, where it becomes much more significant in the forward focusing direction. In addition, photoelectron holography suffers noticeably from multiple-scattering effects. Various reconstruction algorithms were compared for Cu 3p holograms from Cu(001) in Ref. [7]. Photoelectron spectra measured at tens sets of E_K and ϕ over a symmetry-reduced 1/8 total solid angle above the surface in the θ range from 0 to 70° were used to reconstruct the 3D imaging of atoms.

S. Suga and A. Sekiyama, *Photoelectron Spectroscopy*,
Springer Series in Optical Sciences 176, DOI: 10.1007/978-3-642-37530-9_12,

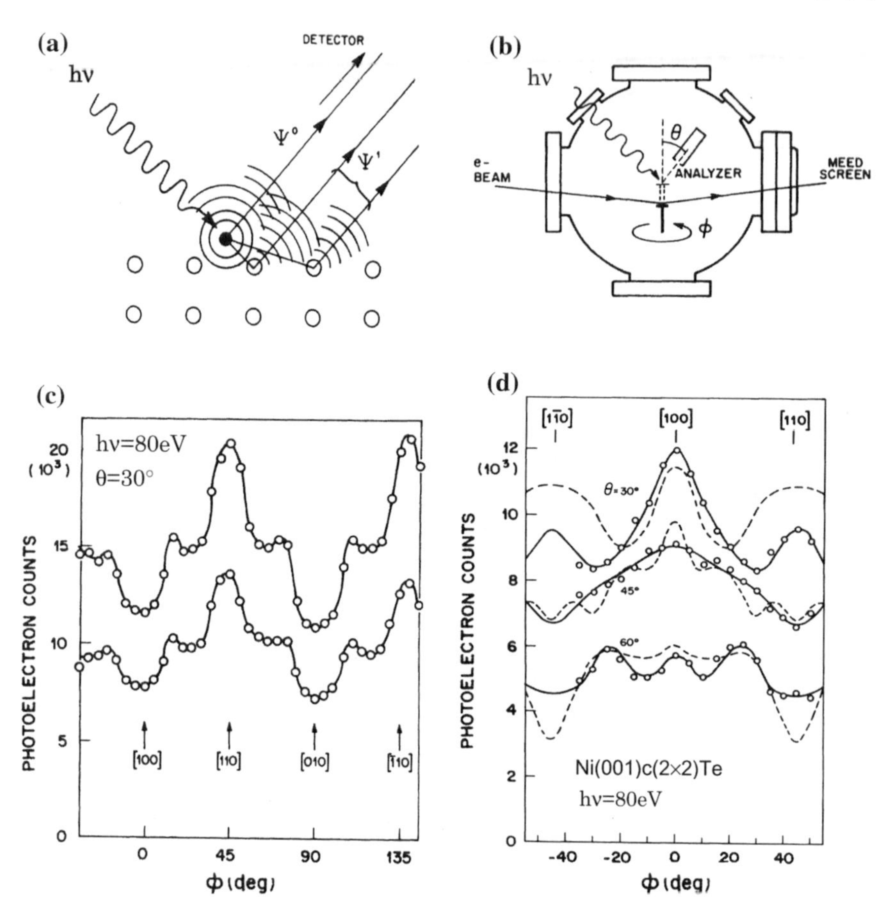

Fig. 12.1 **a** Illustration of photoelectron diffraction (PED) for electrons emitted from a core level of an atom adsorbed on the substrate. Ψ^0 and Ψ^1 represent the waves propagating to the detector directly and indirectly after scattering by neighboring atoms. Diffraction patterns are produced by the interference between the Ψ^0 and Ψ^1 waves. **b** A PED instrument based on an ARPES instrument. Medium-energy-electron diffraction (MEED) is for surface structural analysis. **c** Φ dependence of the Ni 3d valence electron emission from a clean Ni(001) (*upper*) and c(2×2) overlayer Na on it (*lower*). The adsorbed-surface structural information is difficult to be derived from the substrate emission in this case. **d** ϕ dependence of the Te 4d emission from the c(2×2)Te/Ni(001) at $h\nu = 80$ eV. Results at $\theta = 30$, 45 and 60° are shown. *The open circles* (*dashed curves*) represent the experimental results (theoretical calculations) [1]

In order to overcome the forward focusing effects, the differential holography method was employed, where the difference of 2 normalized 3D holograms at different E_K was Fourier transformed [7].

Many holograms must be recorded at different energies in order to do the Fourier transformation and reconstruct the 3D atomic arrangement in the photoelectron holography. As easily recognized, high accuracy measurements of the full angular dependence or high accuracy 2D patterns are necessary for a quantitative analysis in PED. The diffraction patterns can be measured by rotating a small electron analyzer around the sample (Fig. 12.1(b)). However, measuring time can be much shortened and the angular accuracy and angular reproducibility can be much improved by employing a 2D electron analyzer introduced in Sect. 3.4.3. Shortening of the total measuring time is an important requirement for the study of clean surfaces. By employing such an analyzer one can simultaneously obtain the 2D diffraction pattern in a wide solid angle within a short time [8, 9].

We will here demonstrate that single-energy PED pattern is almost enough for the purpose of high resolution surface structural analysis by use of the circular dichroism in PED patterns employing highly circularly polarized (>98 %) soft X-ray light from a helical undulator. The technique used to nonchiral and nonmagnetic materials is reviewed below. Figure 12.2 shows the PED patterns of the Si 2p core level derived electrons from the Si(001) clean surface excited by circularly polarized light with counter clockwise (a) and clockwise (b) helicity at hν slightly above 250 eV [9]. The white regions represent the strong diffraction intensity which are rotated counter clockwise and clockwise depending on the helicity of the excitation light. It is known in the PED that the forward focusing takes place as depicted in Fig. 12.3. In the (110) plane, [115], [113] and [112] forward focusing is expected. However, the real diffraction pattern was found to be azimuthally rotated from these axes as demonstrated in Fig. 12.2. Since the PED excited by linearly polarized light was observed as predicted in Fig. 12.3, the present result was thought to be induced by the circularly polarized light excitation. It was found that the amount of rotation increased when E_K decreased [9].

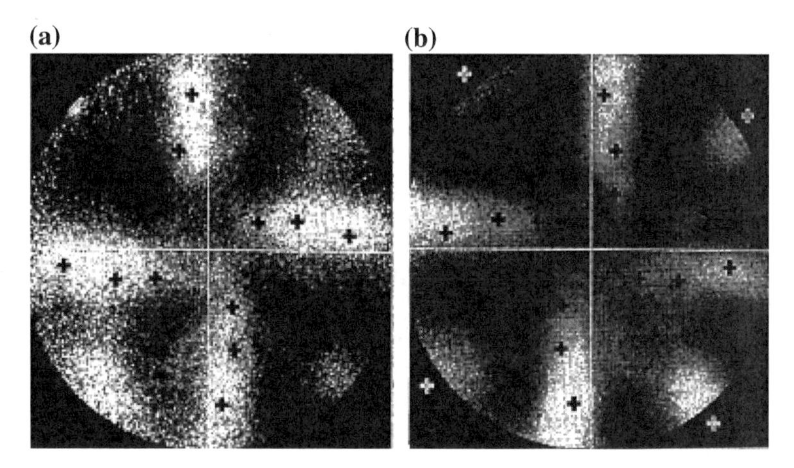

Fig. 12.2 Si 2p core level photoelectron diffraction patterns from the Si (001) surface excited by counter clock-wise (**a**) and clock-wise (**b**) circularly polarized SR. The measurement was done for $E_K = 250$ eV. White regions correspond to strong photoelectron intensity. *The small crosses* show the predicted positions of the forward focusing peaks by the nearest neighbor atoms in the directions of [115], [113], [112] and [101] [9, 10]

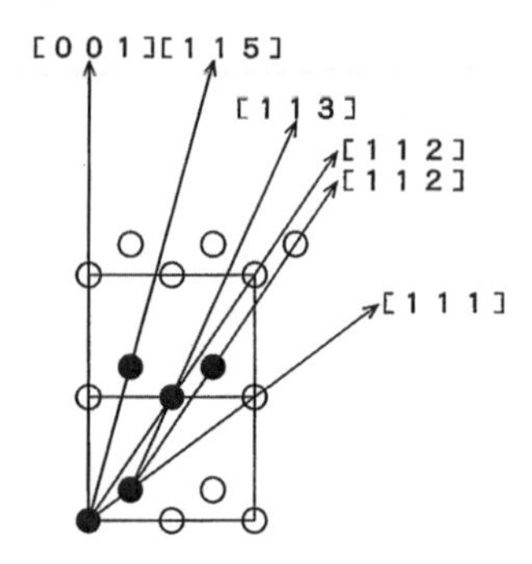

Fig. 12.3 Atomic arrangement of Si in the (110) plane. Forward focusing peaks are expected along the [115], [113], [112], [111] directions for linearly polarized light [9]

The photoelectron wave function at (r, θ, ϕ) (r is the radial coordinate, θ is the polar angle and ϕ is the azimuthal angle with the photoelectron-emitting atom located at the origin) is expressed as

$$\psi \propto e^{ikr}Y_{lm}(\theta, \phi)/r \propto e^{ikr}f(\theta)e^{im\phi}/r, \tag{12.1}$$

where k represents the wave vector of the photoelectron, Y_{lm} is the spherical harmonics with orbital angular momentum l and its z component m. The light is incident from the z axis. For the circularly polarized σ^+ or σ^- light excitation, m is increased or decreased by 1 from the initial state due to the dipole selection rule. For the σ^+ excitation, for example, the dipole allowed transition from the 2p state to the d state takes place from the (l, m) = (1, 1), (1, 0) and (1, −1) states to (2, 2), (2, 1) and (2, 0) states with a relative probability of 6:3:1 when the spin–orbit splitting of the 2p state is neglected. The transition from the 2p state to the s state is much weaker than that to the d state. In fact, the 2p state is split into the j = 3/2 and 1/2 states when the spin–orbit interaction is taken into account. For the σ^+ excitation from the j = 3/2 state, the relative transition probability to the (l, m, spin) = (2, 2, ↑), (2, 1, ↑), (2, 0, ↑) states becomes 6:2:1/3 and to the (2, 2, ↓), (2, 1, ↓) and (2, 0, ↓) states 2:2:1. The difference between the ↑ and ↓ spins can be recognized by spin analysis but cannot be recognized in the conventional photoelectron diffraction in non-magnetic materials.

The locus of the electron (r, θ, ϕ) can be calculated and the following relation is obtained:

$$\phi = -(m/k_{\|}r_{\|}) + C, \tag{12.2}$$

where $k_{\|} = k \cdot \sin\theta$ and $r_{\|} = r \cdot \sin\theta$ and C is a constant. The nearest neighbor atom A is located at **R** from the origin O. Then the angle α between the O–A direction and the electron locus projected into the x–y plane (the plane in Fig. 12.2) is given at $r_{\|} = R_{\|}$ by

$$\tan\alpha = (r_{\|}d\phi/dr_{\|})_{R_{\|}} = m/(k_{\|}R_{\|}) \tag{12.3}$$

If $\phi = 0$ is taken for the O–A direction, the constant C is evaluated as $m/k_{\|}R_{\|}$. The observed deviation angle at infinity is then given by

$$\Delta\phi_{\infty} = C = m/(k_{\|}R_{\|}) \quad (12.4)$$

for the photoelectron with the z-component of the orbital angular momentum m [9, 10]. Namely the forward focusing peak is expected in the direction of $(\theta, \phi + (m/(k_{\|}R_{\|})))$. The angle $\Delta\phi_{\infty}$ is somehow different from α because the locus is not a straight line. The black crosses in Fig. 12.2 show the positions predicted by the above theory for the Si lattice in reasonable agreement with the experimental results. Since both θ and $k_{\|}$ are known from experiment, one can evaluate $R_{\|}$ and R. Thus the surface atomic geometry around the target atom can be evaluated with very high accuracy by this method. This technique is much simpler than the photoelectron holography and immediately applicable to stereoscopic microscopy of the atomic arrangement [10]. Another advantage over the conventional STM for surface nano-structural analysis is, for example, that this PED technique is feasible even when the target atom is embedded below the sample surface.

Even in the field of photoelectron holography, some progress was achieved. In order to reconstruct the 3D atomic arrangement from photoelectron and/or X-ray fluorescence holograms, Fourier transformation was so far utilized requiring many holograms at many E_K or $h\nu$, ϕ and θ. Recently, a new approach named scattering pattern extraction algorithm (SPEA) by means of maximum entropy method (MEM) was proposed to handle a single energy hologram demonstrating its applicability to reconstructing of 58 Au atoms of a Au (001) single crystal [11], where the incident $h\nu$ was 12 keV and the Au L_{α} fluorescence at 9.712 keV was detected with $0 \leq \phi \leq 360°$ and $0 \leq \theta \leq 70°$. By incorporating the crystal translational symmetry into this SPEA-MEM reconstruction algorithm, improvement of the atomic imaging was achieved. This method with considering translational symmetry is applicable to determine the bulk atomic structures around the target atom site or impurity but not straightforward applicable to a system with local distortion around it. An extensive review is given in [12].

Atomic layer dependent magnetism was so far very difficult to be experimentally studied by PES and XAS. Owing to the PED technique with using 2D analyzers, such magnetic information was demonstrated to be obtained by diffraction spectroscopy for Auger electron emission by means of the $h\nu$ dependence of the angular distribution of the Ni LMM ($2p_{3/2}3d3d$) Auger electrons from a Ni wedge ultra-thin film on a Cu (001) surface [13]. The easy magnetization in a Ni film was reported to reorient from the in-plane direction to the surface perpendicular direction beyond 10 ML (it becomes again in-plane beyond 40 ML). Although a rough idea on the depth dependent magnetism might be inferred from the emission angle dependence of Auger electrons, layer-resolved information was difficult to obtain. By virtue of the forward-focusing peaks of Auger electrons for circularly polarized soft X-ray excitation, the mutual configuration between the emitter and scatter atoms can be determined accurately [9]. The light was incident at 45° with respect to the symmetry axis of the analyzer [13]. The typical acquisition time for an Auger diffraction pattern was ~1 s. Scanning of the wedge sample was feasible in the analyzer chamber. The photon beam size was ~0.4 mm

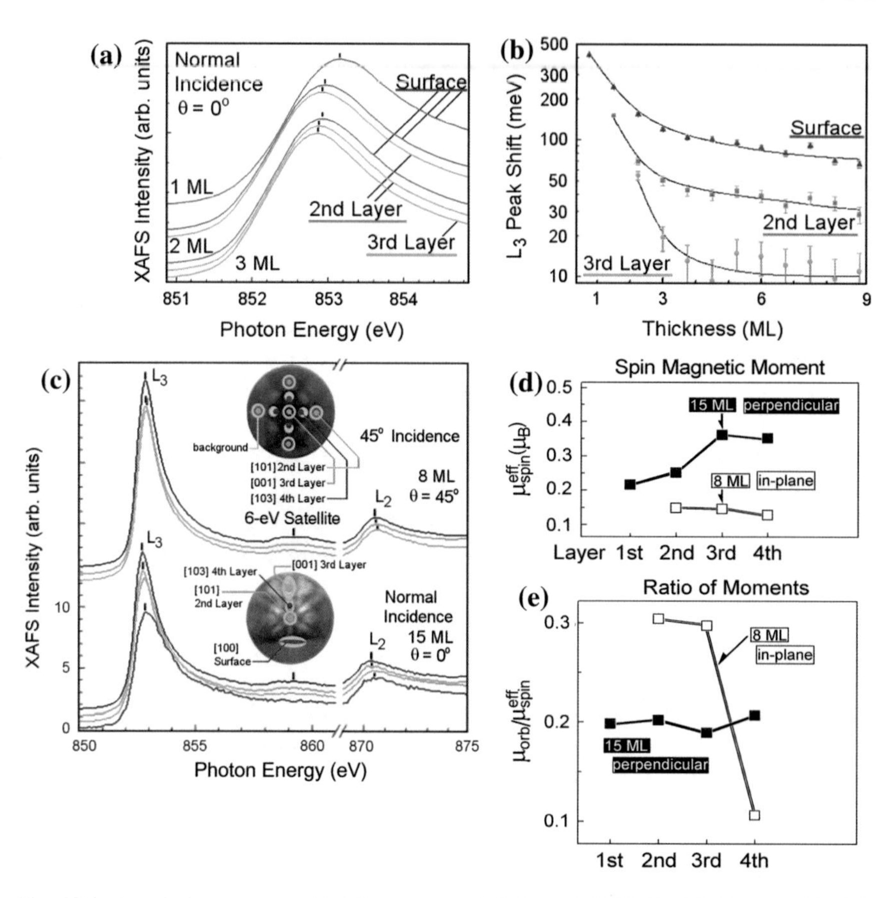

Fig. 12.4 Atomic-layer resolved Ni 2p core level XAS and MCD for Ni films on Cu(001) measured by means of the hν dependence of the Auger electron forward focusing peaks [13]. The spectra of the surface, the 2nd, 3rd and 4th layers were derived from [100] (red), [101] (brown), [001] (green) and [103] (blue) forward focusing peaks in (**c**). XMCD measurements were performed for the 8 and 15 ML Ni films, which were remanently magnetized. XMCD signal was not observed for normal incidence ($\theta = 0°$) but for $\theta = 45°$ in a 8 ML Ni film, clarifying the in-plane magnetization. For a 15 ML Ni film, in contrast, the XMCD was observed for $\theta = 0°$ clarifying the perpendicular magnetization. The spin magnetic moment and the ratio between the orbital and spin magnetic moments are shown in (**d**) and (**e**) [14]

and the wedge length was 5 mm. Fourfold symmetrization was made to get the 2π steradian angular distribution. Slight interlayer mixing between Ni atoms and topmost Cu surface atoms was recognized. Since one can ascribe the individual forward- focusing peaks to [hkl], one can identify the layer to which the emitter atom is contained as long as the film thickness is not too high. From the intensity change of the individual Auger electron forward focusing peaks as a function of the excitation hν, the XAS can be evaluated as shown in Fig. 12.4. The PED was detected by the display type 2D electron analyzer [8]. The XMCD spectrum (Fig. 3(b) of [13]) was then measured as the difference between the σ^+ and σ^-

excitation XAS (Auger electron yield) when the sample was remanently magnetized. The spin magnetic moments derived experimentally for the surface (1st), 2nd, 3rd Ni layers and the 4th layer ($\sim$bulk) are shown in Fig. 12.4(d) [14]. Figure 12.4(e) shows the ratio between the orbital and spin magnetic moments. Although this ratio is almost layer independent for the 15 ML Ni film with perpendicular magnetization, it is much higher for the 2nd and 3rd layer in 8 ML Ni film, suggesting that the orbital magnetic moment is decisive to realize the in-plane magnetization [13, 14]. This method applicable to the subsurface region and shallow interfaces besides the surface layer may also facilitate similar studies of antiferromagnetic materials in future.

The so-called Kikuchi bands are another feature of the photoelectron diffraction in the high hν excitation region. In the case of the photoelectron diffraction in the X-ray region above $\sim$1 keV, the dynamical scattering of electrons from lattice planes provide Kikuchi patterns. A long-range-order dynamical scattering approach for simulating photoelectron diffraction patterns at high E_K up to 20 keV based on quasi-3D translational symmetry of the scattering potential was reported [15]. The predicted patterns were found to show good agreement with experimental results even down to 1 keV.

References

1. N.V. Smith, H.H. Farrell, M.M. Traum, D.P. Woodruff, D. Norman, M.S. Woolfson, B.W. Holland, Phys. Rev. **B21**, 3119 (1980)
2. J.J. Barton, Phys. Rev. Lett. **61**, 1356 (1988)
3. C.S. Fadely, *Synchrotron Radiation Research: Advances in Surface Science* (Plenum Press, New York, 1990)
4. C.S. Fadley, Y. Chen, R.E. Couch, H. Daimon, R. Denecke, J.D. Denlinger, H. Galloway, Z. Hussain, A.P. Kaduwela, Y.J. Kim, P.M. Len, J. Liesegang, J. Menchero, J. Morais, J. Palomares, S.D. Ruebush, E. Rotenberg, M.B. Salmeron, R. Scalettar, W. Schattke, R. Singh, S. Thevuthasan, E.D. Tober, M.A. Van Hove, Z. Wang, R.X. Ynzunza, Prog. Surf. Sci. **54**, 341 (1997)
5. C.S. Fadley, M.A. Van Hove, Z. Hussain, A.P. Kaduwela, R.E. Couch, Y.J. Kim, P.M. Len, J. Palomares, S. Ryce, S. Ruebush, E.D. Tober, Z. Wang, R.X. Ynzunza, H. Daimon, H. Galloway, M.B. Salmeron, W. Schattke, Surf. Rev. Lett. **4**, 421 (1997)
6. C. Fadley, Nucl. Instrum. Meth. **A601**, 8 (2009)
7. S. Omori, Y. Nihei, E. Rotenberg, J.D. Denlinger, S. Marchesini, S.D. Kevan, B.P. Tonner, M.A. Van Hove, C.S. Fadley, Phys. Rev. Lett. **88**, 055504 (2002)
8. H. Nishimoto, H. Daimon, S. Suga, Y. Tezuka, S Ino, I. Kato, F. Zenitani, H. Soezima, Rev. Sci. Instrum. **64**(10), 2857 (1993)
9. H. Daimon, T. Nakatani, S. Imada, S. Suga, Y. Kagoshima, T. Miyahara, Jpn. J. Appl. Phys. **32**, L1480 (1993)
10. H. Daimon, S. Imada, S. Suga, Surf. Sci. **471**, 143 (2001)
11. T. Matsushita, F. Z. Guo, M. Suzuki, F. Matsui, H. Daimon, K. Hayashi, Phys. Rev. **B78**, 144111 (2008)
12. T. Matsushita, F. Matsui, H. Daimon, K. Hayashi, J. Electron. Spectrosc. Rel. Phenom. **177–178**, 195 (2010)

13. F. Matsui, T. Matsushita, Y. Kato, M. Hashimoto, K. Inaji, F. Z. Guo, H. Daimon, Phys. Rev. Lett. **100**, 207201 (2008)
14. F. Matsui, T. Matsushita, H. Daimon, J. Electron Spectrosc. Rel. Phenom. **178–179**, 221 (2010)
15. A. Winkelmann, C.S. Fadley, F.J.G. Abajo, New J. Phys. **10**, 113002 (2008)

Chapter 13
Complementary Techniques for Studying Bulk Electronic States

Nowadays the electronic structure of solids is very effectively probed by the PES and ARPES as discussed so far and complemented by the results of IPES and angle resolved IPES. However, these spectroscopies are often very surface sensitive except for some exceptional cases such as HAXPES, because of the generally short λ_{mp}. On the other hand, the absorption coefficient of the photon with hν higher than several tens eV is in the range of 10^6–10^4 cm^{-1}, which becomes generally decreased with increasing hν except for the sudden increase near a core absorption threshold. The photons in the soft X-ray region, for example, can penetrate more than $\sim$10 nm into a sample, exciting electrons on their paths. In the case of hard X-rays, photons can easily penetrate deeper than 1,000 nm. The hν tunable synchrotron radiation is quite useful to cover the hν regions below and above the core excitation thresholds for absorption spectroscopy. In this regard the photon spectroscopy is a really bulk sensitive method. In order to perform RPES, the starting point is to study the core absorption spectrum. Since the electronic structures near E_F attract wide attention to understand the MIT, Kondo physics, superconductivity and so on, infrared and far-infrared or THz spectroscopies are very useful complementary techniques to photoelectron spectroscopy. Even the momentum information is available in the case of resonance inelastic X-ray and soft X-ray scattering. One should not overlook the potential of photon spectroscopies, which can be performed under such perturbations as the high magnetic field, electric field and pressures. It is highly recommended to combine the PES and ARPES studies with these spectroscopies.

13.1 Core Absorption and Core Fluorescence Spectroscopy

There are several ways to measure the core absorption spectra. The transmittance measurement is a straight way to measure the absorption coefficient. Thin film samples evaporated onto thin rather transparent substrates are sometimes available for core absorption measurements. In this case high quality homogeneous samples

S. Suga and A. Sekiyama, *Photoelectron Spectroscopy*,
Springer Series in Optical Sciences 176, DOI: 10.1007/978-3-642-37530-9_13,

with no pin hole(s) are required. As far as flat specular surfaces are available, optical reflectance measurements are much easier. Kramers–Kronig (K–K) analysis can be applied to the reflectance spectra to derive the optical and dielectric constants as functions of hν. However, the normal incidence reflectivity is generally very low in the soft and hard X-ray regions because the real part of the refractive index is close to 1.0. Then the grazing incidence reflectivity measurement is often performed. It is also empirically known that core absorption spectra are qualitatively similar to the total electron yield (TEY) spectra. This is because the absorption is proportional to the number of core holes, to which the number of low energy secondary electrons produced by the core–hole decay is also proportional. In this case, however, the probing depth is limited by the λ_{mp} of the secondary electrons, which lies in the range of $\sim$10 nm. Still bulk sensitivity comparable to or larger than that of HAXPES is achieved. Another bulk sensitive method for absorption measurements is by means of the fluorescence yield (photon emission yield) measurement, because the photon can escape from a deeper region of the sample as judged from the above mentioned absorption coefficient. In the case of the photon emission under the resonance excitation, however, part of the fluorescence is reabsorbed through the core excitation process known as self-absorption, which not only distorts the spectral shape but also widens the spectral width. When the fluorescence from shallower core levels to the core hole state is measured while scanning hν, bulk sensitive and undistorted core absorption spectrum can be obtained owing to the negligible probability of the self-absorption.

Examples of core absorption spectra at room temperature are shown in Fig. 13.1 for the TM 3p core level ($M_{2,3}$) of TM di-halides evaporated on collodion films [1]. The absorption coefficient free from stray and higher order light from the grating monochromator was evaluated by successive measurements of

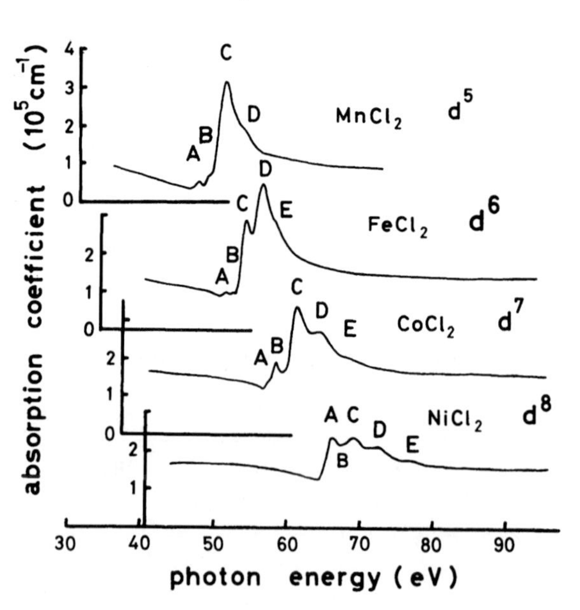

Fig. 13.1 3p core level absorption spectra of transition metal di-halides at room temperature. The occupation number of the 3d level in the ground state is given by d^m in this figure [1]

samples with increased thickness beside using an Al filter. Predominance of the $3p^5 3d^{m+1}$ final state over the $3p^5 3d^m 4s$ states is clear from the increase in the absorption coefficient from $NiCl_2$ to $MnCl_2$. The multiplet structures are quite conspicuous in these spectra and characteristic of the TM species and its valence. They are due to the presence of the incomplete 3d shell and have intra-atomic character as if they are the fingerprint irrespective of the ligand species. The 3p-3d Coulomb and exchange interactions, spin–orbit interaction of the 3p core hole, d–d interactions, crystal field effects as well as the covalency between the 3d and the ligand electrons are taken into account in a theoretical calculation to be compared with these absorption spectral shapes [2].

In the case of high quality flat single crystal surfaces, grazing incidence reflectivity is measured, to which the K–K transformation is applicable. Real and imaginary parts of the dielectric constant (ε_1 and ε_2) and optical constants (refractive index: n and extinction coefficient: k) as well as the energy loss function ($-\mathrm{Im}(1/\varepsilon)$) and the effective electron number (N_{eff}) can be derived. An example is given for NiS_2 in Fig. 13.2 [3]. Many results of reflectivity spectra in the vacuum ultraviolet region were reported [4–6], which were compared with the results of direct absorption spectra.

There are really many results obtained by means of the total electron and/or fluorescence yield methods, since neither thin films nor flat specular surfaces are required. However, quantitative evaluation of the absorption coefficient is not feasible from them. Figure 13.3 shows the results of rare earth 3d core absorption spectra of CeB_6 and PrB_6 obtained by TEY [7]. The probing depth was in the range of 10 nm. The two dominant peak structures were ascribed to the excitation from the $3d_{5/2}$ and $3d_{3/2}$ core levels. The results are similar to those obtained for γ-Ce and PrF_3, suggesting the trivalent nature of the ground states. Then detailed RPES became feasible. The valence band PES spectra were measured at $h\nu$ shown by the vertical lines labelled with letters in the figure. PES spectra excited at $h\nu$ corresponding to C and D were found to be composed of two peaks (Figs. 3 and 4 of [7]) and interpreted as the bonding and antibonding combination of the f^{n-1} and $f^n\underline{L}$ final states, where $\underline{L}$ is the ligand hole in the B 2p band.

One of the results by means of fluorescence yield is shown in Fig. 13.4 for $YbInCu_4$ [8], which shows the bulk valence transitionnear $T_v = 42$ K. The total fluorescence yield (TFY) was obtained by recording the integrated intensity of the scattered light and the partial fluorescence yield (PFY) by measuring the intensity of the Yb $L\alpha_1$ fluorescence ($2p^5 3d^{10} \rightarrow 2p^6 3d^9$ at $h\nu = 7{,}415$ eV) as a function of the excitation $h\nu$. In these spectra, the main structure corresponds to the f^{13} final state and the weak pre-edge shoulder to the $4f^{14}$ final state associated with the Yb^{3+} and Yb^{2+} states, respectively, where the photon excitation is between the 2p core state and the conduction band states with εd and εs character. The splitting between the f^{13} and f^{14} final states is dominated by the Coulomb interaction with the 2p core hole. Here one can clearly notice that all spectral features are much better resolved in the PFY spectrum. The T dependence of the valence could be estimated from PFY by assuming certain spectral line shapes for the Yb^{3+} and Yb^{2+} components. However, the accuracy of the estimated valence value was

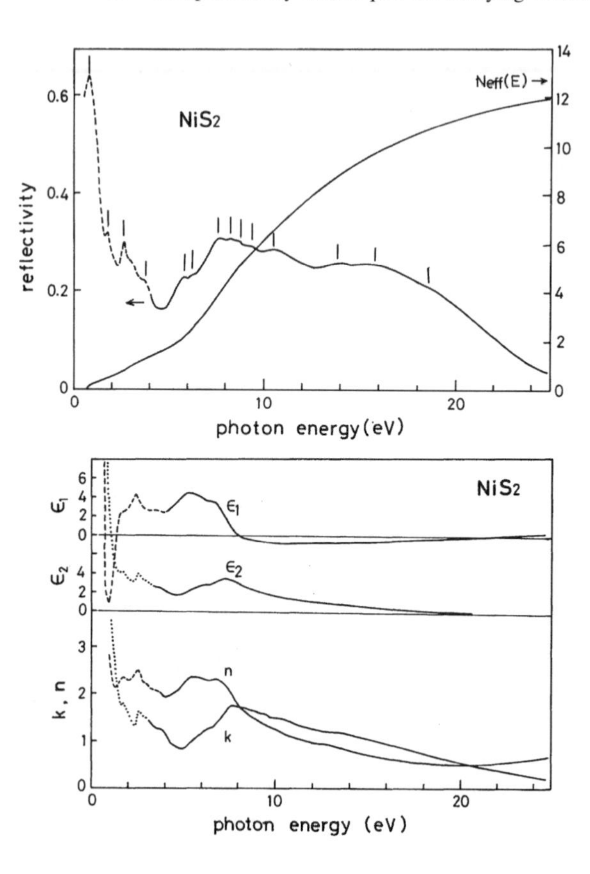

Fig. 13.2 Photoreflectance spectrum of an untreated surface of a NiS_2 single crystal measured at room temperature for an incidence angle of 22.5° (*upper panel*), where N_{eff} derived from this experiment is added. Dielectric constants and optical constants are shown in the *lower panel* [3]

worse than that obtained by HAXPES and resonance inelastic X-ray scattering [8] (RIXS, the technique explained later in Sect. 13.3).

Core level absorption linear dichroism (LD) is very powerful to investigate the orbital symmetry of doped holes and/or partially filled conduction bands, providing fundamental information on the electronic structure in solids. Historically, the orbital symmetry of the intrinsic and doped holes in the high-T_c uprates was established by this technique. Both O 1 s and Cu 2p core absorption spectra of single crystals of LSCO detected by means of fluorescence yield showed that the absorption intensity in the vicinity of the absorption edge became stronger on going from the geometry of **E** // c to **E** ⊥ c, where **E** is the electric field of incident X-rays and c is the direction perpendicular to the layers of cuprates (Fig.2-4 in [9]). From the dipole selection rules, these results indicated the predominant Cu $3d_{x^2-y^2}$ and hybridized O $2p_{x,y}$ characters for the holes in LSCO.

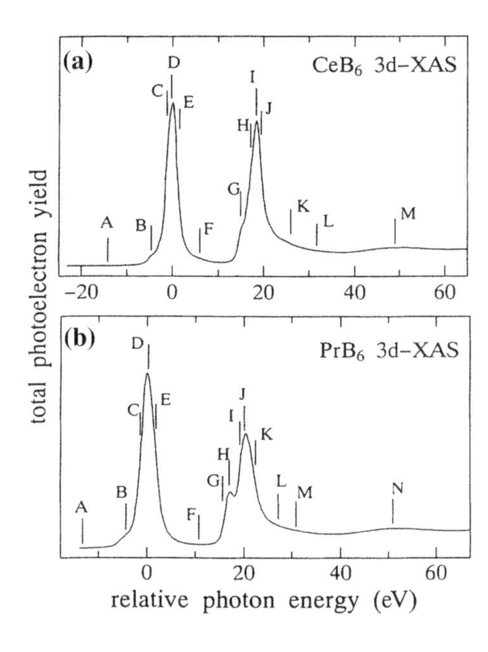

Fig. 13.3 3d core absorption spectra of **a** CeB_6 and **b** PrB_6 measured by means of total electron yield. The clean surfaces were prepared by filing [7]

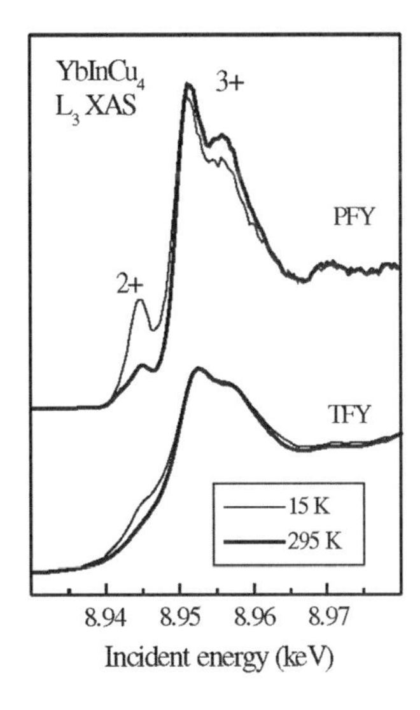

Fig. 13.4 Yb $2p_{3/2}$ core absorption spectra of $YbInCu_4$ above and below the valence transition temperature $T_v = 42$ K obtained by total and partial fluorescence yield (TFY and PFY) [8]

Linear dichroism (LD) in the core absorption was also applied to an organic conductor, the N,N′-dicyanoquinonediimine (DCNQI)-Cu salt by means of TEY [10]. Figure 13.5 shows the N 1s core absorption spectra and their dichroism, where the N atoms are involved in the DCNQI molecule. Sharp peaks at ~399 and ~401 eV and a broad structure around 404 eV are observed in the experimental spectra. The peak at 399 eV is attributed around to a transition from the N 1s core

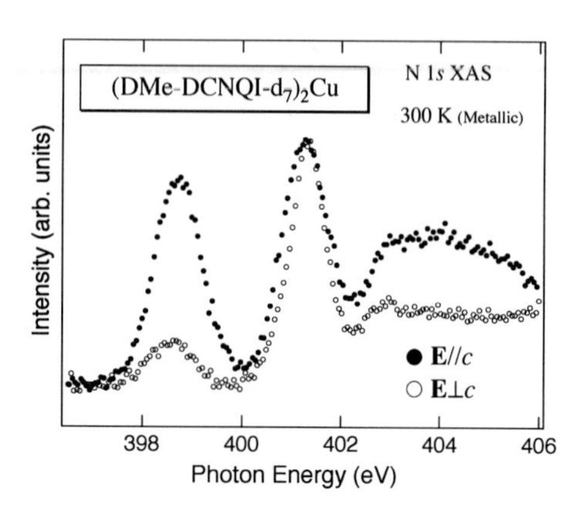

Fig. 13.5 Linear dichroism (LD) in the N 1s core absorption spectra of the quasi-one-dimensional organic conductor DCNQI-Cu [10]

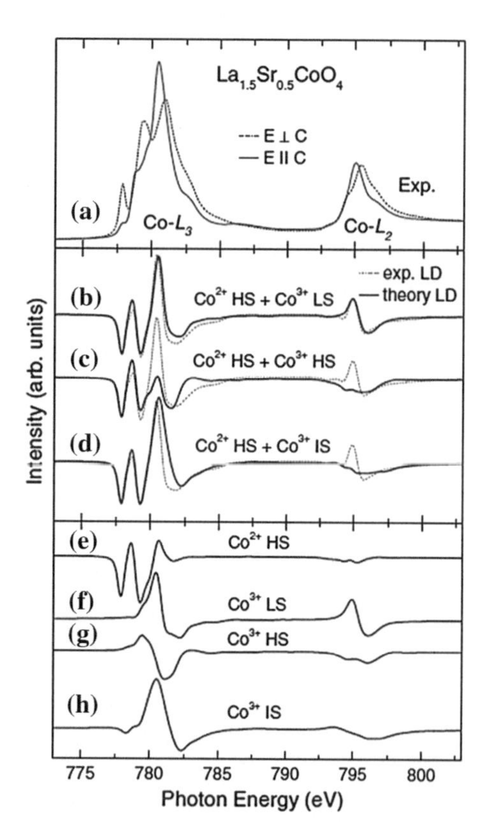

Fig. 13.6 **a** Experimental polarization-dependent Co 2p X-ray absorption spectra of highly insulating $La_{1.5}Sr_{0.5}CoO_4$ for **E** // *c* and **E** ⊥ *c*. **b**, **c** and **d** show the experimental linear dichroism (LD) compared with the theoretical results calculated for scenarios with high-spin (HS) Co^{2+} and either low-spin (LS), HS, or intermediate-spin (IS) Co^{3+} states. The calculated LDs for the HS-Co^{2+}, LS-, HS- and IS-Co^{3+} states are shown in **e**, **f**, **g** and **h** [11]

level to the lowest unoccupied molecular orbital (LUMO) conduction band. Between the **E** // c and **E** ⊥ c spectra, the peak around 399 eV and the broad structure around 404 eV are significantly reduced in the **E** ⊥ c spectrum, reflecting the quasi-one-dimensional character of the LUMO pπ band dispersing along the c axis.

The detailed electronic structure of strongly correlated materials could often be clarified by LD in core absorption. Figure 13.6 shows LD in the Co 2p X-ray absorption spectra (TEY) of $La_{1.5}Sr_{0.5}CoO_4$ compared with results of a configuration interaction cluster model calculation including full atomic multiplets [11]. The $La_{2-x}Sr_xCoO_4$ system is highly insulating for a wide range of x values with anomalously high activation energies for conductivity. The detailed analysis of the LD in Co 2p absorption tells that the Co^{3+} ($3d^6$) ions are in the non-magnetic low-spin (LS) state coexisting with the high-spin (HS, spin S = 3/2) Co^{2+} ($3d^7$) state. In this situation, the charge transfer from the Co^{2+} to Co^{3+} sites could not take place where a so-called spin blockade mechanism is active, naturally explaining the highly insulating nature as well as a remarkable deviation of ordering temperatures between charge (T_{CO} ~ 750 K) and spin (T_{SO} ~ 30 K).

13.2 Infrared and Far-Infrared Spectroscopy

Infrared corresponds to the hν region below ~1.5 eV, which is often called near infrared, mid-infrared and far-infrared with decreasing hν. In infrared spectroscopy, energies are often expressed in the unit of cm^{-1}, where 1,000 cm^{-1} corresponds to 0.124 eV. While infrared spectroscopy can cover only the low hν region, the energy resolution can be quite high. In addition, this technique is rather bulk sensitive. In this respect, infrared spectroscopy is complementary to the high resolution bulk sensitive PES studies near E_F to discuss the details of the electronic structure.

As a typical example, the results on YbB_{12} are shown in Fig. 13.7 [12]. This material is known to be a Kondo semiconductor, where a small energy gap develops at the Fermi level below the sample temperature of ~80 K. The gap magnitude was so far estimated to be in the range of 12–15 meV [13] from electrical resistivity, Hall effect, magnetic specific heat and photoemission experiments [42] of Chap. 9. In this infrared experiment, a disk-shaped sample of 4.5 mm diameter was cut from a single crystal, and mechanically polished for optical measurement. The reflectivity spectra R (hν) were measured in the T range of 8–690 K. R(hν) above 4 meV was measured under a near-normal incidence configuration, using a Fourier interferometer (Bruker IFS 66v) with various combinations of conventional light sources, beam splitters and detectors up to 2.5 eV. R (hν) between 1.3 and 4 meV was measured by using the THz synchrotron radiation source at the BL6B beam line of UVSOR, Okazaki, Japan. The optical conductivity spectra σ(hν) were obtained from a K–K analysis of the measured R(hν) spectra. Although the spectra up to 0.65 eV are not shown here, a dip was induced near 0.1 eV in the R(hν) below 295 K and a broad peak was observed in the σ(hν) spectra in the region of 0.2–0.25 eV with decreasing the T below 400 K. Metallic behavior high temperatures and insulating behavior at low T were thus suggested [12]. In order to discuss the details of the change of the electronic structure near E_F, the T dependence of R(hν) and σ (hν) below 60 meV is shown in Fig. 13.7(a) and (b) below the T of 120 K. Although the spectral

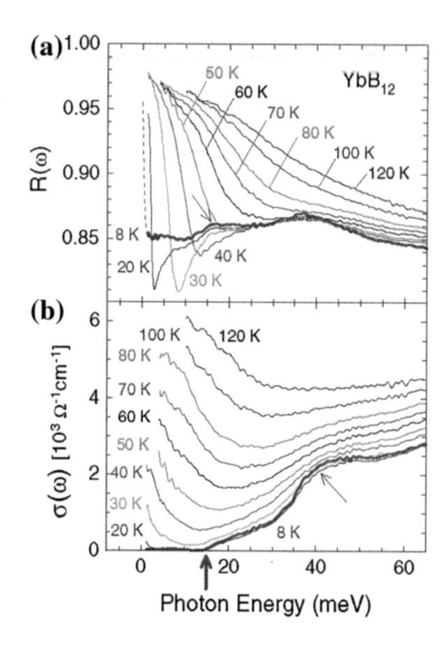

Fig. 13.7 Infrared reflectance spectra R(hν) of YbB_{12} as a function of temperature (**a**). Derived optical conductivity spectra σ(hν) are shown in (**b**). The *arrow* in (**a**) indicates a hump in R(hν). The *black* and *red arrows* in (**b**) indicate a shoulder and the onset in σ(hν) [12]

change is rather monotonous down to 80 K, a faint structure can be recognized at ~40–45 meV below T ~ 70 K in the far infrared σ(hν) spectra in Fig. 13.7(b). With further decreasing the T, the gap formation is seen with leaving a shoulder at ~40 meV.

As the gap develops with decreasing T, the number of free carriers decreases progressively as recognized from the shift of the plasma edge or the Drude tail in R(hν). At 20 and 8 K, the σ(hν) is vanishingly small below 15 meV. This threshold energy was ascribed to the indirect gap energy of the Kondo semiconductor based on the c-f hybridization model [12]. On the other hand, the mid-infrared peak near 0.2–0.25 eV rather commonly observed in other Yb based metals as well was ascribed to the optical transition across the direct gap, whose origin is discussed in detail in Ref. [14]. Indirect transitions usually forbidden in optical transitions might be allowed by second order processes involving phonons and be observable in the far-infrared.

High pressure experiments are also possible in infrared spectroscopy by use of a diamond anvil cell (DAC), though strong photon absorption by diamond takes place near hν = 0.3 eV. Not only the reflectance measurements but also transmittance measurements were performed for the pressure-induced insulator–metal transition in the Mott insulators, TiOBr [15] and TiOCl [16] (Figs. 1 and 2 of [16]). A small piece of sample with a size of ~80 × 80 μm^2 was cut from a single crystal with a thickness ≤5 μm and put into the hole of a steel gasket for the DAC. The pressure dependence of the transmittance and reflectance spectra was measured at room temperature up to 14.5 GPa by use of an instrument equipped with infrared microscope. Part of the measurements was performed at the infrared beam line of the SR source ANKA. The pressure dependent transmittance spectra of TiOBr were measured in the hν region between 0.3 and 2.7 eV for the polarization

E || a and b axes. Pressure dependence of the reflectance spectra were measured in the hν region between 0.03 and 0.93 eV for the polarization **E** || a and b axes. The energy ranges were limited by the normalization method of the spectra by reference signals. The absorption at 0.63 eV for **E** || a and at 1.35 eV for **E** || b were ascribed to the excitations between the Ti 3d levels split by the crystal field. These excitations are infrared active because of the lack of inversion symmetry on the Ti site. The pressure-induced changes were rather similar to those in TiOCl. It was found that the transmittance above ~2 eV was suppressed and shifted to lower hν with the increase of pressure. This behavior was ascribed to the excitation across the charge gap. The transmittance in the whole range between 0.3 and 2.7 eV was drastically suppressed above 14 GPa. On the other hand, the measurement of reflectance spectra was rather difficult at low pressures because the sample was partially transparent and interference fringes induced by the multiple reflections within the sample were observed in the infrared range. Above 10 GPa, such fringes disappeared and R(hν) increased abruptly in the far-infrared and mid-infrared range with slightly different behavior for **E** || a and b axes [15]. Metallization of the sample under pressure was thus clarified. It was also found that TiOBr was more sensitive to the pressure than TiOCl. Between two scenarios proposed to explain these pressure-induced spectral changes in TiOBr, the structural phase transition model was more feasible than the band-width controlled insulator–metal transition model (Mott transition with purely electronic character) according to the X-ray powder diffraction experiment [15]. There are many other results reported on infrared spectroscopy on strongly correlated electron systems under high pressures (for example on V_3O_5 [17] and SmS [18]).

The advantage of infrared spectroscopy compared to PES is not only the applicability of high pressures but also the applicability of external magnetic fields. Namely, infrared spectroscopy can be performed under multiextreme conditions. One interesting example is reported on CeSb [19]. Ce monopnictides have many magnetic phases with complex magnetic structures at low temperatures, under high pressures and high magnetic fields due to the magnetic moments originating from the localized Ce^{3+} $4f^1$ state, where the magnetic interaction mediated by the conduction and valence electrons was thought to play an important role. The hybridization between the Ce 4f and Sb 5p state might be playing such a role in CeSb. Infrared reflectance measurements were performed at the magneto-optical imaging station of BL43IR of SPring-8 in the range of hν = 0.1–1 eV at T between 4 and 70 K, in magnetic fields (B) up to 14 Tesla and with pressure up to 4 GPa on a single crystal sample cut into 200 × 200 × 50 μm^3 and put in a DAC. It was observed under pressures of several GPa that the electrical resistivity at ~30 K increased by one order of magnitude compared with that under ambient pressure. The magnetic phase with this enhancement was the single-layered antiferromagnetic phase AF-1, not present at ambient pressure. The magnetic structure is expressed as +−, where +(−) indicates the magnetic moment of the ferromagnetic layer parallel (antiparallel) to the magnetic field.

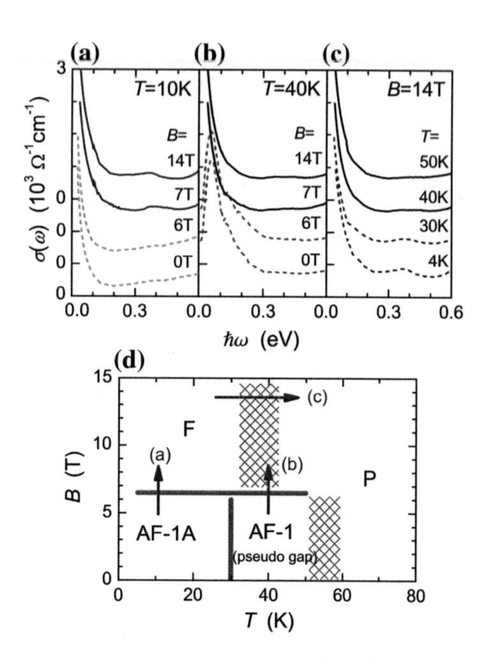

Fig. 13.8 Optical conductivity spectra of CeSb at P = 4 GPa. The magnetic field dependence is shown at 10 and 40 K in (**a**) and (**b**). The temperature dependence is shown at B = 14 Tesla in (**c**). Phase diagram of CeSb at 4 GPa as functions of the magnetic field (B) and temperature (T) (**d**). Solid lines and hatched regions show the first order and second order phase transition [19]

It was also known that the double-layered antiferromagnetic AF-1A phase was stable at low temperatures under ambient pressure and also at high pressures with the ++−−magnetization. In contrast to the metallic behavior of the AF-1A phase, the electronic structure in the AF-1 phase seemed to have a gap as suggested in the temperature dependence of the R(hν) spectra at 4 GPa (Fig. 1 of Ref. [19]), where a phase transition was observed at ~30 K. Metallic behavior was observed at 10 and 70 K under 4 GPa, being judged from R (hν→0) ~ 1. Figure 13.8(a)–(c) shows the magnetic-field (B) and T dependence of the σ (hν) spectra at 4 GPa. In both spectra (**a**) and (**b**), a spectral change was recognized between 6 and 7 Tesla. The spectra below 6 Tesla at T = 40 K in Fig. 13.8(b) are much different from those at 10 K in Fig. 13.8(a) in the sense that the pseudogap is observed in the low hν region, though the gap disappears above 7 Tesla and the spectra become metallic. Such a magnetic-field-induced phase transition from a non-metal to metal was thus observed under high pressures. Although the p-f mixing competes with magnetic interactions and crystal field splitting under ambient pressure, the p-f mixing increases with pressure making the phase diagram simpler than that under ambient pressure. The pseudogap induced in the AF-1 phase under a pressure of ~4 GPa at T ~ 30 K originating from the magnetic band folding effect collapses under a magnetic field >6.5 Tesla [19].

13.3 Resonance Inelastic X-ray Scattering

As already discussed, purely photon-based spectroscopies are much more bulk sensitive having probing depths down to several thousand Å from the surface in many cases in contrast to the electron spectroscopies. However, the momentum information is rather difficult to be obtained by the so far introduced techniques in this chapter. In many studies of materials sciences, the electronic structure near the Fermi level attracts much interest. For metal systems, for example, the Drude tail due to the free electron plasmon will be predominant in the reflectance spectra for very low hν, obscuring the structures due to interband transitions. When the system shows a metal–insulator transition, the infrared reflectance spectroscopy and K–K analysis provide useful information on the optical conductivity spectra, clarifying such phenomena as Kondo-gap, Mott-gap and CDW-gap. Meanwhile, X-ray emission spectroscopy can be applied to probe the occupied valence band states. These photon-in and photon-out method is particularly useful for insulators to which the application of PES is limited by the possible charge up effect. In addition, resonance inelastic X-ray scattering (RIXS) had progressed recently due to its potential of resolving momentum together with the excellent bulk sensitivity down to $\sim$10 μm from the surface [20–25]. Owing to its bulk sensitivity, polished surfaces can be employed for RIXS without deteriorating the reliability of the experiment.

RIXS can be measured in both transmission and reflection mode as illustrated in Fig. 13.9(a). An example of a RIXS set-up employed for the Cu 1s core level RIXS in $CuGeO_3$ and Sr_2CuO_3 [26] is schematically shown in Fig. 13.9(c). The radiation from a planar undulator is monochromatized by a double crystal monochromator and a channel cut crystal and then incident onto the sample. The resolution of the incident hν was better than 300 meV in our set-up at BL19LXU of SPring-8. The horizontal focusing was better than 100 μm on the sample. The momentum difference Δk is defined as the difference between the momenta of the incidence and scattered X-rays. In the present case, Δk was set to be parallel to the long edge of the sample (Fig. 13.9(a)). The sample, analyzer crystal and the slit of the detector should be on a Rowland circle. A 1 m Rowland circle was used for the measurement. A total resolution of 400 meV FWHM was achieved by using a spherically bent Si (553) analyzer crystal. In order to scan Δk, the sample and the whole Rowland circle were slightly rotated by computer control to cover Δk over a few Brillouin zones. Δk was set off the Bragg angle not to saturate the detector. At several Δk the measurement of the scattered X-ray spectrum was performed by rotating the analyzer crystal with translationally moving the detector to satisfy the Rowland circle condition. The energy loss spectrum can be obtained as a function of the excitation energy and Δk.

Here two examples of RIXS on an edge sharing $CuGeO_3$ and a corner sharing Sr_2CuO_3 (Fig. 13.9(b)) with respect to the Cu–O chain are presented in Fig. 13.10(a)–(c) and (d) [26]. The polarization of the light and the Cu–O chain were in fact in the horizontal plane, corresponding to the scattering plane. $CuGeO_3$ has a single chain in the edge-sharing CuO_2 plane configuration with the Cu-O-Cu angle (θ) being 99°,

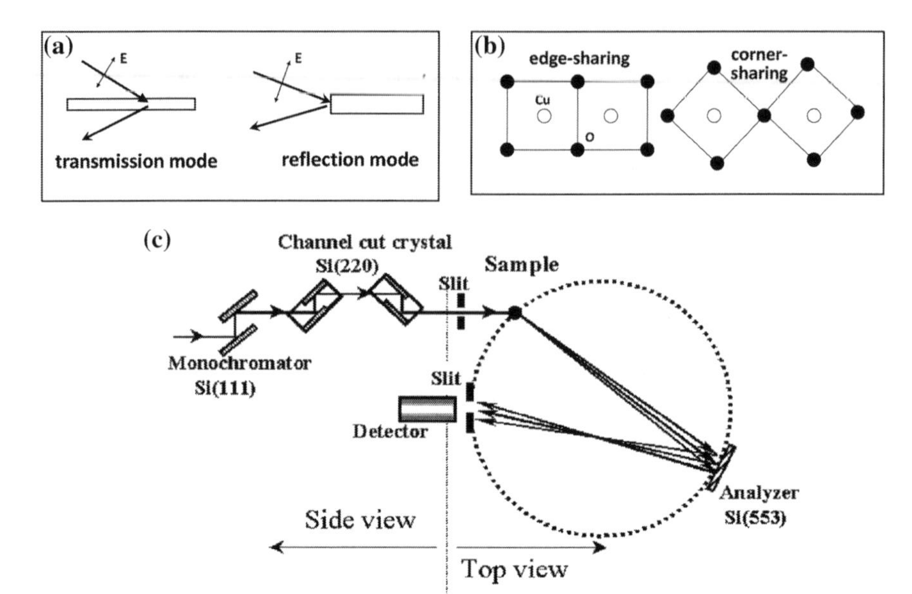

Fig. 13.9 **a** Transmission and reflection modes of resonance inelastic X-ray scattering (RIXS) employed for a $CuGeO_3$ thin film sample and Sr_2CuO_3 [26]. **b** $CuGeO_3$ and Sr_2CuO_3 have edge-sharing and corner-sharing CuO_2 planes, respectively. **c** The optical system employed for RIXS at BL19LXU of SPring-8. The side (*top*) view is shown on the *left* (*right*) side for convenience. The light polarization is perpendicular to the paper plane on the *left* side

where the Cu–Cu chain axis was taken as the x-axis and the CuO_2 plane corresponded to the x–y plane. The $3d_{xy}$ orbital is unoccupied because it has the highest energy among the whole d orbitals. The transfer energy between the neighboring Cu 3d sites via O 2p sites was thought to be very small in this case. On the other hand, Sr_2CuO_3 has a single Cu–O chain in the corner-sharing configuration. The transfer energy was thought to be large and the Cu 3d hole was thought to be in the $3d_{x^2-y^2}$ state. Therefore, very different behavior of charge dynamics was expected.

First the Cu 1s core absorption spectrum was measured by means of fluorescence yield. Then RIXS spectra were measured for a certain Δk at several hν. Cu 1s → Cu 4p excitation was expected according to the dipole selection rules. In a RIXS process, electrons in certain occupied core states are excited to certain unoccupied states while the scattered X-rays are emitted with some energy losses. The photon momentum k is large in the X-ray region and the momentum difference Δk between the incident and scattered photons can easily cover a few BZs. The Δk dependence was measured at several hν.

The experimental results of edge-sharing $CuGeO_3$ are shown in Fig. 13.10(a–c). The inset in (c) shows the Cu 1s absorption spectrum. The quadrupole excitation peak was observed at $h\nu = 8.980$ keV, whereas the main absorption band due to the dipole excitation was rather wide. The RIXS spectra were measured for $\Delta k = 3\pi/c$ at three hν of 8.990, 8.995 and 9.000 keV as shown in (a). Three RIXS

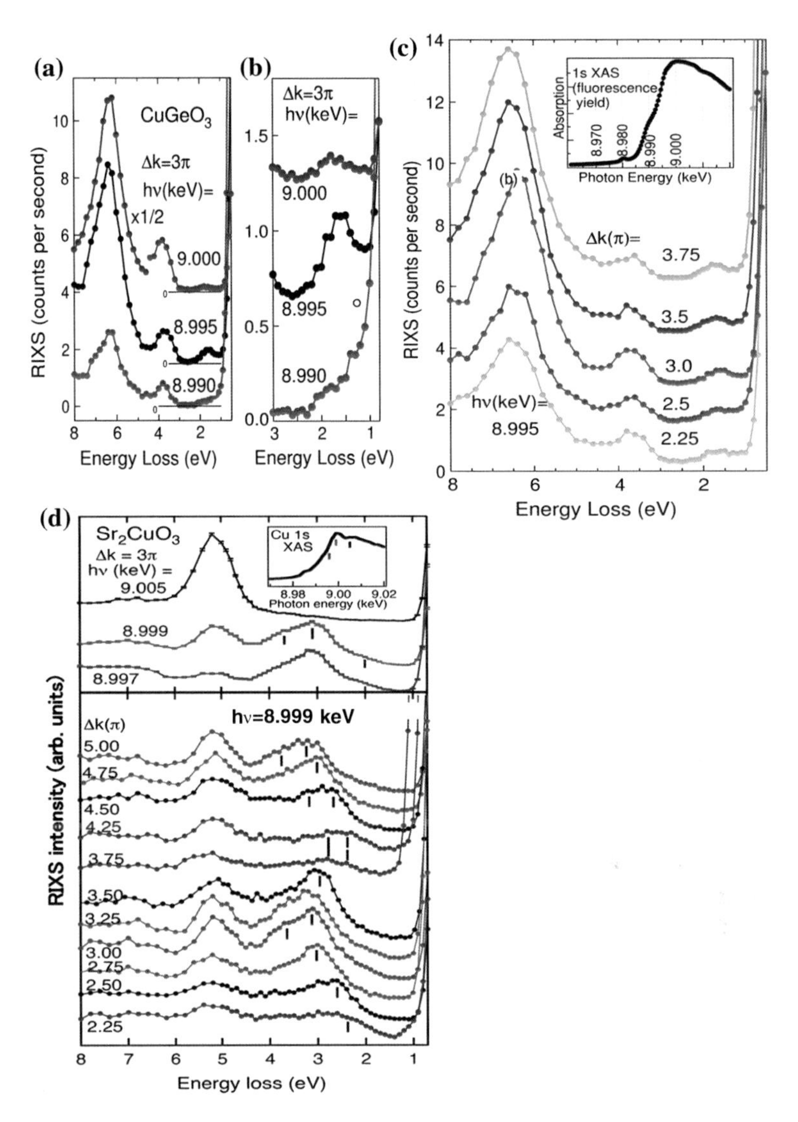

Fig. 13.10 RIXS of edge-sharing $CuGeO_3$ ((**a**), (**b**) and (**c**)) and corner-sharing Sr_2CuO_3 (**d**) at room temperature [26]

structures were observed near 6.3, 3.7 and 1.6 eV. It was recognized that the intensity ratio between the structures at 3.7 and 6.3 eV was the smallest at $h\nu = 9.000$ keV and the structure at 1.6 eV above the smooth tail of the elastic peak is negligible at $h\nu = 8.990$ keV as shown in Fig. 13.10(b). The Δk dependence of the RIXS was then measured at different $h\nu$ (Fig. 13.10(c)). Typical results at $h\nu = 8.995$ keV showed very small dispersion of all RIXS features.

The results of corner-sharing Sr_2CuO_3 are shown in Fig. 13.10(d). The inset shows the Cu 1 s absorption spectrum. Two peak structures were recognized near

8.999 and 9.005 keV in addition to a small absorption hump near 8.985 keV (the quadrupole excitation threshold was near 8.980 keV). Clear differences of RIXS for $\Delta k = 3\pi/b$ were observed at $h\nu = 8.997$, 8.999 and 9.005 keV (upper panel). For excitations at 8.997 and 8.999 keV, three broad energy loss structures were observed at around 3.2, 5.2 and ~6.7 eV. One could further recognize a shoulder near 2 eV for $h\nu = 8.997$ and 8.999 keV and near 3.7 eV for $h\nu = 8.999$ keV. The 3.2 eV structure was dramatically suppressed at $h\nu = 9.005$ keV. RIXS at $h\nu = 8.999$ keV in the lower panel showed a clear Δk dependence of the peak located at 3.2 eV at $\Delta k = 3\pi/b$. Its dispersion had a minimum close to 2.3 eV around $\Delta k = 4\pi/b$ and $2\pi/b$ and a maximum close to 3.2 eV around $\Delta k = 5\pi/b$ and $3\pi/b$, where a shoulder near 3.7 eV was also seen.

There are two energy regions in the Cu 1s absorption spectrum, reflecting different intermediate states in the RIXS processes, namely the low energy region corresponding mainly to the $|1s^1 3d^{10}\underline{L}4p^1\rangle$ states ($\underline{L}$ denotes the hole in the O 2p states), where the Cu 3d hole is transferred to the O 2p state and the higher energy region corresponding mainly to the $|1s^1 3d^9 4p^1\rangle$ state with the Cu 3d hole on the Cu site.

The RIXS results were almost always discussed in comparison with theoretical calculations. The electronic structures calculated for the edge- and corner-sharing CuO_2 planes within a Hartree–Fock (HF) theory by using the random phase approximation, in which the electron correlation effects are perturbatively taken into account [27], were used for comparison. The antiferromagnetic ground state of Mott insulators is well described by the HF theory. Then the charge-excitation energy in Mott insulators could be generally regarded as the antiferromagnetic gap energy within the HF theory. For $CuGeO_3$, two RIXS peaks located near 3.5 and 6.5 eV with very small dispersions of less than 0.2 eV against Δk were predicted, where the spectral weight near 3.5 eV was much smaller than that near 6.5 eV [26]. The Slater-Koster hopping parameters were properly employed. The 3.5 eV peak corresponds to the excitation from the so-called Zhang-Rice singlet (ZRS) [28] to the upper Hubbard band (UHB). The energy loss near 6.5 eV was ascribed to the excitation from the bonding state (BS) (in terms of the electron character) between the Cu 3d and O 2p states to the UHB. Furthermore, the extended d-p model suggested that the intensity ratio between the ZRS → UHB to BS → UHB should be enhanced when the low-energy absorption ($|1s^1 3d^{10}\underline{L}4p^1\rangle$) peak is excited in a qualitative agreement with the $h\nu$-dependent experimental result. On the other hand, the experimentally observed peak at 1.6 eV in the RIXS spectra of $CuGeO_3$ could not be predicted from both theories. This 1.6 eV structure was interpreted as the d–d transition taking place on the same Cu site, as implied from the EELS [29] and soft X-ray O 1s RIXS [30]. Since the $d_{x^2-y^2}$ and d_{xy} orbitals can hybridize via the O 2p orbital in edge-sharing $CuGeO_3$ with $\theta = 99°$, the d–d transition could take place between these states. The strong suppression of this peak at $h\nu = 8.990$ eV in RIXS is because the $|1s^1 3d^{10}\underline{L}4p^1\rangle$ is dominant at this $h\nu$ and the d–d transition cannot take place in the $3d^{10}$ configuration. RIXS in edge-shared Cu oxide systems was later handled by the exact diagonalization of the multiband Hubbard model [31].

The calculated results for Sr_2CuO_3 in the HF theory predicted that both UHB and ZRS had noticeable dispersions caused by the strong hybridization between the Cu 3d and O 2p states and large transfer energies. The dispersions showed π/b periodicity reflecting the antiferromagnetic ground state. The calculated RIXS spectra for typical Δk values of 2nπ/b, (2n + 0.5)π/b and (2n + 1)π/b with integer n showed 2π/b periodicity [26] instead of π/b periodicity, reflecting the partial occupation number of the Cu $3d_{x^2-y^2}$ electrons for each spin component in the band. The 2π/b periodicity was already reported in Sr_2CuO_3 [20] and $SrCuO_2$ [21]. The ZRS → UHB excitation was predicted to have a large dispersion and more than two components (Fig. 3 of [26]). The dispersions of the RIXS features were quantitatively different but qualitatively similar to Fig. 4 of Ref. [27]. It was noticed that its spectral weight shifted towards smaller energies near $\Delta k = 2n\pi/b$ (n = 0, 1, 2...) in agreement with the experimentally observed dispersive feature through 2.3–3.2 eV. In this calculation the ZRS → UHB excitation at $\Delta k = (2n + 1)\pi/b$ had a weak low energy shoulder near the energy loss of 2 eV, which was also consistent with the experimental results. The upper and lower bonding states (BS) between the Cu 3d and O 2p states had also noticeable dispersions, whereas the O 2p non-bonding (NB) band and the middle BS and the Cu 3d lower Hubbard band (LHB) had very small dispersions (Fig.3 of [26]). The small Δk dependence of the 5.2 eV structure in Fig. 13.10(d) was ascribable to the small dispersion of the middle branch of the BS → UHB excitation. The excitation from the NB band and upper BS states was not strong in RIXS because of the little partial occupation number of the Cu $3d_{x^2-y^2}$ electrons. The intensity of the ZRS → UHB excitation was dramatically reduced at $h\nu = 9.005$ keV. This is because this hν corresponds to the intermediate $|1s^1 3d^9 4p^1\rangle$ state and further a Cu 3d excitation requiring the correlation energy U_{dd} is very unfavorable. The distinct d–d excitation is not observed in corner-sharing Sr_2CuO_3 because the hybridization is strong between the Cu $3d_{x^2-y^2}$ and O 2p states and then the hole is not localized on the Cu site. On the other hand, the intensity of the ZRS → UHB excitation relative to the BS → UHB excitation is much stronger than in $CuGeO_3$, reflecting the easy formation of the ZRS in the corner-sharing system.

Cu 1s edge RIXS on La_2CuO_4 was performed and dependences on hν and polarization as well as fine structure in the Cu 4p density of states were discussed [32, 33]. RIXS on $La_{2-x}Sr_xCuO_4$, $La_{2-x}Sr_xNiO_4$ and Nd_2CuO_4 were also reported [34, 35]. Mn 1s edge RIXS was extensively studied including temperature dependence in $La_{1-x}Sr_xMnO_3$, $La_{1-x}Ca_xMnO_3$, $Pr_{1-x}Ca_xMnO_3$ and $Nd_{1-x}Sr_xMnO_3$, where the intersite d–d transitions between Mn atoms depending upon the magnetic order were discussed [36]. For the Yb 2p edge, RIXS was studied in $YbAl_3$, $YbCu_2Si_2$ and $YbInCu_4$ in addition to XAS by means of partial fluorescence yield (PFY–XAS). The results were compared with valence band HAXPES at hν ~ 6 keV performed on scraped surfaces [37]. Through the comparison of the Yb valence values, the importance of suppressing or removal of surface contributions in HAXPES was demonstrated.

The 4f electron properties at the γ–α transition in elemental Ce were studied by the Ce 2p3d RIXS under pressure (0, 1, 2 GPa) to induce the γ–α transition, where the 3d → 2p decay following a resonance excitation close to the Ce 2p → 5d transition was measured [38]. With increasing pressure, the Ce 2p core level XAS main peak ($|4f^1\rangle$ component) was noticeably suppressed while a hump ascribable to the $|4f^0\rangle$ component emerged ~10 eV above. In the 2p3d RIXS, $|4f^2\rangle$ components increased on the γ–α transition and it was demonstrated that the 4f electrons were partly delocalized through band formation with the conduction states, while the electron–electron correlations were reduced on the other hand. The 4f occupation n_f was evaluated as 0.97 in the γ phase ($P = 0.15$ GPa) and 0.81 in the α phase ($P = 2$ GPa). The Kondo temperature T_K was evaluated as 70 K in the γ phase and 1,700 K in the α phase from RIXS.

RIXS in the soft X-ray regime (SX-RIXS) also made much progress recently. In the antiferromagnetic La_2CuO_4 and $CaCuO_2$, high resolution measurements with respect to $h\nu$ and $\mathbf{q}$ at the Cu $2p_{3/2}$ edge allowed the observation of dispersing bimagnon excitations in contrast to the d–d excitations with negligible dispersion [39]. Cu $2p_{3/2}$ core level SX-RIXS was also performed for the underdoped high T_c superconductor $La_{2-x}Sr_xCuO_4$, from which it was suggested that a dynamically inhomogeneous spin state was realized in this system at low temperatures [40]. In the case of SX-RIXS at the Cu 2p edge, the energy resolution down to 120 meV was achieved at the ADRESS beam line in Swiss Light Source [41], facilitating the detailed studies of magnetic excitations as just discussed above. The advantage compared with the inelastic neutron scattering (INS) is the rather high sensitivity of RIXS and the moderate change of the RIXS cross section over the whole BZ in contrast to the low sensitivity of INS around the BZ center (small momentum transfer). Magnetic excitations in 2D quantum antiferromagnets were strongly debated based on the concept of spinons with $S = 1/2$, magnons, and elementary triplet excitation named triplons with $S = 1$, which are three-fold degenerate elementary excitations [42–44]. $Sr_{14}Cu_{24}O_{41}$ is a two-leg quantum spin-ladder compound consisting of two parallel chains (legs) with a transverse exchange coupling. A singlet ground state and dispersive triple excitations (triplons) characterize this material. Figure 13.11 summarizes the Cu $2p_{3/2}$ core level XAS result, the experimental set up, high-resolution Cu $2p_{3/2}$ edge SX-RIXS and the intensity map as a function of momentum [41]. The peak in Fig. 13.11(b) around zero energy loss contains the elastic component as well as unresolved low energy contributions from some phonons and magnetic excitations from the chain. The second peak shows noticeable dispersion with respect respect to q_c ($q_c = 0$ corresponds to the BZ center) as seen from the intensity map in Fig. 13.11(c). In comparison with the results of spectral-density calculations for multi-triplon contributions, the experimental results match very well with the lower boundary of the two-triplon continuum [44]. Since the RIXS intensity is high enough throughout the BZ, the two-triplon spin gap energy was also determined as 100 ± 30 meV. It is thus demonstrated that RIXS is complementary to INS and PES for studying magnetic properties of various materials.

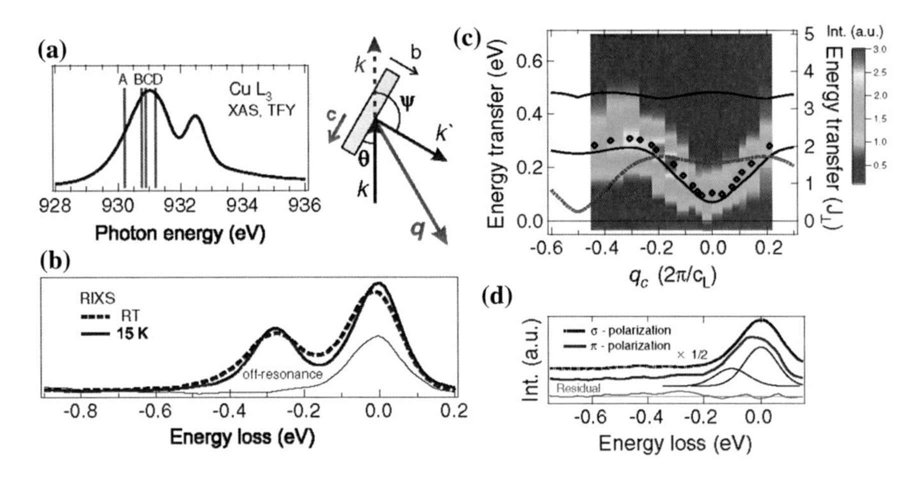

Fig. 13.11 Cu $2p_{3/2}$ RIXS of $Sr_{14}Cu_{24}O_{41}$ [41]. **a** XAS by total fluorescence yield. Typical data were measured at hν of A–D. Single crystals were cleaved ex situ. Mirror like surfaces were obtained. Incident and scattered photons are indicated by **k** and **k**′. **b** and **c**-axes are in the scattering plane. θ and Ψ are the angles of incidence and scattering. Transferred momentum is $\mathbf{q} = \mathbf{k}' - \mathbf{k}$. **b** RIXS excited at the hν of C and at 15 K as well as room temperature. Off resonance RIXS spectrum is added by a thin *solid line*. **c** RIXS intensity map as a function of momentum (q_c) after subtracting the elastic signal. Lower boundary of the two-triplon continuum is given by the open diamond symbols. Calculated one- and two-triplon dispersion curves are given by *dotted* and *full lines* [44]. **d** RIXS spectra measured around the BZ center with $q_c = -0.04*2\pi/c_L$ (where $c_L = 3.93$ Å) for σ (*upper*)- and π (*lower*)-polarized light. The π-polarized spectrum was fitted by two Gaussians

References

1. S. Shin, S. Suga, M. Taniguchi, H. Kanzaki, S. Shibuya, T. Yamaguchi, J. Phys. Soc. Jpn. **51**, 906 (1982)
2. T. Yamaguchi, S. Shibuya, S. Suga, S. Shin, J. Phys. **C15**, 2641 (1982)
3. S. Suga, K. Inoue, M. Taniguchi, S. Shin, M. Seki, K. Sato, T. Teranishi, J. Phys. Soc. Jpn. **52**, 1848 (1983)
4. W. Gudat, E.E. Koch, P.Y. Yu, M. Cardona and C. Penchina, Phys. Status. Solidi. **52** (b), 505 (1972)
5. S. Suga, S. Shin, M. Taniguchi, K. Inoue, M. Seki, I. Nakada, S. Shibuya, T. Yamaguchi, Phys. Rev. **B25**, 5486 (1982)
6. M. Taniguchi, S. Suga, S. Shin, K. Inoue, M. Seki, H. Kanzaki, Solid State Commun. **44**, 85 (1982)
7. S. Suga, S. Imada, H. Yamada, Y. Saitoh, T. Nanba, S. Kunii, Phys. Rev. **B52**, 1584 (1995)
8. C. Dallera, M. Grioni, A. Shukla, G. Vankó, J.L. Sarrao, J.P. Rueff, D.L. Cox, Phys. Rev. Lett. **88**, 196403 (2002)
9. C.T. Chen, L.H. Tjeng, J. Kwo, H.L. Kao, P. Rudolf, F. Sette, R.M. Fleming, Phys. Rev. Lett. **68**, 2543 (1992)
10. A. Sekiyama, A. Fujimori, S. Aonuma, R. Kato, Phys. Rev. B **56**, 9937 (1997)
11. C.F. Chang, Z. Hu, H. Wu, T. Burnus, N. Hollmann, M. Benomar, T. Lorenz, A. Tanaka, H.-J. Lin, H.H. Hsieh, C.T. Chen, L.H. Tjeng, Phys. Rev. Lett. **102**, 116401 (2009)

12. H. Okamura, T. Michizawa, T. Namba, S. Kimura, F. Iga, T. Takabatake, J. Phys. Soc. Jpn. **74**, 1954 (2005)
13. F. Iga, N. Shimizu, T. Takabatake, J. Magn. Magn. Mater. **177–181**, 337 (1998)
14. S. Kimura, Phys. Rev. **B80**, 073103 (2009)
15. C.A. Kuntscher, S. Frank, A. Pashkin, H. Hoffmann, A. Schönleber, S. Smaalen, M. Hanfland, S. Glawion, M. Klemm, M. Sing, S. Horn, R. Claessen, Phys. Rev. **B76**, 241101 (2007)
16. C.A. Kuntscher, S. Frank, A. Pashkin, M. Hoinkis, M. Klemm, M. Sing, S. Horn, R. Claessen, Phys. Rev. **B74**, 184402 (2006)
17. L. Baldassarre, A. Perucchi, E. Arcangeletti, D. Nicoletti, D. Di Castro, P. Postorino, V.A. Sidorov, S. Lupi, Phys. Rev. **B75**, 245108 (2007)
18. T. Mizuno, T. Iizuka, S. Kimura, K. Matsubayashi, K. Imura, H. Suzuki, N. Sato, J. Phys. Soc. Jpn. **77**, 113704 (2008)
19. T. Nishi, S. Kimura, T. Takahashi, Y. Mori, Y.S. Kwon, H.J. Im, H. Kitazawa, Phys. Rev. **B71**, R220401 (2005)
20. M.Z. Hasan, P.A. Montano, E.D. Isaacs, Z.-X. Shen, H. Eisaki, S.K. Sinha, Z. Islam, N. Motoyama, S. Uchida, Phys. Rev. Lett. **88**, 177403 (2002)
21. Y.J. Kim, J.P. Hill, H. Benthien, F.H.L. Essler, E. Jeckelmann, H.S. Choi, T.W. Noh, N. Motoyama, K.M. Kojima, S. Uchida, D. Casa, T. Gog, Phys. Rev. Lett. **92**, 137402 (2004)
22. Y.J. Kim, J.P. Hill, S. Komiya, Y. Ando, D. Casa, T. Gog, C.T. Venkataraman, Phys. Rev. B**70**, 094524 (2004)
23. G. Ghiringhelli, N.B. Brookes, E. Annese, H. Berger, C. Dallera, M. Grioni, L. Perfetti, A. Tagliaferri, L. Braicovich, Phys. Rev. Lett. **92**, 117406 (2004)
24. K. Ishii, K. Tsutsui, Y. Endoh, T. Tohyama, K. Kuzushita, T. Inami, K. Ohwada, S. Maekawa, T. Masui, S. Tajima, Y. Murakami, J. Mizuki, Phys. Rev. Lett. **94**, 187002 (2005)
25. K. Ishii, K. Tsutsui, Y. Endoh, T. Tohyama, S. Maekawa, M. Hoesch, K. Kuzushita, M. Tsubota, T. Inami, J. Mizuki, Y. Murakami, K. Yamada, Phys. Rev. Lett. **94**, 207003 (2005)
26. S. Suga, S. Imada, A. Higashiya, A. Shigemoto, S. Kasai, M. Sing, H. Fujiwara, A. Sekiyama, A. Yamasaki, C. Kim, T. Nomura, J. Igarashi, M. Yabashi, T. Ishikawa, Phys. Rev. B**72**, R 081101 (2005)
27. T. Nomura, J. Igarashi, J. Phys. Soc. Jpn. **73**, 1677 (2004)
28. F.C. Zhang, T.M. Rice, Phys. Rev. B **37**, R3759 (1988)
29. F. Parmigiani, L. Sangaletti, A. Goldoni, U. del Pennino, C. Kim, Z.-X. Shen, A. Revcolevschi, G. Dhalenne, Phys. Rev. **B55**, 1459 (1997)
30. L.-C. Duda, T. Schmitt, A. Augustsson, J. Nordgren, J. Alloys Compd. **362**, 116 (2004)
31. F. Vernay, B. Moritz, I.S. Elfimov, J. Geck, D. Hawthorn, T.P. Devereaux, G.A. Sawatzky, Phys. Rev. **B77**, 104519 (2008)
32. L. Lu, J.N. Hancock, G. Chabot-Couture, K. Ishii, O.P. Vajk, G. Yu, J. Mizuki, D. Casa, T. Gog, M. Greven, Phys. Rev. **B74**, 22509 (2006)
33. A. Shukla, M. Calandra, M. Taguchi, A. Kotani, G. Vanko, S.-W. Cheong, Phys. Rev. Lett. **96**, 077006 (2006)
34. E. Collart, A. Shukla, J.-P. Rueff, P. Leininger, H. Ishii, I. Jarrige, Y.Q. Cai, S.-W. Cheong, G. Dhalenne, Phys. Rev. Lett. **96**, 157004 (2006)
35. J.P. Hill, G. Blumberg, Y.J. Kim, D.S. Ellis, S. Wakimoto, R.J. Birgeneau, S. Komiya, Y. Ando, B. Liang, R.L. Greene, D. Casa, T. Gog, Phys. Rev. Lett. **100**, 097001 (2008)
36. S. Grenier, J.P. Hill, V. Kiryukhin, W. Ku, Y.-J. Kim, K.J. Thomas, S.-W. Cheong, Y. Tokura, Y. Tomioka, D. Casa, T. Gog, Phys. Rev. Lett. **94**, 047203 (2005)
37. L. Moreschini, C. Dallera, J.J. Joyce, J.L. Sarrao, E.D. Bauer, V. Fritsch, S. Bobev, E. Carpene, S. Huotari, G. Vankó, G. Monaco, P. Lacovig, G. Panaccione, A. Fondacaro, G. Paolicelli, P. Torelli, M. Grioni, Phys. Rev. **B75**, 035113 (2007)
38. J.-P. Rueff, J.P. Itie, M. Taguchi, C.F. Hague, J.-M. Mariot, R. Delaunay, J.-P. Kappler, N. Jaouen, Phys. Rev. Lett. **96**, 237403 (2006)

39. L. Braicovich, L.J.P. Ament, V. Bisogni, F. Forte, C. Aruta, G. Balestrino, N.B. Brookes, G.M.D. Luca, P.G. Medaglia, F.M. Granozio, M. Radovic, M. Salluzzo, J. Brink, G. Ghiringhelli, Phys. Rev. Lett. **102**, 167401 (2009)
40. L. Braicovich, J. Brink, V. Bisogni, M.M. Sala, L.J.P. Ament, N.B. Brookes, G.M.D. Luca, M. Salluzzo, T. Schmitt, V.N. Strocov, G. Ghiringhelli, Phys. Rev. Lett. **104**, 077002 (2010)
41. J. Schlappa, T. Schmitt, F. Vernay, V.N. Strocov, V. Ilakovac, B. Thielemann, H.M. Rønnow, S. Vanishri, A. Piazzalunga, X. Wang, L. Braicovich, G. Ghiringhelli, C. Marin, J. Mesot, B. Delley, L. Patthey, Phys. Rev. Lett. **103**, 047401 (2009)
42. K.P. Schmidt, G.S. Uhrig, Phys. Rev. Lett. **90**, 227204 (2003)
43. G.S. Uhrig, K.P. Schmidt, M. Grüninger, Phys. Rev. Lett. **93**, 267003 (2004)
44. K.P. Schmidt, G.S. Uhrig, Mod. Phys. Lett. B **19**, 1179 (2005)

Chapter 14
Surface Spectroscopy by Scanning Tunneling Microscope

The studies of micro and nano materials from the view point of topography and electronic structure are now very important for applications. Some techniques are already explained in Sect. 3.6.4. For topographic imaging of nano-materials, atomic force microscopy (AFM) as well as scanning tunneling microscopy (STM) are widely applied. In the case of magnetic materials, magnetic force microscopy (MFM) and spin-polarized STM techniques are applicable. In MFM, the magnetic stray field is probed by a magnetically covered pyramidal tip [1]. The dipole–dipole force between the magnetic tip and the magnetic sample is measured by detecting the bending of the cantilever. MFM is often performed in atmosphere. UHV MFM facilitates the reduction of the distance between the tip and the sample and improves the spatial resolution, which is well below 100 nm. Although MFM is as powerful for studying magnetic reorientations as scanning electron microscopy with polarization analysis (SEMPA) [2, 3], it can provide no information on the electronic structure. For the studies of electronic structures, scanning tunneling spectroscopy (STS) is applicable.

Nowadays STM is routinely used to study the topography of nano materials. Spatial resolution down to the atomic scale can be achieved. Combining this with STS, one can study the correlation of local structures with local electronic properties. Even the correlation with the local magnetic properties can be studied by means of spin polarized STS. The utilization of relatively narrow surface states helps one to tackle such subjects. An example of ferromagnetic Gd, which was often regarded as a prototype rare-earth metal since it exhibited a half-filled 4f shell with the maximal possible f-shell magnetic moment of $7\mu_B$, is shown first. Figure 14.1(a) shows the photoemission and inverse photoemission spectra (PES and IPES) of a clean Gd (0001) surface at 170 K [1, 4–6]. The T_C of Gd is 292.5 K and the spin-polarized occupied- and unoccupied-surface states with d_{z^2}-like character [7] are clearly observed at $\sim$130 meV below E_F and $\sim$420 meV above E_F. When Gd is evaporated onto W (110) kept at 530 K during the evaporation, isolated Gd islands are grown on a 1 ML Gd covered W substrate [1]. Beside STM, STS was performed at 70 K by means of differential conductivity, dI/dU, measured with a

S. Suga and A. Sekiyama, *Photoelectron Spectroscopy*,
Springer Series in Optical Sciences 176, DOI: 10.1007/978-3-642-37530-9_14,

small voltage modulation at high frequency. The variation of the tunnel current I was detected by a lock in amplifier. dI/dU is proportional to

$$1 + \mathrm{Pt} \cdot \mathrm{Ps} \cdot \cos\theta,$$

where Pt and Ps represent the spin polarization of the tip and the sample and θ is the mutual angle between them. Pt is assumed to be constant. If Ps is assumed to be constant as well, the local orientation θ of the spin of the sample can be measured by this technique.

The d_{z^2}-like Gd (0001) surface states were clearly observed on the Gd (0001) island at ~200 meV below E_F and ~430 meV above E_F as shown by full dots in Fig. 14.1(b). The dI/dU spectrum on the Gd monolayer shows only a broad asymmetric peak near ~300 meV above E_F as shown by empty rectangles. Thus, the topographic contrast dI/dU varies as a function of U. When the sample bias U is positive, the tunneling current I is dominated by the electrons tunneling from the tip to the unoccupied conduction band of the sample. When U is negative, the electrons tunneling from the sample to the unoccupied conduction band of the tip is dominating. Therefore, unoccupied (occupied) surface states of the sample are sensitively probed by dI/dU for positive (negative) U.

In fact, the observed occupied (unoccupied) surface state at U = −220 (+500) meV at 20 K on the Gd (0001) island surface is a majority (minority)-spin polarized state. Namely, exchange-split surface states are observed by STS. A spin-polarized tip with non-reversing spin polarization in the range of interest (U = ±0.5 eV) near E_F, for example Fe, was quite useful for this study [8]. Then spin-polarized tunneling takes place for this surface, confirming the spin-polarization of the sample surface states [9]. The basic principle is the spin-valve effect

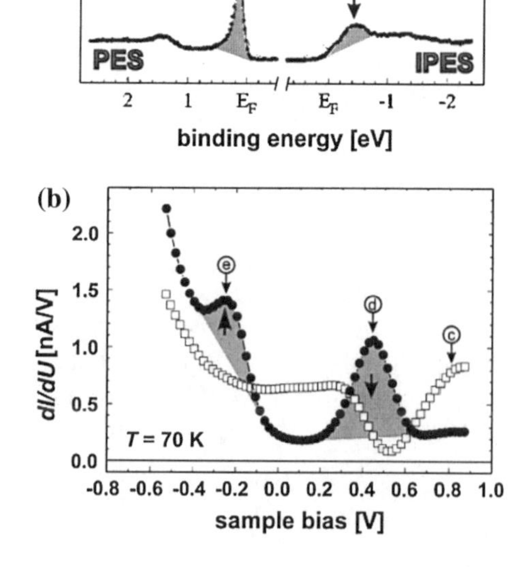

Fig. 14.1 Occupied and empty surface states of Gd(0001) probed by photoemission and inverse photoemission at 170 K (**a**) [4]. dI/dU spectra obtained by scanning tunneling spectroscopy on a Gd (0001) island (*full dots*) and a monolayer Gd (0001) (*empty rectangles*) grown on W (110) at 70 K [1]. Majority spin and minority spin surface states are clearly observed as shown by *arrows* *e* and *d* in Fig. (**b**)

[10]. Namely, a ferromagnetic material has a preferential spin at E_F and the spin must be preserved in the elastic tunneling process and I depends on the mutual magnetization between the tip and the sample. Consequently, dI/dU on a domain with parallel (antiparallel) spin alignment between the sample and the tip increases (decreases), inducing a difference in the magnitude of dI/dU for regions with opposite magnetization in the sample. When the magnetization of the sample (Gd island on W (110)) is reversed by an external filed (±4.3 mT), while keeping the magnetization of the Fe coated W tip unchanged, the above mentioned situation takes place, showing a reversal in the contrast at the majority and minority surface state peak positions. Then the spin polarization of the tunneling current can be evaluated by

$$((\mathrm{dI/dU})\uparrow - (\mathrm{dI/dU})\downarrow)/((\mathrm{dI/dU})\uparrow + (\mathrm{dI/dU})\downarrow).$$

Here ↑ ↓ denote the parallel and antiparallel spin alignment on sample magnetization reversal. Assuming a constant spin polarization of the tip S = 0.44, a spin polarization of about 0.20 (−0.12) is obtained for the majority (minority) part of the surface state. This result at 70 K was about a factor of 2 smaller than the result of spin-polarized IPES [11] at 130 K in Fig. 14.2(b), where qualitatively similar behavior was confirmed. The results shown above were taken by a W tip coated with less than 10 ML thick Fe. If one used a W tip with an Fe coating up to 100–200 ML, tip-magnetization-induced magnetic modification of the sample domain was possible [1]. Even an anti-ferromagnetic tip was used in order to

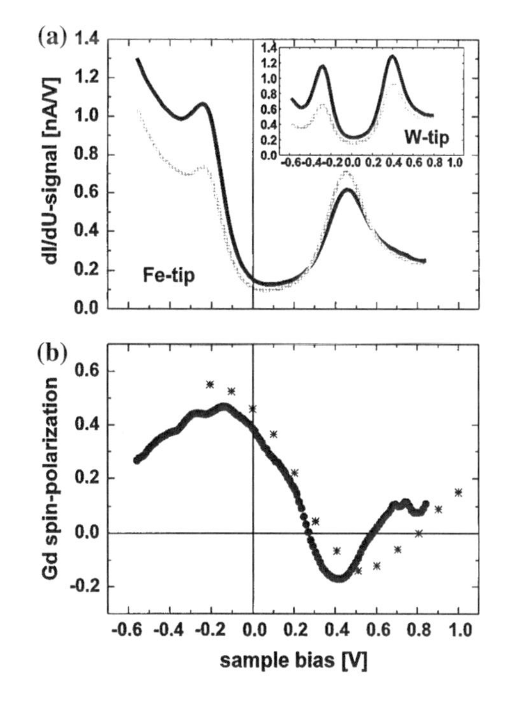

Fig. 14.2 **a** dI/dU spectra at 70 K probed by a 5–10 ML Fe coated W tip on adjacent domains of Gd prepared by depositing 10 ML Gd on a W (110) substrate held at 530 K and those probed by a W tip (inset) [1]. **b** Spin-polarized IPES measured at 130 K (*star marks*) [11] compared with the spin polarization spectra obtained in (**a**) by spin-polarized STS (*full dots*)

reduce the influence of the tip-magnetization on the sample local magnetization. In this experiment, a spatial resolution below 20 nm was demonstrated.

Magnetic memories of ultimate size are desired for modern application. The STM and STS are thought to be very powerful to study such a system by means of their excellent spatial resolution down to the atomic scale. In order to study the excitation dynamics of individual atoms and clusters, inelastic tunneling spectroscopy (ITS) could be performed by use of a low temperature STM [12, 13]. The spin of the tunneling electrons is exchanged with that of the magnetic object. The spin-flip of the tunneling electron changes the spin of the atom or cluster from the ground state with $S_Z = \pm S$ to the excited state with $S_Z = \pm(S - 1)$. The energy of the inelastic spin-flip process E_{sf} corresponds to the energy difference between these two states and can be measured by ITS. When eU reaches the value of E_{sf} with increasing U, the inelastic tunneling channel opens suddenly and I increases. The slight change of I (U) can be identified by a peak in the second derivative d^2I/dU^2. Such an excitation takes place for both positive and negative U. An example of an Fe atom on a Pt (111) substrate is shown in Fig. 14.3 [12].

By subtracting the slight contribution from the substrate, the actual d^2I/dU^2 on an Fe atom was obtained. Since the spectra depend on the individual atoms, the measurement was repeated on many Fe atoms. From the Gaussian fit, the average E_{sf} was evaluated as 5.8 meV. Likewise E_{sf} of a Co atom on Pt (111) was evaluated as 10.3 meV. By use of the atomic manipulation technique by STM, it was

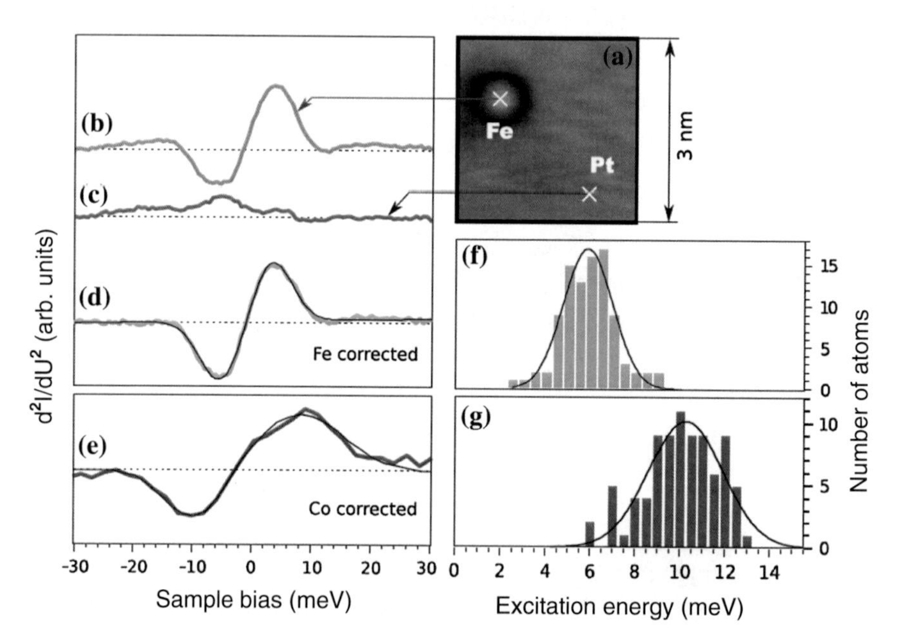

Fig. 14.3 STM and STS of an Fe atom on Pt (111). (**a**) Topography. d^2I/dU^2 spectra on an Fe atom (**b**) and on a Pt (111) substrate (**c**). (**d**) Corrected d^2I/dU^2 spectrum on an Fe atom. (**e**) Corrected d^2I/dU^2 spectrum on a Co atom on Pt (111). Distribution of excitation energies for many atoms on Pt (111) for an Fe (**f**) and a Co (**g**) atom [12]

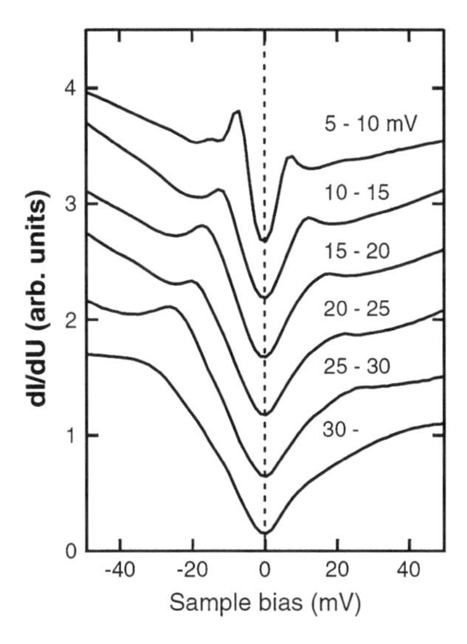

Fig. 14.4 Averaged dI/dU spectra associated with different gap magnitudes in an optimally doped $Ca_{2-x}Na_xCuO_2Cl_2$ superconductor with $T_c = 25$ K. Spatial inhomogeneity is clearly recognized on this surface [17]

possible to prepare numbers of dimers and trimers of Fe and Co. Then d^2I/dU^2 spectra were measured for such systems as well. By considering the modulation voltage ΔU and the thermal resolution, one could evaluate the intrinsic width, which is related to the lifetime τ of the excitation [13]. Thus, the dynamical behavior of the spin-flip excitation in nanomagnetic systems could be studied in detail by STM and ITS.

The STM/STS has also been applied to high-Tc cuprates [14–16]. In situ cleaved very flat stoichiometric surfaces were required for this purpose. Choices of proper samples and low temperature for cleavage are crucial. The differential conductance dI/dU is a measure of the local density of states and a dI/dU map can be obtained at every pixel of the topographic STM image, making spectroscopic mapping [17]. One can change the sample T and also apply even a high external magnetic field [14] for these measurements. An example of such a measurement is shown in Fig. 14.4 in the form of dI/dU spectra for an optimally hole-doped single crystal of the superconductor $Ca_{2-x}Na_xCuO_2Cl_2$ with $T_c \sim 25$ K after in situ cleavage along the (001) plane [17]. On this surface, the so-called checkerboard modulation was clearly observed as in underdoped samples in the dI/dU mapping at a certain energy (for example, +25 meV) at 0.4 K while it was dim in constant current STM topography where a $4a_0 \times 4a_0$ commensurate superstructure with additional $4a_0/3 \times 4a_0/3$ and $a_0 \times a_0$ modulations are observed [16]. Thus, such a modulation was thought to be a robust feature in $Ca_{2-x}Na_xCuO_2Cl_2$. In the region of $|E| < 100$ meV, a V-shaped pseudogap was observed as in Fig. 14.4. Inside this pseudogap, another gap was seen below $|E| \sim 10$ meV with coherence-peak-like features at the gap edges. It was revealed that this low energy gap is spatially inhomogeneous. The experiment at 20 K showed that the gap closed towards T_c. This T dependence and the similarity between this low energy gap and the superconducting gap of $Bi_2Sr_2CaCu_2O_8$ suggests the correlation of this gap with superconductivity [17]. Thus, STM/STS seems to be quite powerful to study the mechanism of high T_c superconductivity on the atomic scale and to complement the information obtained by PES.

STS is well recognized to be a very surface sensitive technique. Although STM and STS are real space imaging and spectroscopy techniques and cannot directly be used to measure the momentum dependence of the excitation energies, it was proposed that the Fourier transform (FT) of the real-space LDOS (local density of states) might be useful to get information on the QP contour at constant energy in **k** space [18]. This approach was thought to be a powerful technique because it enabled one to simultaneously obtain the real-space and momentum-space information on the wave functions, scattering processes and dispersion of QPs. The application of this technique to high T_c cuprates was proposed two decades ago [19]. It was expected that such subjects as (1) QP scattering mechanism, (2) degree of coherence of QP, (3) momentum-space structure or dispersion of QP, (4) relation between the commensurate and incommensurate magnetic signatures, can be tackled. It was known that the QP mean free path of Bi2212 at low T is 2 orders of magnitude below that of optimally doped YBCO, indicating appreciable QP scattering in Bi2212. In addition, strong nanoscale LDOS disorder was observed on the surface between 25 and 65 meV by STM. The samples were cleaved at the BiO plane in cryogenic UHV and immediately transferred into the STM head. High resolution LDOS was measured at 4.2 K with the atomic resolution. The differential tunnel conductance dI/dU was measured at all locations (x, y) in the field of view. Then the 2D map of the LDOS proportional to dI/dU at each E, represented later by $g(\mathbf{r}, \omega)$, was obtained between −6 and −30 meV for the occupied electronic states in a 650 Å region with a resolution of 1.3 Å, where a periodic LDOS modulation was evident. Different spatial patterns and wavelengths were observed at different energies. Then the FT (E, k_x, k_y) of LDOS was obtained [18]. The results of Fourier transform scanning tunneling spectroscopy (FT-STS) applied to Bi2212 are compared with the results of ARPES in Fig. 14.5(a) and (b). The FT-STS results in Fig. 14.5(a) show the locus of scattering $\mathbf{k}_s$ from which the scattering contributing to the LDOS modulation originates. The SC gap function $\Delta(\mathbf{k}_s)$ is displayed in Fig. 14.5(b). Within the employed model, $\Delta(\mathbf{k}_s)$ is equivalent to the momentum dependence of the SC gap. Good correspondence was recognized between FT-STS and ARPES in this case, confirming similar surface electronic structures.

FT-STS was further performed for Bi2212 to discuss the mysterious "pseudogap" excitations at high energies and Bogoliubov QPs at low energies [21–23]. It was found that the low energy excitations were indeed Bogoliubov QPs but they occupied only a restricted region in **k**-space, that shrank rapidly with diminishing hole density. FT-STS was also applied to probe the Fermi surface of strongly inhomogeneous Bi based cuprate superconductors Bi2201. The Fermi surface was found to change on nm length scales, suggesting strong local doping variations [23]. The FT-STS technique was again applied to nearly optimally doped $Ca_{2-x}Na_x$-CuO_2Cl_2 (Na-CCOC) with $T_c = 25$–28 K [24] demonstrating the almost identical SC gap dispersion near the gap node to that of Bi2212 at the same doping level, despite the T_c of Bi2212 being three times larger than that of Na-CCOC.

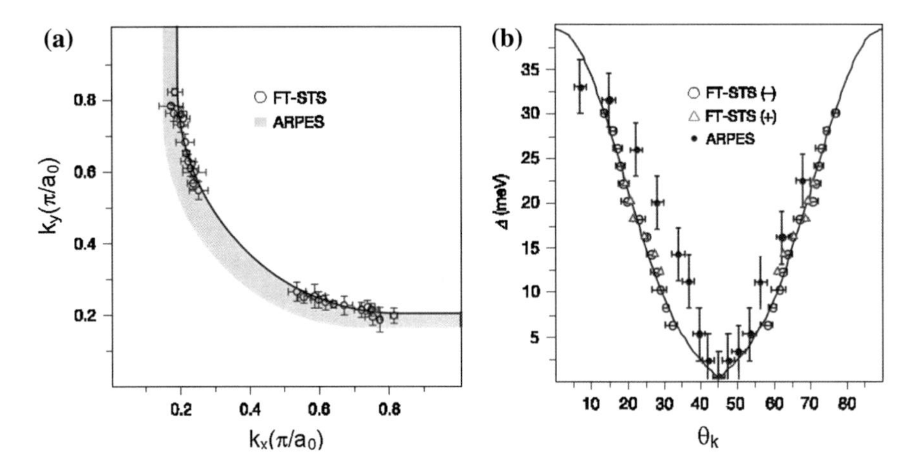

Fig. 14.5 **a** Locus of scattering $\mathbf{k}_s(\omega)$ derived from FT-STS Bi2212 with $T_c = 86$ K. The gray band represents the Fermi surface location from ARPES. **b** Plot of the energy gap $\Delta(\theta_k)$ determined from the filled state measurements (*open circles*) compared with the ARPES results given by the *filled circles* with large error bars. The open triangles (red) show $\Delta(\theta_k)$ determined from the unoccupied state measurements at positive bias [20] (Figures are reprinted by permission from Macmillan Publishers Ltd.)

FT-STS and ARPES in the high T_c cuprates became also known to show contradictory gap functions [25]. Namely, FT-STS results showed a strong deviation of the gap from the $d_{x^2-y^2}$ form. By applying an "octet model" analysis to autocorrelation ARPES, the contradiction was revealed to result from the fact that the octet model did not consider the effects of matrix elements and pseudogap. The canonical $d_{x^2-y^2}$ SC gap around the node was reaffirmed and the fluctuating superconductivity around the node far above T_c was not necessary to explain the QP interference at low energies.

STS could also be applied to detect the QWS in such a system as Pb islands grown on a Cu(111) surface. The energy separation between adjacent empty QWS was found to shrink as the quantum number increased in the case of STS on this system [26]. The origin of such a behavior was discussed in detail based on different models [26, 27].

References

1. M. Bode, M. Dreyer, M. Getzlaff, M. Kleiber, A. Wadas, R. Wiesendanger, J. Phys. Condens. Matter **11**, 9387 (1999)
2. R. Allenspach, M. Spamanoni, A. Bischof, Phys. Rev. Lett. **65**, 3344 (1990)
3. M. Speckmann, H.P. Oepen, H. Ibach, Phys. Rev. Lett. **75**, 2035 (1995)
4. E. Weschke, C. Schüssler-Langeheine, R. Meier, A.V. Fedorov, K. Starke, F. Hübinger, G. Kaindl, Phys. Rev. Lett. **77**, 3415 (1997)

5. D. Li, C.W. Hutchings, P.A. Dowben, C. Hwang, R.T. Wu, M. Onellion, A.B. Andrews, J.L. Erskine, J. Magn. Magn. Mater. **99**, 85 (1991)
6. D. Li, P.A. Dowben, J.E. Ortega, F.J. Himpsel, Phys. Rev. B **49**, 7734 (1994)
7. R. Wu, C. Li, A.J. Freeman, C.L. Fu, Phys. Rev. B **44**, 9400 (1991)
8. R. Wu, A.J. Freeman, Phys. Rev. Lett. **69**, 2867 (1992)
9. M. Bode, M. Getzlaff, R. Wiesendanger, Phys. Rev. Lett. **81**, 4256 (1998)
10. J.C. Slonczewski, Phys. Rev. B **39**, 6995 (1989)
11. M. Donath, B. Gubanka, F. Passek, Phys. Rev. Lett. **77**, 5138 (1996)
12. T. Balashov, T. Schuh, A.F. Takacs, A. Ernst, S. Ostanin, J. Henk, I. Mertig, P. Bruno, T. Miyamachi, S. Suga, W. Wulfhekel, Phys. Rev. Lett. **102**, 257203 (2009)
13. T. Schuh, T. Balashov, T. Miyamachi, T.S. Suga, W. Wulfhekel J. Appl. Phys. **107**, 09E156 (2010)
14. J.E. Hoffman, E.W. Hudson, K.M. Lang, V. Madhavan, H. Eisaki, S. Uchida, J.C. Davis, Science **295**, 466 (2002)
15. M. Vershinin, S. Misra, S. Ono, Y. Abe, Y. Ando, A. Yazdani, Science **303**, 1995 (2004)
16. T. Hanaguri, C. Lupien, Y. Kohsaka, D.-H. Lee, M. Azuma, M. Takano, H. Takag, J.C. Davis, Nature **430**, 1001 (2004)
17. T. Hanaguri, Y. Kohsaka, J.C. Davis, C. Lupien, I. Yamada, M. Azuma, M. Takano, K. Ohishi, H. Takagi, Physica C **460**, 954 (2007)
18. J.E. Hoffman, K. McElroy, D.-H. Lee, K. M Lang, H. Eisaki, S. Uchida, J.C. Davis, Science **297**, 1148 (2002)
19. J.M. Byers, M.E. Flatte, D.J. Scalapino, Phys. Rev. Lett. **71**, 3363 (1993)
20. K. McElroy, R.W. Simmonds, J.E. Hoffman, D.-H. Lee, J. Orenstein, H. Eisaki, S. Uchidak, J.C. Davis, Nature **422**, 592 (2003)
21. Y. Kohsaka, C. Taylor, P. Wahl, A. Schmidt, J. Lee, K. Fujita, J.W. Alldredge, K. McElroy, J. Lee, H. Eisaki, S. Uchida, D.-H. Lee, J.C. Davis, Nature **454**, 1072 (2008)
22. J. Lee, K. Fujita, A.R. Schmidt, C.K. Kim, H. Eisaki, S. Uchida, J.C. Davis, Science **325**, 1099 (2009)
23. W.D. Wise, K. Chatterjee, M.C. Boyer, T. Kondo, T. Takeuchi, H. Ikuta, Z. Xu, J. Wen, G.D. Gu, Y. Wang, E.W. Hudson, Nat. Phys. **5**, 213 (2009)
24. T. Hanaguri, Y. Kohsaka, J.C. Davis, C. Lupien, I. Yamada, M. Azuma, M. Takano, K. Ohishi, M. Ono, H. Takagi, Nat. Phys. **3**, 865 (2007)
25. M. Hashimoto, R.-H. He, J.P. Testaud, W. Meevasana, R.G. Moore, D.H. Lu, Y. Yoshida, H. Eisaki, T.P. Devereaux, Z. Hussain, Z.-X. Shen, Phys. Rev. Lett. **106**, 167003 (2011)
26. M.C. Yang, C.L. Lin, W.B. Su, S.P. Lin, S.M. Lu, H.Y. Lin, C.S. Chang, W.K. Hsu, T.T. Tsong, Phys. Rev. Lett. **102**, 196102 (2009)
27. A. Zugarramurdi, N. Zabala, A.G. Borisov, E.V. Chulkov, Phys. Rev. Lett. **106**, 249601 (2011)

Chapter 15
Outlook

The photoelectron spectroscopy (PES) has progressed very fast in the last 4 decades with great improvement in the instrumentation including the light sources, energy analyzers and detectors. This technique applied to solids was soon recognized as a powerful tool in surface science due to its surface sensitivity, which is facilitated by the short inelastic mean free path of photoelectrons down to $\sim$3 Å. However, this surface sensitivity hindered a lot of studies on the bulk electronic properties of many materials, which have surface electronic structures noticeably different from the bulk electronic structures. Such examples are seen in strongly correlated electron systems.

One approach to overcome this difficulty is to do the measurement at higher hν. Namely, the inelastic mean free path can be beyond 10 Å in many materials for $E_K \geq 700$ eV. Even though the electronic structures in the first layer were noticeably different from those in the bulk, the electronic structures in the second and deeper layers may be less different even when the materials belong to the strongly correlated electron systems. In such cases, detailed hν dependent measurements and/or emission angle dependence may facilitate the depth-resolved evaluation of the different electronic structures. Higher hν/Δhν resolution monochromators with reduced aberration combined with higher brilliance soft and hard X-ray SR undulators is now promoting this frontier.

In the cases of the angle resolved measurements, however, such simple deconvolution in the valence band region is practically impossible. Therefore, high hν SXARPES with high energy and momentum resolution is required to probe bulk band dispersions. Until recently high resolution SXARPES measurements up to $\sim$900 eV were reported. Since the momentum resolution becomes worse only as the square root of the kinetic energy E_K, the angular resolution of modern electron analyzer enables practical band mapping even in this energy region.

In order to realize higher bulk sensitivity, higher hν ARPES is desired. However, it is known that PICSs of many orbitals of atoms decrease very rapidly with increasing the hν beyond 1 keV. Therefore, high photon flux is a necessary condition to perform such experiments. The PICSs of many orbitals are at least two orders of magnitude smaller at 10 keV than at 1 keV but the inelastic mean free

S. Suga and A. Sekiyama, *Photoelectron Spectroscopy*,
Springer Series in Optical Sciences 176, DOI: 10.1007/978-3-642-37530-9_15,

path (λ_{mp}) is almost ten times longer at 10 keV compared to 1 keV. The transmittance of the crystal monochromators is much higher than that of the reflection grating monochromators. If the transmittance of the monochromator is ten times higher at 10 keV than at 1 keV, the counting rate can be comparable. In addition, the longer λ_{mp} means less surface sensitivity, which guarantees longer data acquisition time without serious surface degradation. So far, hard X-ray ARPES (HAXARPES) was reported up to h$\nu \sim$6 keV for W with a high Debye temperature showing band dispersions. If the recoil effects are negligible and the Debye temperature is high enough, HAXARPES will be a promising approach at low temperatures. The hν resolution around 14 keV can be better than 1 meV, being much higher than the hν resolution of around 30–40 meV at 800 eV realized by grating monochromators. Therefore, high resolution HAXARPES at 8–14 keV is a great challenge in the near future for materials with high enough Debye temperature and negligible recoil effects even though the momentum broadening becomes more serious and the available photon flux for such a small Δhν is extremely low.

It is also recognized that λ_{mp} increases again below a few eV of E_K in some materials. Already hν resolution down to 260 μeV was reported near 7 eV by use of a pulsed quasi-cw laser. However, space charge effects cannot always be neglected. Therefore, cw light in the region of h$\nu \sim$11.6–8.4 eV by microwave excited Ar, Kr, and Xe electron cyclotron resonance lamps with hν resolution better than 600 μeV will also be very widely used for low energy ELEPES and ELEARPES. Polarization dependent ELEPES, ELEARPES and HAXPES measurements by use of lasers and SR are very useful to study the orbital-resolved electronic structures as well as matrix element effects. A complete approach will be to do PES and ARPES measurements from several eV up to beyond 10 keV on the same material to reveal and confirm the surface, subsurface and bulk electronic structures.

On the other hand, time resolved photoelectron spectroscopy is in progress after the commissioning of free electron lasers. If the pulse width is 100 fs, for example, the uncertainty principle provides a limited energy resolution of $\sim$6.6 meV. The time resolution will provide dynamic aspects of electronic relaxation if measured by means of coincidence experiments. As for the spatial resolution, the lateral resolution of 20–30 nm is achieved by some measurements. Further improvement in the time- and spatial-resolution will be feasible for the purpose of application.

In this way, the photoelectron spectroscopy is under development even nowadays. Technical developments of fast detectors with parallel data acquisition, high resolutions in energy, momentum and/or spin in the electron analyzers and high resolution and high photon flux in light sources will further promote its frontier, facilitating the detailed studies of electronic and magnetic structures of materials for a wide range of applications.

List of Samples

S. Suga and A. Sekiyama, *Photoelectron Spectroscopy*,
Springer Series in Optical Sciences 176, DOI: 10.1007/978-3-642-37530-9,

Index

S. Suga and A. Sekiyama, *Photoelectron Spectroscopy*,
Springer Series in Optical Sciences 176, DOI: 10.1007/978-3-642-37530-9,

D

E

R

S

X

Z